Lecture Notes in Artificial Intelligence 4387

Edited by J. G. Carbonell and J. Siekmann

Subseries of Lecture Notes in Computer Science

Thomas Barkowsky   Markus Knauff
Gérard Ligozat   Daniel R. Montello (Eds.)

Spatial Cognition V

Reasoning, Action, Interaction

International Conference Spatial Cognition 2006
Bremen, Germany, September 24-28, 2006
Revised Selected Papers

Series Editors

Jaime G. Carbonell, Carnegie Mellon University, Pittsburgh, PA, USA
Jörg Siekmann, University of Saarland, Saarbrücken, Germany

Volume Editors

Thomas Barkowsky
University of Bremen, Department of Computer Sciences
Enrique-Schmidt-Straße 5, 28359 Bremen, Germany
E-mail: barkowsky@informatik.uni-bremen.de

Markus Knauff
Justus-Liebig University Gießen, Department of Psychology
Otto-Behaghel-Strasse 10F, 35394 Giessen, Germany
E-mail: markus.knauff@psychol.uni-giessen.de

Gérard Ligozat
LIMSI-CNRS, Université Paris-Sud
91403 Orsay, France
E-mail: ligozat@limsi.fr

Daniel R. Montello
University of California, Department of Geography,
Santa Barbara, CA, USA
E-mail: montello@geog.ucsb.edu

Library of Congress Control Number: 2007940030

LNCS Sublibrary: SL 7 – Artificial Intelligence

CR Subject Classification (1998): H.2.8, I.2.10, H.3.1, K.4.2, B.5.1

ISSN	0302-9743
ISBN-10	3-540-75665-5 Springer Berlin Heidelberg New York
ISBN-13	978-3-540-75665-1 Springer Berlin Heidelberg New York

This work is subject to copyright. All rights are reserved, whether the whole or part of the material is concerned, specifically the rights of translation, reprinting, re-use of illustrations, recitation, broadcasting, reproduction on microfilms or in any other way, and storage in data banks. Duplication of this publication or parts thereof is permitted only under the provisions of the German Copyright Law of September 9, 1965, in its current version, and permission for use must always be obtained from Springer. Violations are liable to prosecution under the German Copyright Law.

Springer is a part of Springer Science+Business Media

springer.com

© Springer-Verlag Berlin Heidelberg 2007
Printed in Germany

Typesetting: Camera-ready by author, data conversion by Scientific Publishing Services, Chennai, India
Printed on acid-free paper   SPIN: 12173686   06/3180   5 4 3 2 1 0

Preface

This is the fifth volume in a series of book publications featuring basic interdisciplinary research in spatial cognition. The study of spatial cognition is the study of knowledge about spatial properties of objects and events in the world. Spatial properties include location, size, distance, direction, separation and connection, shape, pattern, and so on. Cognition is about the structures and processes of knowledge: its acquisition, storage, retrieval, manipulation, and use by humans, nonhuman animals, and machines. Broadly construed, cognitive activities include sensation and perception, thinking, attention, imagery, attitudes, memory, learning, language, and reasoning and problem-solving; the interaction of these activities with motoric (body movement) and affective (emotional) processing is recognized as critically important, as well. Cognition is typically considered to make up much of the activity of the mind. But though the mind is an expression of the structures and processes of the brain and nervous system, it is also an expression of an organism or agent with a physical body that typically exists in a physical and socio-cultural world.

Researchers study spatial cognition for several reasons. Spatial cognition plays important roles in most of the domains of knowledge and behavior of sentient beings, including activities associated with biological survival, social interaction, cultural practice, and economic exchange. Attempts to describe, predict, and explain the basic components of spatial cognition and their interrelationships stimulate a host of interesting basic questions about how important parts of reality work. Cognition about space and place is an important expression of human-environment or human-earth relationships. What's more, the study of spatial cognition holds the promise of helping to solve many practical problems and improve the lot of humanity, whether by improving information systems, the layout and appearance of built environments, the equitable selection of personnel, or the design of effective and efficient educational programs.

As we stated above, spatial cognition is an interdisciplinary field. That means two things. First is that researchers from a variety of disciplines engage in the study of spatial cognition. Authors and participants at this meeting included representatives from many disciplines, including computer and information science, psychology, geography and cartography, engineering, linguistics, anthropology, architecture, and biology. But this fact only makes the field *multi*disciplinary. The second requirement for *inter*disciplinarity is that the representatives from the different disciplines communicate with each other, with all that this implies, about shared vocabularies, methodologies, and basic philosophical frameworks. Such a truly interdisciplinary spatial cognition community began flowering within the last couple of decades (with a few earlier roots). The meeting at which the papers in this volume were presented provided an example of the emergence of such a community and a contribution to furthering its reality.

Space exists at many scales, and spatial cognition can concern itself with cognition at any or all of these scales. However, a tradition has arisen according to which most research on spatial cognition focuses on spatial scales relevant to human activities over the earth's surface, such as finding one's way, planning trips, and using cartographic maps. Space at this scale is variously known as large-scale, environmental, or geographic space. Spatial cognition researchers want to understand human cognition about the layout of buildings, cities, and regions, not about the layout of atoms or the solar system. However, this scale preference is not universally held, and we expect that insights about cognition at one scale may inform, and be informed by, insights about cognition at other scales. What's more, cognition about space often incorporates multiple scales, as when a person uses a small picture of space—a map—to wayfind in a city.

Historical Roots of Spatial Cognition

The study of spatial cognition has a variety of historical roots. In psychology, the behaviorist Edward Tolman (1948) introduced the term "cognitive map" to explain the behavior of rats that took shortcuts to find goal locations directly, without traversing the routes of the maze. But the developer of one of the earliest successful intelligence tests, Alfred Binet, had written about the psychology of orientation and disorientation several decades before (Binet, 1894). Furthermore, the psychologist C. C. Trowbridge (1913) introduced the idea that people form internal mental representations of the layout of environments, calling them "imaginary maps." The child psychologist Jean Piaget and his colleagues devised the most influential theoretical framework for cognitive development in children, and spatial cognitive development was an essential part of this (Piaget & Inhelder, 1948/1967). The perceptual psychologist J. J. Gibson (1950) developed a theory of perception and activity that prompted cognitive researchers to carefully consider the structure of the world and the structure of information in the patterned energy available to sensory receptors, not just the structure of the mind.

Others besides psychologists contributed to the development of spatial cognition as a field. Trowbridge had actually been influenced by very early work by the geographer, F. P. Gulliver (1908), who wrote about the influence of cartographic map orientation on spatial and geographic learning in school children. This concern with the implications of spatial cognition for spatial learning and education continued throughout the twentieth century. Geographers, planners, and architects in the 1950s and 1960s began to ask questions about the subjectivity of spatial knowledge and its implications for models of spatial behavior that depended on subjective beliefs about spatial layout, including beliefs about distances and directions (e.g., Golledge, Briggs, & Demko, 1969). Much of this got impetus from the writings of Kevin Lynch (1960), an urban planner, who convinced many to look at the pattern of the built environment in terms of its effects on mental representations. Connections among geographers, planners, psychologists, and others became explicit in the 1960s and 1970s with the advent of new journals and conferences (e.g., Downs & Stea, 1973).

Finally, researchers in the middle of the twentieth century began to apply digital computers to understanding a host of phenomena, including the behavior and cognition of humans and other animals. Spatial behavior and cognition was no exception. As early as the 1960s, mathematicians and computer scientists attempted to produce formal and computational models of spatial cognition that were realized on computers (Zadeh, 1965). Researchers in these disciplines also became interested in creating "simulated animals," computational agents that could act or reason intelligently. Researchers in artificial intelligence naturally included action and reasoning in space and place, and about space and place, within the scope of their work (Kuipers, 1978). Some of these researchers focused on creating computational entities—robots—that could act in a coordinated fashion, for example, by moving from place to place without collisions. Other researchers focused on creating computational entities that could reason effectively about space and place, solving problems such as sequencing multiple destinations into one efficient trip or understanding the geometric layout of the world.

Topics in the Study of Spatial Cognition

There are a variety of research topics within the domain of spatial cognition. A first topic concerns the structures and processes of spatial knowledge and reasoning. The cognitive or mental map refers to an internally represented model of the environment; it is a metaphorical term meant to appeal to the idea that something like a cartographic map is stored in the head. However, in many ways the cognitive map is not like a cartographic 'map in the head.' It is not a unitary integrated representation but consists of stored discrete pieces, including representations of landmarks, routes, and regions. The separate pieces are partially linked or associated, often in a hierarchical manner (e.g., when the location of a place is stored in terms of a larger region). Furthermore, spatial knowledge is not well modeled by Euclidean or any other metric geometry, although it clearly expresses spatial information beyond the mere level of topology. The nature of internally represented knowledge in human minds leads to certain patterns of distortions in the way people answer spatial questions. For example, people often believe the distance from place A to B is different than from B to A. Turns are frequently adjusted in memory to be more like straight lines or right angles. At larger scales, most people have somewhat distorted ideas about the sizes and locations of continental land masses on the earth; for example, Southern Europe is thought to be near the equator, when it is actually located at 35-45° north latitude.

A second research topic concerns how cognitive agents acquire spatial knowledge and how the knowledge develops over time. Cognitive agents like humans acquire spatial knowledge directly via sensorimotor systems that operate as they move about the world. They also acquire spatial knowledge indirectly via static and dynamic symbolic media such as maps and images, 3-D models, movies, and language. Researchers are interested in the consequences of different learning media for the nature of acquired knowledge. Spatial knowledge changes over time, through processes of learning and development. Both changes in the child's spatial knowledge and reasoning, and that of an adult visiting a new place for the first time, are of interest to researchers. A widely discussed model of spatial learning in the environment suggests that it develops in a sequence of three stages or elements: landmark knowledge, route knowledge, and survey knowledge. However, a variety of evidence throws into question both the progressive nature of this sequence and the degree to which most people reach the level of survey knowledge.

A third research topic involves questions about how cognitive agents navigate and stay oriented in space. Navigation is coordinated and goal-directed travel through space, and it consists of the two components of locomotion and wayfinding.

Locomotion refers to moving through the local surroundings in coordination with immediately available sensorimotor information, such as when avoiding obstacles and moving toward sensed landmarks. Wayfinding refers to the planning and decision-making that allows one to reach a destination that is not immediately available to sensorimotor systems, such as when choosing efficient routes or orienting to nonlocal features. Wayfinding tasks generally require an internal or external representation of the environment. In order to reach destinations while navigating, agents must establish and maintain orientation—they must know 'where they are,' although the precision and comprehensiveness of this knowledge may not be high. Two broad types of processes are involved in orientation during navigation. One involves recognizing external features such as landmarks, sometimes called piloting. The second involves updating orientation by integrating information about movement speed, direction, and/or acceleration, without reference to recognized features. This is called dead reckoning or path integration.

A fourth research topic concerns how agents use natural language to communicate with each other about space and place. Spatial information is often communicated verbally. People give and receive verbal route directions, read spatial descriptions contained in stories, and increasingly interact with computer systems via verbal queries. There are at least two notable characteristics of the way language expresses spatial information. One is that language expresses mostly nonquantitative or imprecise ("vague") quantitative information about space. Statements about connections and approximate location are more important than precise statements. For example, people say "turn left at the sausage shop" rather than "turn 97° after you have gone 1.3 kilometers." A second characteristic is that interpreting spatial language critically depends on various aspects of the context of the communication. Context is provided by knowledge of who is speaking (or writing), where they are, physical features in the situation, the previous topic of conversation, and so on. The scale of "the umbrella is near the table" is generally understood differently than that of "Belgium is near Germany."

A fifth research topic involves how aspects of spatial knowledge and reasoning are similar or different among individuals or groups. No two individual people know exactly the same things or reason in exactly the same way about space and place. Some people are better at tasks such as wayfinding, learning spatial layouts, or reading maps. In some cases, there may be different ways to think about spatial problems, all of which may be effective. In these cases, we might speak of 'stylistic differences' in spatial cognition rather than skill or ability differences. Researchers are interested in measuring and explaining individual differences. Many factors may be related to variations in spatial cognition: body size, age, education, expertise, sex, social status, language, residential environment, and more. A first goal for research is to measure and document ways these factors might covary with spatial cognition. In addition to describing such covariations, researchers have the goal of distinguishing their underlying causes. Explanations for such covariations are generally quite difficult to determine, however, as one cannot readily do randomized experiments on person characteristics such as age, sex, culture, and activity preferences. Nonetheless, even describing patterns of differences is quite valuable for practical goals such as designing spatial information systems.

A sixth and final topic concerns the neurological structures and processes involved in spatial cognition, an area of research that has advanced considerably with the technical developments of the last couple of decades. Research in this area attempts to answer questions such as how spatial information is encoded in nervous systems, which brain areas process spatial information of different types, how spatial information from different sensory modalities is integrated in the nervous system, and how particular injuries or organic syndromes produce particular deficits in spatial cognition. For example, the role of the hippocampus and other brain structures in spatial cognition (and other domains of cognition) continues to be researched. Recordings of the activity of single brain cells in the hippocampi of rats has revealed the existence of neurons that preferentially fire when the rat is in a particular location, known as place cells. Recordings of cells in other structures that have connections to the hippocampus have uncovered head-direction cells that fire preferentially when the rat is facing in a particular absolute direction in the environment. Clinical studies of organic brain syndromes and injuries in humans have shed light on the neuroscience of spatial cognition in humans, such as specific impairments in aspect of navigational skill following localized brain injuries, a syndrome known as topographical disorientation. Probably the most rapidly advancing research area for studying the neuroscience of spatial cognition is the use of brain-imaging techniques applied to awake and alert human subjects while they reason about or remember spatial information. The most promising of these is functional Magnetic Resonance Imaging (fMRI).

Technologies and the Future of Spatial Cognition Research

A variety of technologies influence spatial cognition, both as a domain of study and a domain of reality. Global Positioning System (GPS) receivers have recently become small and inexpensive, and are increasingly used on an everyday basis by lay persons and specialists alike. A GPS receiver provides information about one's location by geometrically combining distance signals it picks up from several geosynchronous satellites constantly orbiting the earth. This system is part of the technology of automated navigation systems included with many automobiles and increasingly found on cell phones and other portable devices. Several research issues concern how locational information is best displayed or communicated by electronic information systems, and how the availability of such information might change people's experiences and behaviors in space and place. There are a host of spatial cognition questions inspired by efforts to improve the effectiveness and efficiency of Geographic Information Systems (GIS). How can complex geographical information be depicted to promote comprehension and effective decision-making, whether through maps, graphs, verbal descriptions, or animations? How does exposure to new geographic information technologies alter human ways of perceiving and thinking about the world? There are also research issues concerning the way that people do or do not spatialize their understanding of other types of information networks such as the World Wide Web; one often speaks metaphorically of 'navigating' the Web. Undoubtedly one of the most dramatic technological developments with great implications for spatial cognition is the advent of computer-simulated worlds known as virtual environments (virtual reality).

About This Book

This volume contains the 28 contributions that were presented at the international conference *Spatial Cognition 2006*, which was held at the University of Bremen, Germany in September 2006. *Spatial Cognition 2006* was the second international conference organized by the Transregional Collaborative Research Center SFB/TR 8 Spatial Cognition funded by the German Research Foundation (DFG). For this conference, 59 full paper contributions were received, from across the spectrum of research topics within spatial cognition. These submissions were evaluated by an international review committee in a thorough peer review process.

Acknowledgments

The organizers of the conference would like to thank the numerous people who have helped to make Spatial Cognition 2006 a great success. First of all, we thank Eva Räthe, Marion Stubbemann, Inga Buka, Julia Finken, and Martin Köster for their comprehensive organizational and administrative work and their editorial assistance, and we thank Holger Schultheis for maintaining the conference management system. We thank the authors for their careful work in writing their papers and their interesting presentations. We thank the members of the review committee for their careful and timely evaluations of the submitted contributions, as well as for their constructive suggestions. We gratefully acknowledge the financial support of the German Research Foundation (DFG) for their funding of the Transregional Collaborative Research Center SFB/TR 8 Spatial Cognition and the international Spatial Cognition conference series. Last but not least, we thank Alfred Hofmann and his colleagues at Springer for their ongoing support of our spatial cognition book series.

May 2007

Thomas Barkowsky
Markus Knauff
Gérard Ligozat
Daniel R. Montello

Bibliography

Binet, M.A.: Reverse illusions of orientation. Psychological Review 1, 337–350 (1894)
Downs, R.M., Stea, D.: Image and environment. Aldine Publishing Company, Chicago (1973)
Gibson, J.J.: The perception of the visual world. Houghton Mifflin, Boston (1950)
Golledge, R.G., Briggs, R., Demko, D.: The configuration of distances in intraurban space. Proceedings of the Association of American Geographers 1, 60–65 (1969)
Gulliver, F.P.: Orientation of maps. The Journal of Geography 7, 55–58 (1908)
Kuipers, B.: Modeling spatial knowledge. Cognitive Science 2, 129–153 (1978)
Lynch, K.: The image of the city. MIT Press, Cambridge, MA (1960)
Piaget, J., Inhelder, B.: The child's conception of space. Norton, New York (1948/1967)
Tolman, E.C.: Cognitive maps in rats and men. Psychological Review 55, 189–208 (1948)
Trowbridge, C.C.: Fundamental methods of orientation and imaginary maps. Science 38, 888–897 (1913)
Zadeh, L.A.: Fuzzy sets. Information and Control 8, 338–353 (1965)

Organization

Review Committee

Marco Aiello
Elena Andonova
Marios Avraamides
Philippe Balbiani
John Bateman
Thomas Bittner
Heinrich Bülthoff
Wolfram Burgard
Didac Busquets
Kenny Coventry
Ivo Duentsch
Susan L. Epstein
Carola Eschenbach
Sara Fabrikant
Kerstin Fischer
Udo Frese
Merideth Gattis
Sabine Gillner
Janice Glasgow
Gabriela Goldschmidt
Reg Golledge
Hans Werner Güsgen
Steffen Gutmann
Mary Hegarty
Stephen Hirtle
Hartwig Hochmair
Bernhard Hommel
Robin Hörnig
Kathleen Hornsby
Ludger Hovestadt

Barbara Kaup
Alexander Klippel
Werner Kuhn
Lars Kulik
Gerard Ligozat
Jack Loomis
Rainer Malaka
Kim Marriott
Timothy P. McNamara
Daniel R. Montello
Till Mossakowski
Martin Raubal
Jochen Renz
Thomas Röfer
Kerstin Schill
Christoph Schlieder
Cyrill Stachniss
Klaus Stein
Gerhard Strube
Thora Tenbrink
Sabine Timpf
Barbara Tversky
Laure Vieu
Jan Wiener
Stephan Winter
Stefan Wölfl
Mike Worboys
Emile van der Zee
Hubert Zimmer

Related Book Publications

Freksa, C., Knauff, M., Krieg-Brückner, B., Nebel, B., Barkowsky, T. (eds.): Spatial Cognition IV. LNCS (LNAI), vol. 3343. Springer, Heidelberg (2005)

Rodríguez, M.A., Cruz, I.F., Egenhofer, M.J., Levashkin, S. (eds.): GeoS 2005. LNCS, vol. 3799. Springer, Heidelberg (2005)

Meng, L., Zipf, A., Reichenbacher, T. (eds.): Map-based mobile services – Theories, methods and implementations. Springer, Berlin (2005)

Gero, J.S., Tversky, B., Knight, T. (eds.): Visual and spatial reasoning in design III, Key Centre of Design Computing and Cognition, University of Sydney (2004)

Blackwell, A., Marriott, K., Shimojima, A. (eds.): Diagrams 2004. LNCS (LNAI), vol. 2980. Springer, Heidelberg (2004)

Egenhofer, M.J., Freksa, C., Miller, H.J. (eds.): GIScience 2004. LNCS, vol. 3234. Springer, Heidelberg (2004)

Freksa, C., Brauer, W., Habel, C., Wender, K.F. (eds.): Spatial Cognition III. LNCS (LNAI), vol. 2685. Springer, Heidelberg (2003)

Kuhn, W., Worboys, M., Timpf, S. (eds.): COSIT 2003. LNCS, vol. 2825. Springer, Heidelberg (2003)

Barkowsky, T.: Mental Representation and Processing of Geographic Knowledge. LNCS (LNAI), vol. 2541. Springer, Heidelberg (2002)

Egenhofer, M.J., Mark, D.M. (eds.): GIScience 2002. LNCS, vol. 2478. Springer, Heidelberg (2002)

Hegarty, M., Meyer, B., Narayanan, N.H. (eds.): Diagrams 2002. LNCS (LNAI), vol. 2317. Springer, Heidelberg (2002)

Coventry, K., Olivier, P. (eds.): Spatial language: Cognitive and computational perspectives. Kluwer, Dordrecht (2002)

Renz, J.: Qualitative Spatial Reasoning with Topological Information. LNCS (LNAI), vol. 2293. Springer, Heidelberg (2002)

Montello, D.R. (ed.): COSIT 2001. LNCS, vol. 2205. Springer, Heidelberg (2001)

Gero, J.S., Tversky, B., Purcell, T. (eds.): Visual and spatial reasoning in design II, Key Centre of Design Computing and Cognition, University of Sydney (2001)

Habel, C., Brauer, W., Freksa, C., Wender, K.F. (eds.): Spatial Cognition II. LNCS (LNAI), vol. 1849. Springer, Heidelberg (2000)

Habel, C., von Stutterheim, C. (Hrsg.): Räumliche Konzepte und sprachliche Strukturen, Niemeyer, Tübingen (2000)

Habel, C., Werner, S. (eds.): Special issue on spatial reference systems, Spatial Cognition and Computation, vol. 1(4) (1999)

Gero, J.S., Tversky, B. (eds.): Visual and spatial reasoning in design, Key Centre of Design Computing and Cognition, University of Sydney (1999)

Freksa, C., Mark, D.M. (eds.): COSIT 1999. LNCS, vol. 1661. Springer, Heidelberg (1999)

Freksa, C., Habel, C., Wender, K.F. (eds.): Spatial Cognition. LNCS (LNAI), vol. 1404. Springer, Heidelberg (1998)

Egenhofer, M.J., Golledge, R.G. (eds.): Spatial and temporal reasoning in geographic information systems. Oxford University Press, Oxford (1997)

Hirtle, S.C., Frank, A.U. (eds.): COSIT 1997. LNCS, vol. 1329. Springer, Heidelberg (1997)

Burrough, P., Frank, A. (eds.): Geographic objects with indeterminate boundaries, Taylor and Francis, London (1996)

Frank, A.U., Kuhn, W. (eds.): COSIT 1995. LNCS, vol. 988. Springer, Heidelberg (1995)

Frank, A.U., Campari, I. (eds.): COSIT 1993. LNCS, vol. 716. Springer, Heidelberg (1993)
Frank, A.U., Campari, I., Formentini, U. (eds.): Theories and Methods of Spatio-Temporal Reasoning in Geographic Space. LNCS, vol. 639. Springer, Heidelberg (1992)
Mark, D.M., Frank, A.U. (eds.): Cognitive and linguistic aspects of geographic space. Kluwer, Dordrecht (1991)
Freksa, C., Habel, C. (Hrsg.): Repräsentation und Verarbeitung räumlichen Wissens, Informatik-Fachberichte 245, Springer, Berlin (1990)

Table of Contents

Spatial Reasoning, Human-Robot Interaction, and Assistance

Reachability and Dependency Calculi: Reasoning in Network Algebras 1
    Alexander Scivos

The Qualitative Trajectory Calculus on Networks 20
    Peter Bogaert, Nico Van de Weghe, Anthony G. Cohn, Frank Witlox, and Philippe De Maeyer

Qualitative Spatial Representation and Reasoning in the SparQ-Toolbox 39
    Jan Oliver Wallgrün, Lutz Frommberger, Diedrich Wolter, Frank Dylla, and Christian Freksa

Remembering Places in Space: A Human Analog Study of the Morris Water Maze 59
    Sylvia Fitting, Gary L. Allen, and Douglas H. Wedell

The Role of Users' Concepts of the Robot in Human-Robot Spatial Instruction 76
    Kerstin Fischer

Collaborative Assistance with Spatio-temporal Planning Problems 90
    Inessa Seifert

Visuo-Spatial Reasoning and Spatial Dynamics

Dialog-Based 3D-Image Recognition Using a Domain Ontology 107
    Joana Hois, Michael Wünstel, John A. Bateman, and Thomas Röfer

Protein Structure Prediction with Visuospatial Analogy............... 127
    Jim Davies, Janice Glasgow, and Tony Kuo

The Spatial Representation of Dynamic Scenes – An Integrative Approach 140
    Markus Huff, Stephan Schwan, and Bärbel Garsoffky

Modeling Geospatial Events and Impacts Through Qualitative Change 156
    Inga Mau, Kathleen Stewart Hornsby, and Ian D. Bishop

Spatial Concepts, Human Memory, and Mental Reasoning

Preferred Mental Models: How and Why They Are So Important in Human Reasoning with Spatial Relations 175
Marco Ragni, Thomas Fangmeier, Lara Webber, and Markus Knauff

The Spatial and the Visual in Mental Spatial Reasoning: An Ill-Posed Distinction 191
Holger Schultheis, Sven Bertel, Thomas Barkowsky, and Inessa Seifert

Grounded Perceptual Schemas: Developmental Acquisition of Spatial Concepts 210
Amitabha Mukerjee and Mausoom Sarkar

Modeling Human Spatial Memory Within a Symbolic Architecture of Cognition 229
Carsten Winkelholz and Christopher M. Schlick

Updating in Models of Spatial Memory 249
Björn Rump and Timothy P. McNamara

Sensorimotor Interference When Reasoning About Described Environments 270
Marios N. Avraamides and Melina-Nicole Kyranidou

Mechanisms for Human Spatial Competence......................... 288
Glenn Gunzelmann and Don R. Lyon

Navigation, Wayfinding, and Route Instructions

Algorithms for Reliable Navigation and Wayfinding 308
Shazia Haque, Lars Kulik, and Alexander Klippel

Interpreting Route Instructions as Qualitative Spatial Actions 327
Hui Shi, Christian Mandel, and Robert J. Ross

Knowledge Based Schematization of Route Directions 346
Samvith Srinivas and Stephen C. Hirtle

Map Use and Wayfinding Strategies in a Multi-building Ensemble 365
Christoph Hölscher, Simon J. Büchner, Tobias Meilinger, and Gerhard Strube

How Much Information Do You Need? Schematic Maps in Wayfinding and Self Localisation 381
Tobias Meilinger, Christoph Hölscher, Simon J. Büchner, and Martin Brösamle

Wayfinding Strategies in Behavior and Language: A Symmetric and
Interdisciplinary Approach to Cognitive Processes 401
    Thora Tenbrink and Jan M. Wiener

A Spatial Cognitive Map and a Human-Like Memory Model Dedicated
to Pedestrian Navigation in Virtual Urban Environments 421
    Romain Thomas and Stéphane Donikian

Linguistic and Social Issues in Spatial Knowledge Processing

The Influence of Scale, Context and Spatial Preposition in Linguistic
Topology 439
    Anna-Katharina Lautenschütz, Clare Davies, Martin Raubal,
    Angela Schwering, and Eric Pederson

Before or After: Prepositions in Spatially Constrained Systems 453
    Kai-Florian Richter and Alexander Klippel

Discourse Factors Influencing Spatial Descriptions in English and
German 470
    Constanze Vorwerg and Thora Tenbrink

Autobahn People: Distance Estimations Between German Cities Biased
by Social Factors and the Autobahn 489
    Claus-Christian Carbon

Author Index 501

Reachability and Dependency Calculi: Reasoning in Network Algebras

Alexander Scivos

Institut für Informatik,
Albert-Ludwigs-Universität Freiburg
Georges-Köhler-Allee 52,
79110 Freiburg, Germany
scivos@informatik.uni-freiburg.de

Abstract. Reasoning in complex systems of dependencies is important in our highly connected world, e. g. for logistics planning, and for the analysis of communication schemes and social networks. Directed graphs are often used to describe scenarios with links or dependencies. However, they do not reflect uncertainties. Further, hardly any formal method for reasoning about such systems is in use. As it is hard to quantify dependencies, calculi for qualitative reasoning (QR) are a natural choice to fill this gap. However, QR is so far concentrated on spatial and temporal issues. A first approach is the dependency calculus $\mathcal{DC}$ for causal relations [15], but it cannot describe situations in which cycles might occur within a graph. In this paper, refinements of $\mathcal{DC}$ meeting all requirements to describe dependencies on networks are investigated with respect to satisfiability problems, construction problems, and tractable subclassses.

1 Introduction

Reasoning about complex dependencies between events is a crucial task in many applications of our highly partitioned, but widely linked world. Whenever the required answer is a decision or classification, Qualitative Reasoning (QR) is best-suited: It abstracts from metrical details of the physical world and enables computers to make predictions about relations, even when precise quantitative information is not available or irrelevant [3]. QR is an abstraction that summarizes similar quantitative states into one qualitative characterization. From the cognitive perspective, the qualitative method *categorizes* features within the object domain rather than by *measuring* them in terms of some external scale [7]. This is the reason why qualitative descriptions are quite natural for humans.

The two main directions in QR so far are spatial and temporal reasoning. Topics of spatial reasoning comprise topological reasoning about regions [16,6], positional reasoning about point configurations [17], and reasoning about directions [4,9]. For temporal reasoning, either points or intervals are used as basic entities [19,1].

In contrast, this paper elaborates a new direction in QR: reasoning about links and dependencies. For describing linked systems, directed graphs (networks) are established. Network links can represent links in the internet, cash flow, railway connections, road systems, spreading of deseases (for medical analysis), information passed on in an organization, and genetic inheritage. There are many algorithms known for answering specific questions about a given network, e.g. if it is connected, or for finding the shortest path between two entities in it [5]. However, these algorithms assume that the network is defined, i.e. for each pair of entities, it is known how they are linked directly. For finding a network with specific properties and for reasoning with uncertainties, however, no calculus has been established yet.

A network of warehouses with possible delivery routes might illustrate such questions (cf. Fig. 1b). If a network is given, e.g. the following questions can be asked: Can site e deliver to a? Is there a route through f that runs in a cycle? If flawed goods are delivered to site c and to site d, is it possible that they come from a common source?

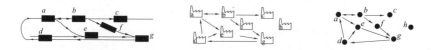

Fig. 1. Examples for networks: Train connections, delivery routes, abstract directed graph

Now assume that only for some pairs, we impose constraints on the relationship between them, but for other pairs, we are undecided: They might be linked or not. For example, is it possible to change the connection between e and g leaving all others as they are so that there is no cycle in the entire system? Or, can a link between e and f be inserted in either direction that is not part of a cyclic route?

Consider a more abstract level of reasoning: Without any other constraints, is it possible to construct a network in which a and b can deliver to each other, and c can be reached from a, but not from b? What follows for the relationship between a and b if d is directly linked to a, and if there is some cyclic path through b and d? This paper aims at giving a formal calculus that answers such questions automatically.

Different kinds of properties will be expressed by different labels.

1.1 Related Work

Of all spatial calculi, the point algebras for linear and partial orders come closest to the network calculi. The linear point algebra PA_{lin} was introduced 1989 by Vilain [19] with the basic labels $\{<, =, >\}$. The general satisfiability problem for PA_{lin} is in **PTIME** [19]. However, real-world problems do not necessarily have linear structures as underlying space. In the nineties, the development of extensions of the linear case into nonlinear structures started to address such problems. Broxvall [2] showed that the constraint satisfaction problem for the point algebra on nonlinear structures PA_{po} with the basic labels $\{\prec, =, \succ, \|\}$ is **NP**-hard. **NP**-hard problems usually have interesting fragments, the so-called tractable subclasses. A fragment of a relational algebra is a subset of relations that is closed under composition and converse. A subclass is called tractable if satisfiability can be decided in polynomial time. Broxvall identified three maximal tractable fragments of PA_{po}. Normally, for these classes the path-consistency method [12] decides satisfiability.

But, these approaches are still too coarse for some applications. For instance, to identify dependencies, it is not only important to state that two points are unrelated, but also to qualify if two points or states have a common ancestor or if there is no such 'decision' point. For this reason, Ragni and Scivos [15] introduced a dependency calculus ($\mathfrak{DC}$) with the basic labels $\{=, \prec, \succ,)($, $\Upsilon\}$ that meets all requirements to describe dependencies in acyclic networks. The network algebras described in this paper are generalizations of these calculi for arbitrary link structures, including networks with cycles.

1.2 Outline

The remainder of this paper is organized as follows: In Section 2, the traditional formalism for graphs and for constraint satisfaction problems is presented. In Section 3, new calculi are introduced from a formal point of view and features of the calculi are described. In Section 4, their computational complexity is investigated. Section 5 summarizes the results and suggests questions for further research.

2 Formalization

2.1 Networks and Conditions

Definition 1. *A* **network** *is a finite directed graph* $\mathfrak{N} = (V, E)$, *where* $V = \{v_1, ..., v_m\}$ *is called* **vertex set** *and* $E \subset V \times V$ *is the set of* **edges** *or* **links***.*

For a pair $(a, b) \in V$, a with respect to b is either

- equal: $a = b$
- preceding: $a \bullet\!\!-b$
- succeeding: $a-\!\!\bullet b$
- linked both ways: $a \bullet\!\!-\!\!\bullet b$
- or distant: $a \,]\![\, b$

From these basic properties, other properties of pairs of vertices can be deduced, such as the **transitive closure** of $\bullet\!\!-, -\!\!\bullet$

$$x \prec y :\leftrightarrow \exists k \geq 0, w_1, \ldots w_k : x \bullet\!\!-w_1 \bullet\!\!- \ldots w_k \bullet\!\!-y,$$
$$x \succ y :\leftrightarrow \exists k \geq 0, w_1, \ldots w_k : x -\!\!\bullet w_1 -\!\!\bullet \ldots w_k -\!\!\bullet y,$$

and the property of being part of a common cycle:

$$x \Diamond y :\leftrightarrow x \prec y \text{ and } y \prec x.$$

The idea will be to label a finite set of properties covering all possible situations.

Reasoning techniques on the labels assure that new assertions about them can be soundly deduced by only knowing these properties.

Example: From $\prec (v, x)$, the converse relation $\succ (x, v)$ can be concluded. Therefore, the formalism of network algebras must be introduced.

2.2 Reasoning in Calculi

As Ligozat and Renz [10] point out, almost all successful calculi are based on a formalism once described by Maddux [11]. Reasoning in such algebras is described in more detail in [14]. Recall the algebraic definition:

Definition 2 (Non-associative Algebra). *A non-associative algebra $\mathcal{R}$ is a tuple $\mathcal{R} = (R, \cup, \cap, \bot, \top, ;, \breve{}, eq)$ such that:*

1. $(R, \cup, \cap, \bot, \top)$ *is a Boolean algebra.*
2. *eq is a constant, $\breve{}$ a unary and ; a binary operation s. t., for any $a, b, c \in R$:*
   $(a\breve{})\breve{} = a$ $\qquad\qquad\qquad\qquad$ $eq; a = a; eq = a$
   $a; (b \cup c) = a; b \cup a; c$ $\qquad\qquad$ $(a; b)\breve{} = b\breve{}; a\breve{}$
   $(a \cup b)\breve{} = a\breve{} \cup b\breve{}$ $\qquad\qquad\quad$ $(a \cap b)\breve{} = a\breve{} \cap b\breve{}$
   $(a; b) \cap c\breve{} = \bot$ *if and only if* $(b; c) \cap a\breve{} = \bot$

Note: A non-associative algebra is a relation algebra in the sense of Tarski [18] if it is associative (i.e. $(a; b); c = a; (b; c)$).

In the case of networks, the elements $a, b, \ldots \in R$ of the non-associative algebra are labels on which constraints for pairs of points can be defined. If $a \cap b = a$, we write $a \sqsubseteq b$. Recall that b is called atomic or basic if $x \sqsubseteq b$ implies $x \in \{\bot, b\}$ and $b \neq \bot$. The set of basic labels is denoted by $B(\mathcal{R})$. In finite algebras like the ones relevant in this paper, each label can be uniquely written as a union of basic labels.

In the following, we will consider calculi in which the set $B(\mathcal{R})$ is jointly exhausive and pairwise disjoint ("JEPD"). Then, the semantic is described by a complete **relabeling function** rel that maps each pair (x, y) of vertices in a network to its unique basic label $rel(x, y) \in B(\mathcal{R})$. For each network, the inverse function rel^{-1} is a homomorphism of the Boolean algebra, especially it respects $\cap, \cup$. The operations of the network algebra mirror the semantic facts:

$$rel(u, u) = eq$$
$$rel(v, u) = (rel(u, v))\breve{}$$
$$rel(u, w) \sqsubseteq rel(u, v); rel(v, w)$$

A non-associative algebra with such a semantic is called a **network algebra**.

In the calculus, we introduce variables for vertices of the (intended) network. Knowledge and requirements are formally described in terms of labels for all pairs of variables.

Definition 3 (Constraint System). *Recall that a* **constraint system** *$\Pi = (X, c)$ over a non-associative algebra consists of a set of variables X and a function $c : X \times X \to R$ that assigns a label to each pair of variables (written as constraints: $x\ c_{xy}\ y$).*

A constraint system where all labels are basic is called a **scenario***.*

Assume that a set of constraints between some points is given. One question might be whether this set is consistent: Is it possible to construct a network in which all the constraints are satisfied? And, what is the computational effort for constructing it?

Definition 4 (Satisfiability). *Let R be a set of relations of a network algebra. An* **interpretation** *of a constraint system $\Pi = (X, c)$ is a pair $(\mathcal{N}, \theta)$ where $\mathcal{N} = (V, E)$*

is a directed graph ("network") and $\theta : X \to V$ maps all variables to vertices of the network. An interpretation **satisfies** the constraint system iff for all variables $x, y \in X$:

$$rel(x^\theta, y^\theta) \sqsubseteq c_{xy}.$$

For a class $\mathcal{C}$ of directed graphs, if $\mathcal{N} \in \mathcal{C}$ satisfies the constraint system Π, it is called a $\mathcal{C}$-model of Π. A constraint system that has a ($\mathcal{C}$-)model is called ($\mathcal{C}$-)**consistent**.

Obviously, if $\bot$ occurs as a constraint, the system cannot be consistent.

Assume that a network algebra is fixed. Typically, reasoning is done to answer the following questions: Is there a model for a given constraint system? What does it look like? What is entailed?

Often, a model has to obey certain conditions. Such conditions are imposed by restricting the models to a class $\mathcal{C}$ of networks. A network is called (cf Fig. 2)

- **strongly connected** if for any two vertices $v \neq w$, $v \prec w$ and $w \prec v$ holds.
- **path-connected** if for any two $v \neq w$, $v \prec w$ or $w \prec v$ holds.
- **acyclic** if the model's network $\mathfrak{N}$ is cycle-free.
- **linear** if the model is a path ($V = (v_1, \ldots, v_n)$, $E = \{(v_i, v_{i+1}) | 1 \leq i < n\}$). Then it is acyclic and path-connected.

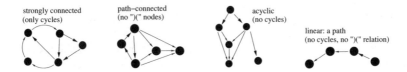

Fig. 2. Some classes of networks: Strongly connected, path-connected, acyclic, and linear

Definition 5 (Reasoning Problems). *For each network algeba and graph class $\mathcal{C}$, there are two types of problems. In both cases, a constraint system Π is given.*

  i. *The constraint satisfiability problem $CSP_\mathcal{C}(\mathcal{R})$: Does Π have a $\mathcal{C}$-model?*
  ii. *The network design problem $NDP_\mathcal{C}(\mathcal{R})$: Find a $\mathcal{C}$-model for Π.*

The size of a problem instance is the number of variables $|X|$.

In an interpretation, a graph is chosen in a way that each variable X is assigned to a vertex, but this assignment is not required to be surjective: Not all elements of the graph must correspond to a variable.

Thus this definition of a satisfiability problem is a generalization of the classical definition of a *Constraint Satisfaction Problem* (**CSP**): For a description consisting of

- a set X of n variables $\{x_1, \ldots, x_n\}$,
- the possible values D_i of variables x_i,
- constraints (sets of relations) over subsets of variables,

is it possible to find an assignment $x_i \mapsto D_i$ satisfying all constraints?

In our case, the possible values D_i are not fixed. We only know that the possible values D_i are part of a member in the class of directed graphs. A CSP is a satisfiability problem with a class that consists of only one given graph.

Lemma 1. *For a class C of directed graphs and set of relations $\mathcal{R}$,*

a. *if $CSP_C(\mathcal{R})$ is in P, the same holds for $NDP_C(\mathcal{R})$, and vice versa.*
b. *if $CSP_C(\mathcal{R})$ is NP-complete, $NDP_C(\mathcal{R})$ is NP-complete, too.*
c. *If P$\neq$ NP, also the reverse direction holds.*

Proof. Obviously, $NDP_C(\mathcal{R})$ is at least as difficult as $CSP_C(\mathcal{R})$.

For a., assume that $CSP_C(\mathcal{R})$ is in P. A PTIME algorithm for $NDP_C(\mathcal{R})$ is given by:

```
for x, y ∈ X:                        (* choose a pair *)
    set possible[x, y] ← false;      (* initialize *)
    set c' ← c_xy                    (* choose a constraint *)
    for {b} ⊑ c':                    (* choose a basic relation *)
        set c_xy ← {b};              (* fix basic relation b *)
        if satisfiable(new constraint system): (* test is in PTIME, by assumption *)
            set possible[x, y] ← true;
            exit inner loop;
    endfor;
    if possible[x, y] = false:
        return not satisfiable;       (* no feasible basic relation found *)
endfor;                              (* take next constraint *)
return the selected scenario.         (* as all constraints are possible *)
```

For b., first assume that $CSP_C(\mathcal{R})$ is NP-complete. If the system is consistent, a scenario can be guessed, i.e. a jointly satisfiable set of basic relations. Therefore, by assumption, a model can be guessed and verified in polynomial time.

c. Assume that $NDP_C(\mathcal{R})$ is NP-complete, but $CSP_C(\mathcal{R})$ is in an easier class C, without loss of generality P$\subseteq C$. Then the algorithm of part a. shows that $NDP_C(\mathcal{R})$ is in C in contradiction to the assumption.

3 The Network Calculi

3.1 The Refined Reachability Calculus $\mathfrak{RRC}$

To answer reasoning questions like the ones posed for the introductory delivery route example, it is necessary to express if two verices are

- directly connected by a link,
- linked by a path (with other sites "in between"),
- part of a possible cyclic path,
- indirectly connected in another way, or
- totally disconnected from each other.

Table 1. The refined reachability calculus $\mathfrak{RRC}$ distinguishes 10 different situations for pairs of vertices. They are based on the direct links and their transitive closures. [+ = yes, - = no].

$\mathfrak{RRC}$ relation	symbol	properties satisfied						
		=	][	•—	≺	—•	≻	•—•
equal	=	+	-	-	-	-	-	-
not being part of a cycle:								
parallel	$\|\|$	-	+	-	-	-	-	-
(indirectly) preceding	≺≺	-	+	-	+	-	-	-
previous	•≺	-	-	+	+	-	-	-
(indirectly) succeeding	≻≻	-	+	-	-	-	+	-
next	≻•	-	-	-	-	+	+	-
being part of a cycle:								
distant-in-cycle	◇	-	+	-	+	-	+	-
previous-in-cycle	◆◇	-	-	+	+	-	+	-
next-in-cycle	◇◆	-	-	-	+	+	+	-
linked both ways	•—•	-	-	-	+	-	+	+

Fig. 3. Two models of the constraint system $x\{≺≺, ◇\}y$, $x\{\|\|, •—•, ◇◆\}z$, $y\{◆◇\}z$. However, it has no models if no additional points are allowed because in any model, there would be a path from x to y, but no direct link from x or z to y.

The non-associative algebra based on $\{=,][, •—, —•, •—•, ≺, ≻\}$ is not JEPD. The JEPD refinement, which we call the **refined reachability calculus** $\mathfrak{RRC}$, distinguishes the 10 basic labels $=,][, ≺≺, •≺, ≻≻, ≻•, ◇, ◆◇, ◇◆, •—•$ explained in Table 1.

Consider an example for an $\mathfrak{RRC}$ constraint system:

$$\Pi = (\{x, y, z\},\ \{x\{≺≺, ◇\}y,\ x\{\|\|, •—•, ◇◆\}z,\ y\{◆◇\}z\})$$

It can be solved as shown in Fig. 3 (left sketch) with an additional vertex v^*:

$$x^\theta ◇ y^\theta,\ x^\theta •—• z^\theta,\ y ◆◇ z^\theta,\ x^\theta ◆◇ v^*,\ v^* ◆◇ y^\theta,\ z^\theta ◆◇ v^*.$$

In most traditional reasoning algebras, omitting some elements from a CSP does not destroy fulfilment. But, in the $\mathfrak{RRC}$ example consistency is lost by omitting v^*, v': A submodel of a model no longer satisfies all its relations.

The answer of the NDP or CSP may change whether new vertices might be added (**open problem**) or not (**confined, or closed problem**). Indeed, Π has no model with surjective θ.

Definition 6 (Confined Problems). *A CSP or NDP is called* **confined** *iff an interpretation $\mathfrak{M} = (\mathfrak{N}, \theta)$ is only accepted as a model if θ is surjective.*

In the confined case, a model of $\Pi = (X, c)$ has at most $|X|$ vertices.

The open problem might have more models than the confined version as the above example illustrates. $\mathfrak{RRC}$ has this feature because its relations $\succ\!\!\!\succ, \prec\!\!\!\prec, \Diamond, \Diamond\!\!\bullet,$ and $\bullet\!\!\Diamond$ (and unions of them) have a special "nonlocal" property:

Definition 7 (Local Relations). *A relation r in a network algebra $\mathfrak{R}$ is called **local** iff for any model $\mathfrak{M} = (\mathfrak{N}, \theta)$ of $\Pi_r := (\{x, y\}, c_{xy} = r)$, each submodel of $\mathfrak{M}$ containing x and y is still a model of Π_r.*

*A relation r in a network algebra $\mathfrak{R}$ is called **weakly local** iff for any model $\mathfrak{M} = ((V, E), \theta)$ of Π_r there is a model $\mathfrak{M}' = ((V, E'), \theta)$ of Π_r with $rel(v^{\mathfrak{M}'}, w^{\mathfrak{M}'}) = rel(v^{\mathfrak{M}}, w^{\mathfrak{M}})$ for all $v, w \in V$ so that each submodel of $\mathfrak{M}'$ containing x^θ and y^θ is still a model of Π_r.*

*A relation r is called **sociable** iff it has the following property: If $\mathfrak{N}$ is a substructure of $\mathfrak{N}^+$, and $\mathfrak{M} = (\mathfrak{N}, \theta)$ is a model of Π_r, $\mathfrak{M}^+ = (\mathfrak{N}^+, \theta)$ is a model of Π_r. All other relations are called **nonlocal**.*

*An algebra, CSP, NDP, or constraint system is called **localic** iff all its relations are weakly local, and **convivial** iff all its relations are sociable.*

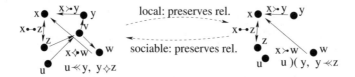

Fig. 4. The idea of local and sociable relations: Local relations are preserved under submodels, sociable ones when the model is enlarged. For instance, $=$ and $\bullet\!\!-\!\!\bullet$ are local and sociable relations as they are independent of all other reachability properties. $\succ\!\!\bullet$ is local, but not sociable (as the pair (x, w) shows). $\{\prec\!\!\!\prec, \Diamond\}$ is sociable, but $\prec\!\!\!\prec$ is nonlocal: In an extension of a model, links to further vertices might close the cycle (as v does for the pair y, z).

Some examples are illustrated in Fig. 4. Each local relation is weakly local. In a localic constraint system, any submodel of a model satisfies all its relations. Adding additional points does not "switch on" satisfiability. In convivial constraint systems, adding additional points may "switch on" satisfiability, but not "switch off" satisfiability.

Traditional calculi are mostly localic. For them, a necessary condition for a consistent CSP is that all its submodels of size $\leq k$ (for some fixed k) are consistent. A popular approach for a tractability proof uses a PTIME algorithm that computes a stronger necessary local condition that is proven to be a sufficient condition for satisfiability.

Recall the following condition used in the famous path consistency algorithm [12].

Definition 8. *A **constraint system** is **(locally) k-consistent** iff for any choice of k variables and any consistent instantiation of $j \leq k - 1$ variables of them, there is a consistent extension to all k variables. 3-consistency is called **path-consistency**.*

The path consistency strategy for solving a localic CSP Π is to apply the operations $\check{}$ and $\circ$ to derive a satisfiability-equivalent k-consistent CSP Π'. Π' can be derived in PTIME, and it is inconsistent if and only if it contains the constraint $\bot$.

(locally) k–consistent

(for any k–tuple: Any model
of its first k−1 points can be extended
to a model of all k points.)

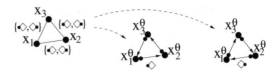

Fig. 5. An example for a locally 3-consistent constraint system. Any instantiation of variables x_1 and x_2 can be completed to a model, similarly for other choices of 2 vertices.

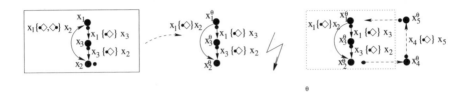

Fig. 6. The constraint system $(\{x_1, x_2, x_3\}, \{x_1\{\blacklozenge, \Diamond\}x_2, x_1\{\blacklozenge\}x_3, x_3\{\blacklozenge\}x_2\})$ is open-satisfiable, as the third sketch shows. It is not locally 3-consistent, but 3-open-consistent: An instantiation of $x_1\{\blacklozenge\}x_2$ is not consistently extendable to the 3 vertex subsystem $\{x_1, x_2, x_3\}$. However, by adding more vertices, it can be extended to a model of the open case.

However, in non-localic algebras, the model can be satisfiable although a submodel is not satisfiable. The notion of (local) k-consistency is not adequate (cf. Fig. 6), so the definition must be adapted.

Definition 9 (Local Consistency for Open CSPs). *A constraint system Π is k-open-consistent iff for each subsystem $\Pi^k \subseteq \Pi$ with k variables holds: Each model of $\Pi^- \subseteq \Pi^k$ can be extended to a model of Π^k.*

Definition 10 (Local Consistency for Confined CSPs). *A subsystem $\Pi^- \subset \Pi$ is called **globally satisfiable** (w. r. t. the constraint system Π) iff there is a satisfiable system Π^+ : $\Pi^- \subset \Pi^+ \subset \Pi$. A constraint system Π is globally k-consistent iff each consistent subsystem $\Pi^- \subset \Pi$ with $|\Pi^-| \leq k$ is globally satisfiable w. r. t. Π.*

Global k-satisfiability of $\mathfrak{RRC}$ is a necessary condition for satisfiabilty. However, generally, neither global k-satisfiability nor k-consistency could be verified in **PTIME**.

Another possible strategy is to find coarser nonlocalic calculi.

3.2 The Coarse Reachability Calculus $\mathfrak{CRC}$

The nonlocality grounds in $\mathfrak{RRC}$'s ability to distinguish between $\blacktriangleleft$ and $\twoheadleftarrow$. By coarsening it, we get a localic calculus: If only reachability is important, the simple **coarse reachability calculus** $\mathfrak{CRC}$ is sufficient. It distinguishes 5 situations as Table 2 shows.

Proposition 1 (Localicity of $\mathfrak{CRC}$). *All $\mathfrak{CRC}$ relations are weakly local. Open $\mathfrak{CRC}$-CSP = confined $\mathfrak{CRC}$-CSP.*

Table 2. The relations of $\mathfrak{CRC}$, its converse and composition. [+ = yes, - = no].

$\mathfrak{CRC}$ relation r	symbol	properties			converse $\breve{r}$	composition with...			
		$=$	$\prec$	$\succ$		$=$	$\parallel$	$\prec$	$\succ$	$\diamond$
equal	$=$	+	-	-	$=$	$=$	$\parallel$	$<$	$>$	$\diamond$
parallel	$\parallel$	-	-	-	$\parallel$	$\parallel$	$\top$	$\{<,\parallel\}$	$\{>,\parallel\}$	$\parallel$
before	$<$	-	+	-	$>$	$<$	$\{<,\parallel\}$	$<$	$\top$	$<$
after	$>$	-	-	+	$<$	$>$	$\{>,\parallel\}$	$\top$	$>$	$>$
cyclic	$\diamond$	-	+	+	$\diamond$	$\diamond$	$\parallel$	$<$	$>$	$\diamond$

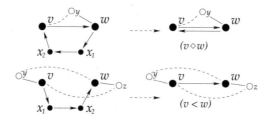

Fig. 7. Examples for substituting a "nonlocal" situation in a model by a "local" one

Proof. (Sketch) For a model $\mathfrak{M}$, an equivalent model $\mathfrak{M}'$ is constructed by adding the transitive closure of the edges. $\mathfrak{M}'$ has the desired property: Reachability between two vertices v^θ, w^θ does not change if additional vertices are omitted, and all $\mathfrak{CRC}$ relations depend on reachability only. A model $\mathfrak{M}$ of an open $\mathfrak{CRC}$-CSP (X, c) is transformed to a model of the closed case by removing all vertices except $\{x^\theta | x \in X\}$ from the equivalent model $\mathfrak{M}'$. Fig. 7 illustrates this process (x_i are the omitted verticses).

3.3 Dependency Calculi

Remember the initial example (cf. Fig. 1): It yields the network

$$\mathfrak{N} = (V = \{a, b, \ldots, h\}, \ E = \{(a, b), (a, e), (b, c), (b, f), \ldots\}).$$

If at c and e trains are delayed, can this have a common cause? If c and e have stored the same product, can it be from the same origin? Do c and e have a common ancestor?

The answer to this question is the dependency property (dep) for vertices v and w:

$$\exists u : u \prec v \text{ and } u \prec w \quad \text{(dep)}$$

If $v \prec w$, $v = w$ or $w \prec v$, (dep) trivially holds for (v, w).
But, (dep) splits the $\mathfrak{RRC}$ relation $\parallel$ into two new relations:

- $v \curlyvee w :\leftrightarrow v \parallel w$ and $\exists u : u \prec v \wedge u \prec w$
- $v \curlywedge w :\leftrightarrow v \parallel w$ and $\neg \exists u : u \prec v \wedge u \prec w$

In Fig. 1, for instance, $c \curlyvee e, c \not\curlyvee h, d \not\curlyvee e$ holds. This leads to the dependency calculus. It has a refined version $\mathfrak{RDC}$ with information about links, reachability, and dependency (cf. Table 3), and a coarse version $\mathfrak{CDC}$ that does not express knowledge about individual links (cf. Table 4). Both calculi are JEPD.

Table 3. The relations of the refined dependency calculus $\mathfrak{RDC}$. [+ = yes, - = no].

$\mathfrak{RDC}$ relation r	symbol	=	][	●≺	≺	●→	≻	●●	dep	converse $r^\smile$
equal	=	+	-	-	-	-	-	-	+	=
parallel:										
unrelated	)(	-	+	-	-	-	-	-	-	)(
forked	Υ	-	+	-	-	-	-	-	+	Υ
comparable:										
earlier	≪	-	+	-	+	-	-	-	+	≫
previous	●≺	-	-	+	+	-	-	-	+	≻●
later	≫	-	+	-	-	-	+	-	+	≪
next	≻●	-	-	-	-	+	+	-	+	●≺
being part of a cycle:										
distant-in-cycle	◇	-	+	-	+	-	+	-	+	◇
previous-in-cycle	◆◇	-	-	+	+	-	+	-	+	◇●
next-in-cycle	◇●	-	-	-	-	+	+	-	+	◆◇
linked both ways	●●	-	-	-	+	-	+	+	+	●●

Table 4. The relations of the coarse dependency calculus $\mathfrak{CDC}$. [+ = yes, - = no].

$\mathfrak{CDC}$ relation	symbol	properties				converse $r^\smile$	composition with ...					
		=	≺	≻	dep		=	)(	Υ	<	>	◇
equal	=	+	-	-	+	=	=	)(	Υ	<	>	◇
unrelated	)(	-	-	-	-	)(	)(	⊤	{)(, Υ, <}	{)(, Υ, <}	)(	)(
forked	Υ	-	-	-	+	Υ	Υ	{)(, Υ, >}	⊤	{Υ, <}	{)(, Υ, >}	Υ
before	<	-	+	-	+	>	<	{)(, Υ, <}	<	⊤	<
after	>	-	-	+	+	<	>	{)(, Υ, >}	{Υ, >}	{=, Υ, <, >, ◇}	>	>
cyclic	◇	-	+	+	+	◇	◇	)(	Υ,	<	>	◇

In both calculi the Υ relation is nonlocal because of the ∃-quantification in the dependency property.

4 Computational Complexity

We have defined the four calculi of Table 5. For the refined calculi, there are open and confined versions of the CSP (or NDP).

How hard is reasoning in the network algebras? By Dijkstra's algorithm [5], checking reachability is possible in **PTIME**. Also, the dependency property (dep) for vertices v, w can be checked in **PTIME**: For each vertex z we need to check if v and w are reachable from it ($2|V|$ checks, each in **PTIME**). Hence, transforming a network (V, E) with a complete link list to its $\mathfrak{RRC}$ description, or to its $\mathfrak{RDC}$ description is feasible in **PTIME**.

What about the other direction? Suppose an $\mathfrak{RRC}$ or $\mathfrak{RDC}$ description is given and the links in the network should be derived. What is the time complexity of their CSP or NDP? Table 6 and 7 list all results that will be proven in this chapter.

Table 5. Overview over the four network calculi

	reachability calculus	dependency calculus
coarse calculi:	$\mathfrak{CRC}$ (5 base relations)	$\mathfrak{CDC}$ (6 base relations)
refined calculi:	$\mathfrak{RRC}$ (10 base relations)	$\mathfrak{CDC}$ (11 base relations)

Table 6. Overview of the computational complexity of the reachability calculus. Proofs for the results with * are sketched in this paper, all others follow as coarser calculi, subcases or refinements.

network algebra	linear	strongly conn.	path-conn.	acyclic	all networks
$B(\mathfrak{CRC})$	PTIME	PTIME	PTIME	PTIME	PTIME
$B(\mathfrak{RRC})$	PTIME	PTIME	PTIME	PTIME	PTIME*
$\mathfrak{CRC}$	PTIME	PTIME	PTIME*	NP-c.*	NP-c.*
open $\mathfrak{RRC}$	PTIME	PTIME	PTIME*	NP-c.	NP-c.
confined $\mathfrak{RRC}$	NP-c.*	NP-c.*	NP-c.	NP-c.	NP-c.

Table 7. Overview of the computational complexity of the dependency calculus

network algebra	linear	strongly conn.	path-conn.	acyclic	all networks
$B(\mathfrak{CDC})$	PTIME	PTIME	PTIME	PTIME	PTIME
$B(\mathfrak{RDC})$	PTIME	PTIME	PTIME	PTIME	PTIME
open $\mathfrak{CDC}$	PTIME	PTIME	PTIME	NP-c.	NP-c.
confined $\mathfrak{CDC}$	PTIME	PTIME	PTIME	NP-c.	NP-c.
open $\mathfrak{RDC}$	PTIME	PTIME	PTIME	NP-c.	NP-c.
confined $\mathfrak{RDC}$	NP-c.	NP-c.	NP-c.	NP-c.	NP-c.

4.1 Reasoning about Scenarios

Proposition 2 ($B(\mathfrak{RRC})$-**NDP and** $B(\mathfrak{RDC})$-**NDP are in PTIME**). *For a given $\mathfrak{RRC}$ scenario or $\mathfrak{RDC}$ scenario, finding a model (or determining inconsistency) is in PTIME, both for the open and for the confined case.*

Proof. Assume that an $\mathfrak{RRC}$ (or $\mathfrak{RDC}$) constraint system Π is given. In both cases, first check 3-open-consistency (called **path consistency**), especially check if "=" is symmetric and transitive. Only if Π is path-consistent, a model (V, E) may exist. Start with $V := X$ and let θ be the identity with the exception that θ maps variables x, y to the same vertex iff c_{xy} is "=". For each pair $x \neq y$, denote which of the properties][, $\bullet\!\!-$, $-\!\!\bullet$, $\bullet\!\!-\!\!\bullet$ $\prec, \succ$ (and (dep)) it should have according to Table 1. The properties][, $\bullet\!\!-$, $-\!\!\bullet$, $\bullet\!\!-\!\!\bullet$ define the edge set E. Because of the path consistency, the $\prec$ and $\succ$ properties are transitive, and in all cases, $x \prec y, y \prec x$ implies $x \Diamond y$.

For the open case, for each $x \prec y$, a new vertex v_{xy} is introduced with $x \bullet\!\!-\! v_{xy}$ and $v_{xy} \bullet\!\!-\! y$ as the only links. (For $\mathfrak{RDC}$, additionally for each $x \curlyvee y$, a new vertex w_{xy} is introduced with $w_{xy} \bullet\!\!-\! x$ and $w_{xy} \bullet\!\!-\! y$ as the only links.)

If a model exists, all direct links have to follow the denoted link properties. Based hereon, collect all vertices reachable in 1 step, in ≤ 2 steps, etc. At most after $n = |V|$

steps, all $\prec, \succ$ relations are known. The task is solved by checking if they coincide with all the denoted properties.

By Lemma 1, also the corresponding CSPs on scenarios are in PTIME. Also, the coarser calculi inherit this property:

Corollary 1. $B(\mathfrak{RRC})$-CSP, $B(\mathfrak{RDC})$-CSP, $B(\mathfrak{CRC})$-CSP, $B(\mathfrak{CRC})$-NDP, $B(\mathfrak{CDC})$-CSP, and $B(\mathfrak{CDC})$-NDP are in PTIME.

Note that this proof also works for restrictions to our four subclasses of networks. Each subclass $\mathcal{C}$ corresponds with a subclass $S_\mathcal{C} \subseteq R$ of the labels: A network is in class $\mathcal{C}$ iff only labels $l \in S_\mathcal{C}$ describe its relations. For example, $S_{\text{str. conn.}} = \{\{\Diamond\}, \{\blacklozenge\}, \{\Diamond\}, \{\bullet\bullet\}\}$ Before the algorithm is applied, the condition has to be checked if all labels comply with the subclass restriction, e.g. if a model is a strongly connected network, only $\{\Diamond\}, \{\blacklozenge\}, \{\Diamond\}, \{\bullet\bullet\}$ can occur.

Hence, all network CSPs and NDPs are in NP because a scenario can be guessed and verified in PTIME. We will see that some of them are NP-hard, others in PTIME. In the following, we will not always investigate both, NDP and CSP as by Lemma 1, it is sufficient to know the complexity of one of the problems.

4.2 Polynomial Reductions

In order to reduce the amount of proofs needed, we first investigate how some problems can be reduced to others.

Lemma 2 (open $\mathfrak{RRC}$-NDP $<_P$ confined $\mathfrak{RRC}$-NDP). *The open $\mathfrak{RRC}$-NDP (also, open $\mathfrak{RDC}$-NDP) can be polynomially reduced to the confined $\mathfrak{RRC}$-NDP (and confined $\mathfrak{RDC}$-NDP, resp.)*

Proof. Let a constraint system $\Pi = (X, \{c_{xy}|\, x, y \in X\})$ be given. The idea is to add new variables for each pair of vertices that might need additional nodes and let $\top$ hold for all new constraints. $\Pi' = (X', c')$ of size $O(|\Pi|^2)$ is constructed whereas

$$X' = X \cup \{z_{xy}|x \neq y \in X\}, \; c'_{xy} = \begin{cases} c_{xy} & \text{;if } x, y \in X \\ \top & \text{;else} \end{cases}.$$

Claim: Π' (as confined NDP) is satisfiability-equivalent to Π (as open NDP).

A (confined) model of Π' is also a model of the open case because $X^\theta \subseteq (X')^\theta$ and the relevant properties (links, reachability,...) among the original vertices X^θ are unchanged.

For each model $\mathfrak{N} = (V, E)$ of the original open NDP instance (X, c), a corresponding model $\mathfrak{N}' = (V', E')$ of the constructed confined NDP (X', c') is obtained by the following process. Let $V' = V \cup (X' \setminus X)$. We trivially extend θ to X'. Then, between vertices of $(X')^\theta$, let the following link assignments E' hold:

For original variables $x, y \in X$, let $(x^\theta, y^\theta) \in E' \iff (x^\theta, y^\theta) \in E$.
For all $x, y \in X$, let $(z_{xy}^\theta, y^\theta) \in E'$
Let $(x^\theta, z_{xy}^\theta) \in E' \iff rel(x^\theta, y^\theta) \in \{\bullet\!\!\prec, \prec\!\!\prec, \blacklozenge, \Diamond, \Diamond\}$,
Let $(z_{xy}^\theta, x^\theta) \in E' \iff rel(x^\theta, y^\theta) \in \{=, \curlyvee\}$

For other pairs, no link is established. This definition lets the relevant properties (reachability, direct links, dependency) be equivalent in both networks as can easily be verified.

Recall that in the coarse reachability calculus $\mathfrak{CRC}$, the open and the confined case are satisfiability-equivalent. Instead of Open $\mathfrak{CRC}$= Confined $\mathfrak{CRC}$, we write $\mathfrak{CRC}$.

Lemma 3 (The Open $\mathfrak{RRC}$-CSP is as Complex as the $\mathfrak{CRC}$-CSP). *Open $\mathfrak{RRC}$-CSP can be polynomially reduced to $\mathfrak{CRC}$-CSP, and $\mathfrak{CRC}$-CSP to Open $\mathfrak{RRC}$-CSP.*

Lemma 3 can be proven by applying a similar construction as in the previous proof.

4.3 Proofs for NP-Completeness

What is the computational complexity of the confined $\mathfrak{RRC}$-NDP?

Theorem 1. *Confined $\mathfrak{RRC}$-CSP (hence, $\mathfrak{RRC}$-NDP) is NP-complete, even when restricted to strongly connected networks.*

The proof is given by a reduction from 3-SAT to the CSP over the fragment

$$\mathcal{F} = \{\{\Diamond\}, \{\blacklozenge\}, \{\Diamond\!\!\!\bullet\}, \{\blacklozenge, \Diamond\!\!\!\bullet\}\} \subset R$$

Assume that an instance of 3-SAT is given with variables $v_1, \ldots v_n$, clauses $c_1, \ldots, c_m$ and literals $l_{jk}(j \leq 3, k \leq m)$.

A corresponding CSP will be constructed in which for each 3-SAT element (each variable, literal, clause), there are two variables which we call a "left" and a "right" one, e.g. for each 3SAT variable v_i, the constructed CSP contains two variables v_i^l and v_i^r. The only possible direct links ("bridges") from left to right vertices will be at the CSP variables v_i^l and v_i^r.

For all v_i, the constraint system includes $v_i^l\{\blacklozenge, \Diamond\!\!\!\bullet\}v_i^r$. If $\blacklozenge$ is realized in an interpretation (i.e. $(v_i^l)^\theta\{\bullet\!\!-\}(v_i^r)^\theta$), the variable v_i is regarded as assigned to "true", otherwise "false" (cf. Fig. 8).

Fig. 9 illustrates the following details: Direct links are demanded between each clause c and all its literals which are linked to the corresponding variable or its negation. There is a path from $(c_j^l)^\theta$ to $(c_j^r)^\theta$ iff at least one of the l_{jk} is true in the assignment corresponding with the choice of $v_i^l\{\blacklozenge, \Diamond\!\!\!\bullet\}v_i^r$ (dashed double-arrow). Whether the vertex $(c_j^r)^\theta$ is reachable from $(c_j^l)^\theta$ thus depends on the link at the "variable" vertices $(v_i^l)^\theta, (v_i^r)^\theta$ in the same way as the satisfiability of a clause depends on the truth assignments.

Fig. 8. The decisive choice between $\blacklozenge$ and $\Diamond\!\!\!\bullet$ occurs only at the "bridges" between "left" and "right" variable vertices

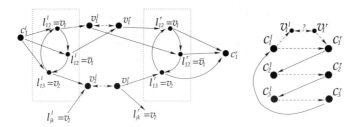

Fig. 9. The details of the construction. Left sketch: The part of the network representing the clause $c_1 = v_1 \vee \bar{v}_1 \vee v_2$. Right sketch: A "top level" view on an instance with 3 clauses. (Solid arrows stand for $\{\blacklozenge\}$, pairs with no arrow or dashed arrow between them are labeled $\{\lozenge\}$. Dashed arrows indicate that there might be a short path (less than 6 direct links), depending on the "bridge" directions).

Direct links $(c_j^r)^\theta \bullet\!\!\prec (c_{j+1}^l)^\theta$ are the only possible connections from one "clause" vertex to any other. Thus, the whole network is cyclic only if each $(c_j^r)^\theta$ is reachable from its corresponding $(c_j^l)^\theta$. Hence the system is satisfiable if and only if there is a variable assignment making each clause "true".

Theorem 2. *Confined $\mathfrak{RRC}$-CSP (hence $\mathfrak{RRC}$-NDP) are NP-hard on linear structures.*

Proof. Recall that NOT-ALL-EQUAL-3SAT is an NP-complete modification of the problem 3SAT whereas an assignment is a solution only if in each clause, at least one literal is assigned *true* and at least one literal is assigned *false* [8].

The proof is a reduction from NOT-ALL-EQUAL-3SAT to the CSP over the fragment

$$\{\bullet\!\!\prec\}, \{\bullet\!\!\prec, \succ\!\!\succ\}, \{\prec\!\!\prec, \succ\!\!\succ\}, \{\bullet\!\!\prec, \succ\!\!\bullet\}, \{\prec\!\!\prec, \bullet\!\!\prec, =, \succ\!\!\bullet, \succ\!\!\succ\},$$

which contains only relations of linear networks.

Let $\mathcal{I}$ be an instance of NOT-ALL-EQUAL-3SAT with variables V and clauses C. We construct an $\mathfrak{RRC}$ instance that contains two supporting variables s_1, s_2, two variables $x_v, x_{\bar{v}}$ for each variable $v \in V$, and three variables y_{j1}, y_{j2}, y_{j3} for the literals in each clause $c_j \in C$.

There are constraints $s_1\{\bullet\!\!\prec\}s_2$, and for all literals $l \in \{v, \bar{v} | v \in V\}$ the two constraints

$$x_l\{\bullet\!\!\prec, \succ\!\!\succ\}s_1; \quad s_2\{\bullet\!\!\prec, \succ\!\!\succ\}x_l$$

demand that in a solution $(\mathfrak{N}, \theta)$, x_l^θ is either directly before s_1^θ (corresponding to a truth assignment of *true*) or directly after s_2^θ (corresponding to an assignment of *false*). Another constraint $x_v\{\prec\!\!\prec, \succ\!\!\succ\}x_{\bar{v}}$ demands that $x_v^\theta, x_{\bar{v}}^\theta$ are on different sides of the couple (s_1^θ, s_2^θ), and therefore correspond to different truth values (see Fig. 10).

Furthermore, if l is the k-th literal in a clause c_j, we add the "truth-value-transporting" constraint $x_l\{\bullet\!\!\prec, \succ\!\!\bullet\}y_{jk}$ that demands that y_{jk}^θ and x_l^θ are on the same side of (s_1^θ, s_2^θ). Finally, for each clause, three constraints

$$y_{j1}\{\prec\!\!\prec, \succ\!\!\succ\}y_{j2}, \ y_{j1}\{\prec\!\!\prec, \succ\!\!\succ\}y_{j3}, \ y_{j2}\{\prec\!\!\prec, \succ\!\!\succ\}y_{j3}$$

are added. For other pairs, no constraints need to be imposed.

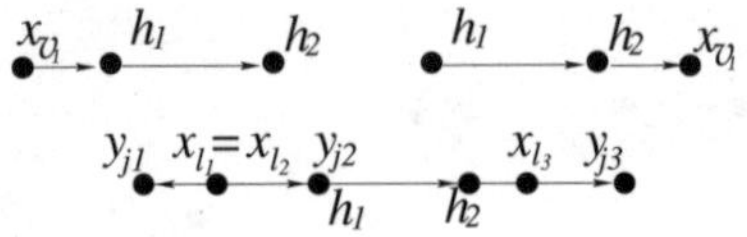

Fig. 10. The three constraints $s_1\{\bullet\!\prec\}s_2$, $x_l\{\bullet\!\prec,\succ\!\succ\}s_1$, $s_2\{\bullet\!\prec,\succ\!\succ\}x_l$ allow two distinct scenarios (left). In the first case, $x_l^\theta\{\bullet\!\prec\}s_1^\theta$, $s_2^\theta\{\succ\!\succ\}x_l^\theta$, in the second case, $x_l^\theta\{\succ\!\succ\}s_1^\theta$, $s_2^\theta\{\bullet\!\prec\}x_l^\theta$. The right sketch shows how the vertices corresponding to a clause may be ordered.

A solution of NOT-ALL-EQUAL-3SAT determines a solution of the constructed CSP as Fig. 10 illustrates.

Vice versa, in a model of the constructed CSP, if l_1, l_2, l_3 are in the same clause, not all $x_{l_1}^\theta, x_{l_2}^\theta, x_{l_3}^\theta$ can be on the same side of $(s_1^\theta s_2^\theta)$ because they would all coincide, and so there would only be two distinct places for the three distinct vertices $y_{j1}^\theta, y_{j2}^\theta, y_{j3}^\theta$.

Corollary 2. $\mathfrak{RDC}$-*NDP and* $\mathfrak{RDC}$-*CSP are* **NP**-*complete, even when restricted to strongly connected, path-connected, linear or acyclic networks.*

Proof. Membership in **NP** is clear: Guess a scenario and check it in **PTIME**. **NP**-hardness follows from the results for $\mathfrak{RRC}$ because $\mathfrak{RDC}$ is a refinement of $\mathfrak{RRC}$.

Theorem 3 (NP-completeness of $\mathfrak{CRC}$-CSP). $\mathfrak{CRC}$-*CSP is* **NP**-*hard for acyclic or arbitrary networks.*

Proof. (Idea) The fragment $\{<,>\}, \{\|\|\}$ is **NP**-hard. This can be shown by a reduction from the **NP**-complete problem BETWEENNESS (described in [8]).

Corollary 3. $\mathfrak{RRC}$-*CSP,* $\mathfrak{CDC}$-*CSP, and* $\mathfrak{RDC}$-*CSP are* **NP**-*complete, in both the open and the confined case, for acyclic or arbitrary networks.*

Proof. For the open case, **NP**-hardness follows because $\mathfrak{RRC}$, $\mathfrak{CDC}$, $\mathfrak{RDC}$ are refinements of $\mathfrak{CRC}$. The closed case follows from Lemma 2.

In most **NP**-hardness proofs of ordered systems and in the ones given here for the reachability calculus, the $\{<,>\}$ or $\{\prec,\succ\}$ relation is responsible for the **NP**-hardness. But, in $\mathfrak{RDC}$ even a fragment without such a label is **NP**-hard:

Theorem 4. *For the fragment* $\mathcal{F} = \{\{)(\}, \{)(, \Upsilon\}, \{\bullet\!\prec\}, \{)(, \bullet\!\prec\}, \{)(, \prec\!\!\prec\}, \{\prec\!\!\prec\},\}$ *of* $\mathfrak{RDC}$, *deciding* $\mathcal{F}$-*CSP is* **NP**-*complete.*

Proof. (Sketch) A reduction from 3SAT to $\mathcal{F}$-CSP is given.

For each 3SAT-element (variables, negated variables, literals, clauses) introduce a vertex. Add another node w. A path from w to the vertex corresponds with the assignment "true". $w\{)(, \bullet\!\prec\}v_i$ and $w\{)(, \bullet\!\prec\}\bar{v}_i$ represent the choice of the truth value. Additional constraints $v_i\{)(\}\bar{v}_i$ demand that only v_i or $\bar{v}_i$ is "assigned" true. "(Negated) variables" are directly linked to "literals" which are linked to the "clauses" they occur in. Label $\bullet\!\prec$ is used for direct links. A clause c_i is indirectly linked to w iff at least one of the literals in the clause is linked to w. Constraints $w\{\prec\!\!\prec\}c_j$ test if the clauses c_j are all assigned true. Hence the constructed system is satisfiable iff the 3SAT instance is satisfiable.

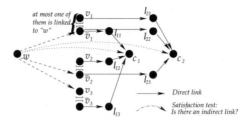

Fig. 11. Reduction from 3SAT to a fragment of the $\mathfrak{RDC}$ calculus. Dashed arrows indicate optional direct links, solid arrows mandatory ones. Dotted arrows stand for the truth test constraints $w\{\nleftarrow\}c_j$. The example represents the 3SAT instance $(\bar{v}_1 \vee v_2 \vee \bar{v}_3) \wedge (v_1 \vee \bar{v}_1 \vee \bar{v}_2)$.

4.4 Tractability Results

However, although the confined case is **NP**-hard, the open CSP is in **PTIME** for some classes of networks.

Theorem 5 (Tractability for path-connected networks). $\mathfrak{CRC}_{path\text{-}conn.}$-CSP is in **PTIME**.

Proof. (Sketch) Let $\Pi = (X, c)$ be given. In path-connected networks, for no pair of vertices the label $\{||\}$ can describe its relation. Hence, all constraints c_{xy} of Π can be substituted by $c_{xy} \cap \{=, <, >, \diamond\}$. Then, construct the satisfiability-equivalent path-consistent CSP by the classical algorithm [12]. Let $\lessgtr := \{=\} \cup \{\diamond\}$.

Claim 1: A constraint system over the coarser calculus with basic labels $\{\lessgtr, <, >\}$ is satisfiability-equivalent to corresponding constraint system of PA_{lin}(with basic labels $\{=, <, >\}$) that is obtained by substituting $\lessgtr$ with $=$.

A PA_{lin} model can be interpreted as a network in which $v\bullet\!\!-\!\!w$ iff $v < w$ in the sense of the partial order. For a model of $\{\lessgtr, <, >\}$, the corresponding PA_{lin} model is found by mapping all vertices of a strongly connected component, i.e. for which $v \lessgtr w$ holds, to the same point, and then applying $\prec$ (i.e. the transitive closure of $\bullet\!\!-\!\!$) as the partial order relation.

Claim 2: A $\mathfrak{CRC}_{path\text{-}conn.}$ constraint system is satisfiability-equivalent to its $\{\lessgtr, <, >\}$ coarsening.

A model of the finer system is trivially a model of the coarser system. Let a model of the coarser system be given. If in all cases in which $v^\theta \diamond w^\theta$ holds, but $v\{=\}w$ is required, v^θ and w^θ are identified, the required relations to other points do not change. If $x^\theta = y^\theta$ holds, but $x\{\diamond\}y$ is required, y can be mapped to a new vertex v^* with $v^* \bullet\!\!-\!\!\bullet x^\theta$, and with all other vertices z^θ, v^* copies the links from x^θ. This construction, applied sequentially for all these cases, satisfies all conditions.

Hence, $\mathfrak{CRC}_{path\text{-}conn.}$-CSP is is solvable if PA_{lin}-CSP is solvable which is known to be in **PTIME**[19].

Corollary 4

1. open $\mathfrak{RRC}_{path\text{-}conn.}$-CSP is in **PTIME**.
2. Open $\mathfrak{RDC}_{path\text{-}conn.}$-CSP, hence also $\mathfrak{CDC}_{path\text{-}conn.}$-CSP is in **PTIME**.

Proof.
1. follows from Lemma 3.
2. In a path-connected solution of a $\mathfrak{RDC}_{path\text{-}conn.}$-instance $\Pi = (X, c)$, for all $x, y \in X$ must hold $x^\theta = y^\theta$, $x^\theta \prec y^\theta$ or $y^\theta \prec x^\theta$. Hence, all labels l can be trimmed to $l \cap \{=, \prec\!\prec, \bullet\!\prec, \succ\!\succ, \succ\!\bullet, \Diamond, \blacklozenge, \Diamond\!\bullet, \bullet\!\bullet\}$ without changing satisfiability. All derived labels are $\mathfrak{RRC}$ labels, hence the solution of the corresponding open $\mathfrak{RRC}_{path\text{-}conn.}$-CSP is a solution of open $\mathfrak{RDC}_{path\text{-}conn.}$-CSP.

Theorem 6 (Tractability of Open $\mathfrak{RRC}$-CSP). *Let $\mathcal{F} \subset R$ be a fragment of sociable $\mathfrak{RRC}$ relations. If an $\mathcal{F}$-constraint system is 3-open-consistent, it is (globally) open-consistent. Open $\mathcal{F}$-CSP and Open $\mathcal{F}$-NDP are in **PTIME**.*

Proof. (Idea) Like for traditional calculi, a path-consistency algorithm [12] generates a satisfiability-equivalent 3-open-consistent constraint system. The precondition of having sociable relations guarantees that a model can be composed of "local" open models satisfying individual labels and hence the system is globally open-consistent iff $O(n^3)$ many subsystems of size 3 are 3-open-consistent. The answer if a subsystem is 3-open-consistent does not depend on the rest of the CSP. Hence each instance is one of finitely many different instance types that can be stored in a table. Hence, each check can be done in linear time.

5 Conclusion and Outlook

The traditional formalism of non-associative algebras and constraint systems can be adapted for network calculi. A label in a network algebra describes how two vertices within a network are related to each other. Unions of basic labels are used when several of them are allowed. A set of variables with such descriptions for pairs of variables form a constraint system. The notion of a solution, traditionally a mapping of variables to elements of a universe, has been extended to finding a structure and a mapping.

Many labels referring to the network structure are nonlocal. Thus, concepts like k-consistency have to be modified. Four different calculi have been presented and investigated in terms of computational complexity. Generally, reasoning in networks is **NP**-complete. For scanarios and some tractable subclasses, reasoning is **PTIME**.

What else should be done? Current research is done on identifying maximal tractable subclasses for all the network algebras. It might be useful to extend the formalism of the network algebra so that properties of single vertices (being a "source" node, being part of a cycle) or properties of triples can be expressed.

References

1. Allen, J.F.: Maintaining knowledge about temporal intervals. Comm. ACM 26(11), 832–843 (1983)
2. Broxvall, M., Jonsson, P.: Towards a complete classification of tractability in point algebras for nonlinear time. In: Jaffar, J. (ed.) CP 1999. LNCS, vol. 1713, pp. 129–143. Springer, Heidelberg (1999)
3. Cohn, A.G.: Qualitative spatial representation and reasoning techniques. In: Brewka, G., Habel, C., Nebel, B. (eds.) KI 1997: Advances in Artificial Intelligence. LNCS, vol. 1303, pp. 1–30. Springer, Heidelberg (1997)
4. Frank, A.: Qualitative Spatial Reasoning with Cardinal Directions. In: Proc. of the 7th Austrian Conf. on AI (1991)
5. Dijkstra, E.W.: A note on two problems in connexion with graphs. Numerische Mathematik 1, 269–271 (1959)
6. Egenhofer, M.: Reasoning about binary topological relations. In: Günther, O., Schek, H.-J. (eds.) SSD 1991. LNCS, vol. 525, pp. 143–160. Springer, Heidelberg (1991)
7. Freksa, C.: Using Orientation Information for Qualitative Spatial Reasoning. In: Theories and Methods of Spatial-Temporal in Geog. Spac. Reasoning
8. Garey, M., Johnson, D.: Computers and Intractability: A Guide to the Theory of NP-Completeness, Freeman, San Francisco (1978)
9. Ligozat, G.: Reasoning about cardinal diretions. J. of Vis. Lang. & Comp. 1(9), 23–44 (1998)
10. Ligozat, G., Renz, J.: What is a Qualitative Calculus? A General Framework. In: Zhang, C., W. Guesgen, H., Yeap, W.-K. (eds.) PRICAI 2004. LNCS (LNAI), vol. 3157, pp. 53–64. Springer, Heidelberg (2004)
11. Maddux, R.: Some varieties containing relation algebras. Trans. AMS 272, 501–526 (1982)
12. Montanari, U.: Networks of constraints: Fundamental properties and applications to picture processing. Inform. Sci. 7, 95–132 (1974)
13. Moratz, R., Renz, J., Wolter, D.: Qualitative spatial reasoning about line segments. In: ECAI (2000)
14. Nebel, B., Scivos, A.: Formal Properties of Constraint Calculi for Qualitative Spatial Reasoning. Künstliche Intelligenz, Heft 4(02), 14–18 (2002)
15. Ragni, M., Scivos, A.: Dependency Calculus: Reasoning in a General Point Algebra. In: Proc. of IJCAI 2005, pp. 1575–1576 (2005)
16. Randell, D., Cui, Z., Cohn, A.: A Spatial Logic Based on Regions and Connection. In: Proceedings KR 1992, pp. 165–176 (1992)
17. Scivos, A., Nebel, B.: Double-Crossing: Decidability and Computational Complexity of a Qualitative Calculus for Navigation. In: Montello, D.R. (ed.) COSIT 2001. LNCS, vol. 2205, Springer, Heidelberg (2001)
18. Tarski, A.: On the calculus of relations. J. of Symb. Logic 6, 73–89 (1941)
19. Vilain, M., Kautz, H., van Beek, P.: Contraint propagation algorithms for temporal reasoning: A revised report. Reasoning about Physical Systems, 373–381 (1989)

The Qualitative Trajectory Calculus on Networks

Peter Bogaert[1], Nico Van de Weghe[1], Anthony G. Cohn[2], Frank Witlox[1], and Philippe De Maeyer[1]

[1] Ghent University, Department of Geography, Krijgslaan 281, 9000 Gent, Belgium
{Peter.Bogaert,Nico.VandeWeghe,Frank.Witlox,
Philippe.DeMaeyer}@UGent.be
[2] University of Leeds, School of Computing, Leeds LS2 9JT, United Kingdom
A.G.Cohn@leeds.ac.uk

Abstract. Moving objects are commonly handled using quantitative methods and information. However, in many cases, qualitative information can be more efficient and more meaningful than quantitative information. A lot of research has been done in generating, indexing, modelling and querying network-based moving objects, but little work has been done in building a calculus of relations between these objects in a qualitative way. In this paper, we introduce a formal definition of how to represent and reason about the relative trajectories of pairs of objects moving along a network.

1 Introduction

In the literature there are two standard approaches when dealing with topological relations between two regions. From the viewpoint of databases, Egenhofer et al. [6] worked out the 9-Intersection Model. Independently, Randell et al. [14] studied the subject from an artificial intelligence point of view resulting in the Region Connection Calculus (RCC). Both approaches reach the same conclusion: a set of eight jointly exhaustive and pairwise disjoint (JEPD) topological relations between two regions without holes. Assuming continuous motion, there are constraints upon the ways these relations change. For example, two objects cannot change their relation from disjoint (DC) to partial overlap (PO) without touching (EC) each other first. These possible changes can graphically be represented by means of a Conceptual Neighbourhood Diagram (CND). CND^s have been introduced in the temporal domain [7], and have been widely used in spatial reasoning, e.g.: for topological relations [14,4]; cardinal directions [5], and for relative orientation [8]. CND^s are typically used for qualitative simulation to predict what will happen in the future. Two relations between entities are conceptual neighbours, if they can be transformed into one another by continuously deforming, without passing another qualitative relation; a CND describes all the possible transitions between relations that can occur [7].

In the real world, most moving objects have a disjoint (DC) relation. A potential problem here is that both the RCC calculus and the 9-Intesection Model can not further differentiate between disjoint objects, nor indeed could any purely topological representation. Moreover, when dealing with moving point objects (MPO's), there are, according to the 9-Intersection Model, only two topological relations between points (i.e. disjoint and meet). Hence these approaches fail to make explicit the level of disjointness of how two or more objects move with respect to each other.. Obvious examples where this type of information is of vital importance is the case of two airplanes and to know whether they are likely to stay in a disjoint relation, if not the consequences can be catastrophic. Therefore, Van de Weghe [16] introduced the Qualitative Trajectory Calculus (QTC). This calculus deals with qualitative relations between two disjoint, moving, point-like objects. Here we want to focus on how QTC can be of use when dealing with moving object in a network situation.

The structure of this paper is as follows. Section 2 gives the definition of QTC which is the basis for the Qualitative Trajectory Calculus on Networks (QTC_N) and describes the usefulness of qualitative relations. Section 3 defines QTC_N and gives an overview of all possible relations and transitions between these relations. Final conclusions and directions for future work are given in section 4.

2 The Qualitative Trajectory Calculus

2.1 Qualitative Relations

Reasoning can be performed with quantitative as well as qualitative information. Typically, when working with quantitative information, a predefined unit of a quantity is used [12]. For example, one could say that a car drives at 30 km/h. In the qualitative approach, continuous information is being quantised or qualitatively discretised by landmarks separating neighbouring open intervals, resulting in discrete quantity spaces [22]. Qualitative reasoning only studies the essence of information, represented as a small set of symbols such as the quantity space $\{-, 0, +\}$ consisting of the landmark value '0' and its neighbouring open intervals '–' and '+'. For example, if one does not know the precise speed of a car and a bicycle, but knows that the speed of the car is higher than the speed of the bicycle, one can label this with the qualitative value '+', meaning that the car is moving faster than the bicycle. One could also say that the bicycle is moving slower than the car, by giving the qualitative value '–' to this relation. Finally, both objects can also move at the same speed, resulting in a qualitative value '0'. One thing is for sure; the speed of a car cannot change from being higher than the speed of the bicycle to being lower than the speed of the bicycle, without passing the qualitative value '0'.

There are a variety of reasons why qualitative reasoning claims their place next to, or complementary to, quantitative reasoning in areas such as Artificial Intelligence and Geographic Information Science. First of all, qualitative knowledge tends to be less expensive than its quantitative counterpart, since it contains less information [8]. Moreover, qualitative data often provide, at an early stage of research, an ideal way to deliver insights in order to identify quickly potential problems that warrant more detailed quantitative analysis [13]. In addition, humans usually prefer to communicate in qualitative categories, supporting their intuition, than using quantitative measures.

2.2 Formal Definition

The Qualitative Trajectory Calculus (QTC) examines changes in qualitative relations between two disjoint, point-like objects. Depending on the level of detail and the number of spatial dimensions, different types of QTC are defined in [16], all belonging to QTC-Basic (QTC_B) [18, 20] or QTC-Double-Cross (QTC_C) [19, 21]. In this section, we focus on QTC-Basic, since this is the basis for defining the Qualitative Trajectory Calculus on Networks (QTC_N).

QTC_B is developed for moving objects in one (QTC_{B1}) or two dimensions (QTC_{B2}). In QTC_{B1}, it is assumed that the movement of two objects is restricted to a line (e.g. two trains moving on a railroad track). In QTC_{B2}, two objects can move freely in a plane (e.g. two ships floating on an ocean). The landmark to describe the qualitative relations in QTC_B is the distance at time t between the two objects. A typical, essential characteristic in both cases of QTC_B is the three character label representing the qualitative movement between two objects. This label represents the following three relationships:

Assume two objects[1] k and l

1. Movement of the first object k, with respect to the position of the second object l at time point t:

   $-$: k is moving towards l:

   $$\exists t_1 (t_1 \prec t \wedge \forall t^- (t_1 \prec t^- \prec t \rightarrow d(k|t^-, l|t) > d(k|t, l|t))) \wedge \\ \exists t_2 (t \prec t_2 \wedge \forall t^+ (t \prec t^+ \prec t_2 \rightarrow d(k|t, l|t) > d(k|t^+, l|t))) \tag{1}$$

   $+$: k is moving away from l:

   $$\exists t_1 (t_1 \prec t \wedge \forall t^- (t_1 \prec t^- \prec t \rightarrow d(k|t^-, l|t) < d(k|t, l|t))) \wedge \\ \exists t_2 (t \prec t_2 \wedge \forall t^+ (t \prec t^+ \prec t_2 \rightarrow d(k|t, l|t) < d(k|t^+, l|t))) \tag{2}$$

   0: k is stable with respect to l (all other cases):

   $$\exists t_1 (t_1 \prec t \wedge \forall t^- (t_1 \prec t^- \prec t \rightarrow d(k|t^-, l|t) = d(k|t, l|t))) \wedge \\ \exists t_2 (t \prec t_2 \wedge \forall t^+ (t \prec t^+ \prec t_2 \rightarrow d(k|t, l|t) = d(k|t^+, l|t))) \tag{3}$$

   $$\exists t_1 (t_1 \prec t \wedge \forall t^- (t_1 \prec t^- \prec t \rightarrow d(k|t^-, l|t) = d(k|t, l|t))) \wedge \\ \exists t_2 (t \prec t_2 \wedge \forall t^+ (t \prec t^+ \prec t_2 \rightarrow d(k|t, l|t) < d(k|t^+, l|t))) \tag{4}$$

---

[1] We introduce the following notation for QTC:
   $x|t$ denotes the position of an object x at time t;
   $d(u,v)$ denotes the distance between two positions u and v;
   $vx|t$ denotes the speed of x at time t;
   $t_1 \prec t_2$ denotes that t_1 is temporally before t_2;
   t^- denotes the time period immediately before t;
   t^+ denotes the time period immediately after t.

$$\exists t_1(t_1 \prec t \wedge \forall t^-(t_1 \prec t^- \prec t \rightarrow d(k\,|\,t^-,l\,|\,t) = d(k\,|\,t,l\,|\,t))) \wedge$$
$$\exists t_2(t \prec t_2 \wedge \forall t^+(t \prec t^+ \prec t_2 \rightarrow d(k\,|\,t,l\,|\,t) > d(k\,|\,t^+,l\,|\,t))) \quad (5)$$

$$\exists t_1(t_1 \prec t \wedge \forall t^-(t_1 \prec t^- \prec t \rightarrow d(k\,|\,t^-,l\,|\,t) > d(k\,|\,t,l\,|\,t))) \wedge$$
$$\exists t_2(t \prec t_2 \wedge \forall t^+(t \prec t^+ \prec t_2 \rightarrow d(k\,|\,t,l\,|\,t) = d(k\,|\,t^+,l\,|\,t))) \quad (6)$$

$$\exists t_1(t_1 \prec t \wedge \forall t^-(t_1 \prec t^- \prec t \rightarrow d(k\,|\,t^-,l\,|\,t) > d(k\,|\,t,l\,|\,t))) \wedge$$
$$\exists t_2(t \prec t_2 \wedge \forall t^+(t \prec t^+ \prec t_2 \rightarrow d(k\,|\,t,l\,|\,t) < d(k\,|\,t^+,l\,|\,t))) \quad (7)$$

$$\exists t_1(t_1 \prec t \wedge \forall t^-(t_1 \prec t^- \prec t \rightarrow d(k\,|\,t^-,l\,|\,t) < d(k\,|\,t,l\,|\,t))) \wedge$$
$$\exists t_2(t \prec t_2 \wedge \forall t^+(t \prec t^+ \prec t_2 \rightarrow d(k\,|\,t,l\,|\,t) = d(k\,|\,t^+,l\,|\,t))) \quad (8)$$

$$\exists t_1(t_1 \prec t \wedge \forall t^-(t_1 \prec t^- \prec t \rightarrow d(k\,|\,t^-,l\,|\,t) < d(k\,|\,t,l\,|\,t))) \wedge$$
$$\exists t_2(t \prec t_2 \wedge \forall t^+(t \prec t^+ \prec t_2 \rightarrow d(k\,|\,t,l\,|\,t) > d(k\,|\,t^+,l\,|\,t))) \quad (9)$$

2. The movement of the second object l, with respect to the position of the first object k at time point t can be described as in 1. with k and l interchanged, and hence:
−: l is moving towards k
+: l is moving away from k
0: l is stable with respect to k
3. Relative speed of the first object k at time point t, with respect to the second object l at time point t:
−: k is slower than l:

$$v_k\,|\,t < v_l\,|\,t \quad (10)$$

+: k is faster than l:

$$v_k\,|\,t > v_l\,|\,t \quad (11)$$

0: k and l are equally fast:

$$v_k\,|\,t = v_l\,|\,t \quad (12)$$

Note that with the introduced three characters, relationships between the two objects can now be described. For example, if object k and object l are both moving towards each other (resulting in a '−' for the first and a '−' for the second character) and object k is moving slower than object l (resulting in a '−' for the third character), this will result in a $(-\,-\,-)_B$ label.

By definition, in QTC_B, there are theoretically 3^3 (27) different relationships. However, in QTC_{B1} only 17 real-life (in theory feasible) possibilities remain. For example, it is impossible for one object to be faster than the other if both objects are not moving (Figure 1: 5a or 5c). Note that in QTC_{B2}, all 27 relations are possible (Figure 2).

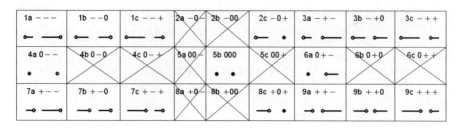

Fig. 1. 17 real-life QTC$_{B1}$ labels[2]

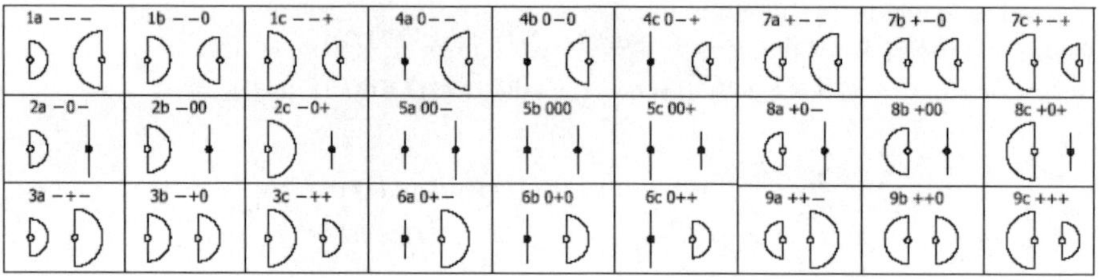

Fig. 2. 27 real-life QTC$_{B2}$ labels[3]

2.3 Theory of Dominance

Adopting the previously defined concept of conceptual neighbours [7] to, *'two trajectory pairs are conceptual neighbours if they can directly follow each other during a continuous movement'*, gives the possibility to create a Conceptual Neighbourhood Diagram (CND) for QTC$_B$.

The construction of the CND is based on the concept of 'dominance space', introduced by Galton [9]. Galton outlined the temporal nature of transitions between qualitative variables, defining some important restrictions concerning the dominance between binary qualitative relations. Central in his theory of dominance are the constraints imposed by continuity. Consider the qualitative distinction between '−', '0' and '+', then a variable capable of assuming any of these three descriptions may change between them. However, a direct change from '−' to '+' and vice versa is impossible, since such a change must always pass the qualitative value '0'. This landmark value '0' only needs to hold for an instant. On the other hand, the '+' of a variable, when changing from '0' to '+' and back to '0', must hold over an interval

---

[2] The left and right dot represent respectively the positions of k and l. A dot is filled if the object can be stationary. The line segments represent the potential object movements. Note that the lines can have different lengths giving the difference in relative speed. The line segments represent whether each object is moving towards or away from the other.

[3] The icons contain line segments with the point object positioned in the middle. The line segment denotes the possibility to move to both sides of the point object. The filled dot represents the case when the object can be stationary. An open dot means that the object cannot be stationary. The icons also contain crescents with the point object in the middle of its straight border. The crescent denotes an open polygon. If a crescent is used, then the movement starts in the dot and ends somewhere on the curved side of the crescent.

[10]. Let us briefly illustrate this point. The issue is that between any two points of a continuous trajectory we can always find, or at least imagine, another intermediate point. In order words, applied to a real number line, between zero and any positive real number, one can always find another positive real number: $0 < 10 < 100$; $0 < 1 < 10$; $0 < 0.1 < 1$; $0 < 0.01 < 0.1$; $0 < 0.001 < 0.01$, etc. It then follows that it is impossible that + only holds over an instant of time. Dual reasoning applies for the change from '0' to '−'. In Galton's [10] terms, we state that '0' dominates '−' and '+', and that '−' and '+' are dominated by '0'.

Based on the concept of dominance, one can construct a *dominance space*. This is a space containing qualitative values with their dominance relations. Figure 3 represents a basic example of the dominance space between the qualitative values '−', '0' and '+', where '−' denotes the set of negative real numbers, '0' is the landmark value, and '+' denotes the set of positive real numbers. It follows from Figure 3 that:

- there is a connection between '−' and '0', and the arrow is in the direction of '−', thus, a transition from '−' to '0' can occur and vice versa, with '0' dominating '−';
- there is a connection between '0' and '+', and the arrow is in the direction of '+', thus, a transition from '0' to '+' can occur and vice versa, with '0' dominating '+';
- there is no direct connection between '−' and '+', thus, a transition from '−' to '+' and vice versa can only occur by passing through '0'.

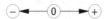

Fig. 3. Dominance space in one dimension

A set of dominance spaces can be combined in order to build composite dominance spaces [10]. In Figure 4a, two *one-dominance*[4] spaces are visualised. A combination of these one-dominance spaces leads to a two-dominance space shown in 4b. The disjunction of the two one-dominance spaces and the two-dominance space leads to an *overall dominance* space in two dimensions represented in Figure 4c.

---

[4] A one-dominance space is a dominance space where all conceptual distances are equal to one. In order to explain the concept of conceptual difference, let us consider three examples:
1. Assume that R_1 and R_2 only differ in one character that can change continuously between both states without passing through an intermediate qualitative value. Then the conceptual distance between R_1 and R_2 is one. Example: $R1 = (000)_{B1}$ and $R_2 = (0+0)_{B1}$ → conceptual distance is one.
2. Assume that R_1 and R_2 only differ in one character that cannot change continuously between both states without passing through an intermediate qualitative value. Then the conceptual distance between R_1 and R_2 is composed of sub-distances. Example: Suppose $R_1 = (0-0)_{B1}$ and $R_2 = (0+0)_{B1}$. Then the conceptual distance between $(0-0)_{B1}$ and $(000)_{B1}$ is one, and the conceptual distance between $(000)_{B1}$ and $(0+0)_{B1}$ is one. Thus, the conceptual distance between $(0-0)_{B1}$ and $(0+0)_{B1}$ is two.
3. Assume that R_1 and R_2 differ in multiple characters. Then the conceptual distance is the sum the sub-distances determined for each individual character (i.e. the Manhattan distance). Example: Suppose $R_1 = (--0)_{B1}$ and $R_2 = (++0)_{B1}$. Then the conceptual distance for the first character is two, and the conceptual distance for the second character is two. Thus, the conceptual distance between $R_1 = (--0)_{B1}$ and $R_2 = (++0)_{B1}$ is four.

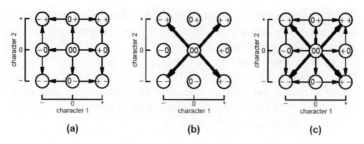

Fig. 4. Combination of dominance spaces

2.4 Conceptual Neighbourhood Diagrams for QTC$_B$

Since QTC$_B$ describes three orthogonal qualitative values we can use the theory of dominance to construct the overall dominance space for three dimensions. Figure 5 shows the one-dominance space for each dimension.

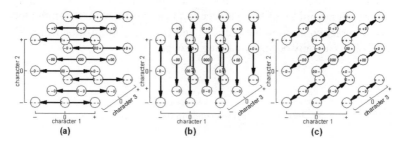

Fig. 5. One-dominance spaces for three dimensions

These three one-dominance spaces can be combined pair wise (Figure 6a), leading to three composite two-dominance spaces as shown in Figure 6b.

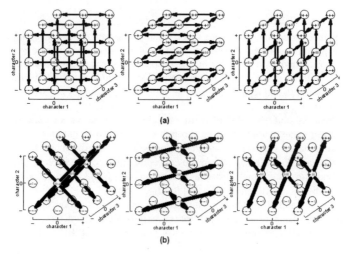

Fig. 6. Construction of two-dominance spaces

Furthermore, the two-dominance spaces can each be combined with their orthogonal one-dominance space (Figure 7a). These combinations all result in the same three-dominance space (Figure 7b).

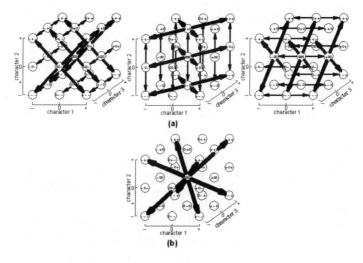

Fig. 7. Construction of the three dominance space

The combination of the three one-dominance spaces, the three two-dominance spaces and the one three-dominance space lead to an overall dominance space which is not visualised since it becomes too complex to display in a two dimensional medium.

A conceptual neighbourhood diagram for QTC_B can now be created by deleting all 'non-existing' transitions between relations (edges in the CND) and by deleting all 'non-existing' relations (nodes in the CND). The CND for QTC_{B2} is equal to the overall dominance space for three orthogonal qualitative values [16], since all relations and all transitions between relations exist. The CND of QTC_{B1} (Figure 8) is different, since we know that only 17 relations are possible, so we delete the ten impossible relations from the overall dominance space as well as all transitions between impossible and (im)possible relations [20].

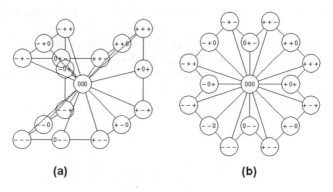

Fig. 8. CND for QTC_{B1}

3 The Qualitative Trajectory Calculus on Networks (QTC$_N$)

According to Moreira et al. [15] there are two types of moving objects: objects that have a completely free trajectory which is only constrained by the dynamics of the object itself (e.g. a bird flying through the sky), and objects that have a constrained trajectory (e.g. a train on a railway track). Note that QTC$_{B2}$ is able to describe the movement of objects which have a free trajectory in two dimensions, and that QTC$_{B1}$ can describe objects which have a constrained (linear) trajectory. The point is that QTC is not able to handle objects that are moving on a set of interconnected linear features such as a network. Consequently, we develop the qualitative trajectory calculus on networks: QTC$_N$.

3.1 Network: Some Definitions

A network, such as a road, rail or river network, is a set of interconnected linear features, which can easily be represented by a connected graph. The graph itself is not a spatial structure, but needs to be embedded in a space or must be 'spatialised' [11]. This can be done by a function which maps each node of the graph onto a location in the defined space, and maps each link of the graph onto a curve segment [11]. In essence, a network is a co-dimensional structure. It is a one-dimensional structure embedded in a two-dimensional or three-dimensional space.

We make the following definitions and assumptions:

- a *graph* is a set of edges, E and nodes, N
- a QTC *network* is a connected graph and a finite set of *objects*
- each edge connects a pair of nodes
- each node has a *degree* which is the number of edges connected to it
- at any time, each object o has a *position* in the graph, which is either a node in N, or is along an edge e in E, in which case the network at t is augmented with an additional *dynamic* node of degree 2 cutting the edge e in two, representing the position of o at t.
- a *path* p from o_1 to o_2 at t is a subgraph of the network at time t, such that every node in p is of degree 2 except two nodes representing the position of o_1 and o_2, which are of degree 1. Thus a path is a sequence of nodes and edges from o_1 to o_2.
- every edge has an associated *length,* which is a positive number
- the length of a path is the sum of the length of the edges in the path
- if M is a path of length $|M|$ and there is no path of length less than $|M|$ between the same two nodes, then M is a *shortest path*
- a *cycle* is a subgraph that contains the same number of edges and nodes, and each node is of degree 2[5].

---

[5] For technical reasons we require that every cycle in a network have at least 3 non dynamic nodes – this can always be achieved by adding further nodes of degree 2 without affecting the overall network topology.

3.2 Formal Definition of QTC$_N$

In QTC$_N$, the landmark used to qualitatively compare the movements of two objects is the shortest path between these objects. This landmark is chosen because of its specificity; a moving object can only approach another object, if and only if it moves along a shortest path between these two objects.

Theorem
A primary object k *on a network can only decrease its distance to a reference object* l *on this network if and only if* k *moves towards* l *along a shortest path.*

Proof:
1. Moving along a shortest path will decrease the distance

Assume a shortest path between k and l is M, and therefore the shortest distance between k and l is $|M|$. If objects (k or l) move along the shortest path M over an infinitesimal unit of distance (ds), they will decrease their distance because ds is not negative.

$$|M| > |M| - ds \qquad (13)$$

2. Moving along any other path (which is not a shortest path) will increase the distance

Assume a shortest path between k and l is M, and therefore the shortest distance between k and l is $|M|$. Any other path N with a length of $|N|$ between k and l, which is not a shortest path, will be longer.

$$|M| < |N| \qquad (14)$$

If k moves along N by a distance ds, where $ds < 0.5\ (|N|-|M|)$, then its distance from l will be $|M| + ds$, since N is not a shortest path. So, if k wants to approach l it must move along a shortest path. □

Using this property, we can state that an object can only approach another object at time t if it does not lie on SP_{kl}^t [6] during t^- and lies on SP_{kl}^t during t^+[7]. An object moves away from another object if it lies on SP_{kl}^t during t^- and does not lie on SP_{kl}^t during t^+. If an object lies on SP_{kl}^t only at t but not during t^- or t^+, or if it lies on SP_{kl}^t throughout $[t^-, t^+]$, then the object will be stationary with respect to the other object (although this relation can only last for an instantaneous moment in time).

We can now reformulate conditions (1) to (12) originating for the construction of the three character label for QTC$_B$ to a QTC$_N$ setting.

1. Movement of the first object k, with respect to the position of the second object l at time point t:

---

[6] SP_{kl}^t denotes the shortest path at time t between objects k and l.

[7] A point p lies on a line L if it is an element of this line ($p \in L$).

−: k moves along the shortest path:

$$\exists t_1 (t_1 \prec t \wedge \forall t^- (t_1 \prec t^- \prec t \rightarrow k|t^- \notin SP_{kl}^t)) \wedge$$
$$\exists t_2 (t \prec t_2 \wedge \forall t^+ (t \prec t^+ \prec t_2 \rightarrow k|t^+ \in SP_{kl}^t)) \tag{15}$$

+: k does not move along the shortest path:

$$\exists t_1 (t_1 \prec t \wedge \forall t^- (t_1 \prec t^- \prec t \rightarrow k|t^- \in SP_{kl}^t)) \wedge$$
$$\exists t_2 (t \prec t_2 \wedge \forall t^+ (t \prec t^+ \prec t_2 \rightarrow k|t^+ \notin SP_{kl}^t)) \tag{16}$$

0: stationary:

$$\exists t_1 (t_1 \prec t \wedge \forall t^- (t_1 \prec t^- \prec t \rightarrow k|t^- \in SP_{kl}^t)) \wedge$$
$$\exists t_2 (t \prec t_2 \wedge \forall t^+ (t \prec t^+ \prec t_2 \rightarrow k|t^+ \in SP_{kl}^t)) \tag{17}$$

$$\exists t_1 (t_1 \prec t \wedge \forall t^- (t_1 \prec t^- \prec t \rightarrow k|t^- \notin SP_{kl}^t)) \wedge$$
$$\exists t_2 (t \prec t_2 \wedge \forall t^+ (t \prec t^+ \prec t_2 \rightarrow k|t^+ \notin SP_{kl}^t)) \tag{18}$$

2. Movement of the second object l, with respect to the position of the first object k at time point t can be described as in Case 1 with k and l interchanged, hence:
−: l moves along the shortest path
+: l does not move along the shortest path
0: stationary
3. Relative speed of the first object k at time point t, with respect to the second object l at time point t:
−: k is slower than l:

$$v_k|t < v_l|t \tag{19}$$

+: k is faster than l:

$$v_k|t > v_l|t \tag{20}$$

0: k and l are equally fast:

$$v_k|t = v_l|t \tag{21}$$

3.3 Possible Relations and Conceptual Neighbours

In order to construct a CND for QTC_N, the possible relations between two objects and the transitions between these relations need to be examined.

If a network is connected, there is always a path between two objects along which they can move towards or away from each other. Since QTC_{B1} describes the movement of objects moving on a line, every binary relation and every transition between these relations stated in QTC_{B1} exists in QTC_N. Every relation in QTC_{B1} can be reached by only changing the speed of the objects. Thus, a transition between relations is triggered by a 'Speed Change' event. Still, due to the co-dimensional nature of a network, there can be additional binary relations and transitions between these relations if the speed is constant. These additional relations and transitions are invoked by two events [2, 17]:

1. a 'Node Pass' event: an object passes a node
2. a 'Shortest Path Change' event: the shortest path between the objects changes.

3.3.1 A Single 'Node Pass' Event

Suppose object m approaches object n (Figure 9a). Using the definition of QTC$_N$, object m will evoke a '−' in the three character label. If m reaches a node in the network with a minimum degree of three, it can either continue its way along a shortest path or it can continue its way on an arc that does not belong to a shortest path. The latter implies that there will be a change in the relation between m and n, because an object can only approach another object if it moves along a shortest path. At the instantaneous moment in time when m passes the node, it will not approach or move away from the other object n (Figure 9b). According to the definition (18), m will evoke a '0' in the three character label. A fraction of time later, m will increase its distance with regard to n, evoking a '+' in the three character label defining the relation between objects m and n (Figure 9c). A 'Node Pass' event can results in a conceptual animation[8] where one of the first two characters in the label changes from '−' to '0' to '+'.

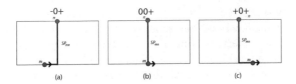

Fig. 9. Example of a transition due to a 'Node Pass' event

In order to have a change in relation due to a 'Node Pass' event certain conditions have to be fulfilled. First of all the degree of the node where the object passes should be at least three. If the degree of the node is less than three and the object moves along the shortest path, an object with positive speed can only continue its way along this shortest path. Secondly, the object causing a 'Node Pass' event must approach the other object. Suppose this object would move away from the other object, it can only continue its way along an arc that does not belong to the shortest path, when it reaches a node. Finally, due to the theory of dominance, a 'Node Pass' event should hold over an interval. Given these three conditions, the transitions caused by a 'Node Pass' event, can be visualised in a CND. Figure 10 gives an overview of all possible transitions between relations due to a single 'Node Pass' event.

---

[8] A conceptual animation is a sequence of QTC relations, following the constraints imposed by qualitative reasoning.

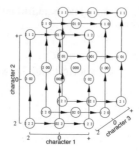

Fig. 10. Possible transitions due to a single 'Node Pass' event

3.3.2 A Single 'Shortest Path Change' Event

We will now illustrate a transition by means of an example shown in Figure 11. Suppose object m lies in between nodes B and C. If m, B and C lie on a cycle, there are at least two paths between n and m. One reaches m via node B another reaches m via node C. In Figure 11a, there is a shorter path via node B (n,A,B,m) and a longer path via node C (n,A,C,m). When m moves away from this shorter path, and therefore moves away from the other object n, m will, according to the definition, evoke a '+' in the three character label. This means the shorter path will be extended and the longer path is shortened. At some moment in time, these two paths will become equally long (Figure 11b). At that instantaneous moment, m will not approach nor move away from n. As a result m will evoke a '0' in the three character label. A fraction of time later, m will move along the newly defined shortest path and therefore decrease its distance compared to the other object, evoking in a '–' in the three character label defining the relation between m and n (Figure 11c). A 'Shortest Path Change' event can result in a conceptual animation where one of the first two characters in the label changes from '+' to '0' to '–'.

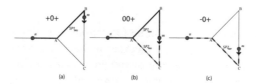

Fig. 11. Example of a transition due to a 'Shortest Path Change' event

Here too three conditions should be satisfied in order to have a change in relation due to a 'Shortest Path Change' event. First of all, at least one of the objects needs to move away from the other object. If both objects are approaching each other, there cannot be a change of the shortest path and thus there cannot be a change in the relation between both objects if the speed remains positive. Secondly, object m needs to lie on a cycle in the network; otherwise m can only be reached by paths using the same immediately proceeding node. In order to have a transition in the relations, the immediately proceeding node of m changes (e.g. the immediately proceeding node of m changes from node B to node C in Figure 11). Finally, due to the theory of dominance, a 'Shortest Path Change' event should hold over an interval. Figure 12 gives an overview of all possible transitions between relations due to a single 'Shortest Path Change' event.

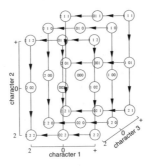

Fig. 12. Possible transitions due to a single 'Shortest Path Change' event

3.3.3 Combination of Events

A transition between the QTC$_N$ relations can be caused by three events. Since these three events ('Speed Change' event, 'Node Pass' event and 'Shortest Path Change' event) occur independently and a 'Node Pass' event or a 'Shortest Path Change' event is caused by only one object, two or more events can occur simultaneously. The result of a transition is the combined transition of the occurring events.

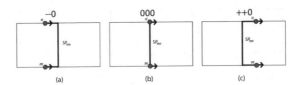

Fig. 13. Example of a transition due to a combined 'Node Pass' event

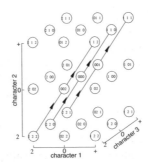

Fig. 14. Possible transitions due to a combined 'Node Pass' event

3.3.3.1 A Combined 'Node Pass' Event

When object m and object n both approach each other, it can be that both objects simultaneously pass a node and therefore create the possibility of a combined 'Node Pass' event (Figure 13). This property induces three additional conceptual animations, resulting in six new transitions as shown in Figure 14.

3.3.3.2 A Combined 'Shortest Path Change' Event

Assume that object m and object n both lie on a cycle within the network. Both objects can lie on two different cycles (Figure 15) or the same cycle (Figure 16). When both objects are moving away from each other, there is a possibility that the node closest to each object along the shortest path in the direction of the other object changes when the shortest path changes. This leads to a combined 'Shortest Path Change' event. This allows three additional conceptual animations resulting in six new transitions (Figure 18a).

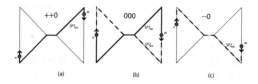

Fig. 15. Example of a transition due to a combined 'Shortest Path Change' event when both objects lie on a different cycle

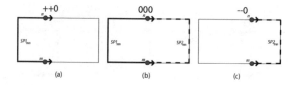

Fig. 16. Example of a transition due to a combined 'Shortest Path Change' event when both objects lie on the same cycle

A combined 'Shortest Path Change' event can also occur when only one object is moving away from the other object. This transition is exemplified in Figure 17. In Figure 17a both objects lie on the same cycle. This means that there are two paths between object n and object m. There is one shorter path (m,A,B,n), and one longer path (m,D,C,n). When n is moving away from m, m is moving towards n and n is moving faster than m, the shorter path will be extended and the longer path will get shorter. At some moment in time, these two paths will become equally long (Figure 17b). At that instantaneous moment, neighter object will approach or move away from each other. As a result both objects will evoke a '0' in the three character label. A fraction of time later, both objects will move along the newly defined shortest path and therefore n will decrease its distance compared to m and m will increase its distance compared to n (Figure 17c). This allows two additional conceptual animations resulting in four new transitions (Figure 18b).

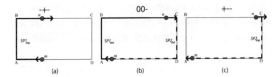

Fig. 17. Example of a transitions due to a combined 'Shortest Path Change' event

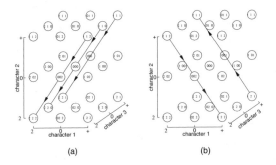

Fig. 18. Possible transitions due to a combined 'Shortest Path Change' event

3.3.3.3 A Combination of a 'Node Pass' Event and a 'Shortest Path Change' Event
Figure 19 illustrates a transition caused by a combination of a 'Node Pass' event and a 'Shortest Path Change' event. This transition occurs when one object passes a node and simultaneously the shortest path changes due to the other. This transition can only occur if the object that passes a node approaches the other object. The other object must then

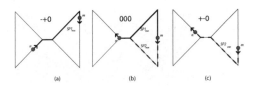

Fig. 19. Example of a transition due to a combined 'Node Pass' and 'Shortest Path Change' event

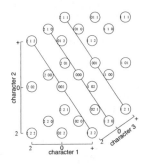

Fig. 20. Possible transitions due to a combined 'Node Pass' and 'Shortest Path Change' event

move away from this object and lie on a cycle of the network. A combination of a 'Node Pass' event and a 'Shortest Path Change' event allows six additional conceptual animations resulting in twelve new transitions as shown in Figure 20.

3.3.3.4 A Combination of a 'Speed Change' Event and a (Combined) 'Node Pass' Event and/or a (Combined) 'Shortest Path Change' Event

Apart from the fact that objects need to move in order for a 'Node Pass' event or a 'Shortest Path Change' event to occur, speed is independent of these two events. Therefore a 'Speed Change' event is also independent of these two events. This means that a 'Speed Change' event can occur simultaneously with a single or a combination of 'Node Pass' events and/or a single or a combination of 'Shortest Path Change' events. An example of a combination of such events is shown in Figure 21. The transitions caused by a combination of a 'Speed Change' event and a 'Node Pass' event and/or a 'Shortest Path Change' event are visualised in Figure 22.

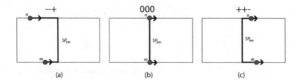

Fig. 21. Example of a transition due to a combination of a 'Speed Change' event and a (combined) 'Node Pass' event and/or a (combined) 'Shortest Path Change' event

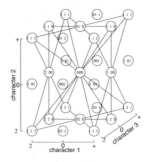

Fig. 22. Possible Transitions due to a combination of a 'Speed Change' event and a (combined) 'Node Pass' event and/or a (combined) 'Shortest Path Change' event

3.4 The Conceptual Neighbourhood Diagram for QTC_N

By combining all possible transitions between relations described in 3.3, an overall CND for QTC_N can be constructed. The overall CND is presented in Figure 23. The CND clearly shows that all of the 27 (3^3) theoretically possible relations exist, but not all of them last over an interval. The ten dashed nodes represent relations that can only exist for an instantaneous moment in time. These relations are the ten none-existing relations in QTC_{B1}. The CND also reveals that in contrast to the CND for QTC_{B2}, not all theoretically possible transitions between relations for QTC_N exist. Out of a possible 98 transitions, 76 remain feasible.

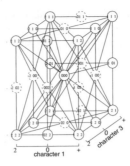

Fig. 23. The Overall CND for QTC_N

4 Conclusion and Further Work

In this paper we defined a Qualitative Trajectory Calculus on Networks (QTC_N). A Conceptual Neighbourhood Diagram (CND) was constructed for QTC_N. Note that QTC_N is more expressive than QTC_{B1}, since ten extra relations exist. However, in QTC_N fewer transitions between relations exist than in QTC_{B2}. In a way QTC_N can be positioned somewhere in between QTC_{B1} and QTC_{B2}. This can be explained by the co-dimensional structure of a network. It is a one-dimensional structure embedded in a two-dimensional or three-dimensional space.

We strongly believe that QTC_N is useful in representing moving objects in the framework of a predefined network. Given that nearly all traffic movements are bounded by a network, QTC_N's application field in Geographic Information Systems for Transportation (GIS-T) seems to offer great potential. We plan to evaluate QTC_N in this domain.

We have presented QTC_N in a relatively informal way concentrating on presenting ideas illustrated with simple examples. In Future work we will fully formalize QTC_N. Moreover, we will construct a composition table [1, 3] for QTC_N.

Acknowledgements

The research work of Nico Van de Weghe and Philippe De Maeyer is funded by the Research Foundation-Flanders, Research Project G.0344.05.

References

1. Allen, J.F.: Maintaining Knowledge about Temporal Intervals. Comm. of the ACM 26(11), 832–843 (1983)
2. Bogaert, P., Van de Weghe, N., De Maeyer, P.: Description, Definition and Proof of a Qualitative State Change of Moving Objects along a Road Network. In: Raubal, M., Sliwinski, A., Kuhn, W. (eds.) Proc. of the Münster GI-Days, Münster, Germany, pp. 239–248 (2004)
3. Cohn, A.G., Bennett, B., Gooday, J., Gotts, N.M.: Representing and Reasoning with Qualitative Spatial Relations about Regions. In: Stock, O. (ed.) Spatial and Temporal Reasoning, pp. 97–134. Kluwer Academic Publishers, Dordrecht (1997)

4. Egenhofer, M., Al-Taha, K.: Reasoning about gradual changes of topological relationships. In: Proc. of COSIT, pp. 196–219 (1992)
5. Egenhofer, M.: Query processing in spatial-query-by-sketch. JVLC 8, 403–424 (1997)
6. Egenhofer, M., Franzosa, R.: Point-set Topological Spatial Relations. International Journal of Geographical Information Systems 5(2), 161–174 (1991)
7. Freksa, C.: Temporal Reasoning Based on Semi-Intervals. Artificial Iintelligence 54, 199–227 (1992)
8. Freksa, C.: Using Orientation Information for Qualitative Spatial Reasoning. In: Frank, A.U., Formentini, U., Campari, I. (eds.) Theories and Methods of Spatio-Temporal Reasoning in Geographic Space. LNCS, vol. 639, pp. 162–178. Springer, Heidelberg (1992)
9. Galton, A.: Towards a Qualitative Theory of Movement. In: Kuhn, W., Frank, A.U. (eds.) COSIT 1995. LNCS, vol. 988, pp. 377–396. Springer, Heidelberg (1995)
10. Galton, A.: A Qualitative Approach to Continuity. In: Amsili, P., Borillo, M., Vieu, L. (eds.) Proc. of the 5th Workshop on Time, Space and Movement (TSM), pp. 17–30 (1995)
11. Galton, A., Worboys, M.: Processes and Events in Dynamic Geo-Networks. In: Rodríguez, M.A., Cruz, I., Levashkin, S., Egenhofer, M.J. (eds.) GeoS 2005. LNCS, vol. 3799, pp. 45–59. Springer, Heidelberg (2005)
12. Goyal, R.K.: Similarity Assessment for Cardinal Directions between Extended Spatial Objects, PhD thesis, University of Maine, USA, pp. 167 (2000)
13. Iwasaki, Y.: Real-World Applications of Qualitative Reasoning: Introduction to the Special Issue. IEEE Intelligent Systems 12(3), 16–21 (1997)
14. Randell, D., Cui, Z., Cohn, A.G.: A Spatial Logic Based on Regions and Connection. In: Nebel, B., Swartout, W., Rich, C. (eds.) Proc. of KR, San Mateo, USA, pp. 165–176 (1992)
15. Moreira, J., Ribeiro, C., Saglio, J.-M.: Representation and Manipulation of Moving Points: An Extended Data Model for Location Estimation. Cartography and Geographic Information Systems 26(2), 109–123 (1999)
16. Van de Weghe, N.: Representing and Reasoning about Moving Objects: A Qualitative Approach, PhD thesis, Ghent University, Belgium, pp. 268 (2004)
17. Van de Weghe, N., Cohn, A.G., Bogaert, P., De Maeyer, P.: Representation of Moving Objects along a Road Network. In: Proc. of Geoinformatics, Gävle, Sweden, pp. 187–197 (2004)
18. Van de Weghe, N., Cohn, A.G., De Tré, G., De Maeyer, P.: A Qualitative Trajectory Calculus as a Basis for Representing Moving Objects in Geographical Information Systems. Control and Cybernetics (accepted for publication)
19. Van de Weghe, N., Cohn, A.G., De Maeyer, P., Witlox, F.: Representing moving objects in computer-based expert systems: the overtake event example. Expert Systems with Applications 29(4), 977–983 (2005)
20. Van de Weghe, N., De Maeyer, P.: Conceptual Neighbourhood Diagrams for Representing Moving Objects. In: Akoka, J., Liddle, S.W., Song, I.-Y., Bertolotto, M., Comyn-Wattiau, I., van den Heuvel, W.-J., Kolp, M., Trujillo, J., Kop, C., Mayr, H.C. (eds.) Perspectives in Conceptual Modeling. LNCS, vol. 3770, pp. 228–238. Springer, Heidelberg (2005)
21. Van de Weghe, N., Kuijpers, B., Bogaert, P.: A qualitative trajectory calculus and the composition of its relations. In: Rodríguez, M.A., Cruz, I., Levashkin, S., Egenhofer, M.J. (eds.) GeoS 2005. LNCS, vol. 3799, pp. 60–76. Springer, Heidelberg (2005)
22. Weld, D.S., de Kleer, J.: Readings in Qualitative Reasoning about Physical Systems, p. 720. Morgan Kaufmann, San Mateo, California (1990)

Qualitative Spatial Representation and Reasoning in the SparQ-Toolbox

Jan Oliver Wallgrün, Lutz Frommberger, Diedrich Wolter, Frank Dylla, and Christian Freksa

SFB/TR 8 Spatial Cognition
Universität Bremen
Bibliothekstr. 1, 28359 Bremen, Germany
{wallgruen,lutz,dwolter,dylla,freksa}@sfbtr8.uni-bremen.de

Abstract. A multitude of calculi for qualitative spatial reasoning (QSR) have been proposed during the last two decades. The number of practical applications that make use of QSR techniques is, however, comparatively small. One reason for this may be seen in the difficulty for people from outside the field to incorporate the required reasoning techniques into their software. Sometimes, proposed calculi are only partially specified and implementations are rarely available. With the SparQ toolbox presented in this text, we seek to improve this situation by making common calculi and standard reasoning techniques accessible in a way that allows for easy integration into applications. We hope to turn this into a community effort and encourage researchers to incorporate their calculi into SparQ. This text is intended to present SparQ to potential users and contributors and to provide an overview on its features and utilization.

1 Introduction

Qualitative spatial reasoning (QSR) is an established field of research pursued by investigators from many disciplines including geography, philosophy, computer science, and AI [1]. The general goal is to model commonsense knowledge and reasoning about space as efficient representation and reasoning mechanisms that are still expressive enough to solve a given task. Qualitative spatial representation techniques are especially suited for applications that involve interaction with humans as they provide an interface based on human spatial concepts.

Following the approach taken in Allen's seminal paper on qualitative temporal reasoning [2], QSR is typically realized in form of calculi over sets of spatial relations (like 'left-of' or 'north-of'). These are called *qualitative spatial calculi*. A multitude of spatial calculi has been proposed during the last two decades, focusing on different aspects of space (mereotopology, orientation, distance, etc.) and dealing with different kinds of objects (points, line segments, extended objects, etc.). Two main research directions in QSR are mereotopological reasoning about regions [3,4,5] and reasoning about positional information (distance and orientation) of point objects [6,7,8,9,10,11,12] or line segments [13,14,15]. In addition, some approaches are concerned with direction relations between extended objects [16,17] or combine different aspects of space [18,19].

Despite this large variety of qualitative spatial calculi, the amount of applications employing qualitative spatial reasoning techniques is comparatively small. We believe that one important factor for this is the following: Choosing the right calculus for a particular application is a challenging task, especially for people not familiar with QSR. Calculi are often only partially specified and usually no implementation is made available—if the calculus is implemented at all and not only investigated theoretically. As a result, it is not possible to "quickly" evaluate how different calculi perform in practice. Even if an application developer has decided on a particular calculus, he has to invest serious efforts to include the calculus and required reasoning techniques into the application. For many calculi this is a time-consuming and error-prone process (e. g. involving writing down large composition tables, which are often not even completely specified in the literature). We think that researchers involved in the investigation of QSR will also benefit from reference implementations of calculi that are available in a coherent framework. Tasks like comparing different calculi with respect to expressiveness or average computational properties in a certain context would clearly be simplified.

To provide a platform for making the calculi and reasoning techniques developed in the QSR community available, we have started the development of a qualitative spatial reasoning toolbox called *SparQ*[1]. The toolbox supports binary and ternary spatial calculi. SparQ aims at supporting the most common tasks—qualification, computing with relations, constraint-based reasoning (cp. Section 3)—for an extensible set of spatial calculi. Our focus is on providing an implementation of QSR techniques that is tailored towards the needs of application developers. A similar approach has recently been reported in [20] where calculi and reasoning techniques are provided in form of a programming library and focuses on algebraic and constraint-based reasoning. SparQ, on the other hand, is a application program that can be used directly and provides a broader range of services. A complementary approach aiming at the specification and investigation of the interrelations between calculi has been described in [21]. There, the calculi are defined in the algebraic specification language CASL. We believe that a toolbox like SparQ can provide a useful interface between the theoretical specification framework and the application areas of spatial cognition, like cognitive modeling or GIS.

In its current version, SparQ mainly focuses on calculi from the area of reasoning about the orientation of point objects or line segments. However, specifying and adding other calculi is simple. We hope to encourage researchers from other groups to incorporate their calculi in a community effort of providing a rich spatial reasoning environment. SparQ is designed as an open framework of single program components with text-based communication. It therefore allows for integrating code written in virtually any programming language, so that already existing code can easily be integrated into SparQ.

---

[1] **S**patial **R**easoning done **Q**ualitatively.

Specifically, the goals of SparQ are the following:

- providing reference implementations for spatial calculi from the QSR community
- making it easy to specify and integrate new calculi
- providing typical procedures required to apply QSR in a convenient way
- offering a uniform interface that supports switching between calculi
- being easily integrable into own applications

The current version of SparQ and further documentation will be made available at the SparQ homepage[2]. In the present text, we will describe SparQ and its utilization. The next section briefly recapitulates the relevant terms concerning QSR and spatial calculi as needed for the remainder of the text. In Section 3, we describe the services provided by SparQ. Section 4 explains how new calculi can be incorporated into SparQ, and Section 5 describes how SparQ can be integrated into applications. Finally, Section 6 contains a case study in which SparQ is employed to compare different calculi with respect to their ability of detecting the inconsistency in the Indian Tent Problem [22].

2 Reasoning with Qualitative Spatial Relations

A qualitative spatial calculus defines operations on a finite set $\mathcal{R}$ of spatial relations. The spatial relations are defined over a particular set of spatial objects, the domain D. In the rest of the text, we will encounter the sets of points in the plane, of oriented line segments in the plane, and of oriented points in the plane as domains. While a *binary calculus* deals with binary relations $R \subseteq D \times D$, a *ternary calculus* operates with ternary relations $R \subseteq D \times D \times D$.

The set of relations $\mathcal{R}$ of a spatial calculus is typically derived from a jointly exhaustive and pairwise disjoint (JEPD) set of *base relations* $\mathcal{BR}$ so that each pair of objects from D is contained in exactly one relation from $\mathcal{BR}$. Every relation in $\mathcal{R}$ is a union of base relations. Since spatial calculi are typically used for constraint reasoning and unions of relations correspond to disjunctions of relational constraints, it is common to speak of disjunctions of relations as well and write them as sets $\{B_1, ..., B_n\}$ of base relations. Using this convention, $\mathcal{R}$ is either taken to be the powerset $2^{\mathcal{BR}}$ of the base relations or a subset of the powerset. In order to be usable for constraint reasoning, $\mathcal{R}$ should contain at least the base relations B_i, the empty relation $\emptyset$, the universal relation U, and the identity relation Id. $\mathcal{R}$ should also be closed under the operations defined in the following.

As the relations are subsets of tuples from the same Cartesian product, the set operations union, intersection, and complement can be directly applied:

Union: $R \cup S = \{\, t \mid t \in R \vee t \in S \,\}$
Intersection: $R \cap S = \{\, t \mid t \in R \wedge t \in S \,\}$
Complement: $\overline{R} = U \setminus R = \{\, t \mid t \in U \wedge t \notin R \,\}$

---

[2] http://www.sfbtr8.uni-bremen.de/project/r3/sparq/

where R and S are both n-ary relations on D and t is an n-tuple of elements from D. The other operations depend on the arity of the calculus.

2.1 Operations for Binary Calculi

For binary calculi the other two important operations are conversion and composition:

Converse: $R^{\smile} = \{\, (y,x) \mid (x,y) \in R \,\}$
(Strong) composition: $R \circ S = \{\, (x,z) \mid \exists y \in D : ((x,y) \in R \land (y,z) \in S) \,\}$

For some calculi, no finite set of relations exists that includes the base relations and is closed under composition as defined above. In this case, a weak composition is defined instead that takes the union of all base relations that have a non-empty intersection with the result of the strong composition:

Weak composition: $R \circ_{weak} S = \{\, B_i \mid B_i \in \mathcal{BR} \land B_i \cap (R \circ S) \neq \emptyset \,\}$

2.2 Operations for Ternary Calculi

While there is only one possibility to permute the two objects of a binary relation which corresponds to the converse operation, there exist 5 such permutations for the three objects of a ternary relation[3], namely [23]:

Inverse: $\mathrm{INV}(R) = \{\, (y,x,z) \mid (x,y,z) \in R \,\}$
Short cut: $\mathrm{SC}(R) = \{\, (x,z,y) \mid (x,y,z) \in R \,\}$
Inverse short cut: $\mathrm{SCI}(R) = \{\, (z,x,y) \mid (x,y,z) \in R \,\}$
Homing: $\mathrm{HM}(R) = \{\, (y,z,x) \mid (x,y,z) \in R \,\}$
Inverse homing: $\mathrm{HMI}(R) = \{\, (z,y,x) \mid (x,y,z) \in R \,\}$

Composition for ternary calculi is defined according to the binary case:

(Strong) comp: $R \circ S = \{\, (w,x,z) \mid \exists y \in D : ((w,x,y) \in R \land (x,y,z) \in S) \,\}$

Other ways of composing two ternary relations can be expressed as a combination of the unary permutation operations and the composition [24] and thus do not have to be defined separately. The definition of weak composition is identical to the binary case.

2.3 Constraint Reasoning with Spatial Calculi

Spatial calculi are often used to formulate constraints about the spatial configurations of a set of objects from the domain of the calculus as a constraint satisfaction problem (CSP): Such a spatial constraint satisfaction problem then consists of a set of variables $X_1, ..., X_n$ (one for each spatial object) and a set of constraints $C_1, ..., C_m$ which are relations from the calculus. Each variable X_i

---

[3] In general, two operations (permutation and rotation) are sufficient to generate all permuations (cmp. [20]). Therefore, not all of these operations need to be specified.

can take values from the domain of the utilized calculus. CSPs are often described as constraint networks which are complete labeled graphs with a node for each variable and each edge labeled with the corresponding relation from the calculus. A CSP is consistent, if an assignment for all variables to values of the domain can be found, that satisfies all the constraints. Spatial CSPs usually have infinite domains and thus backtracking over the domains can not be used to determine consistency.

Besides consistency, weaker forms of consistency called *local consistencies* are of interest in QSR. On the one hand, they can be employed as a forward checking technique reducing the CSP to a smaller equivalent CSP (one that has the same set of solutions). Furthermore, in some cases a form of local consistency can be proven to be not only necessary but also sufficient for consistency. If this is only the case for a certain subset $\mathcal{S} \subset \mathcal{R}$ and this subset exhaustively splits $\mathcal{R}$ (which means that every relation from $\mathcal{R}$ can be expressed as a disjunction of relations from $\mathcal{S}$), this at least allows to formulate a backtracking algorithm to determine consistency by recursively splitting the constraints and using the local consistency as a decision procedure for the resulting CSPs with constraints from $\mathcal{S}$ [25].

One important form of local consistency is *path-consistency* which (in binary CSPs) means that for every triple of variables each consistent evaluation of the first two variables can be extended to the third variable in such a way that all constraints are satisfied. Path-consistency can be enforced syntactically based on the composition operation (for instance with the algorithm by van Beek [26]) in $O(n^3)$ time where n is the number of variables. However, this syntactic procedure does not necessarily yield the correct result with respect to path-consistency as defined above. The same holds for syntactic procedures that compute other kinds of consistency. Whether syntactic consistency coincides with semantic consistency with respect to the domain needs be investigated for each calculus individually (see [27,28] for an in-depth discussion).

2.4 Supported Calculi

As mentioned above, qualitative calculi are based on a certain domain of basic entities: time intervals in the case of Allen's Interval Calculus [2], or objects like points, line segments, or regions in typical spatial calculi. In the following, we will briefly introduce those calculi that are currently included in SparQ and that will be used in the examples later on. A quick overview is given in Table 1 which also classifies the calculi according to their arity (binary, ternary), their domain (points, oriented points, line segments, regions), and the aspect of space modeled (orientation, distance, mereotopology).

FlipFlop Calculus (FFC) and the $\mathcal{LR}$ refinement. The FlipFlop calculus proposed in [9] describes the position of a point C (the referent) in the plane with respect to two other points A (the origin) and B (the relatum) as illustrated in Fig. 1. It can for instance be used to describe the spatial relation of C to B as seen from A. For configurations with $A \neq B$ the following base relations are distinguished: C can be to the **left** or to the **right** of the oriented line going through

Table 1. The calculi currently included in SparQ

Calculus	arity		domain				aspect of space		
	binary	ternary	point	or. point	line seg.	region	orient.	dist.	mereot.
FFC/$\mathcal{LR}$		✓		✓			✓		
SCC		✓		✓			✓		
DCC		✓		✓			✓		
$\mathcal{DRA}_c$	✓				✓		✓		
$\mathcal{OPRA}_m$	✓			✓			✓		
RCC-5/8[4]	✓					✓			✓

A and B, or C can be placed on the line resulting in one of the five relations inside, front, back, start ($C = A$) or end ($C = B$) (cp. Fig. 1). Relations for the case where A and B coincide were not included in Ligozat's original definition [9]. This was done with the $\mathcal{LR}$ refinement [29] that introduces the relations **dou** ($A = B \neq C$) and **tri** ($A = B = C$) as additional relations, resulting in a total of 9 base relations. A $\mathcal{LR}$ relation $rel_{\mathcal{LR}}$ is written as $A, B \ rel_{\mathcal{LR}} \ C$, e.g. $A, B \ \mathbf{r} \ C$ as depicted in Fig. 1.

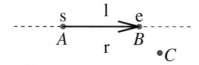

Fig. 1. The reference frame for the $\mathcal{LR}$ calculus, an refined version of the FlipFlop Calculus

Single Cross Calculus (SCC). The Single Cross Calculus is a ternary calculus that describes the direction of a point C (the referent) wrt. a point B (the relatum) as seen from a third point A (the origin). It was originally proposed in [6]. The plane is partitioned into regions by the line going through A and B and the perpendicular line through B. This results in eight distinct orientations as illustrated in Fig. 2(a). We denote these base relations by numbers from 0 to 7 instead of using linguistic prepositions, e.g. 2 instead of *left* as in [6]. Relations 0,2,4,6 are linear ones, while relations 1,3,5,7 are planar. In addition, three special relations exist for the cases $A \neq B = C$ (**bc**), $A = B \neq C$ (**dou**), and $A = B = C$ (**tri**). A Single Cross relation rel_{SCC} is written as $A, B \ rel_{SCC} \ C$, e.g. $A, B \ \mathbf{4} \ C$ or $A, B \ \mathbf{dou} \ C$. The relation depicted in Fig. 2(a) is the relation $A, B \ \mathbf{5} \ C$.

---

[4] Currently only the relational specification is available for RCC, but no 'qualify' module (cmp. Section 3.1).

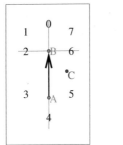
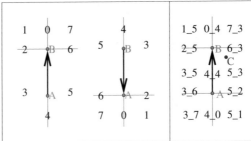

(a) Single Cross Calculus reference frame

(b) The two Single Cross reference frames resulting in the overall Double Cross Calculus reference frame

Fig. 2. The Single and Double Cross reference systems

Double Cross Calculus (DCC). The Double Cross calculus [6] can be seen as an extension of the Single Cross calculus adding another perpendicular, this time at A (see Fig. 2(b) (right)). It can also be interpreted as the combination of two Single Cross relations, the first describing the position of C wrt. B as seen from A and the second wrt. A as seen from B (cf. Fig. 2(b) (left)). The resulting partition distinguishes 13 relations (7 linear and 6 planar) denoted by tuples derived from the two underlying SCC reference frames and four special cases, $A = C \neq B$ (**4_a**), $A \neq B = C$ (**b_4**), $A = B \neq C$ (**dou**), and $A = B = C$ (**tri**), resulting in 17 base relations overall. Fig. 2(b) depicts the relation A, B **5_3** C.

Coarse-grained Dipole Relation Algebra ($\mathcal{DRA}_c$). A dipole is an oriented line segment, e.g. as determined by a start and an end point. We will write $\boldsymbol{d}_{AB}$ for a dipole defined by start point A and end point B. The idea of using dipoles was first introduced by Schlieder [13] and extended in [14].

In the coarse-grained variant of the Dipole Calculus ($\mathcal{DRA}_c$) describes the orientation relation between two dipoles $\boldsymbol{d}_{AB}$ and $\boldsymbol{d}_{CD}$ with the preliminary that A, B, C, and D are in general position, i.e. no three disjoint points are collinear. Each base relation is a 4-tuple (r_1, r_2, r_3, r_4) of FlipFlop relations relating a point from one of the dipoles with the other dipole. r_1 describes the relation of C wrt. the dipole $\boldsymbol{d}_{AB}$, r_2 of D wrt. $\boldsymbol{d}_{AB}$, r_3 of A wrt. $\boldsymbol{d}_{CD}$, and r_4 of B wrt. $\boldsymbol{d}_{CD}$. The distinguished FlipFlop relations are **left**, **right**, **start**, and **end** (see Fig. 1). Dipole relations are usually written without commas and parentheses, e.g. **rrll**. Thus, the example in Fig. 3 shows the relation $\boldsymbol{d}_{AB}$ **rlll** $\boldsymbol{d}_{CD}$. Since the underlying points for a $\mathcal{DRA}_c$ relation need to be in general position the r_i can only take the values **left**, **right**, **start**, or **end** resulting in 24 base relations.

Oriented Point Relation Algebra $\mathcal{OPRA}_m$. The $\mathcal{OPRA}_m$ calculus [11] operates on oriented points. An oriented point is a point in the plane with an additional direction parameter. $\mathcal{OPRA}_m$ relates two oriented points $\boldsymbol{A}$ and $\boldsymbol{B}$ and describes their relative orientation towards each other. The granularity

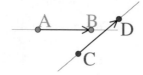

Fig. 3. A dipole configuration: d_{AB} rlll d_{CD} in the coarse-grained Dipole Relation Algebra ($\mathcal{DRA}_c$)

factor $m \in \mathbb{N}$ determines the number of distinguished relations. For each of the two oriented points, m lines are used to partition the plane into $2m$ planar and $2m$ linear regions. Fig. 4 shows the partitions for the cases $m = 2$ (Fig. 4(a)) and $m = 4$ (Fig. 4(b)). The orientation of the two points is depicted by the arrows starting at $\boldsymbol{A}$ and $\boldsymbol{B}$, respectively. The regions are numbered from 0 to $(4m-1)$. Region 0 always coincides with the orientation of the point. An $\mathcal{OPRA}_m$ relation $rel_{\mathcal{OPRA}_m}$ consist of pairs (i, j) where i is the number of the region of $\boldsymbol{A}$ which contains $\boldsymbol{B}$, while j is the number of the region of $\boldsymbol{B}$ that contains $\boldsymbol{A}$. These relations are usually written as $\boldsymbol{A} \;{}_m\angle_i^j\; \boldsymbol{B}$ with $i, j \in \mathcal{Z}_{4m}$[5]. Thus, the examples in Fig. 4 depict the relations $\boldsymbol{A} \;{}_2\angle_7^1\; \boldsymbol{B}$ and $\boldsymbol{A} \;{}_4\angle_{13}^3\; \boldsymbol{B}$. Additional relations describe situations in which both oriented points coincide. In these cases, the relation is determined by the number s of the region of $\boldsymbol{A}$ into which the orientation arrow of $\boldsymbol{B}$ falls (as illustrated in Fig. 4(c)). These relations are written as $\boldsymbol{A} \;{}_2\angle s\; \boldsymbol{B}$ ($\boldsymbol{A} \;{}_2\angle 1\; \boldsymbol{B}$ in the example).

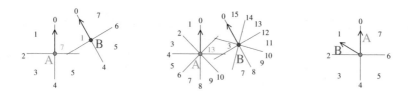

(a) with granularity $m = 2$: $\boldsymbol{A} \;{}_2\angle_7^1\; \boldsymbol{B}$

(b) with granularity $m = 4$: $\boldsymbol{A} \;{}_4\angle_{13}^3\; \boldsymbol{B}$

(c) case where $\boldsymbol{A}$ and $\boldsymbol{B}$ coincide: $\boldsymbol{A} \;{}_2\angle 1\; \boldsymbol{B}$

Fig. 4. Two oriented points related at different granularities

3 SparQ

SparQ consists of a set of modules that provide different services required for QSR that will be explained below. These modules are glued together by a central script that can either be used directly from the console or included into own applications via TCP/IP streams in a server/client fashion (see Section 5). The general architecture is visualized in Fig. 5.

---

[5] $\mathcal{Z}_{4m}$ defines a cyclic group with $4m$ elements.

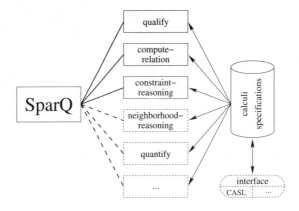

Fig. 5. Module architecture of the SparQ toolbox

The general syntax for using the SparQ main script is as follows:

`$ ./sparq <module> <calculus identifier> <module-specific parameters>`

Example:

`$ ./sparq compute-relation dra-24 complement "(lrll llrr)"`

where 'compute-relation' is the name of the module to be utilized, in this case the module for conducting operations on relations, 'dra-24' is the SparQ identifier for the dipole calculus $\mathcal{DRA}_c$, and the rest are module-specific parameters, here the name of the operation that should be conducted ('complement') and a string parameter representing the disjunction of the two dipole base relations *lrll* and *llrr*[6]. The example call thus computes the complement of the disjunction of these two relations.

Some calculi have calculus-specific parameters, for example the granularity parameter in $\mathcal{OPRA}_m$. These parameters are appended with a '-' after the calculus' base identifier. `opra-3` for example refers to $\mathcal{OPRA}_3$.

SparQ currently provides the following modules:

qualify transforms a quantitative geometric description of a spatial configuration into a qualitative description based on one of the supported spatial calculi

compute-relation applies the operations defined in the calculi specifications (intersection, union, complement, converse, composition, etc.) to a set of spatial relations

constraint-reasoning performs computations on constraint networks

Further modules are planned in future extensions. They comprise a quantification module for turning qualitative scene descriptions back into quantitative

---

[6] Disjunctions of base relations are always represented as a space-separated list of the base relations enclosed in parentheses in SparQ.

geometric descriptions and a module for neighborhood-based spatial reasoning. In the following section we will take a closer look at the three existing modules.

3.1 Scene Descriptions and Qualification

The purpose of the 'qualify' module is to turn a quantitative geometric scene description into a qualitative scene description with respect to a particular calculus. Calculi are specified via the calculus identifier that is passed with the call to SparQ. Qualification is required for applications in which one wants to perform qualitative computations over objects represented by their geometric parameters.

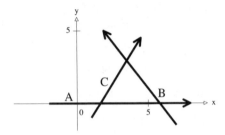

Fig. 6. An example configuration of three dipoles

The 'qualify' module reads a quantitative scene description and generates a qualitative one. A quantitative scene description is a list of base object descriptions (separated by spaces and enclosed in parentheses). Each base object description is a tuple consisting of an object identifier and object parameters that depend on the type of the object. For instance, let us say we are working with dipoles, i.e. oriented line segments. The object description of a dipole has the form '(name x_s y_s x_e y_e)', where name is the identifier of this particular dipole object and the rest are the coordinates of start and end point of the dipole. Let us consider the example in Fig. 6 which shows three dipoles A, B, and C. The quantitative scene description for this situation is:

((A -2 0 8 0) (B 7 -2 2 5) (C 1 -1 4.5 4.5))

The 'qualify' module has one module-specific parameter that needs to be specified:

mode: This parameter controls which relations are included into the qualitative scene description: If 'all' is passed as parameter, the relations between each pair of objects will be determined. If it is 'first2all' only the relations between the first and all other objects are computed.

The resulting qualitative scene description is a list of relation tuples (again separated by spaces and enclosed in parentheses). A relation tuple consists of the object identifier of the relatum followed by a relation and the object identifier of the referent, meaning that the first object stands in this particular relation with the second object. The command to produce the qualitative scene description followed by the result is[7]:

```
$ ./sparq qualify dra-24 all
$ ( (A -2 0 8 0) (B 7 -2 2 5) (C 1 -1 4.5 4.5) )
> ( (A rllr B) (A rllr C) (B lrrl C) )
```

If we had chosen 'first2all' as mode parameter the relation between B and C would not have been included in the qualitative scene description.

3.2 Computing with Relations

The 'compute-relation' module realizes computations with the operations defined in the calculus specification. The module-specific parameters are the operation that should be conducted and one or more input relations depending on the arity of the operation. Assume we want to compute the converse of the dipole relation *llrl*. The corresponding call to SparQ and the result are:

```
$ ./sparq compute-relation dra-24 converse llrl
> (rlll)
```

The result is always a list of relations as operations often yield a disjunction of base relations. In the example above, the list contains a single relation. The composition of two relations requires one more relation as parameter because it is a binary operation, e.g.:

```
$ ./sparq compute-relation dra-24 composition llrr rllr
> (lrrr llrr rlrr slsr lllr rllr rlll ells llll lrll)
```

Here the result is a disjunction of 10 base relations. It is also possible to have disjunctions of base relations as input parameters. For instance, the following call computes the intersection of two disjunctions:

```
$ ./sparq compute-relation dra-24 intersection "(rrrr rrll rllr)"
  "(llll rrll)"
> (rrll)
```

3.3 Constraint Reasoning

The 'constraint-reasoning' module reads a description of a constraint network; this is a qualitative scene description that may include disjunctions and may be inconsistent and/or underspecified. It performs a particular kind of consistency check[8]. Which type of consistency check is executed depends on the first module specific parameter:

---

[7] In all the examples, input lines start with '$'. Output of SparQ is marked with '>'.
[8] The 'constraint-reasoning' module also provides some basic actions to manipulate constraint networks that are not further explained in this text. One example is the 'merge' operation that is used in the example in Section 5 (see the SparQ manual for details [30]).

action: The two consistency checks currently provided are 'path-consistency' and 'scenario-consistency'; the parameter determines which kind of consistency check is performed.

The action 'path-consistency' causes the module to enforce path-consistency on the constraint network using van Beek's algorithm [26] or to detect an inconsistency of the network in the process. In case of a ternary calculus the canonical extension of van Beek's algorithm described in [31] is used. For instance, we could check if the scene description generated by the 'qualify' module in Section 3.1 is path-consistent—which of course it is. To make the test slightly more interesting we add the base relation *ells* to the constraint between A and C; this results in a constraint network that is not path-consistent:

```
$ ./sparq constraint-reasoning dra-24 path-consistency
$ ( (A rllr B) (A (ells rllr) C) (B lrrl C) )
> Modified network.
> ( (B (lrrl) C) (A (rllr) C) (A (rllr) B) )
```

The result is a path-consistent constraint network in which *ells* has been removed. The output 'Modified network' indicates that the original network was not path-consistent and had to be changed. Otherwise, the result would have started with 'Unmodified network'. In the next example we remove the relation *rllr* from the disjunction. This results in a constraint network that cannot be made path-consistent; this implies that it is not consistent.

```
$ ./sparq constraint-reasoning dra-24 path-consistency
$ ( (A rllr B) (A ells C) (B lrrl C) )
> Not consistent.
> ( (B (lrrl) C) (A () C) (A (rllr) B) )
```

SparQ correctly determines that the network is inconsistent and returns the constraint network in the state in which the inconsistency showed up (indicated by the empty relation () between A and C).

In a last path-consistency example we use the ternary Double Cross Calculus:

```
$ ./sparq constraint-reasoning dcc path-consistency
$ ( (A B (7_3 6_3) C) (B C (7_3 6_3 5_3) D) (A B (3_6 3_7) D) )
> Not consistent.
> ( (A B () D) (A B (6_3 7_3) C) (B C (5_3 6_3 7_3) D) (D C (3_7) A) )
```

If 'scenario-consistency' is provided as argument, the 'constraint-reasoning' module checks if a path-consistent scenario exists for the given network. It uses a backtracking algorithm to generate all possible scenarios and checks them for path-consistency as described above. A second module-specific parameter determines what is returned as the result of the search:

return: This parameter determines what is to be returned in case of a constraint network for which path-consistent scenarios can be found. 'First' returns the first path-consistent scenario, 'all' returns all path-consistent scenarios, and 'interactive' returns one solution and allows to ask for the next solution until all solutions have been generated.

Path-consistency is also used as a forward-checking method during the search to make the search more efficient. For certain calculi, the existence of a path-consistent scenario implies consistency. However, this again has to be investigated for each calculus (cmp. Section 2.3). In the following example, we use 'first' as additional parameter so that only the first solution is returned:

```
$ ./sparq constraint-reasoning dra-24 scenario-consistency first
$ ( (A rele C) (A ells B) (C errs B) (D srsl C) (A rser D) (D rrrl B) )
> ( (B (rlrr) D) (C (slsr) D) (C (errs) B) (A (rser) D) (A (ells) B)
    (A (rele) C) )
```

In case of an inconsistent constraint network, SparQ returns 'Not consistent.'. As a future extension, we plan to allow specification of splitting subsets of a calculus for which path-consistency implies consistency. A splitting subset S will be used in a variant of the backtracking algorithm to decide consistency by searching for path-consistent instantiations that only contain relations from S.

4 Specifying Calculi in SparQ

For most calculi inclusion into SparQ should be straightforward. The main action to be taken is to provide the calculus specification. This is done in a Lisp-like syntax. Listing 1.1 shows an extract of the definition of a simple exemplary calculus for reasoning about distances between three point objects distinguishing the three relations 'closer', 'farther', and 'same'.

```
(def-calculus "Relative distance calculus (reldistcalculus)"
    :arity :ternary
    :base-relations (same closer farther)
    :identity-relation same

    :inverse-operation ((same same)
                        (closer closer)
                        (farther farther))
    :shortcut-operation ((same same)
                         (closer farther)
                         (farther closer))
    :composition-operation ((same same (same closer farther))
                            (same closer (same closer farther))
                            (same farther (same closer farther))
                            (closer same (same closer farther))
                            (closer closer (same closer farther))
                            [...]
```

Listing 1.1. Specification of a simple ternary calculus for reasoning about distances

The arity of the calculus, the base relations, the identity relation, and the different operations have to be specified, using lists enclosed in parentheses (e.g. when an operation returns a disjunction of base relations). In this example, the inverse operation applied to 'same' yields 'same', and composing 'closer' and 'same' results in the universal relation written as the disjunction of all base relations. As mentioned in Section 2.2, not all operations are required because some operations are combinations of other operations.

In addition to the calculus specification, it is necessary to provide the implementation of a qualifier function which for an n-ary calculus takes n geometric objects of the corresponding base type as input and returns the relation holding between these objects. The qualifier function encapsulates the methods for computing the qualitative relations from quantitative geometric descriptions. If it is not provided, the 'qualify' module will not work for this calculus.

For some calculi, it is not possible to provide operations in form of simple tables as in the example. For instance, $\mathcal{OPRA}_m$ has an additional parameter that specifies the granularity of the calculus and influences the number of base relations. Thus, the operations can only be provided in procedural form; this means the result of the operations are computed from the input relations when they are required. For these cases, SparQ allows providing the operations as implemented functions and uses a caching mechanism to store often required results.

5 Integrating SparQ into Own Applications

SparQ can also run in server mode which makes it easy to integrate it into applications. We have chosen a client/server approach as it allows for straightforward integration independently of the programming language used for implementing the application.

When run in server mode, SparQ takes TCP/IP connections and interacts with the client via simple plain-text line-based communication. This means the client sends commands which consist of everything following the './sparq' in the examples in this text and can then read the results from the TCP/IP stream.

```
# connect to sparq server on localhost, port 4443
sock = socket.socket(socket.AF_INET, socket.SOCK_STREAM)
sock.connect(('localhost', 4443))
sockfile = sock.makefile('r')

# qualify a geometrical scenario with DRA-24
sock.send('qualify dra-24 first2all ')
sock.send('((A 4 6 9 0.5) (B -5 5 0 2) (C -4 5 6 0))')
scene = readline()      # read the answer
print scene

# add an additional relation (B eses C)
sock.send("constraint-reasoning dra-24 merge")
sock.send(scene + '(B eses C)')
scene2 = readline()     # read the answer
print scene2

# check the new scenario for consistency
sock.send('constraint-reasoning dra-24 path-consistency')
sock.send(scene2)
print readline()        # print the answer
print readline()        # print the resulting constraint network
```

Listing 1.2. Integrating SparQ into own applications: an example in Python

SparQ is started in server mode by providing the command line option --interactive (-i), optionally followed by --port (-p) to specify the port.

```
$ ./sparq --interactive --port 4443
```

If no port is given, SparQ interacts with standard-input and standard-output, i.e., it can be used interactively from the shell.

An example is given in Listing 1.2 which shows a small Python program that opens a connection to the server and performs some simple computations (qualification, adding another relation, checking for path-consistency). It produces the following output:

```
> ( (A rrll B) (A rrll C) )
> ( (A rrll B) (A rrll C) (B eses C) )
> Not consistent.
> ( (B (eses) C) (A () C) (A (rrll) B) )
```

6 A Case Study: Using SparQ for the Indian Tent Problem

In this section we want to demonstrate the application of SparQ to a problem that can be seen as a kind of benchmark in QSR, the so-called *Indian Tent Problem*.

6.1 Indian Tent Definition

The Indian Tent Problem was first discussed by Röhrig [22]. It describes a very simple configuration of points in the plane that is not consistent and can be used to compare spatial reasoning formalisms with respect to their ability to detect this inconsistency.

The Indian Tent consists of a clockwise oriented triangle $\triangle ABC$ and an additional point D (see Fig. 7). The following facts are given: "C is right of AB", "D is left of AB" and "D is right of CB". From these facts follows geometrically that "D is right of CA". Thus, adding the (obviously wrong) fact "D is left of CA" results in an inconsistency.

In the following, we will show how to use SparQ to compare the properties of different calculi. We will model the Indian Tent Problem with three calculi (FFC, DCC and $\mathcal{DRA}_c$) and demonstrate how the constraint-based reasoning abilities of SparQ can be used to gain insights about how well local consistencies like path-consistency or scenario-consistency approximate consistency.

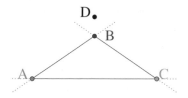

Fig. 7. The Indian Tent Problem

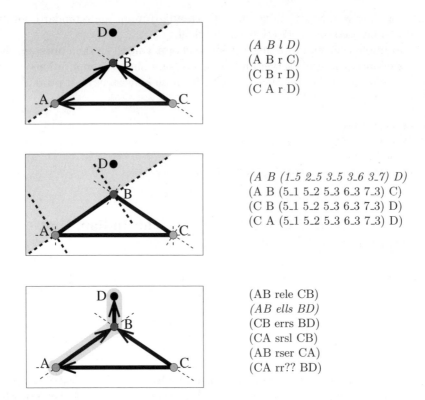

Fig. 8. The Indian Tent problem modeled with the FlipFlop calculus, the Double Cross Calculus and $\mathcal{DRA}_c$. The relation printed in italics refers to the relation depicted by the gray region in the pictures. The '?' is a wildcard in SparQ which stands for an arbitrary valid symbol at this place, i.e., it denotes the disjunction of all possible relations.

6.2 Applying SparQ to the Indian Tent Problem

Fig. 8 shows consistent models of the Indian Tent Problem with FFC, DCC, and $\mathcal{DRA}_c$. For further use, we assume the models to be stored in simple text files with the labels 'FFC_consistent', 'DCC_consistent', and 'DRA_consistent'. We first want to check if we can enforce path-consistency on those models:

```
$ ./sparq constraint-reasoning ffc path-consistency < FFC_consistent
> Unmodified network.
> ( (A B (l) D) (A B (r) C) (C A (r) D) (C B (r) D) )

$ ./sparq constraint-reasoning dcc path-consistency < DCC_consistent
> Unmodified network.
> ( (A B (1_5 2_5 3_5 3_6 3_7) D) (A B (5_1 5_2 5_3 6_3 7_3) C)
    (C A (5_1 5_2 5_3 6_3 7_3) D) (C B (5_1 5_2 5_3 6_3 7_3) D) )
```

```
$ ./sparq constraint-reasoning dra-24 path-consistency < DRA_consistent
> Modified network.
> ( (BD (rlrr) CA) (CB (slsr) CA) (CB (errs) BD) (AB (rser) CA)
    (AB (ells) BD) (AB (rele) CB) )
```

Not surprisingly, SparQ can enforce path-consistency on all three models. Now we replace each last relation specified in Fig. 8 indicating that D is right of CA by the opposite relations '(C A (l) D)', '(C A (1_5 2_5 3_5 3_6 3_7) D)', and '(CA rl?? BD)', respectively. Thus, we create obviously inconsistent configurations which we assume to be stored in the files 'FFC_inconsistent', 'DCC_inconsistent', and 'DRA_inconsistent'. Now these are checked for path-consistency:

```
$ ./sparq constraint-reasoning ffc path-consistency < FFC_inconsistent
> Unmodified network.
> ( (A B (l) D) (A B (r) C) (C A (l) D) (C B (r) D) )

$ ./sparq constraint-reasoning dcc path-consistency < DCC_inconsistent
> Unmodified network.
> ( (A B (1_5 2_5 3_5 3_6 3_7) D) (A B (5_1 5_2 5_3 6_3 7_3) C)
    (C A (1_5 2_5 3_5 3_6 3_7) D) (C B (5_1 5_2 5_3 6_3 7_3) D) )

$ ./sparq constraint-reasoning dra-24 path-consistency < DRA_inconsistent
> Not consistent.
> ( (BD()CA) (CB (slsr) CA) (CB (errs) BD) (AB (rser) CA) (AB (ells) BD)
    (AB (rele) CB) )
```

The $\mathcal{DRA}_c$ model is found to be inconsistent. For FFC and DCC, however, SparQ shows that path-consistency can be enforced for the inconsistent models. For DCC this confirms the results of Röhrig [32]. So we also check the FFC and DCC models for scenario-consistency:

```
$ ./sparq constraint-reasoning ffc scenario-consistency first
    < FFC_inconsistent
> ( (A B (l) D) (A B (r) C) (C A (l) D) (C B (r) D) )

$ ./sparq constraint-reasoning dcc scenario-consistency first
    < DCC_inconsistent
> Not consistent.
```

This result shows that at least for this simple configuration scenario-consistency is sufficient to detect the inconsistency for DCC, while for FFC it is still not sufficient. As this example illustrates, SparQ can be a useful tool for experimentally comparing spatial calculi.

7 Conclusion and Outlook

The SparQ toolbox presented in this text is a first step towards making QSR techniques and spatial calculi accessible to a broader range of application developers. We hope that this initiative will catch interest in the QSR community and will encourage researchers from other groups to incorporate their calculi into SparQ.

Besides including more calculi, extensions currently planned for SparQ are a module for neighborhood-based reasoning techniques [33,15] (e.g. for relaxing inconsistent constraint networks based on conceptual neighborhoods and for qualitative planning) and a module that allows quantification (turning a consistent qualitative scene description back into a geometric representation). This requires the mediation between the algebraic and geometric aspects of a spatial calculus together with the utilization of prototypes. Moreover, we want to include geometric reasoning techniques based on Gröbner bases as a service for calculus developers as these can be helpful, for instance, to derive composition tables [14]. The optimization of the algorithms included in SparQ is another issue that we want to pay more attention to in the future. Finally, we intend to incorporate interfaces that allow to exchange calculus specifications with other QSR frameworks (e.g. [21]).

Acknowledgements

The authors would like to thank three anonymous reviewers for valuable comments and suggestions. This research was carried out at the SFB/TR 8 Spatial Cognition supported by the German Research Foundation (DFG).

References

1. Cohn, A.G., Hazarika, S.M.: Qualitative spatial representation and reasoning: An overview. Fundamenta Informaticae 46(1-2), 1–29 (2001)
2. Allen, J.F.: Maintaining knowledge about temporal intervals. Communications of the ACM, 832–843 (1983)
3. Randell, D.A., Cui, Z., Cohn, A.: A spatial logic based on regions and connection. In: Nebel, B., Rich, C., Swartout, W. (eds.) Principles of Knowledge Representation and Reasoning: Proceedings of the Third International Conference (KR 1992), pp. 165–176. Morgan Kaufmann, San Francisco (1992)
4. Egenhofer, M.J.: A formal definition of binary topological relationships. In: 3rd International Conference on Foundations of Data Organization and Algorithms, pp. 457–472. Springer, Heidelberg (1989)
5. Renz, J., Nebel, B.: On the complexity of qualitative spatial reasoning: A maximal tractable fragment of the region connection calculus. Artificial Intelligence 108(1-2), 69–123 (1999)
6. Freksa, C.: Using orientation information for qualitative spatial reasoning. In: Frank, A.U., Campari, I., Formentini, U. (eds.) Theories and methods of spatio-temporal reasoning in geographic space, pp. 162–178. Springer, Heidelberg (1992)
7. Frank, A.: Qualitative spatial reasoning about cardinal directions. In: Proceedings of the American Congress on Surveying and Mapping ACSM-ASPRS, Baltimore, Maryland, USA, pp. 148–167 (1991)
8. Ligozat, G.: Reasoning about cardinal directions. Journal of Visual Languages and Computing 9, 23–44 (1998)

9. Ligozat, G.: Qualitative triangulation for spatial reasoning. In: Campari, I., Frank, A.U. (eds.) COSIT 1993. LNCS, vol. 716, pp. 54–68. Springer, Heidelberg (1993)
10. Moratz, R., Nebel, B., Freksa, C.: Qualitative spatial reasoning about relative position: The tradeoff between strong formal properties and successful reasoning about route graphs. In: Freksa, C., Brauer, W., Habel, C., Wender, K.F. (eds.) Spatial Cognition III. LNCS (LNAI), vol. 2685, pp. 385–400. Springer, Heidelberg (2003)
11. Moratz, R., Dylla, F., Frommberger, L.: A relative orientation algebra with adjustable granularity. In: Proceedings of the Workshop on Agents in Real-Time and Dynamic Environments (IJCAI 2005) (2005)
12. Renz, J., Mitra, D.: Qualitative direction calculi with arbitrary granularity. [34]
13. Schlieder, C.: Reasoning about ordering. In: Kuhn, W., Frank, A.U. (eds.) COSIT 1995. LNCS, vol. 988, pp. 341–349. Springer, Heidelberg (1995)
14. Moratz, R., Renz, J., Wolter, D.: Qualitative spatial reasoning about line segments. In: Horn, W. (ed.) Proceedings of the 14th European Conference on Artificial Intelligence (ECAI), IOS Press, Berlin, Germany (2000)
15. Dylla, F., Moratz, R.: Exploiting qualitative spatial neighborhoods in the situation calculus. [35], pp. 304–322
16. Billen, R., Clementini, E.: A model for ternary projective relations between regions. In: Bertino, E., Christodoulakis, S., Plexousakis, D., Christophides, V., Koubarakis, M., Böhm, K., Ferrari, E. (eds.) EDBT 2004. LNCS, vol. 2992, pp. 310–328. Springer, Heidelberg (2004)
17. Goyal, R.K., Egenhofer, M.J.: Consistent queries over cardinal directions across different levels of detail. In: Tjoa, A.M., Wagner, R., Al-Zobaidie, A. (eds.) Proceedings of the 11th International Workshop on Database and Expert System Applications, Greenwich, pp. 867–880. IEEE Computer Society Press, Los Alamitos (2000)
18. Sharma, J.: Integrated Spatial Reasoning in Geographic Information Systems: Combining Topology and Direction. PhD thesis, University of Maine (1996)
19. Gerevini, A., Renz, J.: Combining topological and size information for spatial reasoning. Artificial Intelligence 137, 1–42 (2002)
20. Condotta, J.F., Ligozat, G., Saade, M.: A generic toolkit for n-ary qualitative temporal and spatial calculi. In: Proceedings of the 13th International Symposium on Temporal Representation and Reasoning (TIME 2006), Budapest, Hungary (2006)
21. Wölfl, S., Mossakowski, T.: CASL specifications of qualitative calculi. In: Cohn, A.G., Mark, D.M. (eds.) COSIT 2005. LNCS, vol. 3693, Springer, Heidelberg (2005)
22. Röhrig, R.: Representation and processing of qualitative orientation knowledge. In: Brewka, G., Habel, C., Nebel, B. (eds.) KI 1997: Advances in Artificial Intelligence. LNCS, vol. 1303, pp. 219–230. Springer, Heidelberg (1997)
23. Zimmermann, K., Freksa, C.: Qualitative spatial reasoning using orientation, distance, and path knowledge. Applied Intelligence 6, 49–58 (1996)
24. Scivos, A., Nebel, B.: Double-crossing: Decidability and computational complexity of a qualitative calculus for navigation. In: Montello, D.R. (ed.) COSIT 2001. LNCS, vol. 2205, Springer, Heidelberg (2001)
25. Ladkin, P., Reinefeld, A.: Effective solution of qualitative constraint problems. Artificial Intelligence 57, 105–124 (1992)
26. van Beek, P.: Reasoning about qualitative temporal information. Artificial Intelligence 58(1-3), 297–321 (1992)

27. Renz, J., Ligozat, G.: Weak composition for qualitative spatial and temporal reasoning. In: van Beek, P. (ed.) CP 2005. LNCS, vol. 3709, pp. 534–548. Springer, Heidelberg (2005)
28. Ligozat, G., Renz, J.: What is a qualitative calculus? A general framework [34]
29. Scivos, A., Nebel, B.: The finest of its class: The practical natural point-based ternary calculus $\mathcal{LR}$ for qualitative spatial reasoning [35], pp. 283–303
30. Wallgrün, J.O., Frommberger, L., Dylla, F., Wolter, D.: SparQ user manual v0.6. Technical Report 007-07/2006, SFB/TR 8 Spatial Cognition; Universität Bremen (2006)
31. Dylla, F., Moratz, R.: Empirical complexity issues of practical qualitative spatial reasoning about relative position. In: Workshop on Spatial and Temporal Reasoning at ECAI 2004, Valencia, Spain (2004)
32. Röhrig, R.: Repräsentation und Verarbeitung von qualitativem Orientierungswissen. PhD thesis, University of Hamburg (1998)
33. Freksa, C.: Temporal reasoning based on semi-intervals. Artificial Intelligence 1(54), 199–227 (1992)
34. Zhang, C., Guesgen, H.W., Yeap, W.-K. (eds.): PRICAI 2004. LNCS (LNAI), vol. 3157. Springer, Heidelberg (2004)
35. Freksa, C., Knauff, M., Krieg-Brückner, B., Nebel, B., Barkowsky, T. (eds.): Spatial Cognition IV. LNCS (LNAI), vol. 3343. Springer, Heidelberg (2005)

Remembering Places in Space:
A Human Analog Study of the Morris Water Maze

Sylvia Fitting, Gary L. Allen, and Douglas H. Wedell

Department of Psychology, University of South Carolina,
Columbia, SC 29208 USA
fitting@sc.edu

Abstract. We conducted a human analog study of the Morris Water Maze, with individuals indicating a remembered location in a 3 m diameter arena over different intervals of time and with different memory loads. The primary focus of the study was to test a theory of how varying cue location and number of cues affects memory for spatial location. As expected, memory performance, as measured by proximity to the actual location, was negatively affected by increasing memory load, increasing delay interval, and decreasing the number of cues. As memory performance decremented, bias effects increased and were in accordance with the cue-based memory model described by Fitting, Wedell and Allen (2005). Specifically, remembered locations were biased toward the nearest cue and error decreased with more cues. These results demonstrate that localization processes that apply to small two-dimensional task fields may generalize to a larger traversable task field.

Keywords: Categorical coding, environmental cues, Morris Water Maze, spatial cognition, place memory.

1 Introduction

Remembering where objects are located is a common and adaptive activity across species. Several variations on this task may arise from variations in key environmental constraints. For example, the environmental space may be small and two dimensional, such as when remembering a location on the surface of one's computer monitor; moderately large and traversable, such as when remembering where one put one's keys in a room; or much grander in scale and scope, such as when one remembers the location of a residence in the countryside. In addition to the size variable, shape of the environment is a critical factor. For example, several researchers have shown how geometric facets of the environment guide memory for location (Cheng, 1986; Hermer & Spelke, 1994), with some positing that geometric coding of spatial location may constitute a primitive module for processing this information (Gallistel, 1990). The existence of clearly defined paths is another shape constraint likely to influence memory for location, as numerous studies have documented route specific learning (Siegel & White, 1979). In addition to size and shape, the surface features of the environment may be critical. Such surface features may serve as proximal or distal cues for coding memory for location (Egan, 1979).

Clearly it is easier to find one's keys if one remembers placing them next to the lamp (a proximal cue). Or in a large-scale space, one may use the distal cues of surrounding buildings to get a bearing on the location of one's car in a parking lot.

1.1 Memory for Place

In general, the psychological literature posits three distinct varieties of spatial learning and memory, including motor learning, association learning and place learning (Morris & Parslow, 2004; Nadel, 1990; Newcombe & Huttenlocher, 2000). The focus of the present investigation is on place learning, which is distinguished from the other two varieties of spatial memory by a number of characteristics. Place learning involves acquiring memory for spatial locations independent of specific behaviors associated with those locations (Fenton, Arolfo, Nerad, & Bures, 1994; Hebb, 1949; Nadel, 1990; Newcombe & Huttenlocher, 2000; Overman, Pate, Moore, & Peuster, 1996). This type of memory results from the implicit computation of spatial interrelations between distal cues and to-be-remembered locations within a geometric field (O'Keefe & Nadel, 1978; Tolman, 1948). Because orientation is based on external cues, place memory is said to involve an allocentric frame of reference. It features the concept of a geometric field that embeds spatial relations among objects and so is also described as involving a coordinate frame of reference (Newcombe & Huttenlocher, 2000).

Because a geometrical coordinate frame of reference is critical to understanding place learning, researchers have developed a variety of methods to isolate this factor for study. In the animal literature, an important aid in this regard was the development of the Water Maze task by Morris and colleagues (Morris, 1981; Morris, Garrud, Rawlins, & O'Keefe, 1982). In this task, the rat must swim to a hidden platform located in a circular basin that is surrounded by curtains, composing the walls of the environment. Because the spatial field is undifferentiated, the rat cannot use proximal cues within the search space. Furthermore, because the rat's entry point varies across trials, the rat cannot use a simple sequence of motor responses (such as swim 45° to the left for 1 meter). Instead, the presence of distal cues on the curtains affords the rat a way to map the platform's location within this environment in a way that takes advantage of stable configural relationships.

While many researchers have utilized an analog to the Water Maze for studying spatial memory in humans (Astur, Ortiz, & Sutherland, 1998; Astur, Taylor, Mamelak, Philpott, & Sutherland, 2002; Jacobs, Thomas, Laurance, & Nadel, 1998; Lehnung, Leplow, Ekroll, Herzog, Mehdorn, & Ferstl, 2003; Leplow, Lehnung, Pohl, Herzog, Ferstl, & Mehdorn, 2003; Overman et al., 1996; Parslow, Morris, Fleminger, Rahman, Abrahams, & Recce, 2005), several alternative procedures have been developed. One extensively studied paradigm has been to have human subjects remember locations within a small two-dimensional field. For example, Huttenlocher, Hedges and Duncan (1991) asked participants to reproduce the location of a dot in a circular region when no explicit external landmark cues were available. Results indicated that stationary observers tended to impose an implicit polar coordinate system on a circular task field and use geometrically based categories to aid in remembering. The focus of these studies has generally been on the bias in location resulting from the categorical coding of spatial location. Although memory for location in these tasks is generally very good, the systematic biases typically observed imply humans resolve uncertainty of location by moving toward the prototypical location for the appropriate quadrant or category.

1.2 Comparison of Procedures

Because the Morris Water Maze and the dot location task are both used to study place memory, it is instructive to examine similarities and differences between these tasks and corresponding findings. First, the Morris Water Maze task provides evidence for the use of axes extending from visible peripheral cues to organize the field (Morris et al., 1982). In contrast, results from the dot location task are explained in terms of geometric coding, with peripheral cues typically absent in these studies (Huttenlocher et al., 1991). In support of a cue-independent coordinate coding of place, Fitting (2005) demonstrated that including 1 or 3 peripheral cues in the usual dot location task had no effect whatsoever on remembered location. Thus, while the Morris Water Maze has typically been used to study cue-dependent coordinate memory, the dot location task has typically been used to study cue-independent coordinate memory.

This difference in the cue dependency of the representations likely arises from procedural differences in the subject's orientation to the task field across these tasks. Whereas the Water Maze procedure includes varying the entry point to the maze so the initial viewer-based information is not diagnostic, the dot location task fixes the orientation of the viewer to the task field so that viewer-based information may be used and peripheral cues are not critical. Fitting et al. (2005) explored the effect of dynamically varying orientation for the dot location task by including peripheral cues and rotating the task field on the majority of trials. Results indicated that in this dynamic environment, subjects used cues to define category prototypes that biased memory for spatial locations. This result held even for those trials during which the task field was not rotated so that peripheral cues could have been ignored.

Another difference between the tasks lies in the time course of remembering. In the spatial memory task used by Huttenlocher et al. (1991), participants reproduce the location of a dot in a circular region a few seconds after the dot disappears from the display. Thus, this task focuses on spatial coding in short-term memory. In contrast, the classic Morris Water Maze procedure examines place memory across days, a long-term memory task (Morris et al., 1982). The difference in the time course of the tasks raises the possibility that different memory mechanisms may apply to the coordinate-based memory systems for these tasks.

The underlying differences in delay intervals across tasks reflect differences in the research questions these tasks have been used to pursue. Those using the dot location task have been primarily interested in how categorical coding may bias spatial memory (Fitting et al., 2005; Huttenlocher et al., 1991; Huttenlocher, Hedges, Corrigan, & Crawford, 2004; Wedell, Fitting, & Allen, 2007). Short memory durations have been used in this task so that the same subject may be tested on many different locations within the task field, providing a rich data base for modeling the bias in memory. In the category-adjustment model of Huttenlocher et al. (1991), spatial properties of a stimulus specifying its location are represented at two levels, fine-grain and categorical. Categorical coding is posited as the robust product of a relatively rapid process in which location is remembered in terms of its being within a particular categorical region of the response space. A central or salient location within this category acts as a prototype, biasing remembered locations toward it. In the specific task of a stationary observer remembering a location within a circular field, observers appear to impose implicit horizontal and vertical axes on the circle, thus creating quadrants that function as categories, with centrally located points within these quadrants serving as prototypes. Although use of these categorical codes may bias estimates, they may also serve the adaptive purpose of reducing overall error in estimation (Huttenlocher et al., 1991). Fine-grain coding is posited as the product of the process of remembering a location in terms of a geometric coordinate system imposed upon the response field. When the task field is circular, this coordinate system is typically defined by polar coordinates. Fine-grain coding yields metric accuracy, but it may also be more fragile than categorical coding, resulting in a greater reliance on categorical coding with increased memory demands (Haun, Allen, & Wedell, 2005).

In contrast to the goal of studying categorical effects on coordinate memory, the goals addressed by place memory studies using the Morris Water Maze have typically been related to understanding the neurobiology of place learning as distinguished from other varieties of spatial memory. Using these procedures, evidence from neuropsychological and neuroimaging studies in animal and human research indicates that hippocampal areas are uniquely active in place learning (Grön, Wunderlich, Spitzer, Tomczak, & Riepe, 2000; Jarrard, 1993; Morris & Parslow, 2004; O'Keefe & Nadel, 1978, Sutherland & Rudy, 1987). In addition, a number of studies have shown that damage to the hippocampus and related structures in rodents and humans causes impairment in place learning (Astur et al., 1998; Astur et al., 2002; Devan, Goad, & Petri, 1996; Jarrard, 1993; Parslow et al., 2005; Sutherland & Rudy, 1987). Furthermore, psychophysiological studies have shown that principal neurons of the CA1, CA3, and dentate gyrus subfields of the rat hippocampus fire selectively when the animal occupies certain locations in an environment (O'Keefe & Nadel, 1978). Such place cells appear to participate in a distributed and nontopographic map-like representation of the spatial environment (Knierim & Rao, 2003; O'Keefe & Nadel, 1978).

1.3 Human Analog of the Morris Water Maze

We believe that while both the Water Maze and dot location paradigms have provided extremely useful evidence relevant to spatial memory for place, additional insights may be gained by a line of research that combines aspects of these paradigms. The study we describe here builds on a study we conducted that utilized dynamically changing orientation in the dot location task (Fitting et al., 2005). That study showed that the inclusion of rotation trials resulted in participants using a cue-based representation of stimulus locations so that cue locations explained the pattern of bias obtained in memory estimates. The current study addresses two shortcomings in generalizing results of the previous study to procedures more akin to those used in the Morris Water Maze. First, there is the question of whether results from a small space, two-dimensional location task generalize to memory for place in a larger-space environment that is physically traversable. A recent study by Haun et al. (2005) is relevant to this point because they demonstrated that spatial bias pattern in incline estimation and azimuth estimation tasks within a larger-scale environment show similar biases found in the small-scale two dimensional dot location task. Related to this question is whether the use of cues displayed in the small-scale environment that give rise to a more allocentric-based place memory will be similar to cue use when cues are much more peripheral and must be encoded relative to one's location in space, as in the classic Morris Water Maze. A second major point explored in the current study is whether cues may be used as prototypes when location must be remembered for not a matter of seconds, but rather for minutes or even for weeks, durations more consistent with investigations using the classic Morris Water Maze procedure.

To address these questions, we developed a human analog of the Morris Water Maze which consisted of a circular arena 3 m in diameter and 3 m tall. The walls of the arena consisted of black curtains that reached to the floor, with nine different entry points. The floor of the arena was covered in wood chips so there were no discernable proximal cues to use in locating objects. The roof of the arena was undifferentiated as well. Cues were hung on the walls of the arena to provide external reference points for coding the location of objects within the arena.

Note that other researchers have developed human analogs of the Water Maze (Astur et al., 1998; Astur et al., 2002; Jacobs et al., 1998; Lehnung et al., 2003; Leplow et al., 2003; Parslow et al., 2005). One way that our research differs from this previous research is that our primary focus was on cue-based bias, as predicted by our work from the dot location studies. In this regard, the main independent variable was the number (and corresponding locations) of cues. However, because of the extreme differences between the dot location task and the arena task, we included two additional elements designed to affect memory performance. The first of these we call memory load, and it was manipulated between subjects. In the low-load condition, participants were given only one object location to remember. In the high-load condition, they were given three object locations to remember. We predicted better memory when memory load was low. This prediction follows from short-term studies in which having to remember more object locations results in poorer memory for location (Dent & Smyth, 2005). Additionally, it follows from the idea that holding multiple locations in memory could serve to interfere with memory for any given target and hence lead to poorer memory performance. The second manipulation was delay interval, which we manipulated within subjects. The first interval consisted of 10 to 15 minutes after initial encoding, with the delay interval filled with a spatial memory task. The second interval was one week after the first testing, and the third interval was one week after the second testing. As in any memory study, we predicted poorer memory performance with increased delay interval.

There were two main reasons to include the load and delay interval variables. The first of these was simply to document the pattern of learning and forgetting relevant to these variables in the human analog study. Thus, we wanted to see how much one forgets over time and how difficult it is to remember a location when having to remember two other locations in the same space. The second and more important reason for manipulating these variables was to create conditions under which substantial forgetting would take place. Forgetting was necessary because observing bias effects typically requires there to be some disruption of fine-grain memory. With perfect remembering, there would be no way to test hypotheses regarding bias.

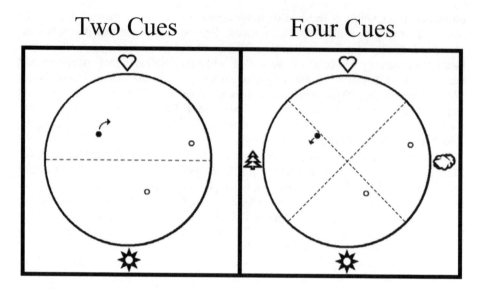

Fig. 1. Schematic illustration of target and cue locations in the arena along with predictions of bias. Dashed lines represent boundaries based on cue determined prototypes. The filled dot shows the target object location, and the open dots show the additional object locations used in the high-load condition. Size of cues and object locations are not to scale.

The cue-based fuzzy-boundary model developed by Fitting et al. (2005) was used to generate predictions regarding the effect of cue manipulation. To illustrate these predictions we present the critical features of the experimental design in Fig. 1. In the two-cue condition (left panel), a heart was located at one end and a sun was located at the other. In the four-cue condition, a tree and a cloud were added at the locations shown. We generated three predictions based on our cue-based fuzzy-boundary model (Fitting et al., 2005). First, we predicted that if cues determine category prototypes, then a bias toward the most proximal cue should be observed for misremembered locations. This prediction is illustrated in Fig. 1 by the arrows, with bias toward the heart in the two-cue condition and toward the tree in the four-cue condition. Thus bias is predicted to be in opposite directions for the two conditions. Note that we only applied this prediction to angular bias, as previous work with the dot location task showed large cue-based effects on angular bias but not on radial bias (Fitting, 2005). Second, our cue-based fuzzy-boundary model predicted that absolute error should decrease with more cues. This is because bias is a key contributor to error, and bias generally increases with distance from the nearest prototypes. With more cues creating more prototypes, the distances to adjacent prototypes tend to decrease, and hence error should correspondingly decrease (indicated in Fig. 1 by differences in the lengths of the arrows in the left and right Panel). Third, we expected bias to be greatest when absolute error was greatest. Thus, we expected bias to be more likely in the long delay-interval conditions and with high load.

A final issue examined in this study was the correlation between measures of place learning and memory on the one hand and measures from tests of small-scale spatial abilities on the other hand. Currently, the relation between place memory and small-scale spatial tasks is simply unclear. We reasoned that if place learning and memory is a unique variety of spatial memory, it should not be closely related to the ability to do spatial tasks that require coding spatial relations in terms of a viewer's axes or in terms of an object's intrinsic axes. However, place memory might be related to the ability to remember the locations of objects in a task field, even when that field is small and viewed by a stationary observer. Consequently, in the current study we predicted that performance on a place-learning task may be significantly correlated with memory performance in our arena task.

2 Method

2.1 Participants

Eighty university students (40 men and 40 women) participated voluntarily in the experiment for research participation credit in undergraduate Psychology classes. The mean age of female participants was 20.0 years (range 18-42 years), and the mean age of male participants was 19.8 years (range 18-26 years). No participants reported visual or motor problems that would have influenced their performance.

2.2 Design

A 2x2x3 mixed factorial design was employed. The between-subjects factors were cue condition, with two levels (two cues or four cues), memory load, with two levels (one target object or three objects), and the within-subjects factor was delay interval, with three levels (immediate, one-week delay, and two-week delay).

Place learning and memory task. The apparatus for the place learning and memory task was a circular arena, 3 m in diameter and 3 m in height located inside a 5x7 m laboratory. The arena was formed by black curtains that had nine equally spaced entry points, which were places where adjacent curtains were not permanently attached to each other. The arena had a ceiling constructed of translucent white cloth that allowed light from four equally spaced light bulbs to illuminate the interior. The floor was covered with wood shavings in which objects could be hidden easily. The curtains, ceiling, and floor provided no reliable basis for specifying locations within the arena. Four pictures (specifically, a tree on the west wall, a heart on the north wall, a cloud on the east wall and a sun on the south wall) or two pictures (the heart on the north wall, and the sun on the south wall) were mounted equidistant from each other on the interior curtains of the arena (see Fig. 1). The pictures were 20 x 25 cm in size, and their distance from the floor was approximately 2 m. Locations to be remembered were designated by toy coins 1.5 cm in diameter and golden in color. The target coin location was on the floor of the arena as shown in Fig. 1, located 42.6° from the 'tree' cue and 47.4° from the 'heart' cue, with a radial value of 78.0 cm. In the high-load condition, the two additional coin locations were on the arena floor as shown in Fig. 1. One object was 75.0° from the 'heart' cue and 15.0° from the 'cloud' cue, with a radial value of 129.4 cm. The other object was 31.0° away from the 'sun' cue and 59.0° from the 'cloud' cue, with a radial value of 72.8 cm.

Performance in the place memory task was scored in terms of three measures. First, absolute error served as a measure of metric inaccuracy. It was defined as the distance in cm from the remembered target location to the actual location. Second, angular bias reflected angular distortion in memory. For this measure, a polar coordinate grid was imposed on the circular arena floor, with radii extending from the center of the arena to the surrounding wall. Angular bias was defined as the signed difference in degrees from the value of the remembered location minus the value of the actual location, with the sign of the angular value indicating the direction of the deviation (negative values represent clockwise bias and positive values represent counterclockwise bias). Third, radial bias was measured in cm by subtracting the radial distance of the actual point from the radial distance of the observed point. A negative radial value indicates a radial bias toward the center of the arena, whereas a positive value indicates a radial bias toward the circumference of the arena.

Spatial ability tests. Participants completed three spatial tests. Two of the spatial tests, the Maze Learning Test (Allen, Kirasic, Dobson, & Long, 1996), assessing spatial-sequential memory, and the Money Standardized Road Map Test (Money, 1976), assessing directional sense, were not assumed to relate to the task at hand and merely served as filler tasks. The third task, the Building Memory Test (Ekstrom, French, Harmen, & Dermen, 1976) was designed to assess spatial-visual memory as a small-place memory task. The objective of the test was to learn the position of buildings in a street map. After studying a map for three minutes, participants were asked to indicate on a test page the location of 12 buildings. The format was multiple-choice, with six possible response choices for each location. Four minutes were allowed for retrieval. The test had two parts, each with a different map and set of 12 test items. Maximum possible score was 24.

2.3 Procedure

Each participant was involved in three data collection sessions separated by one-week intervals. The first session involved a learning phase and a test phase. The second and third sessions involved a test phase only.

At the beginning of the first session, the experimenter met the participant outside the room containing the arena, obtained informed consent, and provided instructions. For the learning phase, the participant was seated in a wheelchair and was blindfolded to interfere with the use of visual information outside the arena for orientation. Then the experimenter wheeled the participant to the arena following a circuitous route to reduce the use of vestibular or kinesthetic information for orientation. After wheeling the participant to the first entry point (entry point number 1) the experimenter instructed the participant to remove the blindfold and stand up. Then, inside the arena the experimenter led the participant to the predetermined hiding place(s) and hid the coin(s) under the wood-shavings. The participant was told to remember the hiding place(s) exactly because he or she would later have to indicate the hiding place(s) with a marker. The participant was given as long as needed to memorize the hiding place of the coin(s). No participant needed more than 1 minute. After he or she had memorized the hiding place(s), the blindfold was replaced, and the participant was wheeled out from the room containing the arena by a circuitous route. This exit ended the learning phase.

Outside the room, the participant was administered the Building Memory Test as a filler task, followed by the test phase of the arena procedure. The experimenter wheeled the blindfolded participant back into the room containing the arena and, following circuitous routes each time, stopped at three different entry points (entry point number 7, 1, and 5). As during the learning phase, at each entry point the participant was instructed to remove the blindfold and stand. Inside the arena, the participant was instructed to orient himself or herself, walk from the entry point position directly to the hiding place or series of hiding places, and to indicate the hiding place(s) by placing a small red plastic disk to mark the exact location. By instructing the participant to walk directly from the entry point position to the hiding place or series of hiding places, the participant was restricted in his/her use of strategy and could not rely on any physical measurement that he/she might have used during the learning phase. After placing the marker(s), the participant came back to the entry point of that trial, sat in the wheelchair, and replaced the blindfold. The experimenter recorded the responses quickly using a grid in place under the wood shavings. After the third test trial, the experimenter wheeled the participant outside the room containing the arena. This concluded the initial session.

Participants returned one week later for the second session. Initially, they were administered the Maze Learning Test. Then, they were seated in the wheelchair, blindfolded, and conveyed into the room containing the arena. The events comprising the second test phase were the same as those in the first session, except that a different sequence of entry sites was involved (entry point number 1, 2, and 6). The second session ended with the participant being wheeled from the room after the third entry point.

Participants returned two weeks after the initial session for the third session. Initially, they were administered the Road Map Test. Then, testing proceeded as in the first and second sessions, except that a different sequence of entry sites was involved (entry point number 8, 3, and 1). The third session ended with the participant being wheeled from the room after the third entry point. Participants were then debriefed and thanked for their participation.

3 Results

3.1 Overview of Analyses

Data from the place memory task were analyzed in a series of analyses of variance (ANOVA's). The primary analyses were performed using as the dependent measures the mean absolute error, angular bias, and radial bias, from the three trials on each session. Sex of participant was not integrated in any of these analyses because preliminary analyses did not reveal any significant effects or interactions involving this variable ($p > .05$). We report analyses only for the target location, as it was common to all conditions.

For the variable absolute error and radial bias, if participants had a score of more than two standard deviations above the mean for that condition then the data point was replaced by the mean for that condition. Outlying data points for the variable angular bias were identified by an error of 90° or greater. This method was used in order to eliminate cases that indicated misplacement into an adjacent quadrant. In the two cue condition, out of 120 data points (40 participants with one mean estimate in each of 3 sessions) 8 data points (6.7%) were replaced for absolute error, 6 (5%) for radial bias, and 5 (4.2%) for angular bias. In the four cue condition out of 117 data points (39 participants with one mean estimate in each of 3 sessions) 4 data points (3.4%) were replaced for absolute error, 6 (5.1%) for radial bias, and 0 (0%) for angular bias. No effects for trials or memory load were noted regarding outlying data points. For all statistical analyses, the significance level was set at .05.

3.2 Analyses of Absolute Error (in cm)

Fig. 2 illustrates how absolute error for the target location varied as a function of delay interval and load. The principal analysis of absolute error was a 2 (Cue Condition) x 2 (Memory Load) x 3 (Delay Interval) mixed ANOVA performed on mean absolute error per session for the target object location. Results indicate a main effect of number of cues, $F(1, 75) = 4.07$, $p = .047$, $MSe = 490.11$, reflecting greater accuracy when more peripheral cues were available, $M_{4\text{-Cues}} = 31.21$ ($SD = 11.70$) and $M_{2\text{-Cues}} = 36.98$ ($SD = 14.88$). The analysis also showed a significant main effect of memory load, $F(1, 75) = 6.52$, $p = .013$, $MSe = 490.11$, reflecting greater accuracy in the low-load condition, $M_{\text{Low}} = 30.44$ ($SD = 12.72$) and $M_{\text{High}} = 37.77$ ($SD = 13.62$). The Cue x Load interaction did not reach conventional levels of significance in our

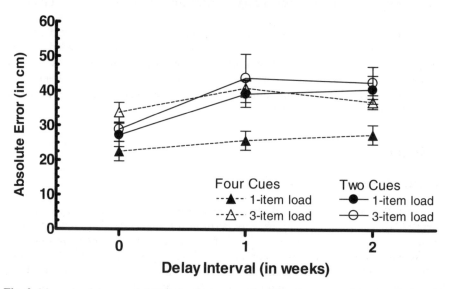

Fig. 2. Mean absolute error (±SEM) for the target object location across delay-interval session for each load condition

analysis[1]. The main effect of delay interval was significant, $F(2, 150) = 13.99$, $p < .001$. The trend analysis revealed a significant linear trend, $F(1, 75) = 20.50$, $p < .001$, along with a quadratic trend, $F(1, 75) = 7.91$, $p < .01$. Post hoc analyses indicated that participants produced less error when tested immediately after acquisition trial than after a one-week delay ($p < .001$) or after a two-week delay ($p < .001$). No difference was noted between a one-week and a two-week delay. There was also a significant Cue x Delay interaction, $F(2, 150) = 3.40$, $p = .036$, $MSe = 155.15$, reflecting similar accuracy when tested immediately after acquisition trials, but higher accuracy in the four-cue condition after one and two week delays relative to the two cue condition (combined across load condition).

To test for trial effects, an additional analysis was conducted adding the three trials within each session as a within-subjects factor. Results for a 2 (Cue Condition) x 2 (Memory Load) x 3 (Delay Interval) x 3 (Trial) mixed ANOVA indicated a significant main effect of trials, $F(2, 150) = 11.41$, $p < .001$, $MSe = 152.41$, reflecting greater accuracy for the first and second trial, $M = 33.29$ ($SD = 14.46$) and $M = 32.25$ ($SD = 13.62$), respectively, compared to the third trial with $M = 37.06$ ($SD = 15.80$). Further, no significant interactions were noted with trial.

3.3 Analyses of Angular Bias (in degrees)

Fig. 3 illustrates how angular bias for the target location varied as a function of delay interval and load. A parallel 2x2x3 ANOVA was conducted on angular bias. Results indicate an overall effect of number of cues, $F(1, 75) = 13.87$, $p < .001$, $MSe = 1278.77$, reflecting a bias effect in opposite directions for the different cue conditions, $M_{2\text{-Cues}} = 12.55$ ($SD = 20.51$) and $M_{4\text{-Cues}} = -4.60$ ($SD = 21.23$). This opposite bias effect is consistent with the cue-based fuzzy-boundary prediction of bias toward the nearest cue. The main effect of delay interval was not significant, but there was a significant Cue x Delay interaction, $F(2, 150) = 11.06$, $p < .001$, $MSe = 137.55$. This interaction was due to much stronger bias effects for the one and two week delays than for the no delay condition.

In a more detailed analysis of bias effects, one-sample t-tests were conducted at each interval for each cue and load condition. No significant bias effects were noted in any of the groups when tested immediately after the acquisition trial, indicating good short-term spatial memory for all four groups. Unbiased memory, with no significant effects at any of the three delays, was further noted when four cues were available and only one target location had to be remembered, indicating stable memory across time. However, in the four-cue high-load condition, memory for the target location was significantly biased in the one-week delay interval toward the nearest available environmental cue ($t = -2.66$, $p = .015$). For the two-cue condition, both high and low memory load groups demonstrated significant bias effects toward the nearest available cue for the one-week

---

[1] Although the Cue X Load interaction looks strong in Fig. 2, it was not significant when we conducted analyses on the untransformed scores. However, it was statistically significant when the ANOVA was conduced on the log transformation of the error scores, $F(1, 75) = 5.16$, $p = .026$. This interaction reflected the significantly reduced absolute error for the target in the four-cue low-load condition compared to the other three conditions. However, a test of violations of sphericity was significant for the log transformed scores but not for the untransformed scores, so this result must be viewed with caution.

delay ($t = 3.45$, $p = .003$ and $t = 2.25$, $p = .037$, respectively) and for the two-week delay ($t = 3.04$, $p = .007$ and t = 2.89, $p = .010$, respectively). Note that all significant tests of bias were in the direction predicted by the cue-based fuzzy-boundary theory (Fitting et al., 2005).

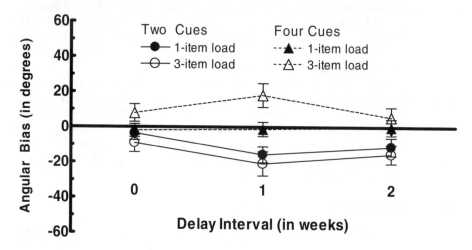

Fig. 3. Mean angular bias (±SEM) for the target object location across delay interval session for each load condition

3.4 Analyses of Radial Bias (in cm)

Fig. 4 illustrates how radial bias for the target location varied as a function of delay interval and load. A parallel 2x2x3 ANOVA was performed on mean radial bias. Results indicate an overall load effect, $F(1, 75) = 16.92$, $p < .001$, $MSe = 448.88$, reflecting greater bias toward the circumference in the low-load condition relative to the high-load condition, $M_{Low} = 9.97$ ($SD = 12.62$) and $M_{High} = -1.43$ ($SD = 11.88$). No other effects reached conventional levels of significance.

3.5 Correlation Analysis

Pearson correlations were conducted on the paper and pencil Building Memory Test of spatial location and the three measures from the human arena memory task (after reverse scoring angular bias in the four-cue condition so that positive values in both conditions indicated a bias toward the nearest environmental cue). There were no significant correlations between the small-scale memory task and any of the large-scale performance measures. However, correlations between absolute error and angular bias scores were significant for each session and averaged $r = .47$, $p < .001$. This finding supports our prior predictions that bias is a contributor to error. No significant correlation with absolute error was noted for radial bias.

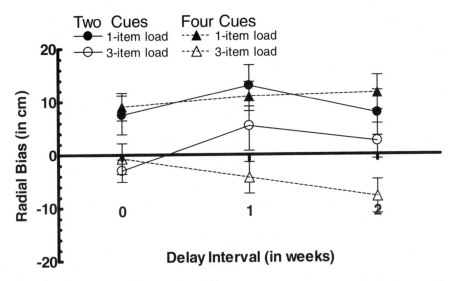

Fig. 4. Mean radial bias (±SEM) for the target object location across delay interval session for each load condition

Finally, we investigated the stability of individual differences in the arena task performance by computing Cronbach's alpha across three delay intervals. Cronbach's alpha was 0.74, indicating relatively stable individual differences in spatial memory. Thus, although stable individual differences were found, these were not predicted by the small-scale test.

4 Conclusion

We conducted a human analog study of the Morris Water Maze using a traversable arena. It is important to note that our version of the Water Maze analog differed from the typical Water Maze procedure and human analog versions of that procedure used in the past in several ways. First, in our procedure there was only one learning trial, which consisted of the individual freely exploring the area and considering ways to remember the exposed target. In the typical procedure, the target is hidden and the individual initially finds it through free exploration, and then additional learning trials are recorded. Second, animal studies and human studies, such as those reported by Leplow et al. (2003), typically track the path of the individual searching for the hidden object so that dependent variables are based on the movement record, recording proximity to the target, movement away from the target, time to find the target and other process measures. These measures are related to a third fundamental difference between our study and the classic Water Maze procedure, which is that in our study the individual does not receive feedback on the location of hidden target; in the classic procedure such feedback is given when the individual finds the target. These differences arose from our attempt to blend aspects of the small scale place memory paradigm of Huttenlocher et al. (1991) with the navigation-based procedure of the Water Maze. Our design then allowed us to use dependent measures more suitable for assessing memory representation of the location of the target and associated biases.

In our study, we manipulated three main variables: number of cues, number of targets (or load), and delay interval. Manipulation of these variables provided us with the tools to assess basic memory performance in this traversable and mostly undifferentiated space. Not surprisingly, memory accuracy (inversely related to absolute error) decreased with increase in delay interval, increase in number of targets to remember, and decrease in number of cues. Despite decrements in performance related to these variables, the overall performance of participants demonstrated robust memory for spatial location. Even after a two week delay with no feedback, participants placed the target on average within about 37.7 cm of the actual target location (inside the 300 cm diameter task field). In the most advantageous of task conditions (one-item to be remembered using four cues), the target was placed within about 27.3 cm of its true location after a two week delay. The forgetting curves, linear with a minor quadratic component, indicate that most of the forgetting is within the first week, as there was no significant increase in absolute error from week 1 to week 2. Thus, within this task our subjects were capable of accurate spatial memory after a minimal encoding session.

The lack of performance differences between one and two week delays is subject to different interpretations. One possibility is that forgetting asymptotes after one week, as indicated in this study. Alternatively, the one-week trial may have served as a retrieval and rehearsal opportunity that bolstered memory for the hidden coin location(s) and thus obscured any additional forgetting that may have occurred. A research design that manipulates delay between subjects so that participants have no intermediate retrieval trials would address this issue. However, the advantage of the reported experiment which utilized a within-subjects manipulation of delay is that it provided a much more powerful test of the delay factor than a comparable between-subjects design.

An important innovation of the current study as compared with previous human analog studies of the Morris Water Maze was our focus on predicting bias in spatial memory. We tested whether theories of spatial memory bias established in a small two-dimensional space (Huttenlocher et al., 1991; Fitting et al., 2005) would translate to the larger traversable space in our arena apparatus. In particular, we tested predictions generated from our cue-based fuzzy-boundary model of spatial memory (Fitting et al., 2005). These predictions were supported in three ways. First, angular bias was in the predicted direction: The remembered location shifted toward the heart cue in the two-cue condition and away from the heart cue (toward the tree cue) in the four-cue condition. This is consistent with cues being used as category prototypes and uncertainty about location being resolved toward the prototype, as proposed by the category-adjustment model (Huttenlocher et al., 1991). A second way in which model predictions were supported was the increase in angular bias and absolute error for the two-cue versus the four-cue condition. The cue-based fuzzy trace model predicts that error and bias increase as distance to the nearest adjacent cues increase (Fitting et al., 2005). Adjacent cues were more closely spaced in the four-cue condition and so error was reduced. A third way in which the model was supported was in the relationship between error and bias. The model predicts that bias will increase as memory errors increase. Thus, the fact that bias was only significant at the 1 and 2 week delay intervals in which absolute error was greatest is consistent with the model. Furthermore, absolute error significantly correlated with predicted bias, consistent with the idea that the greater the error in memory, the more likely the error will be resolved in a biased direction.

Although some small effects on radial bias were observed, these were not the focus of the investigation. The significant load effect on radial bias is consistent with the radial prototype being influenced by the existence of other radial locations, possibly averaging across the radial locations in order to form a radial prototype. Such speculation would need to be systematically tested in future research.

The correlational analyses indicated stable individual differences in our arena task. These were not related to sex differences and were not predicted by a small-scale paper and pencil spatial memory task. Thus, one area of future investigation would be to determine what small-scale tasks predict performance in the larger-scale spatial memory task.

In summary, the present study provided evidence for the impact of the three investigated basic factors on human place memory: environmental cue availability, delay interval, and memory load. The most important implication of the present research is that place memory in a large-scale situation may be governed by an imposed coordinate system, with errors resolved toward the nearest available peripheral cue in the environment, as demonstrated in previous small-scale tasks (Fitting, 2005). Although further investigation using a variety of cue and spatial locations is needed, these results support the idea that cues are used as prototypes, with memory for locations biased toward the nearest available environmental cue and the magnitude of bias being a function of distance to the available cue. While the test environment used in our study was traversable, its size was necessarily small and comparable to that used in other Water Maze analog tasks (Leplow, Höll, Zeng, & Mehdorn, 1998). As such, our test area would be considered a vista space (Montello, 1993), which may not necessarily afford active exploration for building a representation. However, we believe this study provides the groundwork for future tests of bias in large scale, traversable environments.

Acknowledgement. The authors gratefully thank Lena Fitting Kourkoutis for helping with the data transformation.

References

Allen, G.L., Kirasic, K., Dobson, S.H., Long, R.G.: Predicting environmental learning from spatial abilities: An indirect route. Intelligence 22, 327–355 (1996)

Astur, R.S., Ortiz, M.L., Sutherland, R.J.: A characterization of performance by men and women in a virtual Morris water task: A large and reliable sex difference. Behavioural Brain Research 93, 185–190 (1998)

Astur, R.S., Taylor, L.B., Mamelak, A.N., Philpott, L., Sutherland, R.J.: Humans with hippocampus damage display severe spatial memory impairments in a virtual Morris water task. Behavioural Brain Research 132, 77–84 (2002)

Cheng, K.: A purely geometric module in the rat's spatial representation. Cognition 23(2), 149–178 (1986)

Dent, K., Smyth, M.M.: Verbal coding and the storage of form-position associations in visual-spatial short-term memory. Acta Psychologica 120, 113–140 (2005)

Devan, B.D., Goad, E.H., Petri, H.L.: Dissociation of hippocampal and striatal contributions to spatial navigation in the water maze. Neurobiology of Learning and Memory 66, 305–323 (1996)

Egan, D.: Testing based on understanding: Implications from studies of spatial ability. Intelligence 3, 1–15 (1979)

Ekstrom, R.B, French, J.W., Harmen, H.H., Dermen, D.: Manual for kit of factor-referenced cognitive tests. Educational Testing Service, Princeton, NJ (1976)

Fenton, A.A., Arolfo, M.P., Nerad, L., Bures, J.: Place navigation in the Morris water maze under minimum and redundant extra-maze-cue conditions. Behavioral and Neural Biology 62, 178–189 (1994)

Fitting, S.: Memory for spatial location: Cue effects as a function of field rotation. Master's Thesis, Department of Psychology, University of South Carolina, Columbia, SC (2005)

Fitting, S., Wedell, D.H., Allen, G.L.: Memory for spatial location: Influences of environmental cues and task field rotation. In: Cohn, A.G., Mark, D.M. (eds.) COSIT 2005. LNCS, vol. 3693, pp. 459–474. Springer, Heidelberg (2005)

Gallistel, C.R.: The organization of learning. MIT Press, Cambridge (1990)

Grön, G., Wunderlich, A.P., Spitzer, M., Tomczak, R., Riepe, M.W.: Brain activation during human navigation: gender-different neural networks as substrate of performance. Nature Neuroscience 3, 404–408 (2000)

Haun, D.B.M., Allen, G.L., Wedell, D.H.: Bias in spatial memory: a categorical endorsement. Acta Psychologica 118, 149–170 (2005)

Hebb, D.O.: The organization of behaviour. Wiley-Interscience, New York (1949)

Hermer, L., Spelke, E.S.: A geometric process for spatial reorientation in young children. Nature 370(6484), 57–59 (1994)

Huttenlocher, J., Hedges, L.V., Corrigan, B., Crawford, L.E.: Spatial categories and the estimation of location. Cognition 93, 75–97 (2004)

Huttenlocher, J., Hedges, L.V., Duncan, S.: Categories and particulars: Prototype effects in estimating spatial location. Psychological Review 98, 352–376 (1991)

Jacobs, W.J., Thomas, K.G.F., Laurance, H.E., Nadel, L.: Place learning in virtual space II. Topographical relations as one dimension of stimulus control. Learning and Motivation 29, 288–308 (1998)

Jarrard, L.E.: On the role of the hippocampus in learning and memory in the rat. Behavioral and Neural Biology 60, 9–26 (1993)

Knierim, J.J., Rao, G.: Distal landmarks and hippocampal place cells: effects of relative translation versus rotation. Hippocampus 13(5), 604–617 (2003)

Lehnung, M., Leplow, B., Ekroll, V., Herzog, A., Mehdorn, M., Ferstl, R.: The role of locomotion in the acquisition and transfer of spatial knowledge in children. Scandinavian Journal of Psychology 44, 79–86 (2003)

Leplow, B., Höll, D., Zeng, L., Mehdorn, M.: Spatial orientation and spatial memory within a 'locomotor maze' for humans. In: Freska, C., Habel, C., Wender, K.F. (eds.) Spatial Cognition. LNCS (LNAI), vol. 1404, pp. 429–446. Springer, Heidelberg (1998)

Leplow, B., Lehnung, M., Pohl, J., Herzog, A., Ferstl, R., Mehdorn, M.: Navigational place learning in children and young adults as assessed with a standardized locomotor search task. British Journal of Psycholog 94, 299–317 (2003)

Money, J.: Manual: A standardized road-map test of direction sense. Academic Therapy Publications, San Rafael, California (1976)

Montello, D.R.: Scale and multiple psychologies of space. In: Frank, A.U., Campari, I. (eds.) COSIT 1993. LNCS, vol. 716, pp. 312–321. Springer, Heidelberg (1993)

Morris, R.G.M.: Spatial localization does not require the presence of local cues. Learning and Motivation 12, 239–260 (1981)

Morris, R.G.M., Garrud, P., Rawlins, J.N.P., O'Keefe, J.: Place navigation impaired in rats with hippocampal lesions. Nature 297, 681–683 (1982)

Morris, R.G., Parslow, D.M.: Neurocognitive components of spatial memory. In: Allen, G.L. (ed.) Human Spatial Memory: Remembering where, pp. 217–247. Lawrence Erlbaum Associates, Inc, London (2004)

Nadel, L.: Varieties of spatial cognition: Psychobiological considerations. In: Diamond, A. (ed.) Annals of the New York Academy of Sciences, vol. 608, pp. 613–636. Academy of Sciences, New York (1990)

Newcombe, N.S., Huttenlocher, J.: Making space: The developmental of spatial representation and spatial reasoning. MIT Press, Cambridge, MA (2000)

Overman, W.H., Pate, B.J., Moore, K., Peuster, A.: Ontogeny of place learning in children as measured in the radial arm maze, Morris search task, and open field task. Behavioral Neuroscience 110(6), 1205–1228 (1996)

O'Keefe, J., Nadel, L.: The hippocampus as a cognitive map. Clarendon Press, Oxford (1978)

Parslow, D.M., Morris, R.G., Fleminger, S., Rahman, O., Abrahams, S., Recce, M.: Allocentric spatal memory in humans with hippocampal lesions. Acta Psychologica 118(1-2), 123–147 (2005)

Siegel, A.W., White, S.H.: The development of spatial representations of large-scale environments. In: Reese, H.W. (ed.) Advances in child development and behavior, vol. 10, pp. 9–55. Academic Press, New York (1975)

Sutherland, R.J., Rudy, J.W.: Configural association theory: The role of the hippocampal formation in learning, memory, and amnesia. Psychobiology 17, 129–144 (1987)

Tolman, E.C.: Cognitive maps in rats and men. Psychological Review 55, 189–208 (1948)

Wedell, D.H., Fitting, S., Allen, G.L.: Shape effects on memory for location. Psychonomic Bulletin & Review 14, 681–686 (2007)

The Role of Users' Concepts of the Robot in Human-Robot Spatial Instruction

Kerstin Fischer

University of Bremen, FB10 – Sprach- und Literaturwissenschaften, 28334 Bremen
kerstinf@uni-bremen.de

Abstract. Spatial instructions are always delivered for a particular communication partner. In this paper I investigate the the role of users' concepts of their communication partner in human-robot interaction by analysing the spatial language choices speakers make in three comparable corpora with three different robots. I show that the users' concepts of their artificial communication partner is only mildly shaped by the appearance of the robot, and thus that users do not mindlessly use all clues they can get about their communication partner in order to formulate their spatial instructions. Instead, spatial instruction in human-robot interaction also depends on the users' models of the communication situation, as well as on external variables, such as gender.

Keywords: Spatial Language, Human-Robot Interaction, Linguistic Variation.

1 Introduction

The problem I address in this paper is the role of speakers' concepts of their communication partner in verbal spatial instruction. Speakers have a range of different ways of referring to spatial configurations at their disposal, and in a particular situation they have to decide for one variant over another. Among many other possibilities, speakers may refer to the spatial scenario in the following figure in different ways, for instance:

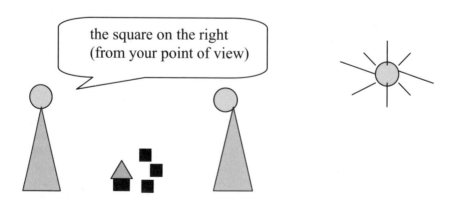

There are however multiple possibilities to refer to this, and to similar, spatial scenes, for instance:

Table 1. Spatial Instruction Strategies

the square on the right (from your point of view)	relative & partner-oriented
the square on the left (from my point of view)	relative & speaker-oriented
the square in front of the house	landmark-based & intrinsic
the square west of the house	landmark-based & extrinsic
the square in the middle	group-based
go straight, - a bit more to the right – stop	path-based

While these are not all possibilities for referring to spatial configurations, by far, they are the most relevant for our scenario discussed in the following. For a detailed overview of the spatial choices see Tenbrink (2005).

In this paper I address the conditions for this linguistic variation in spatial instruction focusing on the role of the communication partner. The question is thus, depending on whom speakers are talking to, which linguistic choices will they make?

Previous research on spatial language in dialogue has shown partner-specific effects particularly in the domain of perspective taking. Schober (1993) reports that if speakers had to imagine a communication partner, they exclusively relied on the partner's perspective, although Hermann and Grabowski (1994) have shown that this requires more cognitive effort for the speaker. If they had a real communication partner, they mostly used their own perspective. This is in line with results by von Stutterheim & Kohlmann (1998) who also found that in a situation in which the addressee could give feedback, speakers did not adapt their perspective. Schober (1998) attributes the difference to speakers' knowledge about the success of their instructions due to feedback. If their communication partners signal to them that understanding is successful, speakers may stick to the mode of instruction which needs the least collaborative effort (Clark & Wilkes-Gibbs, 1986). In case they do not know whether the communication partner understands them, they adjust to the partner's perspective in order to guarantee understanding in the absence of feedback. In Schober's (1995) study, however, speakers often take the perspective of their communication partner although there is the possibility of feedback. Moreover, after exchanging roles as instructor and instructee, speakers try to take the other's perspective as often as these have taken theirs. Finally, speakers' use of spatial instructions independent of particular viewpoints increases in these dialogues. Schober (1998) therefore argues that speakers are generally orienting towards the principle of ease in communication as a collaborative effort.

While these results point to partner-specific effects due to interaction, there are also results that show that the users' recipient design, their adapting their utterances to the expected needs of their communication partner (Sacks et al. 1974), influences the spatial choices made. Schegloff (1972), for instance, has shown that speakers may carry out extensive membership analyses in order to formulate a spatial description, that is, they try to determine the categories their addressee belongs to in order to infer the level of knowledge they can rely on in their spatial description. Similarly, Fischer and Bateman (2006) show that in human-robot communication speakers ask clarification questions particularly on these issues. Speakers' attribution of knowledge to their communication partner is furthermore apparent in Schober (2005) who shows that depending on the spatial capabilities of their addressees, speakers adapt to their communication partners. Moreover, Fischer (2005) shows that the choice of perspective in human-robot interaction depends crucially on what speakers believe the robot to be able to do. A linguistically skilful robot obviously creates a different, more complicated image than a robot that does not display any linguistic knowledge. The speakers' choice of perspective in these dialogues therefore depended essentially on the users' partner model.

What is still open, however, is the exact nature of the relationship between partner model, concept of the situation, external variables, and the spatial language chosen. My procedure here will be to analyse corpora of human-robot dialogues that differ only with respect to a single variable, namely the robot appearance. Human-robot dialogues, in contrast to natural dialogues among humans, have the advantage that the role of the addressee, the robot, can be investigated in much more detail than it can be done with humans since we can control the situations completely and vary only one variable at a time. Robots can moreover be manipulated far more easily than humans can be. Human-robot interaction, in contrast to human-to-human communication, allows therefore the identification of the parameters that speakers take into account when choosing their language for their particular communication partner (Fischer, 2003).

2 Method and Data

In the framework of the I1-project in the SFB/TR8 in Bremen, we elicit human-robot interaction data in scenarios that differ only with respect to single parameters. On the basis of the data elicited, we determine the influence of each parameter on the users' linguistic choices, and identify correlations in users' choices in order to predict behaviour on the basis of non-intrusive user models. The corpora used in this study are dialogues in which only the robot appearance is varied, and all other factors are kept constant. In the methodology employed here (Fischer, 2003), the comparability and control of the situation is achieved by keeping the robot behaviour constant. That is, the robot's behavioural (and, in other corpora, verbal) output is based on a fixed schema which is the same for all dialogues and across corpora. In this way, not only are all speakers confronted with exactly the same situation. The methodology also allows us to investigate speakers' concepts of their (artificial) communication partner. Because of the fixed schema that is causally unrelated to the speakers' actual linguistic behaviour, the speakers' sense-making efforts cannot be attributed to particular features or misbehaviours of a particular robot, but they have to be understood as arising from the speakers' own cognitive models of the situation, including their communication partner.

For example, if the speaker after a miscommunication (which is of course frequent if the robot does not really react to the speakers' utterances) uses a different descriptive term for the goal object, she displays that she holds the problem to be due to the robot's limited lexicon. If she switches to another reference system, she displays that she believes the reference system to be possibly inappropriate in this situation. That is, the original utterance and its reformulation serve as solutions to the same communicative task, and the second choice is based on an analysis what may have gone wrong in the previous turn, depending on the speaker's concept of the communication partner. Admittedly, a single instance of a particular linguistic choice may be attributed to many other factors as well; it is the systematicity and the repeatedness both for the same speaker and between speakers that allow the conclusion that we are observing the users' strategies depending on their concepts of their communication partner. To conclude, the design allows the identification of the explanatory models speakers build up to make sense out of the human-robot situation (Fischer, 2003).

Finally, the frequent impression of miscommunication encourages speakers to employ more than one linguistic strategy. If speakers are immediately successful, they have been found to stick to this strategy (Moratz et al., 2001, Fischer, 2006). This would however yield very uninteresting data to us. Instead, the speakers' reformulations are particularly revealing concerning the participants' concepts of the communication partner and their understanding of the situation.

Fig. 1. Aibo

Fig. 2. Scorpion

The dialogues discussed here differ only with respect to the robot addressed. The first robot used was Sony's Aibo, a commercially available robot which looks like a little dog (see Fig. 1). The second robot has been built by Frank Kirchner and colleagues at the University of Bremen (e.g. Spenneberg and Kirchner, 2002) and looks like a metal spider (see Figure 2). The third robot used is another commercially available experimental robotic platform, the Pioneer (see Fig. 3).

Procedure. The task users had to fulfil was to verbally instruct the robot to move to particular goal objects, pointed at by the experimenter. The objects, as well as the robot, were placed on the floor in front of the participants in various spatial configurations, changed during the experiment. Pointing was used in order to avoid prompting the participant with particular spatial descript-ions. Most tasks involved a single goal object, one task involved a sequence of several objects to which the robot had to go. After two tasks, which took up about two thirds of the time per dialogue, the experimenter told each speaker that she was allowed to refer to the objects directly. This prompt was meant to investigate the speakers' willingness to change their instructions, given their previous experience with the robot. The robot was steered by a student employee behind a screen according to a fixed schema of robot behaviours (Fraser & Gilbert, 1991). After the recording, speakers filled out a questionnaire in which they were asked whether they had believed that they were talking to a real robot, which all of them acknowledged.

Fig. 1. Pioneer

Participants. Participants were 66 German students from the University of Bremen of whom 12 interacted with Aibo, 21 with Scorpion, and 33 with Pioneer. Each participant interacted with just one robot, going through all task, which are defined by the various spatial configurations that we tested. Thus, all received prompts after the second task, and all filled out questionnaires at the end of the session. The instruction was simply to make the robot move to the objects pointed at by the experimenter. The prompt was 'oh, by the way, you may refer to the objects directly'.

The data elicitation conditions can thus be summarised as follows:

- 66 German human-robot dialogues
    - 12 dialogues with Aibo, a dog-like pet robot
    - 21 dialogues with Scorpion, a huge metal insect (e.g. Spenneberg and Kirchner, 2002)
    - 33 dialogues with Pioneer, a car- or box-like robot
- Task: to instruct the robot verbally to move to particular goal objects
- Robot behaviour: according to fixed schema of behaviours, independent of the speakers' utterances ('Wizard-of-Oz' scenario)
- after two tasks (after about 10 minutes of interaction), speakers were prompted that they can use object-based descriptions

The statistical analysis ANOVA was carried out using the Statistica software package.

Data Coding. The users' linguistic behaviour in the first two tasks, before the prompt, were coded in the following way:

- different types of **spatial instruction**, in particular, path- versus object-based instructions; that is, the speakers' linguistic behaviour in the first two tasks, before the prompt, was analysed with respect to consistent use of path-based instructions, such as '*move left*', consistent use of goal-based instructions in which the participants name an object, such as '*go to the left bowl*', or mixed usage;
- different degrees of **emotional expression** as evidenced by the use of interjections, partner-oriented signals and displays of interpersonal relationship, like *my friend*;
- different types and amounts of **feedback**, for instance, whether the speaker used feedback signals, like *okay*, explicit evaluations, like *good job*, and the use of character traits, e.g. *good boy*;
- different assumptions of competence; here, we coded the amount of **structuring cues** (see also Fischer & Bateman, 2006), distinguishing between implicit, for instance, *now*, and explicit, for instance, *first of all, the next step,* structuring cues as well as task overviews. Moreover, we coded different types of **intensifiers**, in particular, colloquial, for example, *slightly,* versus technical, for instance, *30 degrees*, and metaphorical, for example, *sharp*, intensification. Finally, we investigated the abstractness of **object descriptions**, for example, whether users employed abstract terms, such as *object* or *obstacle*, or whether they used basic-level terminology, for instance, *bowl*;
- different displays of relationship between the communication partners, for instance, with respect to **anthro- and zoomorphisation**, **politeness**, e.g. the use of *please* and *thank you*, **reference to the robot**, for instance as *robot* and *it* or as *he*, and the **sentence mood** chosen, in particular, whether no verb is used or whether the verb is in the infinitive and thus no assertion about the relationship is made, or whether the imperative is chosen or other linguistic forms, for example, the declarative or complex modal constructions;
- different intonation contours, indicating either different degrees of certainty and assertiveness or different degrees of interactivity. That is, rising **intonation contours** have been suggested to indicate lack of certainty (Lakoff, 1972), or to function as a device to increase interactivity and to encourage the partner to contribute to the conversation (Fishman, 1983). Thus, spatial instructions were coded for whether prosodic delivery was overwhelmingly with rising, falling, or level intonation contour, or whether it was mixed;
- different displays of the concept of the human-robot interaction situation as evidenced by the **dialogue beginning**, in particular, whether the speakers begin the interaction with a greeting or whether they immediately proceed to an instruction.

3 Hypotheses

There is ubiquitous research showing that speakers orient at their communication partners' appearance when formulating their utterances. Not only Schegloff's (1972) findings on membership analysis make us expect that speakers will use every clue possible to create models of their artificial communication partners that help them make their linguistic choices; also Roche (1989), for instance, has shown that appearance is a crucial factor in intercultural communication. Moreover, results by Nass and colleagues (Nass and Moon, 2000; Nass and Brave, 2005) show that users transfer knowledge from the source domain of human social interaction to human-computer interaction. All these findings make us expect that the spatial language directed at robots that differ in appearance will differ considerably. The hypothesis investigated here is thus that the users' spatial instruction strategies should crucially depend on their concepts of their artificial communication partner, which in turn should be influenced by the appearance of the robot. The analysis should thus concentrate not only on the spatial language features themselves, but also on evidence of the users' concepts of their communication partner as apparent from other linguistic choices. The prediction, based on previous findings, is thus that the linguistic features outlined will be distributed differently in the different corpora.

Besides the prediction that the users' linguistic behaviour will differ significantly in general, we may furthermore try to identify in more detail what the speakers' concepts are determined by. In particular, we may ask whether the users' attributing properties to robots takes place on the basis of analogies from the source domains of the robots, i.e. the animal kingdom and, for example, toy cars. If this is the case, it can be predicted that with respect to the features analysed, users will ascribe more competence to a more technically appearing system, here the Pioneer, compared to robots whose source domains are from the animal kingdom. Correspondingly, spatial instruction strategies, structuring cues, intensifiers, and object-naming strategies may be distributed differently for the three robots, such that Pioneer is trusted with a higher competence and Aibo and Scorpion with a lower degree of competence.

In contrast, those features pointing to stronger interpersonal and emotional relationships between speaker and robot should be higher for Aibo and possibly the Scorpion than for Pioneer. Thus, if speakers make use of properties of the source domains, one may expect more zoomorphisation for Aibo and Scorpion than for Pioneer, more instances of personifying references to the robot for Aibo and Scorpion than for Pioneer, more feedback, more emotional expression, more interactivity with Aibo and Scorpion, as evidenced by the prosodic delivery of the instruction, and more interpersonal information in the sentence mood and in the dialogue beginning.

4 Results

4.1 General Results

The first result is that without extra information, speakers consistently used only partner-centred instructions; that is, they exclusively took the robot's perspective, sometimes even getting up from their chair to turn into the direction which the robot was facing in order to identify the reference frame from the robot's point of view. Exceptions are the few goal object-based descriptions, in which the reference frame was often group-based. Thus, the spatial instruction style found was in 42.4% of all cases exclusively, and in 77.2% mostly, path-based. After the prompt, that speakers may also refer to the objects directly, however, 79% switched to object-based descriptions, most of which were group-based, which shows that speakers generally prefr this instruction strategy, yet did not initially believe it to be useful in the human-robot situation. The spectrum of spatial strategies taken throughout the dialogues comprises:

path-based	*go right*
goal-based	*the bowl on the right* (meaning the robot's right)
group-based	*to the bowl in the middle* (meaning in the middle of a group of objects)

4.2 The Role of the Robot

The analysis of variance shows that there are significant differences with respect to **spatial instruction** strategies between the corpora (F=11.08; df=2; p < 0.002). In particular, post hoc analyses show that whereas spatial instruction for Aibo is mainly path-based, the numbers of object-based instructions is considerably higher for Pioneer and Scorpion than for Aibo (see Figure 4).

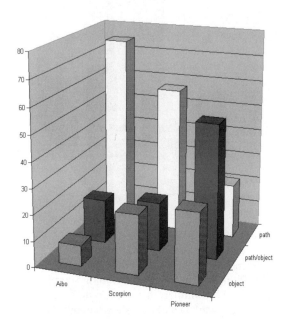

Fig. 2. Percentages of Spatial Instruction Strategies Used for the three Robots

The t-test (one-tailed) reveals that users employed significantly more higher-level (object-based) spatial instructions for Pioneer than for either Aibo or Scorpion. That is, Aibo and Scorpion are addressed with similar, very basic constructions, whereas Pioneer is addressed in a more elaborate way. Details are shown in the Table 2:

Table 2. Differences in Spatial Instruction

robot	Aibo	Scorpion	Pioneer
Aibo	-	n.s.	p<0.003
Scorpion		-	p<0.07
Pioneer			-

Furthermore, the choice of **intensifiers** differed for the different robots (F=3.877; df=3; p < 0.03). The four different types of intensifiers identified in the corpora were **technical**, such as *30 degrees to the right*, **colloquial**, such as *slightly*, **metaphorical**, such as *sharp*, or **none** at all (see Fig. 5).

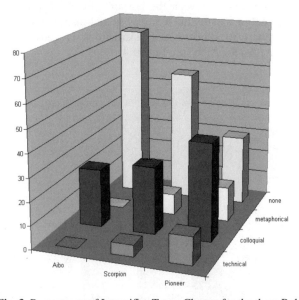

Fig. 3. Percentages of Intensifier Types Chosen for the three Robots

Post-hoc analysis by means of the t-test shows that users employed significantly more intensification for Pioneer than for either Aibo or Scorpion. That is, again Aibo and Scorpion are addressed in similar, more restricted ways, whereas Pioneer is addressed in a more elaborate way. Details are shown in the Table 3:

Table 3. Differences in Intensification

robot	Aibo	Scorpion	Pioneer
Aibo	-	n.s.	p=0.09
Scorpion		-	n.s.
Pioneer			-

Furthermore, the **prosodic delivery** was revealed to be significantly (F=15.1488; df=1; p < 0.01) different for the three robots. In particular, post-hoc analysis shows that speakers delivered their spatial instructions for Aibo with falling intonation contours, while for Pioneer they employed rising intonation contours, see Table 4. Scorpion is somewhat in the middle, since it also differs significantly from the Pioneer, but with a statistical tendency for a difference with Aibo as well.

Table 4. Differences in Intonation Contours

robot	Aibo	Scorpion	Pioneer
Aibo	-	p=0.11	p<0.0002
Scorpion		-	p<0.003
Pioneer			-

Unfortunately, this result is difficult to interpret due to the multifunctionality of intonation in discourse. Thus, a rising intonation contour may express uncertainty, in this case supporting the hypothesis that speakers orient more to the 'devicehood' of the more mechanically appearing robots Scorpion and Pioneer, suggesting that users are uncertain about communicating with such technical devices. However, rising intonation also has the function of encouraging interactive negotiation, and thus it could be argued that speakers are more interactive with Pioneer and Scorpion. Finally, also the source domain hypothesis cannot be ruled out completely since with dogs we are used to use falling intonation contours.

The most striking result however is that contrary to expectations, regarding all other variables investigated none turned out to be significantly different. In particular, none of the variables **emotional expression, feedback, structuring cues, object abstractness, zoomorphisation, politeness, reference to the robot, grammatical mood** or **greeting** yielded statistical differences between the corpora. This is particularly surprising for the amount of zoomorphisation, which should be higher for zoomorph robots like Aibo and Scorpion, yet speakers turned out to use similar conceptualisations for the Pioneer, for example:

(1) P110: den mag er besonders gern nicht? -- (laughter), den hat er zum fressen gern, (laughter) (breathing) [*this one he likes best, doesn't he? (laughter), he could eat it alive (laughter)*] [ol4,box2]

Thus, many factors related to the attribution of competence and emotionality regarding interactional management and interpersonal relationship were not influenced by the appearance of the robots.

4.3 The Role of External Variables: Gender

Investigating the corpora further for other relevant factors reveals that classical sociolinguistic variables are of some influence. The speaker variable gender turned out to be the most relevant regarding some of the variables suspected to be involved in different displays of relationship between human and robot. Thus, reference to the robot (F=5.3216; df=1; p < 0.03), politeness (F=5.7493; df=1; p < 0.03) and emotionality (F=3.3604; df=3; p < 0.02) were significantly influenced by the speakers' sex, such that men referred to the robot less often in general, and less often in an zoomorphised way, i.e. by using he or boy, for instance. Furthermore, men were found to use fewer interjections and contact signals and fewer politeness items, such as *thank you* or *please*.

Moreover, a linguistic feature showing different degrees of competence ascribed to the robot, the choice of object-naming strategies, was also found to be essentially gender-specific (F=5.3722; df=2; p < 0.01). In particular, women used more concrete object descriptions, such as *bowl, glass, pot*, instead of abstract terms, such as *object* or *obstacle*.

4.4 The Role of Participants' Concepts of the Communication Situation

The users' choice of dialogue beginning, which is suspected to indicate the speakers' concept of what the particular human-robot interaction situation consists in (see Fischer 2006ab), proved to be influential for the choice of structuring cues (F=3.12678; df=3; p < 0.04). That is, speakers who greeted the robot in some way or other were more inclined to use structuring cues, such as *next, then*, or even to give descriptions of the spatial scene, e.g. *vor Dir sind drei blaue Schälchen zu sehen – fahr mal zu dem mittleren. – blauen Schälchen* [*in front of you are three small blue bowls. why don't you drive to the middle – blue bowl.*], than those who immediately start the dialogue with a spatial instruction. Thus, the use of a greeting at the dialogue beginning correlates with attributions of higher linguistic, perceptual, and cognitive competence.

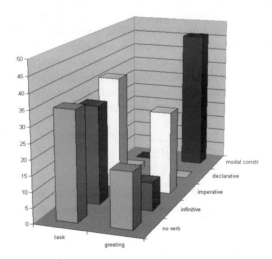

Fig. 4. Percentages of Sentence Mood Used by Speakers who Begin the Dialogues with a Greeting versus with a Task-oriented Instruction

Moreover, the users' dialogue beginnings also correlate with their choice of sentence mood (F=6.426; df=2; p < 0.003). That is, speakers who greet the system are more likely to use a verb in their spatial instruction (instead of using only spatial terms, for instance, *right, straight*) and to employ grammatical mood that expresses an interpersonal relationship between speaker and hearer, such as the imperative, the declarative mood or even complex modal constructions, such as *why don't you* in the example above. In contrast, speakers who begin their dialogues with a spatial instruction are more likely to use the infinitive, the imperative, or no verb at all (see Figure 6).

5 Discussion

Human-robot interaction constitutes for most speakers a very unusual communication situation, and thus it could have been expected that speakers use all information available to them from the robot to design their utterances for their unfamiliar communication partner. Accordingly, Nass and Brave (2005) have found that speakers employ the cues given by the voice of human-computer interfaces to infer properties of their communication partner. They propose that people mindlessly transfer knowledge about human social actors to human-computer interfaces (see also Nass and Moon 2000). Starting from this hypothesis I expected that users in our human-robot dialogues would transfer many properties of the robot source domain to the verbal interaction. However, this was the case only to a very limited degree. Indeed the spatial instruction strategies were adapted to the particular robot such that the pet robot Aibo and the Scorpion were addressed in a very low-level style of spatial instruction and with no or only colloquial intensification. Moreover, Aibo and Scorpion were found found to be addressed with falling intonation, in contrast to the Pioneer, which may correspond to the fact that one orders dogs to go somewhere, rather than to ask them; however, it may also be related to uncertainty in the communication with artificial communication partners, Pioneer being more 'robotic' in its appearance than Aibo.

However, if the source domain was what mattered the most, a Scorpion would probably not be hypothesised to be as intelligent as a dog. Indeed, Scorpion was addressed with instructions at the same level as Aibo and with similar amounts of technical and metaphorical means of intensification.

Moreover, due to the fact that Aibo is a pet robot, emotional expression and zoomorphisation would have been expectable to a higher degree if users really mindlessly transferred properties of human-dog communication to human-Aibo interaction. However, the most important influencing factor here was gender, not the difference between Aibo versus Scorpion or Pioneer.

Finally, if the source domains mattered that much, also other linguistic properties should have been influenced by the appearance of the robots. Instead, only three linguistic properties, spatial instruction, intonation contour and intensification, were found to differ significantly between the robots, whereas most features relevant in the interpersonal relationship and the attribution of competence to the robot were not affected.

The explanation for these findings, I propose, can be found in the fact that in the dialogues investigated here, speakers are talking to an artificial communication partner more than to anything else. The hypothesis would thus be that instead of relying on cues from the source domains *doghood*, *scorpionhood* or *technical devicehood* for the three robots, I want to suggest that the speakers in fact reacted to different degrees of artificialness of the robots, which in turn influences the definition of the human-robot situation in general.

Moreover, in the English data recorded in the project under the same conditions, none of the linguistic features investigated differed significantly for the three robots (Fischer, forthcoming), not even instructional strategy and intensification. This may be partly due to the fact that there we used the same speakers for all three robots, although recording took place at least three months apart. However, the lack of differences fits the overall picture very well, indicating a rather weak role of the appearance of the robot. However, also in the English data, the linguistic choices users made correlated significantly with the dialogue beginnings, as it was found for the data analysed here with respect to structuring cues and grammatical mood. Thus, structuring cues and sentence mood, variables that had been suspected to be related to ascribed competence of the communication partner and the interpersonal relationship respectively, have been found to be determined by the way speakers conceptualise the communication situation as it is apparent from their way of opening the dialogue. This is in line with previous findings in human-robot and human-computer communication, which show that the speakers' own perception of the communication situation determines the linguistic properties of their utterances, e.g. the use of clarification questions (Fischer and Bateman 2006), the reaction to situations of misunderstanding (Fischer 2006a, Fischer and Wilde 2005), the prosodic delivery of utterances (Fischer 2006a), as well as cooperative and uncooperative conversational strategies (Fischer 2006a). With respect to these linguistic strategies it was found that they can be significantly related to the users' conceptual models of the communication situation as either managing a technical tool or as pretending to enter a conversation with the artificial communication partner (see also Fischer 2006b).

Considering the relationship between partner model, concept of the situation, external variables, and the spatial language chosen, we can conclude that all three play a role in the formulation of spatial utterances. While the concepts evoked by the particular robots interact with concepts of the human-robot communication situation in general, and in particular with the concept of the robot as human-like or tool-like, in creating a model of the partner with respect to its capabilities, external factors, here gender, also play a decisive role with respect to interpersonal factors of human-robot interaction.

References

Clark, H.H., Wilkes-Gibbs, D.: Referring as a collaborative process. Cognition 22, 1–39 (1986)

Fischer, K.: Linguistic Methods for Investigating Concepts in Use. In: Stolz, T., Kolbe, K. (eds.) Methodologie in der Linguistik. Frankfurt a.M.: Peter Lang (2003)

Fischer, K.: Discourse Conditions for Spatial Perspective Taking. In: Proceedings of the Workshop on Spatial Language and Dialogue, WoSLaD, Delmenhorst (October 2005)

Kerstin, F.: What Computer Talk Is and Isn't. Human-Computer Conversation as Intercultural Communication. Saarbr (2006a)

Fischer, K.: The Role of Users' Preconceptions in Talking to Computers and Robots. In: Proceedings of the Workshop on How People Talk to Computers, Robots, and other Artificial Communication Partners, Delmenhorst (April 21-23, 2006) (2006b)

Kerstin, F.: (forthcoming): Talking for an Addressee. Alignment, Recipient Design and Interaction in So-called Simplified Registers. Habilitation Thesis, University of Bremen

Fischer, K., Wilde, R.: Methoden zur Analyse interaktiver Bedeutungskonstitution. In: Ueckmann, N. (ed.) Von der Wirklichkeit zur Wissenschaft: Aktuelle Forschungsmethoden in den Sprach-, Literatur- und Kulturwissenschaften. Hamburg: LIT (2005)

Fischer, K., Bateman, J.A.: Keeping the initiative: An empirically motivated approach to predicting user-initiated dialogue contributions in HCI. In: Proceedings of EACL 2006, Trento, Italy (2006)

Fishman, P.: Interaction: The work women do. Social Problems 25, 397–406 (1978)

Fraser, N., Gilbert, G.N.: Simulating Speech Systems. Computer Speech and Language 5, 81–99 (1991)

Theo, H., Joachim, G.: Sprechen. Psychologie der Sprachproduktion. Heidelberg: Spektrum (1994)

Lakoff, R.: Language and Woman's Place. Oxford University Press, Oxford (1975)

Moratz, R., Fischer, K., Tenbrink, T.: Cognitive Modelling of Spatial Reference for Human-Robot Interaction. International Journal on Artificial Intelligence Tools 10(4), 589–611 (2001)

Nass, C., Brave, S.: Wired for Speech: How Voice Activates and Advances the Human-Computer Relationship. MIT Press, Cambridge, MA (2005)

Nass, C., Moon, Y.: Machines and mindlessness: Social responses to computers. Journal of Social Issues 56(1), 81–103 (2000)

Jörg, R.: Xenolekte. Struktur und Variation im Deutsch gegenüber Ausländern. Berlin/New York, de Gruyter (1989)

Sacks, H., Schegloff, E.A., Jefferson, G.: A Simplest Systematics for the Organization of Turn-Taking for Conversation. Language 50, 696–735 (1974)

Schegloff, E.A.: Notes on a Conversational Practice: Formulating Place. In: Sudnow, D.N. (ed.) Studies in Social Interaction, pp. 75–119. The Free Press, New York: MacMillan (1972)

Schober, M.F.: Spatial Perspective-taking in Conversation. Cognition 47, 1–24 (1993)

Schober, M.F.: Speakers, Addressees, and Frames of Reference: Whose Effort Is Minimized in Conversations about Locations? Discourse Processes 20, 219–247 (1995)

Schober, M.F.: How Addressees Affect Spatial Perspective Choice in Dialogue. In: Olivier, P., Gapp, K.-P. (eds.) Representation and Processing of Spatial Expressions, pp. 231–245. Erlbaum, London (1998)

Schober, M.F.: Spatial Dialogue between Speakers with Mismatched Abilities. In: Proceedings of the Workshop on Spatial Language and Dialogue, WoSLaD, Delmenhorst (October 2005)

Spenneberg, D., Kirchner, F.: Scorpion: A Biomimetic Walking Robot. In: Robotik 2002. VDI-Bericht, vol. 1679, pp. 677–682 (2002)

Thora, T.: Localising objects and events: Discoursal applicability conditions for spatiotemporal expressions in English and German. PhD Thesis, University of Bremen (2005)

von Stutterheim, C., Kohlmann, U.: Selective Hearer-Adaptation. Linguistics 36(3), 517–549 (1998)

Collaborative Assistance with Spatio-temporal Planning Problems

Inessa Seifert

SFB/TR 8 Spatial Cognition, Universität Bremen, Germany
seifert@sfbtr8.uni-bremen.de

Abstract. The paper describes a collaborative assistance approach with spatio-temporal planning, which requires user's active participation in the problem solving task. The proposed collaborative assistance system operates on a region-based representation structure, which allows for partial specification of constraints at different levels of granularity. Weakly specified constraints contribute on the one hand to high computational complexity when generating alternative solutions and on the other hand to large solution spaces. The paper introduces Partial Order, Neighboring Regions and Partial Order of Neighboring Regions heuristics, which allow for pruning of significant parts of the search space, and produce hierarchical structuring of the solution space. Resulting hierarchical organization of the solution space reflects human mental processing of geographic information. To reduce cognitive load during observation of solution space, filtering of certain aspects, set-oriented structuring and case-based reasoning approaches are introduced.

1 Introduction

Nowadays, geographic information systems (GIS) provide meta-information about geographic regions and locations and make it possible to associate specific properties, such as points of interest (e.g., national parks, museums or sightseeing attractions) with the topological data, i.e., how locations are connected with each other. Such information provides a basis for assistance with spatial problems like spatio-temporal planning, i.e., planning of an individual journey or a city tour.

When planning a journey through an unfamiliar environment, travelers have to make a decision on *what* they are going to do, i.e., specify a set of activities. Along with the question *what*, they have to decide *where* the activities take place. Since the most of the interesting places and attractions are distributed around a country or a city, the corresponding locations have to be grouped together, to reduce the time for traveling from one place to another. Furthermore, journeys are usually constrained in time, so that journey planners have to fix the durations of their activities and put them into a feasible temporal order. Yet, especially in the early planning stage the information on *what*, *when* and *where* is known only partially and is available at different levels of granularity.

Planning individual journey involves not only dealing with spatial and temporal constraints, but also traveler's background knowledge, personal preferences, moods or even emotions. Common approaches utilize user modeling and personalization techniques, for generation of personalized tour plans (e.g., McGinty & Smith, 2002). However, the main disadvantage of the personalized tour generation is an exclusion of user's active participation in the process of planning. In doing so, the generated tours are difficult or even impossible to change.

The paper describes a collaborative assistance approach with spatio-temporal planning, which requires user's active participation in the problem solving task (Schlieder & Hagen, 2000). Since spatio-temporal planning task is now shared between an artificial assistance system and a user, the problem domain is separated into hard spatio-temporal constraints, for example specific locations and temporal order of activities, and soft constraints, for example, personal preferences. An assistance system supplies a user with alternative solutions, which fulfill the specified hard constraints. Depending on the knowledge available at the beginning of a planning process, user's input can be underspecified. Dealing with a weakly specified user input, we face two problems: (a) underspecified spatio-temporal constraints contribute to a high computational complexity when generating alternative solutions; (b) the resulting solution space is considerably large, so that observation of all possible generated solutions becomes a cognitively demanding task (Knauff et al., 2002).

In order to attack the computational complexity of spatial planning problems, the state of the art artificial spatial systems utilize hierarchical representation of spatial knowledge (e.g., Chown, 2000; Kuipers, 2000; Leiser & Zilberschatz, 1989). On the other hand, series of psychological experiments have shown that mental processing of geographical information is also hierarchical (Hirtle & Heidorn, 1993; Tversky, 1993; Hirtle, 1998). In our recent work (Seifert et al., 2007), we have introduced a region-based representation structure, which aims at providing correspondence between hierarchies in human mind and a spatial assistance system. Due to the hierarchical organization, the proposed region-based representation structure allows for specification of spatial constraints at different levels of granularity, and facilitates the generation of alternative solutions.

In the scope of this paper, the conceptual model of the proposed region-based representation structure is going to be refined. The paper addresses the cognitively motivated principles for structuring of the spatio-temporal problem domain as well as of resulting solution space. The proposed principles aim at reducing the cognitive load during collaborative search for a solution.

The paper describes a prototypical implementation of the region-based representation structure, which is demonstrated by means of an example for planning of an individual journey to the US state California. The example setting shows how the introduced cognitively motivated structuring principles can be applied to a large-scale environment. Furthermore, the paper illustrates how users can specify constraints at different levels of granularity and which kind of solutions can be generated as a result.

To deal with weakly specified user's input, several heuristics are introduced, which allow for pruning significant parts of the problem space. The paper includes an evaluation of the prototypical implementation of a PROLOG-based constraint solver, considering underspecified constraints, proposed heuristics, computing time, and number of alternative solutions. Subsequently, set-oriented principles for structuring of the solution space will be introduced.

Furthermore, the region-based representation structure allows for case-based reasoning. In doing so, a user can be provided with an already assembled tour (a precompiled solution), which can be modified at different levels of granularity and easily adapted to her personal preferences.

The paper ends with an outlook, as well as a set of open question, which require further interdisciplinary research in the field of spatial cognition.

In the next section I'm going to outline the characteristics of the spatio-temporal problem domain and put them into the terminology of computer science/artificial intelligence.

2 Characteristics of the Spatio-temporal Problem Domain

Spatio-temporal planning aims at selecting activities and putting them into a feasible order under consideration of temporal and spatial constraints. Along with the hard spatio-temporal constraints planning an individual journey encompasses soft constraints like personal preferences, interests, moods and emotions. Usually, concrete requirements on the course of a journey are not known from the very beginning, but rather become formed during the process of planning. Furthermore, journey plans are often ambiguous, i.e., many things are left un- or underspecified, in order to account for further changes in the future. In the following, I refer to the recent explorations studies, which address the problems of tourists in unfamiliar environments (Brown, 2007). The analysis of the studies is focused on the *pre-visiting* of unfamiliar environments, and is taken from the reasoning and knowledge representation point of view.

2.1 Initial Problem State

Trips to foreign countries are planned by using external media like traveling guides, various information resources on the Internet and geographic maps. The details on *what* and *where* are often available at different levels of granularity, for example, sightseeing in specific parts of a city or a country, hiking in national parks, skiing in mountains, or swimming at a see coast. The question *when* is expressed qualitatively, i.e., more as a feasible temporal order of activities (e.g., sightseeing before swimming at a see coast), and less in precise points in time.

2.2 Underspecified Spatial and Temporal (Hard) Constraints

Usually, journeys are constrained in time. Temporal constraints include the duration of the whole journey as well as durations and temporal order of activities, which take place during the journey. Spatial constraints encompass particular places or regions and the distances between them. Travelers usually allocate the days they spend to some particular locations, leaving many things unspecified, for example, when exactly they visit each location and which kinds of activity, in the following denoted as activity types, they pursue there.

2.3 Soft Constraints

Soft constraints, like personal preferences, interests, moods or emotions, play a crucial part in the selection of appropriate locations and activity types (i.e., *where* and *what* to do). Also constraints which consider diversity of locations are very individual: traveler's preferences may vary from seeing as many tourist attractions as possible, to preferring so called insider tips, which are situated in the most cases aside of the major tourist attractions (i.e., *where*). Temporal constraints together with specific activity types also may depend on individual preferences, like shopping on Saturdays.

2.4 Goals Are Open-Ended

Traveling is for leisure. There are no hard criteria for an optimal solution, no overall goal, it is open ended. There are no particular optimization criteria, since walking in the streets and enjoying the "street life" may be one of the most enjoyable parts of a journey. Finally, most of the tourist plans are 'good enough' and deliberately ambiguous so that they can take into account future contingencies.

2.5 Clustering of Multiple Target Locations

In order to avoid spending too much time for traveling between places, journey planners group together attractions which are close together, or pick up an area with more than one potential facility, for example, restaurants.

2.6 Case-Based Reasoning

Dealing with a huge amount of the information accessible from various information resources, travelers often help themselves by copying plans and tours of those people, who have already been to that places, and refine them with their own modifications.

2.7 Planning a Journey Is an Ill-Structured Problem

Since there is no well-defined initial state, no well-defined optimization function, and no particular goal state, planning an individual journey through a foreign country is an ill-structured problem (Simon, 1973).

3 A Collaborative Assistance Approach

The term of collaborative assistance has been introduced by Schlieder & Hagen (2000). A collaborative approach allows for computational support with ill-structured problems, where constraints can be formalized only partially. A collaborative assistance system accompanies a mental problem solving process, by sharing it between a human and a computational constraint solver. To enable such joint problem solving process, the problem domain is separated into hard and soft constraints. An assistance system generates a set of solutions, which fulfill the hard constraints. Subsequently, a user examines the generated solution space and applies her personal criteria regarding the solution quality, i.e., soft constraints, which can not be outsourced to a computational constraint solver. An assistance system provides users with operations, which allow for refinement and modification of constraints, in order to obtain new improved solutions. Finally, the joint problem solving task can be seen as an interaction cycle (see Figure 1).

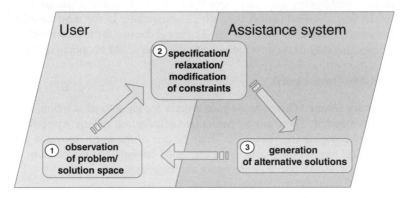

Fig. 1. Interaction cycle

The interaction cycle starts with the observation of the problem domain (1) and subsequent specification of initial constraints (2), which are used by an assistance system for generation of alternative solutions (3). Depending on how many and how precise the initial constraints are specified, the resulting solution space may become considerably large. In order to facilitate mental co-processing during collaborative problem solving, the assistance system should take into consideration the specifics of mental processing of the domain knowledge. During the observation of the solution space, a user can make a decision on her personal preferences regarding the quality of solution. In doing so, the solutions space becomes again a problem space (1). A user specifies new constraints, or modifies and relaxes the existing constraints (2), so that an assistance system can generate new improved solutions (3).

From the characteristics of the spatio-temporal problem domain, introduced in the previous section, the requirements on collaborative spatio-temporal planning assistance can be summarized as follows. The assistance system should:

1. Provide a representation of an unfamiliar large-scale environment, which facilitates the observation of the problem space.
2. Allow for input of underspecified spatio-temporal constraints at different levels of granularity,
3. Generate alternative solutions,
4. Provide users with skilful operations which facilitate the observation of the solution space and allow for identifying the specifics of a particular solution or sets of solutions,
5. Allow for refinement of the solution space by introducing further constraints, i.e., constraint specification,
6. Allow for relaxation and modification of constraints of a pre-compiled spatio-temporal configuration in order to obtain new solutions, i.e., case-based reasoning.

In the following, a conceptual model of the region-based representation structure is going to be introduced together with the cognitively motivated principles for structuring of unfamiliar large-scale environments.

4 Region-Based Representation Structure

In our recent work (Seifert et. al., 2007), we introduced a region-based representation structure, which represents a spatial hierarchy, consisting from locations, activity regions[1], super-ordinate regions, nodes and paths.

The problem space is structured hierarchically into regions using the cognitive phenomena of *regionalization* (Montello, 2003) and *region connectivity* (Wiener, 2004), to facilitate mental co-processing during shared spatio-temporal planning task.

When planning an individual journey, travelers have to make a decision on (a) how-long a journey should take, (b) how many places in a foreign country they are going to visit, (c) which means of transportation they are going to use for moving from one location to another, (d) what kind of activities should be planed in which locations, and (e) how long they are going to stay in some particular location, or how much time should be planed for a specific activity type.

Referring to the definition by Simon (1981), "... *solving a problem simply means representing it so as to make the solution transparent*", the problem space should resemble the decision space of a human reasoner. In other words, the representation structure of the problem domain should correspond to the spatio-temporal planning problem, which has to be solved in a collaborative way.

4.1 Spatial-temporal Constraint Satisfaction Problem

A spatio-temporal planning problem can be handled as a constraint satisfaction problem (CSP), which includes duration of a journey, a set of activities, and temporal order of activities. Each activity has a duration, an activity type and a spatial assignment. A solution for a spatio-temporal planning problem is defined as a feasible temporal order of activities, which fulfill the following constraints:

1. Activities do not overlap with each other in time,
2. The sum of durations of activities together with temporal costs required to overcome distances between spatial assignments should not exceed a given temporal scope of a journey.
3. Spatial assignments of activities are different from each other.

To allow for partial specification of constraints, the corresponding activity attributes are handled in the following way. Duration of each activity remains mandatory, whereas activity type and spatial assignment are optional. Furthermore, spatial assignments can be specified at different levels of granularity: e.g., using a specific location, an activity region, or a super-ordinate region.

---

[1] Regions addressed in (Seifert et. al., 2007) as thematic regions are denoted in this paper as activity regions, to provide a stronger association with specific activity types.

4.2 Structuring Principles

The proposed structuring principles aim at producing a hierarchical region-based representation structure of an unfamiliar environment, which corresponds to temporal structure of a journey. Since the resulting representation of a problem space should resemble a human decision space, the granularity levels of the spatial hierarchy should contain elements, which possess specific semantics, i.e., are meaningful for a traveler. In the following, I'm going to start with the lowest level of the region-based hierarchy.

Location: is a one dimensional point that represents an area, which is associated with one or more activity types. The physical extension of an area depends on to the duration of a planed activity and on the velocity of traveler's locomotion within the area. A location has an attribute, which stands for semantic importance of the location in comparison to other locations associated with a same activity type. As an example for measurement of semantic importance of a location can serve information on population of the area, it's historical or cultural attractions.

Activity region: contains a set of locations, which are associated with a specific activity type. In this vein, an activity region represents a set of optional locations, where a planed activity can take place. Multiple activity regions can also share the same connectivity node and overlap with each other.

Connectivity node: is a semantically important location, which binds other locations within an activity region with an optimal connectivity cost.

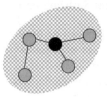

Fig. 2. Locations within an activity region connected by a node

A connectivity cost is computed from an average temporal distance cost to the corresponding locations and depends on a velocity of traveler's locomotion.

Path: connects different connectivity nodes and has a distance cost, which is represented as a temporal interval. Each intersection of path segments is also modeled as a node, which carries no additional connectivity costs.

Super-ordinate region: contains several activity regions (see Figure 3).

When structuring a large-scale environment into super-ordinate regions, specifics of mental processing of geographical large-scale environments has to be taken into consideration.

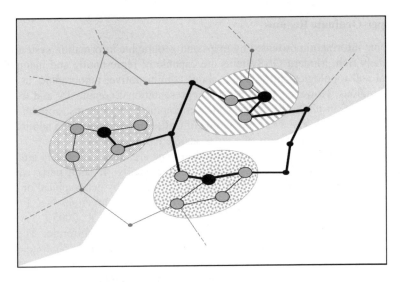

Fig. 3. Two super-ordinate regions with activity regions connected by paths between the corresponding connectivity nodes

Fig. 4. Example partitioning of California

4.3 Super-Ordinate Regions

Geographic information provided by maps and geographic information systems (GIS) is extremely rich. Modern GI-Systems are capable of representing and manipulating of spatial and topological information about administrative, topographic, as well as thematic regions. However, to reduce the representational complexity and to convey the structure of an unfamiliar large-scale geographic environments, popular traveler guides like the "Vis a Vis, Traveling Guide of California", combine the administrative regions into super-ordinate parts of a large-scale environment.

Such global parts divide a large-scale geographic area into a relatively small number of regions, which can be easily processed mentally. If such parts cannot be mapped to administrative regions, they are usually labeled by geographical or climatic specifics of the environment in combination with cardinal directions (see Figure 4), like North Coast, Central Cost, South Coast, etc. (cf., Lyi Y. et al., 2005).

The Figure 4 illustrates an example for partitioning of California. The super-ordinate regions resemble the topographic properties of Californian landscapes: coast regions, mountain regions, a valley and a desert. To resolve the ambiguity of the regions, which share the same topographical properties, an administrative description, e.g., Sierra Nevada and cardinal directions are used, e.g., Northern Mountains, North Coast, Central Coast, South Coast.

4.4 Representation of the Relations Between the Structural Elements

The region-based representation structure consists from locations, activity regions, super-ordinate regions, connectivity nodes and paths. The relations between the structural elements are described as follows.

Neighbor of: neighboring super-ordinate regions are related with each other by a "neighbor of" relation, for example, the super-ordinate region Desert is a "neighbor of" the South Coast, Sierra Nevada and the Central Coast.

Part of: Since locations are part of an activity region, and an activity region is a part of super-ordinate region, the corresponding spatial hierarchy of regions is represented as a partonomy. Such partonomies allow for representing and reasoning about qualitative topological relations between regions with rough boundaries (Bittner & Stell, 2002). The Figure 5 illustrates an example partonomy consisting of locations, which belong to the Redwood National Park: Berry Glenn, Orick and Big Lagoon. The Redwood National Park is situated at the North Coast of California.

Activity type of region: Each activity region is associated with a specific activity type. For example a Redwood National Park is associated with a possibility for hiking.

Connectivity node of region: Connectivity nodes, which are associated with activity regions, have specific connectivity costs. Whereas nodes, which represent intersections of path segments, are assigned to super-ordinate regions and don't have additional connectivity cost.

Edge: An edge connects two nodes, i.e. path segments, and has a distance cost. The relations between the structural elements of the region-based representation described in this section are realized as logical predicates in the constraint programming language PROLOG.

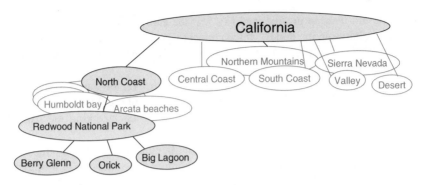

Fig. 5. Example partonomy of California

4.5 Specification of Spatial and Temporal Constraints

Spatio-temporal constraints are specified by a set of activities and a temporal scope of a journey. The Figure 6 illustrates the activity data structure:

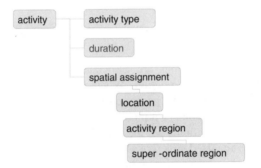

Fig. 6. Activity data structure

Each activity contains assignments for activity type, duration, as well as a spatial assignment, which can be specified at different levels of granularity: location, activity region or a super-ordinate region. The duration of an activity is mandatory, whereas activity type and a spatial assignment are optional.

5 Generation of Alternative Solutions

Generation of alternative spatio-temporal configurations encompasses instantiations of all feasible assignments of the corresponding underspecified attributes on the granularity level of activity regions. To obtain the shortest path between the connectivity nodes of the corresponding activity regions, we utilize a standard shortest path algorithm (Dijkstra); this may be enhanced in the future by using more sophisticated routing algorithms. Feasible solution should fulfill the following requirements (1), activities do not overlap in time, and (2), i.e., the sum of durations of activities and distance costs between the nodes, which connect specified, or instantiated activity regions, does not exceed the given temporal scope of a journey. (3) Resulting activity regions are different from each other.

Since the relations between the elements of the region-based representation structure are implemented as logical predicates, the generation of alternative solutions involves instantiation of underspecified facts using the backtracking mechanism of constraint programming language PROLOG.

Depending on the knowledge available at the beginning of the planning process, initial constraints can be either weakly specified, allow for generation of a single solution, or overconstrained. If the specified constraints do not allow for generation of a feasible solution, the constraints have to be explicitly relaxed by the user. In the other case, a weakly constrained initial state results in a large solution space and in time consuming computation.

The PROLOG-based constraint solver instantiates possible values of underspecified activity regions using the data stored in a database. For a faster generation of the spatio-temporal configurations the next section describes three heuristics, which can be applied to prune significant parts of the search space by using the relations between the super-ordinate regions.

5.1 Partial Order

The underspecified activity regions are associated with the corresponding super-ordinate regions by a "part-of" relation. When possible values of underspecified activity regions are instantiated, the partial order heuristic "remembers" the order of super-ordinate regions, and hinders jumping back to super-ordinate regions, which have already been visited. In doing so, the partial order heuristic produces automatically a hierarchically organized structure of the resulting solution space. Since mental processing of geographic information is hierarchical, solution space, organized in a hierarchical manner, is easier to observe. However, special cases like round trips have to be taken into account to allow travelers to come back to the place where their journey has started.

5.2 Neighboring Regions

Nearest neighbor (NN) is a well known heuristic, which is utilized to attack the computational complexity of Traveling Sales Man Problems (Golden et. al., 1980). Using (NN), a search algorithm looks for a node with a minimal distance to a most recent node visited to get a good approximation for an optimal path between multiple destinations to be accessed. Since a feasible solution of the given spatio-temporal constraint satisfaction problem doesn't require search for an optimal configuration of activity regions, the proposed neighboring regions heuristic utilizes the "neighbor of" relation between super-ordinate regions. The Neighboring Regions allows for visiting any new super-ordinates but doesn't prohibit going back to a visited region neighbored to the currently visited super-ordinate region.

5.3 Partial Order of Neighboring Regions

The heuristic is a combination of Partial Order and Neighboring Regions heuristics. In this case, the heuristic "remembers" the visited super-ordinate regions, prohibits jumping back and allows moving only to a neighboring super-ordinate region. Compared to the first two heuristics, it is the strongest heuristic.

6 Evaluation of the Prototype

The Table 1 provides an overview about the test data, which has been used for evaluation of the prototypical implementation. For the evaluation of the performance of the region-based representation structure number of activity regions, nodes and edges is relevant. Therefore, test data considering locations has been omitted.

Table 1. Test data

structural element	amount
super-ordinate region	7
activity region	82
node	89
edge	253
activity type	4

The procedure for generation of the alternative spatio-temporal configurations is written in PROLOG. However, for the implementation of a comfortable graphical user interface a JAVA-based client is needed. To simulate a possible behavior of a JAVA-based GUI, the PROLOG-based generation procedure is called from a client by using a PROLOG-vendor specific JAVA-Interface.

The following table contains the description of an example underspecified user input: a journey of 14 days, 3 days of sightseeing, 3 days of swimming, 3 days of hiking and 3 days for alpine ski, without specification of spatial assignments. The rest of the time, 2 days, is planed for traveling from one place to another.

Table 2. Underspecified user input

activity	activity type	duration	activity region	super-ordinate region
1	sightseeing	3 days	*unspecified*	*unspecified*
2	swimming	3 days	*unspecified*	*unspecified*
3	hiking	3 days	*unspecified*	*unspecified*
4	alpine ski	3 days	*unspecified*	*unspecified*

The Table 3 demonstrates applied heuristics, computing time required for generation of alternative solutions, and the number of the generated solutions. 'NONE' stands for no heuristics applied, 'PO' stands for Partial Order, 'NB' stands for NeighBoring regions, and finally 'PONB' stands for Partial Order of NeighBoring regions heuristic.

Table 3. Applied heuristics, resulting solutions, and computing time

heuristic	solutions	computing time in seconds
NONE	520	148
PO	326	109
NB	456	33
PONB	278	24

The demonstrated evaluation of the performance of the proposed heuristics allows for analysis of user's input. Such analysis determines the degree of underspecified constraints. In this vein, the assistance system can apply a suitable heuristic automatically and give users feedback about how much time the generation of the resulting solutions may cost. Yet, another opportunity to avoid time consuming computation of alternative solutions is letting a user mark the regions and places that she finds interesting, before the generation starts. Taking into account only pre-selected regions and location, the search space and the time for generation of alternative solutions can be significantly reduced.

6.1 Partial Specification of Spatial Assignments

In the following example, the constraints are represented as a set of activities with partially specified attributes. The activities together with partially specified spatial assignments have to fit into a specified temporal scope of a journey. For example, travelers are going to spend 5 days for visiting locations at the Central Coast of California. They plan two activities of 2 days each. The rest of the time is planed for driving by car from one place to another. The first activity includes a specified activity region: San Francisco Beaches, which combines optional locations at San Francisco Bay, together with a corresponding activity type: swimming. The second activity includes 2 days of sightseeing. Since our travelers wish to see the Central Coast, the spatial assignment of the second activity is specified at a granularity level of the Central Coast. The Table 4 summarized the described set of activities:

Table 4. Partially specified spatial assignments

activity 1		activity 2	
activity type	swimming	activity type	sightseeing
activity region	San Francisco beaches	activity region	*unspecified*
super-ordinate region	Central Coast	super-ordinate region	Central Coast
duration	2 days	duration	2 days

Considering a temporal scope of 5 days, 2 alternative solutions are generated using the available test data. The first solution proposes a trip from San Francisco Beaches to Big Sur, via San Jose, Castroville and Monterey. The second solution proposes to stay in the area of the city of San Francisco, and enjoy the sightseeing there.

Each of the generated solution consists of activity regions, which combine several smaller locations, where activities can take place, for example San Francisco Beaches and Big Sur, the super-ordinate regions, here the Central Coast, and a set of nodes connecting the activity regions. A user has now choice either to travel from one region, San Francisco Beaches, to another, Big Sur, or to stay during her journey in the area of San Francisco.

6.2 Partial Specification of Spatial Assignments and Activity Types

Another example illustrates the case, where the second super-ordinate region is specified, but no specific activity type is given. The corresponding set of partially specified activities is the following:

Table 5. Partially specified activity types and spatial assignments

activity 1		activity 2	
activity type	swimming	activity type	*unspecified*
activity region	San Francisco Beaches	activity region	*unspecified*
super-ordinate region	south coast	super-ordinate region	Northern Mountains
duration	2 days	duration	2 days

The first alternative solution proposed to drive from San Francisco Beaches, via Sacramento, Lakehead, and stay in one of the places near Lake Shasta for swimming. The second solution leads from San Francisco Beaches to Mount Shasta, also via Sacramento and Lakehead, and proposes alpine skiing.

The two solutions differ from each other in the activity regions Shasta Lake and Mount Shasta as well as in the corresponding activity types 'swimming' and 'alpine skiing'.

7 Structuring of the Solution Space

The evaluation of the prototypical implementation of the region-based representation structure has shown that underspecified constraints require on the one hand much computing time, and on the other hand contribute to large solution spaces (see Table 3 for details). To reduce the cognitive load during observation of the solution space, the assistance system should allow for a compact representation of the solutions space and provide users with skilful operations, which allow for comparison of alternative solutions or sets of solutions.

7.1 Filtering of Aspects

Providing users with specific filters that allow concentrating on certain aspects, of the spatio-temporal problem domain gives travelers an opportunity to assess the quality of specific solutions and specify further constraints in the next interaction step, i.e. marking specific super-ordinate regions and, in this way, to reduce the search space during subsequent generation of new solutions. Therefore, aspectualization is a suitable method to attack the complexity of geographic information (Freksa & Barkowsky, 1996).

7.2 Clustering of Quasi Identical Solutions

Usually, in an early route planning stage a particular order of activities with unspecified activity types is not crucial. Therefore, the assistance system should allow travelers to observe sets of solutions, for example, combining quasi identical solutions which encompass the same activity regions or super-ordinate regions, leaving out different orders of the corresponding activities.

7.3 Top-Down Approach

Since human processing of geographic information is hierarchical, the assistance system should provide operations of zooming in and zooming out in the spatial region-based hierarchy. Specification of partial order considering super-ordinate regions, provides so called top-down approach and allows a user to step into the spatial hierarchy introducing more and more detailed constraints, by navigating from coarse to fine regions.

7.4 Bottom-Up or Case-Based Reasoning Approach

In opposite to the top-down approach, the assistance system can provide users with already assembled tours, which can be successively enriched with modifications which reflect users' personal preferences, as well as relaxations of the initially proposed spatial assignments or activity types at a higher granularity level. Since modification of constraints takes place sequentially, the generation of new solutions doesn't require much computing time and the resulting number of alternatives is relatively small.

8 Outlook and Future Work

The paper presented an approach for a collaborative assistance with spatio-temporal planning, which involves user's active participation in the planning process. Since the problem solving task is shared between a human and an artificial assistance system, mental cognitive capacity as well as specifics of mental processing of information about large-scale environments has to be taken into consideration. In order to establish an adequate dialog between a human and a computational constraint solver the cognitive phenomena of regionalization and region connectivity are used for organizing and representing of spatial knowledge. The proposed region-based representation structure of the problem domain allows for specifying constraints at different levels of granularity and dealing with underspecified constraints. The paper described the principles for structuring of large-scale environments, which aim at linking together spatial and temporal structure of a spatio-temporal planning problem. Depending on the knowledge available at the beginning of the collaborative planning process, spatial assignments as well as activity types can be partially specified. The underspecified constraints require much computing time and contribute to large solution spaces. In the scope of the paper, several heuristics are introduced, which involve the relation between the super-ordinate regions: Partial Order, Neighboring Regions and Partial Order of Neighboring Regions, which allow for pruning of significant parts of the search space.

Series of psychological experiments conducted in a virtual reality lab have shown that humans can solve spatial problems like navigation in regionalized environments more efficiently then in environments without a pre-defined region-based structure (Wiener, 2004). However, it is still an open question addressed to empirical behavioral research, whether similar heuristics are used by humans, when dealing with the similar spatio-temporal planning problems.

The paper introduced first ideas towards cognitively motivated principles for structuring of large-scale environment. The resulting spatial hierarchy serves not only for reducing of the computational complexity of the spatio-temporal planning problem, but also provides travelers with semantically meaningful information. However, a formal model of the cognitively motivated structuring principles is a matter of future work, which requires further research not only on how humans learn and structure knowledge about large-scale environments, but also how such knowledge can be successfully conveyed.

Another open issue is a graphical user interface, which should visualize the solution space and provide user's with sophisticated operations. Such operations should allow for navigation in the solution space and comparison of alternative solutions.

Acknowledgments

I gratefully acknowledge financial support by the German Research Foundation (DFG) through the Transregional Collaborative Research Center SFB/TR 8 Spatial Cognition (project R1-[ImageSpace]). I also want to thank Thora Tenbrink, Kai-Florian Richter and Falko Schmidt, who provided me with the valuable feedback on the first draft of the paper. Special thanks to Zhendong Chen and Susan Träber, who worked on the implementation of the prototype, searched and maintained the test data. I also want to thank two anonymous reviews, whose comments helped to improve the quality of the paper.

References

Bittner, T., Stell, J.G.: Vagueness and Rough Location. Geoinformatica 6, 99–121 (2002)
Brown, B.: Working the problems of Tourism. Annals of Tourism Research 34(2), 364–383 (2007)
Chown, E.: Gateways: An approach to parsing spatial domains. In: ICML 2000. Workshop on Machine Learning of Spatial Knowledge (2000)
Freksa, C., Barkowsky, T.: On the relation between spatial concepts and geographic objects. In: Burrough, P., Frank, A. (eds.) Geographic objects with indeterminate boundaries, London: Taylor & Francis, pp. 109–121 (1996)
Golden, B., Bodin, L., Doyle, T., Stewart, W.: Approximate travelling salesman algorithms. Operations Research 28, 694–711 (1980)
Hirtle, S.C.: The cognitive atlas: using GIS as a metaphor for memory. In: Egenhofer, M., Golledge, R. (eds.) Spatial and temporal reasoning in geographic information systems, pp. 267–276. Oxford University Press, Oxford (1998)

Hirtle, S.C., Heidorn, P.B.: The structure of cognitive maps: Representations and processes. In: Gärling, T., Golledge, R.G. (eds.) Behavior and environment: Psychological and geographical approaches, pp. 170–192. North-Holland, Amsterdam (1993)

Knauff, M., Schlieder, C., Freksa, C.: Spatial Cognition: From Rat-Research to Multifunctional Spatial Assistance Systems. Künstliche Intelligenz, Heft 4/02, arendtap Verlag, Bremen (2002)

Kuipers, B.: The spatial semantic hierarchy. Artificial Intelligence 119, 191–233 (2000)

Leiser, D., Zilbershatz, A.: The Traveller: A computational model of spatial network learning. Environment and Behavior 21, 435–463 (1989)

Lyi, Y., Wang, X., Jin, X., Wu, L.: On Internal Cardinal Direction Relations. In: Cohn, A.G., Mark, D.M. (eds.) Proceeding of Spatial Information Theory 2005. LNCS, pp. 283–299. Springer, Heidelberg (2005)

Ginty, Mc., Smyth.: Shared Experiences in Personalized Route Planning. In: Proceeding of AAA Symposium on (2002)

Montello, D.R.: Regions in geography: Process and content. In: Duckham, M., Goodchild, M.F., Worboys, M.F. (eds.) Foundations of Geographic Information Science, Taylor & Francis, pp. 173–189. London (2003)

Seifert, I., Barkowsky, T., Freksa, C.: Region-Based Representation for Assistance with Spatio-Temporal planning in Unfamiliar Environments. In: Gartner, G., Cartwright, W., Peterson, M.P. (eds.) Location Based Services and TeleCartography. Lecture Notes in Geoinformation and Cartography. Springer, Heidelberg (2007)

Schlieder, C., Hagen, C.: Interactive layout generation with a diagrammatic constraint language. In: Freksa, C., Habel, C., Wender, K.F. (eds.) Spatial cognition II - Integrating abstract theories, empirical studies, formal methods, and practical applications, pp. 198–211. Springer, Heidelberg (2000)

Simon, H.A.: The Structure of Ill Structured Problems. Artificial Intelligence 4 (1973)

Simon, H.A.: The Sciences of Artificial, 2nd edn. MIT Press, Cambridge (1981)

Tversky, B.: Cognitive maps, cognitive collages, and spatial mental models. In: Frank, A., Campari, I. (eds.) Spatial information theory, pp. 14–24. Springer, Heidelberg (1993)

Vis a Vis, Traveling Guide of California, in German: Kalifornien, Dorling Kindersley, Duncan Baird Publishers, London (2004)

Wiener, J.M. (2004) (PhD Thesis, University of Tübingen, Germany): Places and Regions in Perception, Route Planning, and Spatial Memory (2004)

Dialog-Based 3D-Image Recognition Using a Domain Ontology

Joana Hois[1], Michael Wünstel[1], John A. Bateman[1], and Thomas Röfer[1,2]

[1] SFB/TR8 Spatial Cognition, Universität Bremen,
Postfach 330 440, 28334 Bremen, Germany
{joana,wuenstel}@informatik.uni-bremen.de, bateman@uni-bremen.de
http://www.sfbtr8.spatial-cognition.de/
[2] DFKI Lab Bremen, Safe and Secure Cognitive Systems,
Robert-Hooke-Straße 5, 28359 Bremen, Germany
thomas.roefer@dfki.de
http://www.dfki.de/

Abstract. The combination of vision and speech, together with the resulting necessity for formal representations, builds a central component of an autonomous system. A robot that is supposed to navigate autonomously through space must be able to perceive its environment as automatically as possible. But each recognition system has its own inherent limits. Especially a robot whose task is to navigate through unknown terrain has to deal with unidentified or even unknown objects, thus compounding the recognition problem still further. The system described in this paper takes this into account by trying to identify objects based on their functionality where possible. To handle cases where recognition is insufficient, we examine here two further strategies: on the one hand, the linguistic reference and labeling of the unidentified objects and, on the other hand, ontological deduction. This approach then connects the probabilistic area of object recognition with the logical area of formal reasoning. In order to support formal reasoning, additional relational scene information has to be supplied by the recognition system. Moreover, for a sound ontological basis for these reasoning tasks, it is necessary to define a domain ontology that provides for the representation of real-world objects and their corresponding spatial relations in linguistic and physical respects. Physical spatial relations and objects are measured by the visual system, whereas linguistic spatial relations and objects are required for interactions with a user.

1 Introduction

Dialogs between humans and robots about natural objects in a visual scene have always required interdisciplinary research. Different components from linguistics and psychology for generating and comprehending natural dialogs, on the one side, have to be combined with image processing and knowledge representation for a formal representation of the respective objects, on the other.

In this paper, we introduce a system that analyzes a partial 3D-view of an office, in particular with a view toward desktops and the configurations of objects that are found. These 3D-images of real world scenarios are taken by a laser scanner. The images are processed by a visual component that segments and classifies the occurring objects (object types). Based on this analysis, the user is invited to correct those object classifications, so that it is possible to ask about the objects with regards to their spatial relations afterwards. The language used in the dialog and for object representation is English. This linguistic component is mainly divided into two phases, which we call the *training phase* and the *action phase*. During the training phase, the system enlists the assistance of the user for determining the types of those objects that could not be classified unambiguously by the visual component. The resulting object types from this dialog are transferred into a domain ontology, which offers a knowledge structure and additional information about the domain. During the action phase the user can ask about objects by referencing their spatial relations. Questions about objects that have a certain spatial relation to other objects and questions about object types are currently supported.

This capability enables our system to provide a basis for applications in diverse contexts. For example, it can be extended for assigning a robot instructions about activities it has to perform, while referring to objects via their spatial positions and relations. Another possibility is to generate a scene description of the 3D-view, possibly for blind people. Our main focus at present, however, is to develop this system for examining and improving the following three mutually supportive aspects:

- **Cognitive Vision.** The field of computer vision is very domain specific in general. The area of application ranges from medical to technical inspection tasks. The objective of cognitive vision is to expand the vision system's abilities by employing cognitive properties, which are seen as being focused on these individual domains. For the perception system developed, this means that the system is able to deal with uncertainties and can learn. It may offer more than one possible solution and this can be interpreted either within the subsequent user dialog or by ontology-based deduction.
- **Linguistic User Interaction.** A human-computer dialog can improve the performance of a vision module by involving the user in the object classification and by letting the user ultimately determine an object's type. If there are objects that could not be classified after 3D-object segmentation and perception, the user can specify the type. Moreover, the dialog component also provides the possibility to ask questions about objects in the scene with respect to their spatial relations. In this case, the system has to clarify which physical spatial relations and objects, measured by the vision system, agree with the spatial relations and objects linguistically expressed by the user.
- **Domain Ontology.** In order to combine a natural language component with an image recognition system, we need to consider the formal definition of the recognized objects that are to be referenced linguistically. In addition to those objects, the spatial relations that hold between them also need to be represented in a formal structure. This structure should not only provide a distinct representation of the concepts that are used by the user but also

support reasoning strategies in order to deal with user requests and to improve the object classification. We will introduce a domain ontology guided by an ontological library for linguistic and cognitive engineering, which seems to be particularly appropriate to our problems while also offering a solid basis for further development.

There are few approaches that try to combine all of these issues. For example, in [16] a system is introduced that generates linguistic descriptions of images with references to spatial relations. Their emphasis is on relating linguistic expressions of spatial relations to the physical alignment of objects using fuzzy sets. In contrast to this work, we also want to support a reverse dialog, in which the user can use expressions of spatial relations and the system has to detect them. In addition, the use of a formal specification of concepts by means of a domain ontology seems to be more promising, as it directly supports reasoning strategies and provides a fundamental structure for linguistically expressed spatial relations as well as for the representation of spatial reference systems. Furthermore, no aspects of cognitive vision in their system are considered.

In [20] a cognitive vision system is presented that generates natural language descriptions from traffic scenes. This framework uses also a fuzzy logic formalism to present schematic and conceptual knowledge, and it especially aims to explore the destination of objects moving in the scene. Apart from the different domain that is used in this approach compared to ours, the system in [20] scans the input images to detect vehicles and derive further information about the scene, for example the position of lanes. Instead of searching for specific objects in the scene, our system segments all objects independently of their types. Therefore, we have to deal with multiple object types and even with objects that are not classifiable, which is also a reason why we access a domain ontology. But generally, the system described in [20] is more concerned with the tracking of objects and their motion than with their classification and their relative spatial positions.

The system *DESCRIBER*[1] is also a system that combines spoken language and object recognition and is able to generate phrases about objects in a scene. For this purpose, the system is trained by a data record of synthetic scenes paired with spoken descriptions about the objects in the scene, their colors, size, shape, and relative spatial relations. In contrast to [16] and [20], this system is also combined with a robotic system that supports the detection of objects corresponding to the description in novel spoken phrases [22]. The vision component of the robot uses two color video cameras for the recognition of scenes. Although the application of this system is similar to our system, the realization is very different: We are using 3D images taken by a laser, and do not need to calculate 3D information of a scene on the basis of two 2D images. Moreover, instead of recognizing objects on the basis of their color, we have developed a vision component which perceives objects not only by physically measured values but also by using additional background knowledge and techniques of cognitive vision. And instead of learning spatial relations via respective data records, our

---

[1] http://www.media.mit.edu/cogmac/projects/describer.html

system uses general spatial calculi for generating linguistic expressions of spatial relations.

2 Cognitive Computer Vision

The field of cognitive vision has the intention of transferring ideas from cognition into computer vision. Cognitive vision can therefore be seen as a special domain within the field of computer vision. Its goals are described in [6]:

> "The term cognitive vision has been introduced in the past few years to encapsulate an attempt to achieve more robust, resilient, and adaptable computer vision systems by endowing them with a cognitive faculty: the ability to learn, adapt, weigh alternative solutions, and even the ability to develop new strategies for analysis and interpretation."

The realization of these goals can be achieved, e.g., by a virtual commentator which transforms the visual information acquired into a textual scene description [21]. To obtain such a commentator many of the claims of a cognitive vision system have to be realized. As we want to improve our object recognition system by the above cited approach, the system is motivated by it and complies with its respective claims.

Moreover, we see textual scene description as a contribution to the range of techniques by which the abilities of a system can be tested and evaluated. These aspects are therefore also part of this work and are integrated into our object recognition system: The system is supposed to be expandable, offer alternative solutions and make them available for formal (textual) interpretation.

2.1 The Object Recognition System ORCC

The recognition system ORCC[2] (see Fig. 1) identifies and classifies objects within an indoor environment and strives for the long term goal for a scene interpretation. The system comprises modules based on the fields of computer vision, cognitive vision and computer graphics. The central component, computer vision, involves a range of methods from density segmentation up to bounding-box calculation. The module cognitive vision contains techniques that build on the results of the computer vision part. Typical cognitive capabilities supported here are the ability of learning to classify objects preferably independent from particularly designated "special" features. Thus objects are classified according to their underlying functionality (or applicability) not by some special characteristic. A table, for example, is defined by a horizontal plane within a certain height interval. The goal therefore is not to recognize a special class of objects but categories in the sense of [21]. The results are processed in a way to use them for a textual description of the scene. The third module, the computer graphics component, is used to visualize the scene and the result of the speech interaction.

---

[2] Acronym for **O**bject **R**ecognition using **C**ognitive **C**omputing.

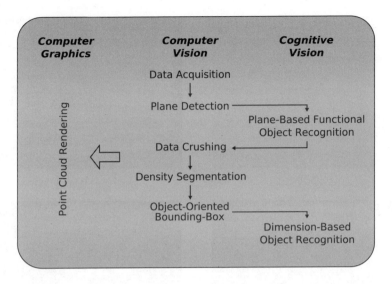

Fig. 1. ORCC System

2.2 Components of the ORCC System

As the system uses the raw data of a real scene several features are necessary, described as follows:

Sensor. The laser sensor mounted on a pan-tilt unit provides depth information from a viewpoint about 1.35 m above the ground (see Fig. 2(a)). The system has a horizontal resolution of 0.5° and the scanner is also tilted in 0.5° steps. The accuracy of the depth estimation is about 1.0 cm.

Plane Detection. For plane detection a normal vector based algorithm is used [23]: The normal vector in every data point is calculated using its corresponding surroundings. Only if the local quality of the normal vector (given by the eigenvalues within a Singular Value Decomposition step) and neighbor quality (given by the comparison with neighboring normals) falls within an appropriate range of tolerance it is used. The segments themselves are obtained via a region growing algorithm (see Fig. 2(b)).

Plane-Based Functional Object Recognition. The task of this non-final classification step is to find objects that are necessary for the segmentation of an object (see Fig. 2(c)). The object models used here hold direct attributes like *orientation* or *size* and relational attributes like *distance to* or *deviation of the orientation from the horizontal plane*. A tabletop (the plane of the tabletop), on the one hand, is located within a certain interval above the floor, has a size that ranges within a certain interval, and its orientation is horizontal. On the other hand, a wall has a certain distance (ideally zero) to the ceiling and is oriented

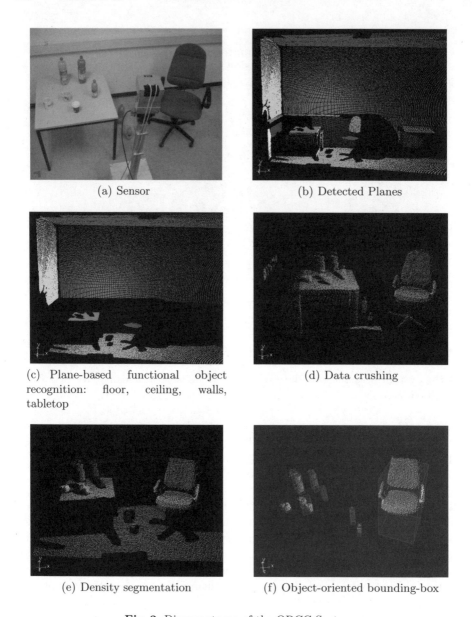

Fig. 2. Diverse stages of the ORCC System

vertically. A common approach for the modeling of such problems is to use Bayesian networks [13]. The structure of the network is given by the different features of the objects. The result for each object is a multivariate Bayesian function. The main and the variance values are determined analytically and are stored within an XML file.

Data Crushing. In the overall approach there is no search for specific objects but an estimation of segmented objects. To be able to segment the objects based on their three-dimensional point distribution, it is necessary to remove structural elements of the scene, such as walls (see Fig. 2(d)). After the removal of the space enclosing points (objects), the predominantly interesting points are obtained.

Density Segmentation. The segmentation step detaches objects from their surroundings. In some approaches, special models (e.g. cars) are determined within a scene [1]. But because we also need to be able to segment unknown objects, a point density approach is chosen. As the objects are predominantly positioned on planes like tabletops or floors, this information is used for the density segmentation (see Fig. 2(e)).

Object-Oriented Bounding-Box Calculation. The calculation of the object-oriented bounding box (see Fig. 2(f)) is a preliminary step of the dimension-based object recognition procedure. It is separated from the latter as it has to deal with tail points, which appear at edges; these are due to special characteristics of the scanning equipment [17].

Dimension-Based Object Recognition. The dimensions of an object category are described by naïve, continuous Bayesian networks. Every object class corresponds to a single Bayesian Network that is parameterized by its mean and variance values. These values are determined analytically using the values from a training set. Given the dimensions of an object resulting from the step before, the grade of membership to each class is determined. The object is classified as belonging to the class with the highest probability value, although competing alternatives are also given [25].

3 Representation of Objects and Spatial Relations

For our purposes we need to represent the scene in terms of its occurring objects and their relations, which requires explicit taxonomical and partonomical structures. This representation is also affected, however, by the linguistic interaction intended between the user and the system. Hence, the structure has to take into account linguistic expressions that possibly correspond with the scene together with their meanings. To cover all these criteria, we decided to make use of an ontological structure that directly supports our requirements.

Ontologies are nowadays widely used in diverse applications with different granularities and with varying formal considerations. They are used to clarify the concepts being modeled and the relations holding between them. They also provide a reasoning mechanism by the ontological structure itself. However, there is still no standard conceptual modeling guideline, although there are series of principles by which ontologies can be modeled and classified [8]. Ontologies have also raised attention since the beginning of the Semantic Web development

[3], which is intended to provide a navigable structure aligned by its semantic concepts. It is also a main focus to ensure re-usability and sharing of this ontologically defined knowledge.

In the past, several research efforts have been made to represent space and spatial relations using insights from ontological development (see [2] for an exhaustive overview). The work reported here applies one of these efforts in order to examine its practical usability and the necessity for further extensions. For computability reasons, we use description logic for the development of our domain ontology.

3.1 Domain Ontology Modeling

We have created a domain ontology that consists not only of objects that can be recognized by the ORCC system or that are known to our image recognition tools, but also of general objects that can be present in an office scene and that we have already explored in earlier office domain ontologies [12]. Its structure is guided by an ontology framework with an emphasis on linguistic and cognitive aspects, namely DOLCE [15].

Originally developed as a part of the WonderWeb project[3], DOLCE was designed as a foundational ontology. As it is strongly influenced by natural language and human commonsense, it seems a promising basis for our current work. Its main concepts are divided into four categories, to wit *Perdurants*, *Endurants*, *Qualities*, and *Abstracts*. Perdurants describe entities that unfold in time (e.g., an event of some kind), whereas Endurants describe entities that are wholly present at each point in time (e.g., a cup, a glass, a desk, etc.). Qualities inhere in entities, so every entity can have certain qualities, even Qualities themselves. The value of these qualities are expressed by entities of the *Abstract* concept, called *quale*. A quale describes a position of an individual quality in a *quality space*, which corresponds to *conceptual spaces* described in [9].

Objects in ordinary office scenes are types of the concept Endurant, more precisely of its subconcept *Physical Object*. Such concepts are defined by having direct spatial qualities and as being wholly present in time. The object types we have modeled in our domain ontology are hierarchically structured by means of their functions, as emphasized in [5]. This functional bias is of particular importance: In the long term, we are planning to connect object recognition with possible robot actions that depend on 'functional roles' an object can have. For instance, additional information about a mug, as some kind of a drinking vessel, is that it can be filled with liquid, an instance of *Amount Of Matter*. Therefore mugs, cups, coffeepots, bottles are subsumed under drinking vessel; staplers, hole punchers, scissors are subsumed under office-supply, etc. This information can then be derived and used for performing possible actions, such as drinking. In addition, a human-computer interaction can linguistically refer to such actions.

---

[3] http://wonderweb.semanticweb.org

In our domain ontology, we provide a fundamental data set for object types that are common in an office. If the system detects an object type that cannot be classified and the user suggests a type that is not known to the ontology, the system will handle this by classifying this concept simply as a physical object. If the user refers to this object later by using a specific type that is already known from the domain ontology, the system will refine the respective object type and for this scene the user can refer to this object with both types.

3.2 Spatial Relations

Spatial relations between objects in a scene are used for the linguistic user dialog during the *action phase* (see section 4), in which the user may use relations like "Show me the object that is behind the book" and the system has to infer what "behind something" means and which book the user is referring to, regarding the current object configuration of the scene.

To describe spatial relations in linguistic terms, we have to consider the different linguistic frames of reference that lead to these terms and that are employed by the user. An essential result from linguistic and psychological research concerning linguistic expressions of spatial relations is the general difference between three main frames of reference, named *instrinsic*, *relative*, and *absolute* [14].

In the first case, spatial relations between two objects are described by the position of one object (the *referent*) relative to an intrinsic orientation of another object (the *relatum*). In relative reference systems, the relative position of a referent to its relatum is described from the viewpoint of a third position (the *origin*). Finally, if the description of spatial relations between two objects is made with respect to some externally fixed direction, we speak of an absolute reference system. For a detailed discussion of reference systems and their literature, as well as extensive examples of linguistic terms involving spatial relations, see [24].

With the current status of our system, we consider only projective linguistic expressions of spatial relations corresponding to a relative reference system. As absolute reference systems do not seem to be employed for indoor office environments in our western culture [19], this restriction is well motivated and therefore we do not take into account absolute relations at present. The integration of intrinsic references is also left to future work, although this is important because intrinsic references are very common in natural language [11].

As indicated above, a spatial relation comprises physical objects each being origin, relatum, or referent. In DOLCE, a Physical Object has an individual spatial (and temporal) quality that defines the physical position of the object in space (and time) [15]. The quale of this spatial quality is a spatial region in geometric space. Although the properties of the geometric space adopted is not further defined in DOLCE, they are intended to represent the absolute value of the spatial position of one physical object. They are not supposed to express spatial relations that belong to linguistic terms and describe assignments of abstract concepts to certain physical conditions between physical objects. Hence, spatial

116   J. Hois et al.

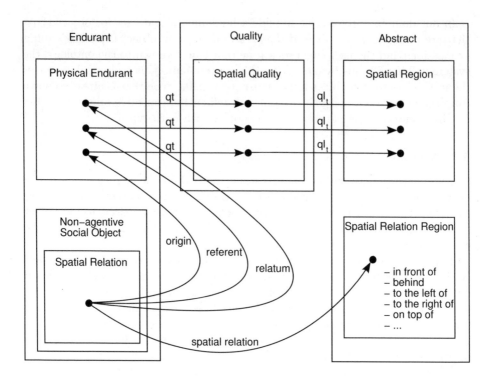

Fig. 3. Spatial relations

relations addressed in this paper are not represented in the existing ontological framework yet.[4]

In order to deal with this central area of spatial conceptualization, we are exploring a novel definition of spatial relations as instances of the DOLCE category *Non-agentive Social Object* that relate the objects to their relation. This is because Non-agentive Social Objects in DOLCE are defined as entities that are generically dependent on a community of agents. In this sense linguistic terms of spatial relations are in this sense socially constructed and maintained. We call this entity *Spatial Relation*. The values, it can take, are subsumed under the abstract concept *Spatial Relation Region*. Instances of Spatial Relation take as arguments three Physical Endurants distinguished by being origin, relatum, or referent, and one Spatial Relation Region, the value of the spatial relation. This is illustrated in Fig. 3. Even though we are currently only considering relations in terms of a relative reference system, this representation also covers intrinsic and absolute reference systems. In this case origin and relatum are equal.

---

[4] Note: In DOLCE, there are also *places*, e. g. "the underneath of a table", defined as a *Feature* of objects. But this should not be confused with spatial relations, like "something *is* underneath a table", because such places are regarded as "parasitic entities" that constantly depend on their *hosts*; they do not directly act as physical spaces for other objects.

In our application, a distinguished viewpoint of the user (the origin), on which the spatial relations of the objects depend, has to be declared. As the user's viewpoint is equal to the position of the laser with respect to the scene, we can use an initial instantiation for the laser in our domain ontology as origin. Although it is possible for the user to have a position different from the laser, we are currently constrained to the laser's viewpoint as this is where the scene is scanned from and displayed to the user. This is also done with respect to the scope of our application: A robot getting instructions from users that share the same viewpoint toward the scene from the position of the robot, when users are not in the same room as the robot or when they are located at the same position as the robot (for example, in case of a wheelchair).

Starting from this origin, we calculate the relative position of objects and integrate them into the domain ontology. The calculation is based on the values measured by the ORCC system, that is the orientation and the distance according to the laser. The assignment of the linguistic terms of spatial relations is oriented by *heterogeneous non-overlapping acceptance areas*, introduced in [10]. The system assigns appropriate linguistic spatial terms by calculating the angles between two objects (referent and relatum) with respect to the laser sensor (origin). Term assignments in our application conform with the distribution of angles and their linguistic projective terms (left, right, in front of, behind) that is described in [18], regarding only relative references. This is taken at present for simplicity, as it will be for further investigations to show how we can best integrate the representation of diverse spatial calculi in DOLCE and how this can be related to linguistic expressions of spatial relations.[5]

Generally, our domain ontology has the following major advantages that contribute to the system: 1. It represents all relevant information in our system about object representation and spatial relations shared by the dialog and the classification component. 2. Spatial relations that are expressed in natural language can be directly accessed due to the ontological representation of spatial relations. 3. As it is developed on the basis of a fundamental ontology, it shall easily be combined with ontologies guided by the same fundamental basis. 4. This structure also eases extending spatial relations for other reference systems.

Although the classification of objects is not directly influenced through this domain ontology, the quality of the dialog can be improved, as additional information about possible constellations of objects in an office scene also forms part of our domain ontology. This information is helpful for the user dialog, as described in the next section, but not vital. To infer new knowledge from the ontology, the system is using the tableaux reasoner Pellet.[6]

---

[5] Our approach is to extend the notion of the axiomatization of simple spatial regions involving single physical objects in terms of geometric spaces. Our abstract spatial relation quality regions, for example, correspond more to axiomatizations of spatial calculi involving orientation, such as Freksa's Doublecross calculus [7].

[6] http://www.mindswap.org/2003/pellet

4 Linguistic Dialog

With the current status of the system's dialog, we are offering rather basic user interactions, as our aim is not yet to develop sophisticated natural language dialogs that provides multiple variants for similar meanings. We will demonstrate the representation of, and referencing to, physical objects and their spatial relations represented in the domain ontology on the basis of the dialogs instead. In particular, our linguistic component is part of the realization of the cognitive vision claims, mentioned above. An improvement of this component is considered below (see section 6).

Generally we divide our human-computer interaction with the system into two phases: In the first phase (the *training phase*), after the scene has just been scanned, the system generates the ontological instances of the objects as far as they are identified by the image recognition system. If an object was not identified, or was identified as several types, the system tries to identify the object by consulting the user. Thus, objects that were classified unambiguously by the ORCC system are directly instantiated into the domain ontology together with their spatial relations to surrounding objects. Objects with ambiguous classifications have to be determined by the user during the training phase.

To be as non-intrusive as possible, the system analyzes the ambiguously classified objects to see whether they are most likely to be of a specific type by ontological reasoning. For some concepts, we can assume that objects of the same kind happen to be close to each other, like books, bottles, or mugs. However, it is crucial to emphasize that we do not restrict the concepts in this way. A bottle does not necessarily need to be nearby other bottles. We just note that they can be usually grouped together. In case of an unclassified object that is surrounded by three objects of one type, the system asks the user if the unclassified object (which is marked in the image during this dialog) is also of that type (or, if not, what else). This technique does not bother the user with many different and unlikely suggestions, but only the most probable one. Hence, the domain ontology already facilitates a simpler, more controlled style of interaction and it improves the convenience of the system. A similar assumption is that certain kinds of objects are also often close to each other, like a mouse is usually near to a computer, a keyboard, or a laptop, and not near other mice, which gives us a similar reasoning strategy as just described. Such modeled background knowledge is usually related to the functionality of objects. The probability value of an object type that is calculated by the ORCC system is also a hint for prioritizing the list of possible types. But we do not omit types that have very low priority values as these may still prove to be relevant later. In either case, users are always free to define object types as they like.

In addition to each request of the system about possible object types, the referring object is marked in the image. Hence, the user can determine the respective type directly looking at the image. The system will consult the user for clarification only in case an object was classified ambiguously. After the type has been determined, an instance of the object is integrated into the domain ontology, including spatial relations to surrounding objects, as described in the previous section. In Fig. 5 an example of a training phase dialog is shown.

In the second phase (the *action phase*), the user can either request specific types of objects by referencing spatial relations to other objects, or can request one or several objects, that have a certain relation to a specifically described object. Although the latter request is just a "show me"-operation, this can be easily extended to other, more application-relevant instructions, such as "give me", "take", or "do something with".

Our system can handle only a few linguistic queries at present, such as "Show me the object that is to the right of the mug!" or "What is the object behind the mug?". After syntactically processing the user's request, the ontological knowledge about relative positions between different objects takes effect. Depending on the spatial relations that the user expressed in the request, the system tries to derive the correct event. Currently linguistic expressions, such as "to the right/left of", "on (top of)", "in front of", and "behind", are supported. If the user, for example, asks for "all objects" that have a certain position relative to a specific object (the relatum), all possibilities are looked up transitively in the domain ontology. In case the request is about a specific object with an indication of its relative position, the ontological instantiation of this type and its spatial relations are explored. An example of an action phase dialog is shown in Fig. 6.

As we have already mentioned, this human-user interaction should be seen primarily as a first approach toward one possible application of the combination between image recognition, linguistic references and ontological modeling of spatial relations. There are certainly other fields of application which include functional aspects of objects, like assigning a task to a robot to perform certain actions depending on the objects' functionalities, or different interaction scenarios, such as generating a description of the scene. This will be investigated in future developments. The results of our current experiment serve mainly as a basis for investigating the representation of spatial relations, formally as well as linguistically, which are evaluated in the next section.

5 First Results

To show first results of our system's processing sequence, an example scene is analyzed in this section.

Fig. 4 shows a 3D scene (4(a)) in which a part of an office with its typical objects is illustrated[7], along with the corresponding 2D pictures (4(b) and 4(c)). The position of the laser indicates the viewpoint toward the scene. Initially the system has detected six objects unambiguously and correctly: The floor (ID: 721), a table (ID: 725), two bottles (ID: 741, 742), a book (ID: 747), and a chair (ID: 752). Hence, the respective objects and their corresponding spatial relations are integrated into the domain ontology relating to this scene. Six remaining objects have been detected by the ORCC system, which finally have to be determined by the user. The resulting dialog is shown in Fig. 5.

---

[7] An ID of each object is inserted for a comprehensible reference in the text.

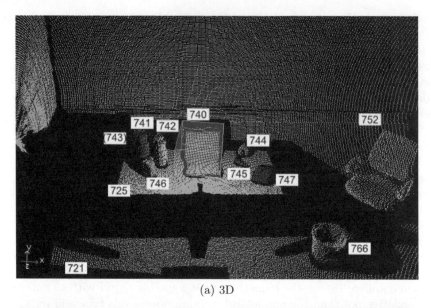

(a) 3D

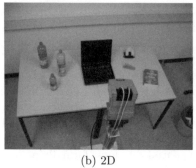

(b) 2D

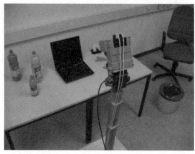

(c) 2D

Fig. 4. Example of a scanned scene (3D and 2D)

5.1 Example of a Training Phase Dialog

In this dialog, the system asks the user about the remaining objects. It suggests the different types that the visual component has classified for each object. The results of this classification are shown in Table 1. Additionally to each question the respective object is marked in the scene. Beginning with the first question ("Is the marked object (id:740) a/an laptop, or briefcase, or something else?", the marking of the object is shown in Fig. 4(a)), the object with ID 740 is ambiguously classified as a laptop, a bin, or a briefcase (see Table 1). One of the additional kinds of information about possible spatial constellations of objects, mentioned above, is that laptops are usually on a table and bins on a floor. This relation is represented in our domain ontology as an optional attribute: Objects of type laptop are related to an object of type table by the spatial relation *on*

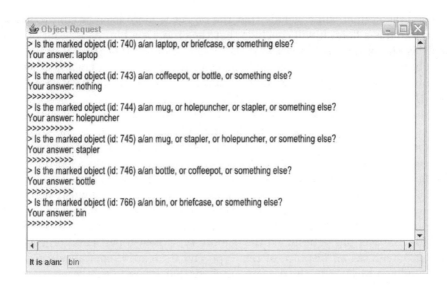

Fig. 5. Dialog during training phase

Table 1. Result of the ORCC analysis of the suggested object types for each object (the bold types are the correct object types)

ID	Classification & Probability		
721	floor (0.99)		
725	tabletop (0.96)		
740	laptop (0.58)	bin (0.15)	briefcase (<0.01)
741	bottle (<0.01)		
742	bottle (0.06)		
743	coffeepot (0.61)	bottle (<0.01)	*Segmentation failure*
744	mug (0.71)	holepuncher (0.50)	stapler (0.09)
745	mug (0.33)	stapler (0.16)	holepuncher (0.09)
746	bottle (0.23)	coffeepot (0.02)	
747	book (0.61)		
752	chair (0.15)	bin (<0.01)	stapler (<0.01)
766	laptop (0.25)	bin (0.18)	briefcase (<0.01)

(top of) while objects of type bin are related to an object of type floor. As we mentioned above, these are not strict conditions for being a laptop or a bin, but optional ones, and they are considered during the training phase.

For this reason, the system seeks for and detects the Spatial Relation with ID 740 as referent that has the relation *on (top of)* and analyzes the object type of the corresponding relatum. Such "on (top of)" relations are based on the results from the ORCC system, which calculates the vertical alignment of objects on the basis of density segmentation (see Section 2.1). If the relatum is a table, the system will conclude that the object is more likely a laptop and if the relatum is a floor, it will conclude that the object is a bin. As a result of this analysis, the system omits to suggest the less likely object type. In case of ID 740, the object is located on a table and therefore the system does not ask whether the object might be a bin. The inverse case is shown in the question about the bin (ID: 766).

This example demonstrates that additional information about typical office scenes, which is represented in the domain ontology, can help to simplify the questions. Although the user finally has to determine the type of the object, the system avoids suggesting unlikely object types. This behavior does not only aim for dialog simplification but also for a more sophisticated user interaction, as the user is not irritated by implausible suggestions. This directly binds, therefore, ontological domain information, the formal scene representation, and the visual information in order to improve the overall capabilities of the system as a whole.

After the user has determined the definite type of each particular object, the system integrates an instance of this type and its relative spatial positions to surrounding objects into the domain ontology that refers to this scene.

One of the six remaining, unclassified objects (ID 743) does not actually occur in the original scene (cf. 4(b)). Sometimes the visual component segments further, not existing objects because "shadows" of other objects are detected by the system as being objects themselves. In this case, the system applies its general object classification (described in section 2.1) to the object. The object with ID 743 is classified as a coffeepot or bottle. As it is surrounded by two definitely identified bottles, no further ontological information affects the question about this object ("Is the marked object (id: 743) a/an coffeepot, or bottle, or something else?"). As the user is aware that this object does not exist, the answer is "nothing", which means that the system does not instantiate an object at all (we can see this in the user request below). This demonstrates that the integration of the user via a dialog can support clarifying the identification of a scene in difficult cases.

After all, depending on the answers of the user, the system has instantiated all objects as well as their spatial relations in the domain ontology referring to this scene and can make use of this in dialogs during the action phase, which is described in the following section.

5.2 Example of an Action Phase Dialog

An example of a dialog during the action phase is shown in Fig. 6. The user can ask questions about the objects in the scene by using spatial relations. For instance, the request "what are the objects to the right of the stapler?" lists the corresponding objects with their relative positions. In the system's answer the book on the table (compare 4(b)) is mentioned.

The answer to the question about the objects behind the stapler lists the hole puncher, which also matches with the image in Fig. 4(a). Depending on the frame of reference and the acceptance areas that we introduced in section 3, the system specifies that the bin is the only object in front of the chair

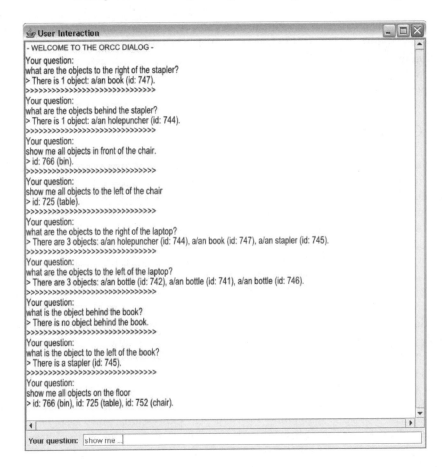

Fig. 6. Dialog during action phase

and the table is to the left of the chair from the user's viewpoint, which also seems intuitive in combination with Fig. 4(a). In the case that the system has to list all objects to the left of the laptop, it only refers to the objects that were instantiated during the training phase. Thus, it lists three bottles that actually exist in the image (Fig. 4(b)) without the additional object (Fig. 4(a), ID 743) that were segmented by the visual component. Again accordings to the 2D image in Fig. 4(b), the system lists the hole puncher, the book, and the stapler as the objects that are to the right of the laptop.

Another result from the use of acceptance areas and the relative frame of reference is that there are no objects behind the book and the stapler is the object to the left of the book according to the system. It is important in this case to remember the viewpoint of the user, from which the relative positions are calculated. It may also be usual for the user to refer to the stapler or the hole puncher as being behind the book in respect of their alignments on the table.

The planned integration of additional frames of reference, such as the intrinsic reference required here, will incorporate such possibilities.

Finally the question about all objects that are on the floor gets the reply from the system containing the bin, the table, and the chair, which again matches correctly with Fig. 4(b).

6 Conclusion and Future Work

In this paper, we have argued that visual systems can be supported by linguistic user dialogs as well as a domain ontology. We have presented a cognitive visual system that analyzes a real world 3D image. A first approach toward the representation of spatial relations between physical objects in an ontological structure was introduced and we have discussed the resulting human-user interaction dialogs.

Further research aspects include the incorporation of representations of spatial calculi, for instance, RCC-8 [4] or Doublecross [7]. Also the computation of responding to user requests during the action phase that ask only for *one* object has to be refined as it currently only considers distance and relation but not proximity to spatial relation axes. The differences in the linguistic expressions of the resulting spatial relations will be examined and evaluated empirically for their suitability for differing tasks. The representation of spatial relations in ontologies still remains a crucial concern. Further considerations have to be made with respect to current efforts in ontological engineering. The interrelation between representing spatial relations depending on spatial calculi in ontologies and spatial relations depending on linguistic expressions must also be made clearer. To this end, we believe that the combination described here of functionally interpreted perceptual visual input with a formal domain ontology offers an excellent foundation for subsequent research on the perception, grounding, and use of spatial descriptions in interactions between artificial agents and humans.

One of our next foci is on including further linguistic frames of reference. Linguistic terms according to intrinsic references seem to be very promising and will be integrated into the system, especially in connection with a detection of objects' intrinsic orientation within the ORCC system. We will also consider *internal relations*, such as "the object in the back corner of the room" — in contrast to *external relations* described in intrinsic, relative, and absolute reference systems. Such extensions are also essential to clarify user request ambiguities. The possibility of different viewpoints from the laser will also be important. We plan to extend this in the long term to multi-robot scenarios.

An extensive evaluation of our system is also to be done. This depends mostly on the configuration of larger data records that we have to prepare. Moreover, we are planning to carry out experiments with our system and humans to investigate the system's performance and quality. The possibilities of the user interaction will also be extended in order to support this evaluation as required.

Acknowledgements

The Collaborative Research Center for Spatial Cognition (Sonderforschungsbereich Transregio SFB/TR8) of the Universität Bremen and the Universität Freiburg is funded by the Deutsche Forschungsgemeinschaft (DFG), whose support we gratefully acknowledge.

This work has been especially supported by the SFB/TR8 subprojects A2-[ThreeDSpace] and I1-[OntoSpace].

References

1. Anguelov, D., Taskar, B., Chatalbashev, V., Koller, D., Gupta, D., Heitz, G., Andrew Y. N.: Discriminative learning of markov random fields for segmentation of 3D range data. In: IEEE International Conference on Computer Vision and Pattern Recognition (CVPR), San Diego, California (June 2005)
2. Bateman, J.A., Farrar, S.: Spatial Ontology Baseline. SFB/TR8 internal report I1-[OntoSpace] D2, Collaborative Research Center for Spatial Cognition, University of Bremen, University of Freiburg, Germany (2004)
3. Berners-Lee, T., Hendler, J., Lassila, O.: The Semantic Web. Scientific American 284(5), 34–43 (2001)
4. Anthony, G., Cohn, A.G., Bennett, B., Gooday, J., Gotts, N.M.: Qualitative spatial representation and reasoning with the region connection calculus. GeoInformatica 1(3), 275–316 (1997)
5. Coventry, K.R., Carmichael, R., Garrod, S.C.: Spatial prepositions, object-specific function and task requirements. Journal of Semantics 11, 289–309 (1994)
6. Auer, P., et al.: A Research Roadmap of Cognitive Vision, ECVision: European Network for Research in Cognitive Vision Systems (19.04.2005), http://www.eucognition.org/ecvision/research_planning/ECVisionRoadmapv5.0.pdf
7. Freksa, C.: Using orientation information for qualitative spatial reasoning. In: Frank, A.U., Campari, I., Formentini, U. (eds.) Spatio-Temporal Reasoning, pp. 162–178 (1992)
8. Gómez-Pérez, A., Fernández-López, M., Corcho, C.: Ontological Engineering with examples from the areas of Knowledge Management, e-Commerce and the Semantic Web. Springer, Heidelberg (2004)
9. Gärdenfors, P.: Conceptual Spaces: The Geometry of Thought. A Bradford Book. MIT Press, Cambridge (2000)
10. Hernández, D.: Qualitative Representation of Spatial Knowledge. In: Hernández, D. (ed.) Qualitative Representation of Spatial Knowledge. LNCS, vol. 804, Springer, Heidelberg (1994)
11. Herskovits, A.: Language and Spatial Cognition: an interdisciplinary study of the prepositions in English. Studies in Natural Language Processing (1986)
12. Hois, J., Schill, K., Bateman, J.A.: Integrating Uncertain Knowledge in a Domain Ontology for Room Concept Classifications. In: Bramer, M., Coenen, F., Tuson, A. (eds.) The Twenty-sixth SGAI International Conference on Innovative Techniques and Applications of Artificial Intelligence. Research and Development in Intelligent Systems, Springer, Heidelberg (2006)
13. Krebs, B., Burkhardt, M., Wahl, F.M.: Integration of Multiple Feature Detection by a Bayesian Net for 3D Object Recognition. Mustererkennung, 143–150 (1998)

14. Levinson, S.C.: Space in Language and Cognition. Cambridge University Press, Cambridge (2003)
15. Masolo, C., Borgo, S., Gangemi, A., Guarino, N., Oltramari, A.: Ontologies library (final). WonderWeb Deliverable D18, ISTC-CNR, Padova, Italy (December 2003)
16. Matsakis, P., Keller, J., Wendling, L., Marjamaa, J., Sjahputera, O.: Linguistic description of relative positions in images. IEEE Transactions on Systems, Man and Cybernetics, Part B 4(32), 573–588 (2001)
17. Meyer, A.: Merkmals- und formbasierte 3D-Objekterkennung für Büroszenen. Diplomarbeit, Universität Bremen (2005)
18. Moratz, R., Tenbrink, T.: Spatial reference in linguistic human-robot interaction: Iterative, empirically supported development of a model of projective relations. Spatial Cognition and Computation 6(1), 63–106 (2006)
19. Moratz, R., Tenbrink, T., Bateman, J.A., Fischer, K.: Spatial knowledge representation for human-robot interaction. In: Freksa, C., Brauer, W., Habel, C., Wender, K.F. (eds.) Spatial Cognition III. LNCS (LNAI), vol. 2685, pp. 263–286. Springer, Heidelberg (2003)
20. Nagel, H.-H.: Steps toward a Cognitive Vision System. AI Magazine 25(2), 31–50 (2004)
21. Nagel, H.-H.: Cognitive Vision Systems (CogViSys) (31.08.2001), http://cogvisys.iaks uni-karlsruhe.de/homepage_CogViSys_V3B.html
22. Roy, D., Gorniak, P., Mukherjee, N., Juster, J.: A Trainable Spoken Language Understanding System For Visual Object Selection. In: International Conference of Spoken Language Processing (2002)
23. Stamos, I., Allen, P.K.: 3-D Model Construction using Range and Image Data. In: Proceedings of the IEEE Conference on Computer Vision and Pattern Recognition CVPR 2000, pp. 531–536. IEEE, Los Alamitos (2000)
24. Tenbrink, T.: Semantics and Application of Spatial Dimensional Terms in English and German. SFB/TR8 internal report I1-[OntoSpace], Collaborative Research Center for Spatial Cognition, University of Bremen, University of Freiburg, Germany (2005)
25. Wünstel, M., Röfer, T.: A Probabilistic Approach for Object Recognition in a Real 3-D Office Environment. In: WSCG'2006 Posters Proceedings (2006)

Protein Structure Prediction with Visuospatial Analogy

Jim Davies, Janice Glasgow, and Tony Kuo

School of Computing, Queen's University,
Kingston, ON, K7L 3N6, Canada
jim@jimdavies.org, janice@cs.queensu.ca, kuo@cs.queensu.ca

Abstract. We show that visuospatial representations and reasoning techniques can be used as a similarity metric for analogical protein structure prediction. Our system retrieves pairs of α-helices based on contact map similarity, then transfers and adapts the structure information to an unknown helix pair, showing that similar protein contact maps predict similar 3D protein structure. The success of this method provides support for the notion that changing representations can enable similarity metrics in analogy.

1 Introduction

It is well known that the right representation greatly facilitates reasoning [2] and there is a growing recognition of the need for intelligent architectures to accommodate a diversity of representations [30].

The guiding theory of our research is that changing representations allows reasoners to see similarities in one representation type that might be difficult to detect in another. For example, functional representations of a human face and the front of a car may have very little semantic overlap. In this research we focus on visuospatial representations. In our example, representing the headlights and eyes as circles, and the grill and mouth as a centrally-located hole allows connections to be drawn between these components.

As people often have visuospatial experiences when solving problems [6,11,31], an important step in establishing our theory described above is to computationally show that visuospatial representations can be used to solve a variety of problems. In this paper we provide support for this notion in the domain of protein structure prediction, an example of a complex problem-solving domain. We will describe the problem, and then how our system, *Triptych*, uses visuospatial reasoning on image representations to solve it.

1.1 Protein Structure Prediction

A primary goal of molecular biology is to understand the biological processes of macromolecules in terms of their physical properties and chemical structure. Since knowing the structure of macromolecules is crucial to understanding their functions, and all life crucially depends on protein function [24], an important part of molecular biology is understanding the three-dimensional (3D) structure of proteins.

Proteins are composed of one or more chains of amino acid residues(so called because part of the amino acid is released when it forms a chain). The description of which residues appear and in what order is the protein's "primary structure". According to the laws of chemistry, the chains twist, fold, and bond at different points, forming a complex 3D shape. Subchains form regular "secondary structures", the two main types being α-helices and β-strands. The three-dimensional structure of a chain is its "tertiary structure", and the overall protein shape (which may involve several chains) is known as its "quaternary structure". A major unsolved problem for the biological sciences is to be able to reliably predict the quaternary structure from the primary. This, at the highest level, is our problem domain.

Approaches to protein structure prediction vary from those that apply physical principles to those that consider known amino acid sequences and previously determined protein structures. Many of the latter use what is known as "homology" as a similarity metric. In this context homology is the similarity of two amino acid sequences. Our work also falls in the latter category, but rather than using primary structure directly, we compare contact maps.

Contact maps. A *distance map*, D, for a protein with n amino acid residues is an $n \times n$, symmetric array where entry $D(a_i, a_j)$ is the distance between residue a_i and residue a_j, generally calculated at the coordinates of the C_α (carbon-alpha) atoms for the residues. The order of the amino acids on the axes is the same as their order on the chain. Given a distance map D, we compute a *contact map* C for the protein as a symmetric, $n \times n$ array such that:

$$C(a_i, a_j) = \begin{cases} 1, & \text{if } D(a_i, a_j) < t; \\ 0, & \text{otherwise.} \end{cases}$$

where t is a given threshold value (in our work this theshold is 10Å). There exists a contact between residues a_i and a_j if and only if they are within a given distance t of one another in the protein structure. Figure 1 illustrates image representations for a distance map and a contact map reconstructed from the Protein Data Bank (PDB) [3].

Researchers have considered various approaches for the process of predicting contact maps for a protein from its primary sequence and structural features; these are primarily based on neural network-based methods [12]. While results from this work are encouraging, it still results in maps that contain a large degree of noise. Thus we carry out our initial experiments on idealized maps generated from the PDB. Future work will include prediction of structure from predicted contact maps.

A contact map is a translational and rotational invariant, visuospatial representation that captures some of the protein's relevant structural information. Our general hypothesis is that visual processing on contact maps enables effective retrieval of similar structures, even if sequence homology is ignored. Contact

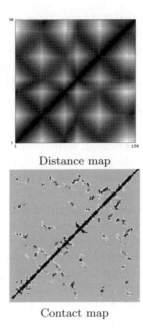

Distance map

Contact map

Fig. 1. Distance map and contact map for the protein Bacterioferritin (Cytochrome B1). The axes represent the residues of the protein starting from the N terminus (bottom left corner). In the distance map, darker colors correspond to closer distances. For the contact map, black areas correspond to values of 1, where residues are in contact (within 10Å of one another). Secondary structures are easily recognizable in a contact map: α-helices appear as thick bands along the main diagonal; β-sheets appear as thin bands parallel and perpendicular to the main diagonal.

maps provide a "fingerprint" that can be used to efficiently compare proteins to find ones with similar substructures. We will refine this hypothesis when we describe our implementation.

2 Analogy Applied to Protein Structure Prediction

Rather than working with whole proteins, we are working with pairs of α-helices. At the highest level, each time Triptych runs it takes as input: 1) the contact map for the unknown (target) helix pair, and 2) a memory of known helix pair structures and contact maps. The final output consists of a location in space (x, y, z coordinates) of each amino acid residue in the target helix pair.

Analogical problem solving is founded on the premise that similar problems have similar solutions. Experiences are retrieved, mapped, and reused during problem solving. Aaronson et al. [1] suggest that analogical reasoning is particularly applicable to the biological domain, partly because biological systems are often homologous (rooted in evolution). As well, biologists often use analogy, where experiments are designed and performed based on the similarity between features of a new system and those of known systems. Analogical and case-based reasoning has previously been applied to a number of problems in molecular biology; an overview of these systems can be found in [27].

Our system retrieves and adapts protein data from the PDB in order to construct potential 3D structural models for our target helix pair. These models are evaluated in terms of domain knowledge and the "best" structures will ultimately be used as building blocks at the next level of model building.

We retrieve similar α-helix pair contact maps and adapt the known structures to predict alignments for the unknown structures.

To predict the alignment of helices in 3D space, we consider helix pair contact maps, C_{s_m,s_n}, corresponding to pairs of helices (s_m, s_n) which have greater than four contacts.[1] This map is the subarray of C such that the rows of C_{s_m,s_n} correspond to the amino acid residues in secondary structure s_m and the columns correspond to the residues in secondary structure s_n. These maps need only be defined for contacts along and below the diagonal of the helix pair contact map, as the map for pair (s_m, s_n) is equivalent to that for (s_n, s_m). Note, that unlike the protein contact map the contact maps for pairs of helices are not generally symmetric. The images in Figure 2 illustrate a contact maps for pairs of α-helices.

The *retrieve* task returns a list of retrieved helix pairs, ordered according to similarity. The similarity metric is a visual similarity between source and target contact maps. The *adapt* module transfers structure information from the top retrievals (called the "sources") and modifies the information according to the specifics of the target.

3   Implemented Modules: Retrieval and Adaptation

Our focus is on predicting the alignment, or relative location, in 3D space of α-helix pairs given the contacts between their residues.

Retrieval Module. For each query map C_{s_m,s_n} we retrieve helix pairs with contact maps most similar to C_{s_m,s_n}.

A similarity measure for comparing the query contact map with maps generated from structures in the PDB was derived using techniques from machine vision, where we consider the black regions to be the image within the array. We were less concerned about the dimensions of the map, than what it looked like in terms of shape and location of black regions (regions which contain contacts). For example, Figure 2 illustrates three different maps for pairs of helices, where maps (a) and (c) are considered similar to one another, and (b) is different from the other two.

First we blur the images using Gaussian smoothing [19]. This is often done to remove unwanted details and noise. Contacts are treated as black points, and points surrounding them are turned some shade of gray depending on their distance from the nearest contacts. The grayscale tone is determined by a Gaussian

---

[1] If there are fewer than five contacts between two secondary structures it is difficult to determine their orientation from their contacts.

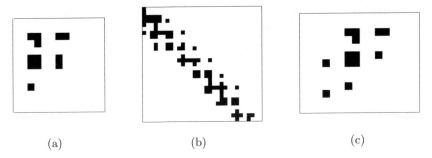

Fig. 2. Illustration of similar, (a) and (c), contact maps and a map (b) that is dissimilar to the other two. (b) shows the sub-contact $C_{Helix-6, Helix-8}$ map for a pair of helices in protein Bacterioferritin. Since the diagonal band shows contacts that extend from the beginning of one and end of another, to the end of one and beginning of another, we can discern that the helices are oriented anti-parallel to one another.

distribution where the contacts are the means. The maps are then morphed using a technique called *closing*, which removes low-valued points but keeps the rest of the image intact [19].

The retrieval of similar contact maps involves a two-tiered approach. Given a query contact map, the first tier uses three general content descriptors to cull the dataset of dissimilar contact maps: quadtrees, color and edge distributions, and gray-level co-occurrence matrices.

Quadtrees have been successfully applied to image compression, comparison, and classification. The quadtree [38] is a hierarchical data structure used to represent images. For an image, a two-dimensional region is recursively decomposed into quadrants where each quadrant is a node in the quadtree.

Color distribution [37] is a common feature used in image retrieval. Pixel color values are put into a histogram form: colors are discretized and counted and placed in bins. Global histogram representation has the drawback of loss of location, shape, and texture information. As a result images retrieved based on similar color distributions may not be semantically related.

Edge detection [42], and the features that can be extracted from it, is commonly used as a content descriptor of images. In this work we use the Canny edge detection method [5]. The Gaussian smoothing was necessary for this step to work, as it uses gradients and cannot be applied to binary images. Our measure of similarity based on edge detection involves comparing histograms showing the frequency of edges with angles of $0°$, $45°$, $90°$, and $135°$.

A statistical method that considers the spatial relationship of pixels, the gray-level co-occurrence matrix (GLCM) [21] is a texture analysis method from which various statistical features can be extracted. Each entry *(i, j)* in the GLCM corresponds to the number of occurrences of the pair of gray levels i and j which are a distance d apart in the original image. For example, if d is 1, then GLCM entry (1, 2) will contain the number 4 if there are four instances of gray value 1 adjacent to gray value 2 in the original image. In analysis, the GLCM are normalized so the histogram or features extracted can be compared.

A committee of these general content descriptors is used in the first tier of retrieval. Quadtrees vectors were generated from the binary, smoothed, and morphed contact maps. The color and edge distributions and gray level co-occurrence matrices were obtained from the smoothed contact maps. The committee results in a set of contact maps which are present in the retrievals of two or more general content descriptors. We determined empirically that 100 retrievals for each descriptor is sufficient. The results of the committee are then used in the second tier of retrieval.

For the second tier, the Jaccard's distance [25] was calculated between each contact map from the first tier and the query map. Because the maps vary in size, a sliding window approach was used to determine the best matching regions between the query and the contact maps from the first tier. The best mapping regions also provide registration of residues for evaluation using RMSD (Root-Mean-Squared Distance), the standard measure of distance for protein structures. The best 25 retrievals were then selected from the 100 as the final set of contact maps to be returned.

Adaptation Module. For each query contact map the retrieval process returns potential helix pairs from the PDB, ranked in order of estimated similarity. For each query map, the adaptation phase transfers the structure information from the highest-ranking structures to the input helix pair.

Transferring locations requires a mapping function – that is, a set of alignments that determine which residues in the target structure map to which residues in the retrieved source structure. This is achieved by first aligning the contact maps so that the mean cell location of contacting amino acid residues in the retrieved structure aligns with the mean cell location of contacting residues in the target. Then all amino acid residues in the target structure that have corresponding residues in the source structure are given the coordinate information from these residues. Since the registration may not feature a great overlap of the maps, usually there remain some target residues with no coordinates (i.e., no corresponding residue in the known structure to transfer over). Since α-helices tend to have a consistent structure, the missing coordinates are filled in using general domain knowledge. Specifically, each turn of an α-helix is estimated at 5.4 Å along the helix axis and each turn at 5 Å across. Using this information and the helix axis, calculated from the filled-in locations, our system is able to infer these unmatched residue locations. This is textbook Biochemistry knowledge. Figure 3 illustrates the portions of the helices that are determined through our mapping function and those constructed from domain knowledge (grown area).

Given this implementation and our overall hypothesis, our refined hypothesis is that analogy using contact map similarity can effectively generate accurate protein substructure predictions. We applied the retrieval and adaptation

Fig. 3. In this figure the lower helix is the target and the upper is the source. The dotted gray circle represents the mapping area. The locations of the target amino acid residues for which there are no cooresponding source residues are inferred based on the known geometry for helices. These "grown" areas are represented with the dotted black line.

components of Triptych to a set of 61 proteins, mostly all α chains, retrieved from the PDB.[2]

4 Results

For each protein, we computed the distance map, contact map and secondary structure contact map. From the contact maps, we were able to derive 422 maps that described contacts for pairs of helices.

Table 1. The retrieval results of the committee on 422 unique queries when the top N out of 100 are returned as the final set of contact maps

N	Mean	Std	MeanBest	Rank
100	1.8604	0.8035	0.5259	7.5
50	1.6498	0.6447	0.5303	7
25	1.3944	0.5077	0.5506	5
10	1.1919	0.4166	0.6034	3

The results of the retrieval process for 422 unique test queries are shown in Table 1. N is the number of helix pairs retrieved; *Mean* describes the average RMSD for the queries and *Std* is the average standard deviation. *Mean Best* and *Rank* describe the average best RMSD and its median rank within the final set of contact maps. The results suggest the following: 1) as N, the number of retrieved helix pairs, decreases the average RMSD of the final set of contact maps improves, 2) the *Mean Best* represents the best structure match and worsens as

---

[2] The proteins were 1a0aA, 1a1z_, 1a28A, 1acp_, 1afrA, 1aj8A, 1akhA, 1akhB, 1am9A, 1aoiA, 1aoiB, 1arv_, 1auiB, 1auwA, 1bbhA, 1bcfA, 1bgp_, 1bh9A, 1bh9B, 1bu7A, 1bvb_, 1c52_, 1cc5_, 1cem_, 1cktA, 1clL_, 1cpq_, 1csh_, 1cy5A, 1d9cA, 1dceB, 1dpsA, 1ea1A, 1eerA, 1eteA, 1fce_, 1fgjA, 1ft1B, 1furA, 1gakA, 1hcrA, 1hnr_, 1hryA, 1huuA, 1hyp_, 1kx2A, 1lbd_, 1lfb_, 1lis_, 1lmb3, 1mhyD, 1neq_, 1pbwA, 1pru_, 1rzl_, 1tc3C, 1tx4A, 1uxc_, 2af8_, 2hddA, and 2ilk_.

Table 2. Experimental results when considering the adaptation of the top N results. RMSD denotes the mean of the best scores for each of the 422 input helix pairs for the top N retrievals.

n	RMSD
1	3.6668
5	2.2667
10	1.8814
25	1.5286
50	1.3921
100	1.3011
200	1.2507
422(all)	1.2426

N decreases, and 3) as N increases from 25 to 50 to 100, the *Mean Best* does not change substantially.

Further examination of the 100 retrievals using the committee determined that 65.40% of the 422 queries have its best RMSD fall within the top 10 retrievals, 83.18% within the top 25 and 96.45% within the top 50. Thus, a final set of contact maps consisting of the top 25 retrievals from a set of 100 seems to be the best balance between a low average RMSD over all the retrievals and a low RMSD for the average best retrieval. This ensures all the retrievals are similar to the query and contains the best match in $\sim 83\%$ of the helix pairs.

Using the results of the retrievals module, we evaluated the adaptation method by comparing the *predicted* locations of the residues to the *actual* locations, as given in the Protein Data Bank (PDB) in terms of RSMD. The results when considering the top N retrievals, for $N = 1, 5, 10, 25, 50, 100, 200$, and 422 are presented Table 2. These results suggest that we converge to a good solution when considering about the top 50 solutions.

The RMSDs presented are acceptably accurate in the biochemistry literature. See [41] for an empirical study with an RMSD of 1.6.

Note that the retrieval scores for the *Mean Best* (in terms of RMSD distance between the correct and predicted structures) are less than the adaptation scores (which reported the distance between the retrieved structures and the correct structure). The reason for this is that the retrieval scores are based on the RMSD of only the regions of the helices in contact with each other. The adaptation method extends the helices beyond the regions of contact based on biochemical knowledge, affording more opportunity for error.

5 Related Work

Previous methods for the recovery of 3D structure from distance contact maps are mostly based on distance geometry and stochastic optimization techniques, though none look specifically at prediction of helix pair structures. Nilges et al. [32] applied distance maps and dynamical simulated annealing to determine the 3D structure of proteins. More recently Vendruscolo et al. [40] proposed a dynamic approach that generates a structure that has a contact map similar to the query contact map.

The issue of visual knowledge in analogy and case-based reasoning has attracted the attention of researchers in several areas. Below we relate our work to some analogical problem solving systems that use visuospatial knowledge.

Previous visual analogy work in molecular biology domains include visualizing crystallographic data at different resolutions [16,17,22,28,29], in drug design [4,15], and in in-vitro fertilization [26]. Perner has applied visual analogy to image segmentation of CT images [34], HEp-2 cell images [33,35], and the identification of fungi [36].

Analogy with spatial reasoning has been applied to non-bioinformatics domains as well. FABEL [14] is an example of a system that adapts diagrammatic cases in the domain of architectural design. In FABEL, the source diagram specifies the spatial layout of a building or similar structure. FABEL adapts source diagrams by extracting and transferring specific structural patterns to the target problem. It uses domain-specific heuristics to guide pattern extraction and transfer.

REBUILDER [18] is an analogical reasoner that does retrieval, mapping, and transfer of software design class diagrams. The diagrams are represented structurally, not visuospatially, however. This means that, for example, that the connection is between two nodes is more important than the length and direction of that connection. That is, REBUILDER works with a different level of visual abstraction, a level at which only the structural relationships between visual elements, such as connectedness, are relevant to the task. Determining the right level of visual abstraction for visual case-based problems requires additional research. The choices made by REBUILDER depend largely on the specific domains in which they operate. In REBUILDER's domain of software design class diagrams, only the structural relations appear to be important.

FAMING [10] makes analogies with physical mechanism parts. FAMING uses the Structure-Behavior-Function (SBF) ontology to describe the cases. The structure is described in terms of a metric diagram (a geometric model of vertices and connecting edges), a place vocabulary (a complete model of all possible qualitative behaviors of the device), and configuration spaces (a compact representation of the constraints on the part motions). Shape features can involve two objects, expressing, for example, one part's ability to touch another part. Human designers are necessary for FAMING's processing. The designer chooses which cases and functions should be used, which dimensions the system should attempt to modify, and which shape features should be unified. It uses qualitative kinematics to propose design solutions for the desired function following the designer-suggested idea. Though not described as a visuospatial system, the important parts of physical mechanisms of the sort FAMING uses inevitably contain much knowledge that could be construed as spatial or visual.

Visual analogy has been used for cognitive modeling as well. DIVA [7] is an analogical mapper that uses visuospatial representations, using the Java Visual

Object System. It does no transfer of problem solutions and uses the ACME architecture for mapping [23]. MAGI [13] takes visual representations and uses the Structure-Mapping Engine [9] to find examples of symmetry and repetition in a single image through analogy.

Non-visual case-based problem-solving systems, such as CHEF [20] and PRODIGY [39] provide interesting points of comparison regarding the transfer process. CHEF is a case-based reasoner that transfers and adapts cooking recipes from a source to a target. The Prodigy case-based reasoning system implements the theory of Derivational Analogy [39]. It models transfer using memories of the justifications of each step, allowing for adaptation of the transferred procedure. Traces, called "derivations", are scripts of the steps of problem solving, along with the justifications for why the steps were chosen over others.

The Galatea system [8] uses only visuospatial representations of problem-solving procedures and transfers a source solution to a target solution. By using a sufficiently abstract visual language it is able to transfer problem-solving procedures between semantically distant analogs. The work on Galatea also supports the notion that visuospatial representations are useful for problem-solving.

6 Discussion

Though not based on a cognitive model, Triptych shows how visuospatial reasoning can facilitate problem solving in a another complex domain, building the case for the value of visual and spatial representations and reasoning for intelligent systems in general.

The theory behind this work is that changing representations can provide novel similarity insights. In this work we use contact maps and treat them as binary images, applying image processing techniques to them to retrieve similar protein substructures. This is in contrast with, for example, Jurisica et al. [28], who retrieve based on generated attributes. In the adapt module, the information transferred is purely spatial. The success of this method for α-helix pair structure prediction provides preliminary support for this theory, in that generated visuospatial representations can provide a means to find similarity. Future work will compare the results of contact map retrieval to sequence homology retrieval to investigate in exactly which conditions contact map similarity (representing visuospatial representations) is superior to the non-visual homology similarity metric.

In this paper we also described and demonstrated the applicability of the analogy methodology to the problem of secondary structure alignment from contact maps. Our hypothesis was that analogy using contact map similarity can effectively generate accurate protein substructure predictions. Triptych retrieves protein substructures based on visual similarity of contact maps. Initial results suggest that the retrieve and adapt phases are successful in finding similar contact maps in the PDB and modifying these to predict the alignment of pairs of helices, supporting this hypothesis. The advantage and novelty of our approach lies in its use of multiple sources of knowledge, including existing structural knowledge from the PDB, expert and text book knowledge (as used in the helix extension), as well as knowledge mined from the database. Once the viability of the approach is shown to be effective with idealized contact maps, the predicted, error-prone contact maps can be used as input.

Acknowledgements

Funding provided by: the Natural Science and Engineering Research Council (Ottawa); Institute for Robotics and Intelligent Systems (Ottawa); Protein Engineering Network Center of Excellence (Edmonton).

References

1. Aaronson, J.S., Juergen, H., Overton, G.C.: Knowledge discovery in genbank. In: Hunter, L., Searls, D., Shavlik, J. (eds.) Proceedings of the First International Conference on Intelligent Systems for Molecular Biology, pp. 3–11. AAAI Press, Stanford, California, USA (1993)
2. Amarel, S.: On representations of problems of reasoning about actions. In: Michie, D. (ed.) Machine Intelligence 3, vol. 3, pp. 131–171. Elsevier/North-Holland, Amsterdam, London, New York (1968)
3. Berman, H.M., Westbrook, J., Feng, Z., Gilliland, G., Bhat, T.N., Weissig, H., Shindyalov, I.N., Bourne, P.E.: Protein data bank. Nucleic Acids Research 28, 235–242 (2000)
4. Biname, J., Meurice, N., Leherte, L., Glasgow, J., Fortier, S., Vercauteren, D.P.: Use of electron density critical points as chemical function-based reduced representations of pharmacological ligands. Journal of Chemical Information and Computer Science 44, 1394–1401 (2004)
5. Canny, J.: A computational approach to edge detection. IEEE Transactions on Pattern Alanysis and Machine Intelligence 8(6), 769–798 (1986)
6. Casakin, H., Goldschmidt, G.: Expertise and the use of visual analogy: Implications for design education. Design Studies 20, 153–175 (1999)
7. Croft, D., Thagard, P.: Dynamic imagery: A computational model of motion and visual analogy. In: Magnani, L., Nersessian, N.J. (eds.) Model-Based Reasoning: Science, Technology, & Values, pp. 259–274. Kluwer Academic: Plenum Publishers, New York (2002)
8. Davies, J., Goel, A.K.: Visual analogy in problem solving. In: Nebel, B. (ed.) Proceedings of the International Joint Conference for Artificial Intelligence 2001, pp. 377–382. Morgan Kaufmann Publishers, Seattle, WA (2001)
9. Falkenhainer, B., Forbus, K.D., Gentner, D.: The structure-mapping engine: Algorithm and examples. Artificial Intelligence 41, 1–63 (1990)
10. Faltings, B., Sun, K.: FAMING: supporting innovative mechanism shape design. Computer-aided Design 28(3), 207–216 (1996)
11. Farah, M.J.: The neuropsychology of mental imagery: Converging evidence from brain-damaged and normal subjects. In: Stiles-Davis, J., Kritchevsky, M., Bellugi, U. (eds.) Spatial Cognition– Brain bases and development, pp. 33–59. Erlbaum, Hillsdale, New Jersey (1988)

12. Fariselli, P., Olmea, O., Valencia, A., Casadio, R.: Prediction of contact maps with neural networks and correlated mutations. Protein Engineering 14(11), 835–843 (2001)
13. Ferguson, R.W.: Magi: Analogy-based encoding using regularity and symmetry. In: Ram, A., Eiselt, K. (eds.) Proceedings of the Sixteenth Annual Conference of the Cognitive Science Society, Atlanta, GA, pp. 283–288. Lawrence Erlbaum Associates, Mahwah (1994)
14. Gebhardt, F., Voss, A., Grather, W., Schmidt-Belz, B.: Reasoning with Complex Cases. Kluwer Academic Publishers, Dordrecht (1997)
15. Glasgow, J., Epstein, S.L., Meurice, N., Vercauteren, D.P.: Spatial motifs in design. In: Proceedings of the Third International Conference on Visual and Spatial Reasoning in Design (2004)
16. Glasgow, J.I., Conklin, D., Fortier, S.: Case-based reasoning for molecular scene analysis. In: Janice, I. (ed.) Working Notes of the AAAI Spring Symposium on Case-Based Reasoning and Information Retrieval, pp. 53–62. AAAI Press, Menlo Park, California (1993)
17. Janice, I., Glasgow, S., Fortier, D., Allen, F.: Knowledge representation tools for molecular scene analysis. In: Proceedings of the 28th Annual Hawaii International Conference on System Biotechnology Computing Track (January 1995)
18. Gomes, P., Seco, N., Pereira, F.C., Paiva, P., Carreiro, P., Ferreira, J.L., Bento, C.: The importance of retrieval in creative design analogies. In: Bento, C., Cardoso, A., Gero, J. (eds.) Creative Systems: Approaches to Creativity in AI and Cognitive Science. Workshop program in the Eighteenth International Joint Conference on Artificial Intelligence, Acapulco, Mexico, pp. 37–45 (August 2003)
19. Rafael, C.G., Richard, E.W.: Digital Image Processing. Addison-Wesley, New York (1992)
20. Hammond, K.J.: Case-based planning: A framework for planning from experience. Cognitive Science 14(4), 385–443 (1990)
21. Haralick, R.M., Shanmugam, K., Dinstein, I.: Textural features for image classification. IEEE Transactions on Systems, Man and Cybernetics SMC-3(6), 610–621 (1973)
22. Hennessy, D., Buchanan, B., Subramanian, D., Wilkosz, P.A., Rosenberg, J.M.: Statistical methods for the objective design of screening procedures for macromolecular crystallization. Acta Crystallogr D Biol Crystallogr 56(Pt 7), 817–827 (2000)
23. Holyoak, K.J., Thagard, P.: The analogical mind. American Psychologist 52(1), 35–44 (1997)
24. Hunter, L.: Life and its molecules. AI Magazine 25(1), 9–22 (2004)
25. Jaccard, P.: Nouvelles recherches sur la distribution florale. Bulletin de la Société Vaudoise des Sciences Naturelles 44, 223–270 (1908)
26. Jurisica, I., Glasgow, J.I.: Extending case-based reasoning by discovering and using image features in in-vitro fertilization. In: SAC 2000. ACM Symposium on Application Computing, Biomedical Computing- special session on biomedical applications of knowledge discovery and data mining, CITO. Villa Olmo, Italy (March 2000)
27. Jurisica, I., Glasgow, J.I.: Applications of case-based reasoning in molecular biology. AI Magazine 25(1) (2004)
28. Jurisica, I., Rogers, P., Glasgow, J.I., Collins, R.J., Wolfley, J.R., Luft, J.R., DeTitta, G.T.: Improving objectivity and scalability in protein crystallization: Integrating image analysis with knowledge discovery. Intelligent Systems in Biology, Special Issue of IEEE Intelligent Systems, 26–34 (2001)

29. Jurisica, I., Rogers, P., Glasgow, J., Fortier, S., Collins, R., Wolfley, J., Luft, J., DeTitta, G.T.: Integrating case-based reasoning and image analysis: High-throughput protein crystallization domain. In: Proceedings of the Innovative Applications of Artificial Intelligence (IAAI01), Seattle, pp. 73–80. IRIS, CITO (August 2001)
30. McCarthy, J., Minsky, M., Sloman, A., Gong, L., Lau, T., Morgenstern, L., Mueller, E.T., Riecken, D., Singh, M., Singh, P.: An architecture of diversity for commonsense reasoning. IBM Systems Journal 41(3), 530–539 (2002)
31. Monaghan, J.M., Clement, J.: Use of computer simulation to develop mental simulations for understanding relative motion concepts. International Journal of Science Education 21(9), 921–944 (1999)
32. Nilges, M., Clore, G.M., Gronenborn, A.M.: Determination of the three-dimensional structures of proteins from interproton distance data by dynamical simulated annealing from a random array of atoms. FEBS Lett. 229, 129–136 (1988)
33. Perner, P.: Image analysis and classification of hep-2 cells in flourescent images. In: Fourteenth International Conference on Pattern Recognition, pp. 1677–1679. IEEE Computer Society Press, Los Alamitos (1998)
34. Perner, P.: An architecture for a cbr image segmentation system. Journal of Engineering Application in Artificial Intelligence 12(6), 749–759 (1999)
35. Perner, P.: Why case-based reasoning is attractive for image interpretation. In: Case-Based Reasoning Research and Development, pp. 27–44. Springer, Heidelberg (2001)
36. Perner, P., Gunther, T., Perner, H.: Airborne fungi identification by case-based reasoning. In: Ashley, K.D., Bridge, D.G. (eds.) ICCBR 2003. LNCS, vol. 2689, pp. 73–79. Springer, Heidelberg (2003)
37. Smith, J.R., Chang, S.F.: Quad-tree segmentation for texture-based image query. In: Proceedings of the second ACM international conference on Multimedia, pp. 279–286 (1994)
38. Sullivan, G.J., Baker, R.L.: Efficient quadtree coding of images and video. IEEE Transactions on Image Processing 3(3), 327–331 (1994)
39. Veloso, M.M.: Prodigy/analogy: Analogical reasoning in general problem solving. In: Wess, S., Richter, M., Althoff, K.-D. (eds.) Topics in Case-Based Reasoning. LNCS, vol. 837, pp. 33–52. Springer, Heidelberg (1994)
40. Vendruscolo, M., Kussell, E., Domany, E.: Recovery of protein structure from contact maps. Folding and Design 2, 295–306 (1997)
41. Wang, J., Stieglitz, K.A., Kantrowitz, E.R.: Metal specificity is correlated with two crucial active site residues in escherichia coli alkaline phosphatase. Biochemistry 44(23), 8378–8386 (2005)
42. Won, C.S., Park, D.K., Park, S.J.: Efficient use of mpeg-7 edge histogram descriptor. Electronics and Telecommunications Research Institute Journal 24, 23 (2002)

The Spatial Representation of Dynamic Scenes – An Integrative Approach

Markus Huff, Stephan Schwan, and Bärbel Garsoffky

Cybermedia Research Unit, Knowledge Media Research Center
Konrad-Adenauer-Straße 40, 72072 Tübingen, Germany

Abstract. This paper addresses the spatial representation of dynamic scenes, particularly the question whether recognition performance is viewpoint dependent or viewpoint invariant. Beginning with the delimitation of static and dynamic scene recognition, the viewpoint dependency of visual recognition performance and the structure of the underlying mental representation are discussed. In the following, two parameters (an easy to identify *event model* and salient *static features*) are identified which appeared to be accountable for viewpoint dependency or viewpoint invariance of visual recognition performance for dynamic scenes.

1 Representing Dynamic Scenes Spatially

Research addressing the question whether mental representation and therefore visual recognition performance for real world events is viewpoint dependent or viewpoint invariant has a long tradition in cognitive psychology. After watching a real world event from a certain viewpoint, *viewpoint dependency* is defined when the recognition performance impairs with increasing angular difference between the viewpoints in the learning and the test phase. Otherwise, when the recognition performance does not depend on the angular difference between learning and test phase, viewpoint invariant recognition performance is observed. Summarizing the viewpoint debate of the 1990s in which evidence for both viewpoint invariant (Biederman, 1987; Biederman & Bar, 1999) and viewpoint dependent (Tarr, 1995) recognition performance for static objects was found; Hayward (2003) concluded that both sides acknowledge an influence of viewpoint. "We propose that human object recognition may be thought of as a continuum in which the most extreme exemplar specific discriminations require exclusively viewpoint dependent mechanisms, while the most extreme categorizations recruit exclusively viewpoint invariant mechanisms." (Tarr, 1995; p. 1503) Results regarding the spatial representation of *static scenes* were quite similar. In addition to a number of studies showing that the spatial representation of static scenes was viewpoint dependent (McNamara, 2003; Diwadkar & McNamara, 1997; Simons & Wang, 2002) yet others found that under certain circumstances static scene recognition can also be viewpoint independent (e.g. Evans & Pezdek, 1980).

But what about dynamic scenes? Do they rely on mechanisms similar to those involved in the recognition of static objects and static scenes? In general, dynamic scenes consist of multiple objects undergoing certain movement patterns within a static frame. Typical examples are scenes containing events like tactical moves in a soccer or basketball game. We conceive events in a relatively broad and basic manner when two or more objects show regular movement patterns (Schwan & Garsoffky, in press). The term event does not necessarily imply complex activities like *making a bed* (Zacks, Tversky, & Iyer, 2001) or *opening a door* (Worboys & Hornsby, 2004). Therefore we use the term similar to scholars of event perception (Gibson, 1979) and not to scholars of activity theory (Newtson, 1973). Although we were using quite short dynamic scenes as stimulus material (8 vs. up to 640 seconds as used by Zacks et al., 2001) our definition of *event* is not incompatible with the one of Zacks et al. (2001). While watching an event (in the meaning of Zacks) from a greater distance (as it was the case in the study with basketball scenes, Garsoffky, Huff, & Schwan, 2007), the movements of a complex activity were perceived only as moving objects. While, over the last few years, static scenes have been investigated in many studies, dynamic scenes and the underlying cognitive processes have only been examined in few experiments. Why is it so important to differentiate *dynamic scene recognition* from *static scene recognition*? First, a dynamic scene continuously changes its relational properties (Garsoffky, Schwan, & Hesse, 2002). The spatial relations between the objects and in relation to the observer are not constant over time. This implies that, unlike watching a static scene, the observer does not see a stable object constellation over a certain period of time. A second point is that dynamic scenes typically are not repeated in the same fashion, whereas static objects or scenes can be encountered on multiple occasions.

These differences have some implications for expected visual recognition performance for dynamic scenes on the one hand, and for the underlying cognitive processes and properties of the mental representation on the other. Garsoffky et al. (2002) posed three theoretical models, each of which makes different predictions concerning the visual recognition of dynamic scenes: the *static-scene* model, the *dynamic-event* model, and the *film-form* model. These models are described in more detail below.

1.1 Static-Scene Model

The static-scene model predicts *viewpoint dependent* recognition performance. As research on static scene recognition has shown, visual recognition performance is highest if the viewpoint used in the test phase is the same as that used in the learning phase (Diwadkar & McNamara, 1997). As the difference between these two viewpoints increases, recognition performance decreases. This is due to alignment processes which attempt to match the novel viewpoint with the viewpoint from the learning phase. According to McNamara (2003) there is therefore ample evidence indicating that spatial memories of static scenes are orientation-dependent. Assuming that static and dynamic scenes are processed in similar ways, recognition performance for dynamic scenes should also be optimal when learning and test phase viewpoints are identical. Further, recognition performance should decrease with increased angle differences between learning and test phase viewpoints. This model also includes the possibility of multiple-view representations given that a scene is shown from different viewpoints in the learning phase. The viewpoint dependency of recognition performance should be relative to those viewpoints initially seen.

1.2 Dynamic-Event Model

The second model proposed by Garsoffky et al. (2002) is the dynamic-event model. This model assumes *viewpoint independent* visual recognition performance. It is based on the assumption that dynamic scenes constitute a special type of event, and that they are processed accordingly. Research on events has shown that due to the high informational density of unfolding circumstances, information about objects, layout, topology, and progression of the event is extracted (Posner & Keele, 1968) giving way to a more abstract mode of representation. This abstraction process should also hold for the spatial properties of dynamic scenes and ought in turn to result in a more viewpoint independent representation.

1.3 Film-Form Model

The film-form model proposed by Garsoffky et al. (2002) also predicts a *viewpoint dependent* recognition performance. This model, originally developed by film makers, assumes that the mental representation of a dynamic event is aligned with the viewpoint of the *establishing shot* depicting the scene. The establishing shot is the first referential section at the beginning of a filmic event. The scene is first presented as a whole from a considerable distance. Each of the subsequent viewpoints is integrated into this frame. According to this model the mental representation of dynamic scenes should therefore be viewpoint dependent. Interestingly this model is partly compatible with the *spatial reference model* proposed by Shelton and McNamara (2001, 2004). They proposed that learning and remembering the spatial layout of a static scene involves interpreting the given structure in terms of a spatial reference system. They suggested that this process is analogous to determining the "top" of a figure; in effect, conceptual "north" is assigned to the layout, creating privileged directions in the environment (Mou & McNamara, 2002). The frame of reference for this interpretation is selected using egocentric and environmental cues, such as viewing orientation and alignment processes, respectively. Due to this model the resulting visual recognition performance is viewpoint dependent relative to the reference vector.

2 Empirical Overview

Over the course of the last three years the authors have conducted eleven laboratory recognition studies addressing the issue of *viewpoint dependency* in visual recognition performance with dynamic scenes. The first five studies incorporated complex, naturalistic types of dynamic scenes; namely short sequences from soccer games (experiment 1 to 3, Garsoffky, Schwan, & Hesse, 2002), short sequences

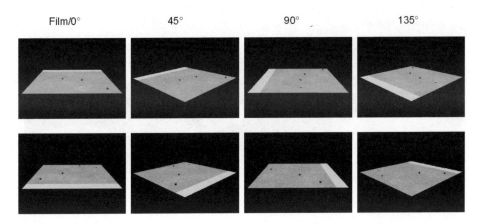

Fig. 1. In this figure the used viewpoints in the test phase are illustrated. If the dynamic scene was presented from 0°, the viewpoint deviation in the test phase was 0°, 45°, 90° and 135°.

from basketball games (experiment 4, Garsoffky, et al., 2007) and traffic scenes (experiment 5, Huff, Garsoffky, & Schwan, unpublished). In experiment 6 to 11 more artificial material was used (a kind of *ball race*, Garsoffky, Schwan, & Hesse, 2004; Huff, unpublished). All experiments were based on the same design scheme, employing a within-subjects design, and measuring recognition performance using a *visual recognition* paradigm. After twice observing a dynamic scene in the learning phase, participants in the test phase were shown a number of pictures recorded from different viewpoints at different points of time (see figure 1). For each picture the participant had to decide by key press whether he/she had already seen the depicted scene.

Each experiment contained one control condition, in which the dynamic scene was presented without modification or intervention (e.g. film-cuts, supplemental verbal descriptions, etc.). All further conditions were experimental conditions containing certain variations which were expected to weaken the viewpoint dependency effect. As it turned out, even the control conditions, on account of different underlying stimulus materials, showed substantial differences with regard to the viewpoint dependency effect. This will be the main focus of the present paper.

Soccer games. In experiments 1 to 3 we used already existing animations of regular sport events; namely soccer episodes from the 1998 world championships. Each episode contained a successful shot at goal from one of the two teams. Since they were extracted from existing material, there was no possibility of controlling distractor items. For this reason, only *hits* (correct answers to target items in a visual recognition test) could be analyzed.

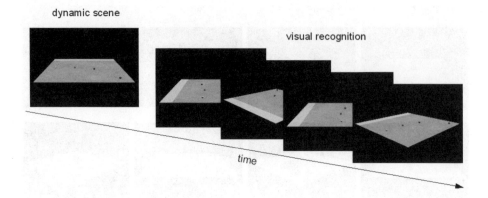

Fig. 2. This figure illustrates the experimental recognition paradigm. After twice presenting a dynamic scene, participants were shown several video stills depicting either the target scene or a distractor scene from varying viewpoints at different points of time.

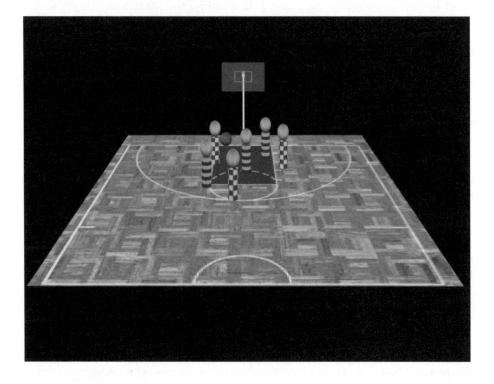

Fig. 3. This figure shows an example of a basketball game used as stimulus material

Fig. 4. This figure shows an example of a traffic scene depicting an accident on a highway

Basketball plays. In experiment 4 we used simulations of typical moves from basketball games as stimulus material[1] (Garsoffky et al., 2007). Each event showed players from two teams, distinguishable by the colors and patterns of their shirts (see figure 3) and ended with one team scoring a basket. It was attempted to design the events in a manner as realistic as possible. In this case it was possible to control distractor items used in the visual recognition test, because each target item had a corresponding distractor item. Therefore calculation of *sensitivity* and *response bias* measures in terms of the signal detection theory (Green & Swets, 1966) was possible, as was the case for each of the following experiments.

Traffic scenes. In study 5 traffic scenes were used (Huff et al., unpublished). The dynamic scenes, depicting an accident on a highway, consisted of multiple vehicles (cars, trucks, motorcycles) driving on a road with three lanes (see figure 4). Along the road there were several prominent objects (e.g. huge trees, a building, etc.). The distractor scenes involved changing the starting position of one of the cars responsible for causing the accident at the end of the scene.

---

[1] Examples of the stimulus materials mentioned in this paper are available at http://www.iwm-kmrc.de/cybermedia/sc2006/

Fig. 5. This figure illustrates the ball races used in the current studies. Four different colored balls could be seen moving across a quadratic plane on parallel courses towards the white finishing line.

Ball races. Experiments 6 to 11 used stimulus material of a more artificial nature (e.g. Huff, Garsoffky, & Schwan, 2006; see figure 5). Four different colored balls could be seen moving across a quadratic plane on parallel courses towards a specific target (either a kind of soccer goal or a white finishing line). Each ball had a different starting position and a distinctive movement characteristic (constant velocity, accelerating or decelerating).

3 Empirical Results Concerning the Recognition of Dynamic Scenes

The experiments investigating the viewpoint dependency of visual recognition performance were analyzed with regard to the dependent measures *hits*, *sensitivity* and *response bias*.

3.1 Recognition Performance Regarding Hits

In all experiments viewpoint dependent recognition performance was observed regarding the dependent variable *hits* (see figure 6). Recognition performance was highest when the viewpoint in the test phase was the same as that used in the learning phase. It was therefore assumed that the mental representations responsible for dynamic scene recognition have similar properties to those mental representations responsible for static scene recognition (*static scene model*). That is, the viewpoint seen in the initial learning phase became part of the mental representation of the dynamic scene.

However, the hit rate only describes reactions to target items (video stills depicting the already seen dynamic scene). Reactions to distractor items - representing the other half of the reactions - are ignored by this measure. Hence different causes could account for this effect. For example the effect could imply a conservative response bias; participants respond less often with "yes" when angular difference between the viewpoints in the learning and the test phase is greater, independent of the type of recognition item (target or distractor). Although experiments 1 to 3 (Garsoffky et al., 2002) yielded some support for a viewpoint dependent mental representation, the experimental setup of these experiments precluded the calculation of more refined measures e.g. *sensitivity* measure and *response bias*. In the following experiments, this particular problem was solved by using modeling software to design the stimulus material. From now on it became possible to construct specific distractor scenes, incorporating all the properties of the original scene excepting a modification of the movement characteristic of a single object. The matching of target and distractor items was now possible and consequently the calculation of sensitivity.

3.2 Recognition Performance Based on Signal Detection Measures

Altogether, eight empirical studies were conducted with stimulus material allowing the calculation of *sensitivity* and *response bias* measures according to the signal detection theory (Green & Swets, 1966). In three studies, using ball races as stimulus material, viewpoint dependency with regard to the sensitivity measure was found. In contrast, the sensitivity measure of five experiments demonstrated viewpoint invariant recognition performance. This was the case both for realistic and artificial stimulus material. In experiment 4 using *basketball games* (Garsoffky et al., 2007), in experiment 5 using *traffic scenes* (Huff, et al., unpublished), and in three studies based on the more artificial ball races, no viewpoint dependency effect with regard to the sensitivity measure was found; visual recognition performance was independent of the viewpoint used in the test phase.

Additionally, in all eight empirical studies there was a main effect for viewpoint deviation regarding the dependent variable *response bias*. Explicitly, this means that the participants' reactions tended to be more conservative, the greater the deviation of the viewpoint in the test phase from that in the learning phase (see also figure 6).

Basketball scenes

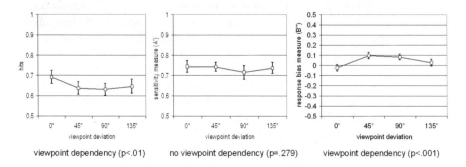

Ball races with information about the hidden structure

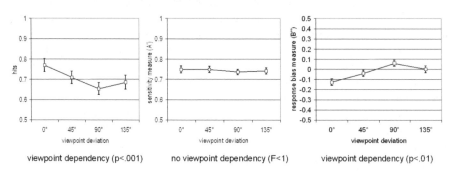

Ball races without information about the hidden structure

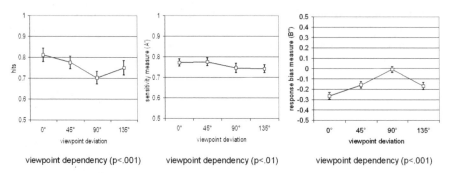

Fig. 6. This figure illustrates the findings of the reported studies. As typical examples the basketball scenes (first row; Garsoffky et al., 2007), the ball races with information about the hidden structure (second row; Huff, et al., 2006), and the ball races without information about the hidden structure (third row; Huff, unpublished) are diagrammed. The first diagram in each row shows the viewpoint dependency effect for *hits*. The second one displays the *sensitivity* measure (a viewpoint dependency effect was only found with artificial material without information of the hidden structure as displayed in the last row). The third diagram in each row shows the viewpoint dependency for the *response bias* measure.

4 An Integrative Account of the Findings

In summary, the findings of the eleven experiments indicate that viewers do not necessarily process dynamic scenes in a viewpoint dependent way, and that a more differentiated theoretical account is therefore required. On the one hand, viewpoint dependent recognition performance in terms of *hits* was found in all studies. On the other hand, the pattern of results concerning *sensitivity* proved more complex, with three studies demonstrating an influence of viewpoint on this specific variable and six others showing none. Furthermore, in all studies an influence of viewpoint was found for *response bias*. Therefore, the finding of a viewpoint dependency of *hits* must at least partially be attributed to corresponding variations in *response bias* (see also figure 6). Contrastingly, the influence of viewpoint on *sensitivity* appears to be moderated by additional factors. First, viewpoint dependency of sensitivity was found for artificial material (i.e. ball races), but not for naturalistic dynamic scenes. Second, even in the context of ball races, viewpoint dependency diminished when the movement patterns of the ball races were easily characterized by a sequence of unfolding events.

In relating visual recognition performance to the attributes of the underlying dynamic scenes, it can be conclusively concluded that participants tend to represent a dynamic scene in a viewpoint independent manner if the scene itself offers cues which them help to form an abstract mental representation. This can be explained by the fact that an abstract mental representation of a dynamic scene is more economic in the usage of memory on the one hand, and easier to handle on the other. A closer analysis of the findings indicates two parameters of dynamic scenes which could help to foster the development of a viewpoint independent representation: an easy to identify *event model* and *salient static features*.

4.1 Event Model

An *event model* is a parsimonious description of an unfolding event in terms of its gist. Examples include the concatenation of circumstances finally leading to an accident in the traffic scenes shown from an exterior point of view, or the strategic moves of players on the offensive team in the basketball scenes (Garsoffky et al., 2007). Since event models represent a highly economic characterization of a dynamic scene, we assume that viewers base their mental representation on such event models whenever possible. According to the findings of our studies, at least two options as to how such event models are induced can be identified. First, a dynamic scene can possess certain *inherent* movement patterns which are easily identifiable. Second, even if no such obvious event features exist, an event model representation can be *induced* by having a verbal description accompany the event.

Inherent Event Model. Dynamic scenes containing an evident event structure offer cues for the construction of an abstract, gist-like mental representation of the scene. We assume that such a representation does not include the original viewpoint and that viewers are therefore able to recognize the scene from different viewpoints with equal accuracy.

This is compatible with the assumption of the dynamic event model approach (see 1.2). For example Jenkins, Wald, & Pittenger (1978) presented participants a sequence of slides depicting a certain activity. In the subsequent recognition test containing (1) slides not shown in the learning phase, but depicting the presented activity and (2) slides portraying a distractor activity. Results showed that the false alarm rate in response to (1) was highest, suggesting that participants had developed a mental representation which abstracted from the pictorial details of the selected slides (see also Hannigan & Reinitz, 2001).

Similar processes seem to be responsible for the corresponding finding of the present studies. For example, the *basketball games* (Garsoffky et al., 2007) contained typical strategic moves often seen on TV; two teams could be seen playing on a basketball court; the offensive team was in possession of the ball, the defensive team trying to obtain the ball. Each move ended with the offensive team scoring a basket. Instead of separately encoding each player (up to six) and his individual movements, it is possible to encode two teams (an offensive and a defensive team) together with their respective moves in time. As a result the event model could sound like: "The red team possess the ball and moves toward the basket which is defended by the blue team. As the red team is entering shooting range, a basked was made." In this description, irrelevant details are lost and the representation is not bound to a specific viewpoint. Thus, the resulting mental representation enables viewpoint independent visual recognition performance. It is arguable that such an effect only holds for highly realistic stimulus material with high familiarity – all individuals have at some point seen a basketball or soccer match on TV. Our empirical results, however, further showed viewpoint independent recognition performance to occur for artificial dynamic scenes (ball races) with an evident structure, for example if one ball dominated the others on account of being the fastest and passing all others (Huff, unpublished).

Induced Event Model. Even if the stimulus material is less obviously structured, participants are still able to adapt an event model, if they are explicitly informed about the existence of hidden structures. This was carried out in two studies (Huff, unpublished) in which participants were instructed, prior to the experiment, to focus their attention on possible event structures. As a result no viewpoint dependent recognition performance with regard to sensitivity was observed, indicating that viewers were able to develop an appropriate event model.

4.2 Static Features

A further attribute of dynamic scenes which appears to be important for *viewpoint dependent* versus *viewpoint invariant* recognition performance are salient static features within a dynamic scene. There are two functions which such salient static features can serve. First, they can indicate the spatial orientation of the

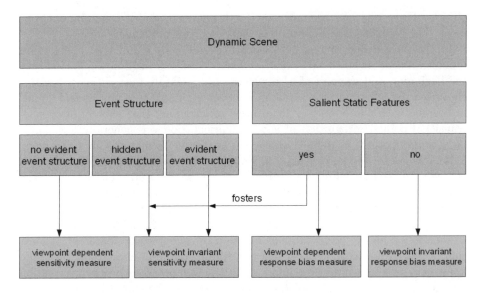

Fig. 7. According to the theoretical approach in this paper a dynamic scene can contain an event structure and salient static features. This figure illustrates this approach and displays the predictions regarding the viewpoint dependency for *sensitivity* and *response bias* measures.

scene (e.g. the basket of a basketball scene) and second, they can structure the dynamic scene into discreet sub-events (e.g. the lines of the basketball court or the tree beneath the road in the traffic scene).

The *orientational* function of salient static features can explain the tendency of participants to react more conservatively, the larger the deviation of the viewpoint between the learning and the test phase. All eleven experiments contained such static features and accordingly showed a viewpoint dependent response bias. In particular, such a static feature allows the viewer to readily identify the spatial orientation of a recognition item without having to take the depicted event into account. We therefore assume that the abstract event model description of a dynamic scene is supplemented by an additional "tag", which – based on the salient static features – indicates the initial orientation of the scene.

Additionally the *structuring* function of salient static features fosters the development of a viewpoint invariant mental representation of dynamic scenes. In this context, there is evidence from two lines of research. In the first, Easton and Sholl (1995) and Sholl and Nolin (1997) proposed a model of spatial representation and retrieval regarding static scenes. This model contains two sub-systems. The first, the *self-reference system*, codes *self-to-object relations* in the form of body-centered coordinates. That is, the body itself becomes part of the mental representation including terms like *left-right*, *front-back*, and *up-down*. The encoded mental representation is *viewpoint-dependent*. A second system, the so-called *object-to-object-system*, codes the spatial relations between the enlisted objects. These relations are not coded with regard to the surrounding environment. The resulting representation is therefore *viewpoint-independent*.

In the second line of research, Gernsbacher (1985) found that abstraction processes occur at so-called *transition* points. Such transition points indicate the beginning and the end of a sub-event, and facilitate abstraction processes. He varied the left-right orientation of pictures within picture stories and found, in keeping with the boundary hypothesis of encoding, that shortly after such transition points, recognition of left-right orientation was generally low. Similarly, *static features* as defined above can also serve as markers for structuring a dynamic scene into sub-events. For example, the point at which a car changes lane can be used to define the beginning of a new sub-event. There is evidence that at these points, processes of abstraction from visual details take place (Gernsbacher, 1985). It would therefore seem reasonable to assume that such points play an important role in the development of event models. Nevertheless, as mentioned above, (*orientational* function of salient static features) at least some surface information concerning the overall spatial orientation of the scene appears to be retained. In sum, if indeed abstraction processes occur at transition points, then the orientation of the original scene is also stored in combination with the abstracted dynamic content.

Taken together, the two functions of salient static features can contribute to explaining the pattern of results found in the present studies. The *orientational* function seems to be responsible for the conservative response bias, and the *structuring* function appears to aid viewpoint invariant recognition performance. If a dynamic scene contains salient static features, these can facilitate the encoding of a viewpoint invariant mental representation of a dynamic scene in two ways. First, by enabling the encoding of an *object-to-object* system (McNamara, 2003). Second, by making certain points (so called transition points) available at which abstraction processes can take place (Gernsbacher, 1985). In contrast, the absence of such *salient static features* forces the viewer not only to encode a *self-reference* system, but also makes it more difficult to *structure* the dynamic scene into sub-events.

4.3 Summary

To summarize, several parameters were identified in this integrative approach which are thought to influence the viewpoint dependency of visual recognition performance for dynamic scenes (as summarized in figure 7). In particular, the perceptibility of the event structure of a given dynamic scene determines the development of a viewpoint invariant event model. If such an event structure is perceivable to the viewer, either because it is inherently salient or because it is cued by additional information, an abstract mental representation is developed. Only if no such structure is evident will viewers tend to develop a viewpoint dependent representation. This process of developing an event model is further facilitated by the existence of salient static features within the dynamic scene. Such static features also allow the abstract event model description of a dynamic scene to be supplemented by an additional "tag" which indicates the initial orientation of the scene and in turn leads to a viewpoint dependent response bias.

5 Discussion

In this paper an attempt was made to integrate the findings of eleven experiments adressing the spatial properties of dynamic scene recognition. Two major points were identified. The first concerns the issue of *measurement*. It was shown that the notion of viewpoint dependent recognition is inherently ambiguous on account of the fact that it can relate to *hits*, *sensitivity* or to *response bias*. It is therefore important to analyze *all* reactions of participants. While *hits* described only one half of the reactions, signal detection measures (sensitivity and response bias) were able to describe all reactions. In the reported studies it was also shown that *viewpoint dependency* of sensitivity and response bias are largely independent of one another.

The second major point of the current studies relates to the characteristics of the dynamic scenes. In this paper we identified two determinants which could be responsible for viewpoint dependent or viewpoint invariant encoding of dynamic scenes. These characteristics, in particular a *perceivable event structure* and the *existence of salient static features*, are assumed to condition whether a dynamic scene is encoded in the sense of the *static-scene* or the *dynamic-event* model (Garsoffky et al., 2002). Based upon our empirical results, it is evident that viewers tend to encode viewpoint invariant mental representations whenever possible. If a dynamic scene contains elements such as an *event model* or *salient static features*, then a mental representation can be encoded which is abstract and which facilitates viewpoint invariant recognition performance.

Acknowledgments

This research was supported by grants from the Deutsche Forschungsgemeinschaft (DFG). The authors thank three anonymous reviewers for their helpful comments on this work.

References

Biederman, I.: Recognition-by-components: A theory of human image understanding. Psychological Review 94, 115–147 (1987)

Biederman, I., Bar, M.: One-shot viewpoint invariance in matching novel objects. Vision Research 39, 2885–2899 (1999)

Christou, C.G., Bülthoff, H.H.: View dependence in scene recognition after active learning. Memory & Cognition 27(6), 996–1007 (1999)

Diwadkar, V.A., McNamara, T.P.: Viewpoint dependence in scene recognition. Psychological Science 8, 302–307 (1997)

Easton, R.D., Sholl, M.J.: Object-array strucure, frames of reference, and retrieval of spatial knowledge. Journal of Experimental Psychology: Learning, Memory & Cognition 21, 483–500 (1995)

Evans, G.W., Pezdek, K.: Cognitive mapping: Knowledge of real-world distance and location information. Journal of Experimental Psychology: Human Learning and Memory 6, 13–24 (1980)

Garsoffky, B., Huff, M., Schwan, S.: Changing viewpoints during dynamic events. Perception 36, 366–374 (2007)

Garsoffky, B., Schwan, S., Hesse, F.W.: Viewpoint dependency in the recognition of dynamic scenes. Journal of Experimental Psychology: Learning, Memory, & Cognition 28, 1035–1050 (2002)

Garsoffky, B., Schwan, S., Hesse, F.W.: Does the viewpoint deviation effect diminish if canonical viewpoints are used for the presentation of dynamic sequences? In: Forbus, K., Gentner, D., Regier, T. (eds.) Proceedings of the 26th Annual Conference of the Cognitive Science Society, pp. 428–433. Erlbaum, Mahwah, NJ (2004)

Gernsbacher, M.A.: Surface information loss in comprehension. Cognitive Psychology 17, 324–363 (1985)

Gibson, J.J.: The ecological approach to visual perception, Boston, Houghton Mifflin (1979)

Green, D., Swets, J.: Signal detection theory and psychophysics. Wiley, Oxford (1966)

Hannigan, S.L., Reinitz, M.T.: A demonstration and comparison of two types of inference-based memory errors. Journal of Experimental Psychology: Learning, Memory, and Cognition 27, 931–940 (2001)

Hayward, W.G.: After the viewpoint debate: Where next in object recognition? Trends in Cognitive Sciences 7, 425–427 (2003)

Huff, M.: Verbal Influences on the recognition of dynamic scenes (Unpublished manuscript)

Huff, M., Garsoffky, B., Schwan, S.: The viewpoint dependency of traffic scenes (Unpublished manuscript)

Huff, M., Garsoffky, B., Schwan, S.: The influence of the serial order of visual and verbal presentation on the verbal overshadowing effect of dynamic scenes. In: Proceedings of the 28th Annual Conference of the Cognitive Science Society, pp. 1539–1544. Lawrence Erlbaum, Mahwah (2006)

Jenkins, J.J., Wald, J., Pittenger, J.B.: Apprehending pictorial events: An instance of psychological cohesion. In: Savage, C.W. (ed.) Perception and cognition: Issues in the foundations of psychology, Minneapolis, pp. 129–163. University of Minnesota Press (1978)

McNamara, T.P.: How are the locations of objects in the environment represented in memory? In: Freksa, C., Brauer, W., Habel, C., Wender, K. (eds.) Spatial cognition III: Routes and navigation, human memory and learning, spatial representation and spatial reasoning, pp. 174–191. Springer, Heidelberg (2003)

Mou, W., McNamara, T.P.: Intrinsic frames of reference in spatial memory. Journal of Experimental Psychology: Learning, Memory, & Cognition 28, 162–170 (2002)

Newtson, D.: Attribution and the unit of perception of ongoing behavior. Journal of Personality and Social Psychology 28, 28–38 (1973)

Posner, M.I., W.,, S, K.: On the genesis of abstract ideas. Journal of Experimental Psychology 77, 353–363 (1968)

Schwan, S., Garsoffky, B.: Event segmentation and memory. In: Shipley, T., Zacks, J. (eds.) Understanding events: How humans see, represent, and act on events, Oxford University Press, New York (in press)

Shelton, A.L., McNamara, T.P.: Systems of spatial reference in human memory. Cognitive Psychology 43, 274–310 (2001)

Shelton, A.L., McNamara, T.P.: Orientation and perspective dependence in route and survey learning. Journal of Experimental Psychology: Learning, Memory, & Cognition 30, 158–170 (2004)

Sholl, M.J., Nolin, T.L.: Orientation specificity in representations of place. Journal of Experimental Psychology: Learning, Memory, and Cognition 23, 1494–1507 (1997)

Simons, D.S., Wang, R.F., Roddenberry, D.: Object recognition is mediated by extraretinal information. Perception & Psychophysics 64, 521–530 (2002)

Tarr, M.J.: Rotating objects to recognize them: A case study on the role of viewpoint dependency in the recognition of three-dimensional objects. Psychonomic Bulletin & Review 2, 55–82 (1995)

Worboys, M., Hornsby, K.: From objects to events: GEM, the geospatial event model. In: Proceedings of the 3rd International Conference on GIScience, Phelps Hall: NCGIA (2004)

Zacks, J., Tversky, B., Iyer, G.: Perceiving, remembering, and communicating structure in events. Journal of Experimental Psychology: General 130(1), 29–58 (2001)

Modeling Geospatial Events and Impacts Through Qualitative Change

Inga Mau[1], Kathleen Stewart Hornsby[2], and Ian D. Bishop[1]

[1] Department of Geomatics, University of Melbourne,
Parkville, Victoria 3010, Australia
i.mau@pgrad.unimelb.edu.au,
i.bishop@unimelb.edu.au
[2] Department of Geography, University of Iowa, Iowa City, IA, 52242, USA
kathleen-stewart@uiowa.edu

Abstract. This paper presents a qualitative formal framework to model the impact of an event. An event could be a forest fire or a flood, for example, that results in a discernable change such as a reduced vegetation height. This framework provides a qualitative classification of impacts in order to reason about events and their impacts. The underlying conceptual model distinguishes between *immediate* and *delayed* impacts. Based on this distinction, a set of basic types of impacts are differentiated, in particular *abrupt* and *gradual* impacts. We analyze how the temporal relation between an event and an impact can be used to capture combinations of impact types, called *evolving* impacts. To link event-impact relations spatially, this work introduces the concept of qualitative impact maps that represent the extent of an impact type. The combination of qualitative impact maps with event-impact inference rules enables the identification of events that are likely to underlie these impacts. The application potential of this approach is demonstrated via a case study based on vegetation change and event data for a nature reserve near Melbourne, Australia. This study shows how the model can support decision making for planning and management in nature reserves.

1 Introduction

The representation of change is a central topic in geographic information science (GIS), since few domains if any are completely static. Instead, the properties or attributes of geospatial objects can evolve or the location of objects can vary over time. A number of approaches have focused on the description of identity changes of geospatial objects [13,14,24,25] or on spatial and non-spatial attribute changes [16,4,9]. In many cases, changes are the result of *events*. A fire event in a nature reserve, for example, burns trees and bushes resulting in an abrupt decrease in vegetation height in the reserve. The event of a traffic accident on an interstate highway could result in an impact, such as a traffic jam, or lane closures, or both. We refer to a change caused by an event as an *impact*.

This paper proposes a qualitative classification of impacts that abstracts relevant information about attribute change avoiding unnecessary detail. Vegetation changes, for example, are usually detected by analyzing an enormous amount of vegetation data. Representing the actual changes as qualitative impact types allows us to reason about impacts and events from a more cognitive perspective. A categorization of impact types assigns a small number of qualitative categories to represent the complex changes that occur to phenomena as a result of an event. The aim is to depict spatio-temporal information in a comprehensible way that supports effective decision-making. The classification of impacts and in particular their temporal relations to events is a major focus of this paper. Based on the identification of different impact types domain-specific inference tables and impact maps can be used to link event types and impact types causally and spatially. The goal of this work is to provide a model that supports the identification of likely events based on changes that have occurred in a geospatial domain, and also has the potential to help predict changes based on events that have happened.

The rest of the paper is organized as follows: in section 2, related work on event-based modeling and modeling impacts for dynamic geospatial domains is discussed. Section 3 presents a qualitative categorization of impacts based on a set of temporal event-impact relations. In order to capture more semantics associated with impacts, this typology is extended to consider additional types of impacts including abrupt and gradual impacts as well as combinations of these basic impact types. A formalization of gradual and abrupt impacts is developed and impact maps are introduced to relate the concepts spatially. In section 4, the developed model of impact categorization is applied to a case study of events and related vegetation change in a nature reserve near Melbourne, Australia. A prototype implemented in ArcMap/ArcObjects shows a user interface for querying event-impact relations in the nature reserve based on the qualitative impact types.

2 Modeling Events and Impacts

Currently, most geographic information systems (GIS) have limited ability to trace and analyze changes over time [14,16,18]. Most conventional GISs are *snapshot-based,* where only the data corresponding to the time when the data is recorded is represented. These snapshot-based approaches typically support analyses based on time-slices of data and do not offer the possibility of analyzing changes in the past or predicting future outcomes [12].

An *event-based* spatio-temporal system offers new opportunities for analyzing and understanding changes in spatial data. Integrating events into information systems facilitates a deeper understanding of spatio-temporal relationships by enabling the user to relate changes to the events underlying these changes, for example, the disappearance of a pond can be related to a drought occurring in that area [3, 25]. Where an *occurrent,* for example, a storm or a forest fire, refers to a happening in the real world that unfolds itself through a period of time [11], in this work, an *event* refers to the representation of an occurrent in a spatio-temporal data model. We treat occurrents as discrete events that can be clearly distinguished from each other, for example the representation of an individual forest fire in a certain area at a certain time. Characteristics of events are their uniqueness and countability [8]. The domain focus for this work is modeling events relating to vegetation changes. For example, burning, clearing, and flooding are typical occurrents that are experienced in managing a nature reserve or national park and that result in changes in the vegetation. This work is extended to treat the change that occurs subsequently to an event, referred to as an *impact*. The set of event-impact relations that is distinguished provides support for identifying events from impacts that have occurred and also provides a basis for the converse case, i.e., predicting impacts that are possible based on events that have happened.

2.1 Modeling Events

Early work on incorporating events in geographic information systems considered, for example, a raster-based approach for modeling spatio-temporal changes through time [19] as well as approaches to monitor and analyze successive states of spatial entities [5] where such state changes are modeled as events. An event-based approach was explored for date-stamping data separately instead of assigning a time to each snapshot and then organizing the data in chronicles that comprise data about numerous specific events relating to different objects [9]. These chronicles are then used to make inferences about the state of the world at a given time or to construct at least a partial snapshot. A more recent focus on event-based modeling for dynamic geospatial domains gives events the same status as objects in the data model. The corresponding event-based geospatial data model extends typical object-based queries to allow querying of dynamic systems by modeling object-event and event-event relationships [25] and more recently has been extended to include, for example, event-process and process-process relationships [10]. Different kinds of event-based semantics, such as significant or noteworthy events, are often associated with a dynamic geospatial domain and also require modeling support [6].

2.2 Modeling Impacts on Geospatial Objects

In order to model relationships between occurrents in the real world and the changes that result from these occurrents, we distinguish between events and subsequent associated impacts in a geospatial domain. An impact involves one or more objects changing from one state to another where the *state* of an object refers to its attribute values. For example, a land parcel has the attribute *vegetation_type* and the attribute value *forest*. An event such as a forest fire results in the vegetation type of the land parcel changing from *forest* to *heathland*. This change in the value of an attribute corresponds to the impact of the forest fire event. In reality, the trees are burnt and the vegetation structure and composition is altered. This alteration is mimicked in the model such that the state described by the vegetation type of the land parcel is changed. Another example corresponds to the building of a dam on a river where *dam_building* is the event and the impact is a change in the value of the attribute, *water_depth*, modeling the accumulating water on one side of the dam.

An important topic related to the modeling of impacts of events is that of causal relations between events and objects in dynamic spatio-temporal systems. A number of researchers have examined the topic of causality. One such study focused on modeling causality by considering the spatial overlap and temporal order of events [3]. This approach helps to identify events and the relationships among them by searching for spatial and temporal patterns using statistical and additional methods of data observation and interpretation. However, causal inferences are still left to the discretion of the user. Causal relationships are often closely related or implicit in temporal relationships and have been investigated by temporal logics to support temporal reasoning. The interval calculus [2] and event calculus [15] are early examples of temporal logics that model events as temporal objects and support inferences about temporal relationships. Causality has also been examined in models that represent spatio-temporal phenomena and relationships. For example, a generic model has been proposed for representing causal links within spatio-temporal GIS using an extended entity-relationship formalism [1]. Using this model, causal relationships between events, objects and agents are modeled and conditions applied.

A more recent approach investigates causal relations between events and states, where state refers to either an object's state or the state of a process [10]. A process, for example the flow of traffic, can have different states such as being slow or fast. Possible causal relations between events and states are *cause*, *initiate*, *terminate*, *enable*, *perpetuate*, *allow* and *prevent*. For example, an event *causing* another event is defined as the successor event *being the result of* the predecessor event. In this work, we focus on event-impact relations, in particular where events result in changes to the attributes of objects.

3 Categories of Impacts

Events can have different kinds of impacts on objects. We present a qualitative categorization of impacts beginning with the case of single events having single impacts. This categorization is developed based on a set of temporal event-impact relations where events and impacts are first modeled as intervals, and then where events are modeled as instants. This typology is extended by considering different types of impacts including abrupt and gradual impacts. These types of impacts are analyzed with respect to qualitative as well as quantitative attributes. Finally, a model for evolving impacts is developed based on combinations of basic impact types that occur successively.

3.1 Modeling Temporal Event-Impact Relations

In the vegetation domain, impacts following events like fires or clearings usually involve vegetation changes that are durative and only in exceptional cases instantaneous. These impacts last for a certain time period and are not finished in a moment of time, for example a clearing changes a forest to heathland, which then slowly regrows into a forest. For this reason, in this work we focus on modeling impacts using an interval-based perspective. Depending on the treatment of events (modeled as either durative or punctual), different event-impact relations can be distinguished and in the sections that follow different classes of temporal event-impact relations are derived systematically.

3.1.1 Events and Impacts as Intervals

When events are assumed to be durative and are modeled as intervals, then any of Allen's [2] thirteen possible temporal interval relations can be applied to describe possible event-impact relations. However, event-impact relations are assumed to conform to three basic conditions:

(1) the event starts before the impact starts, or alternatively,
(2) the event starts at the same time as the impact starts, and
(3) no event can finish later than the impact.

The first two conditions mean that an impact that results from an event cannot start before the event begins, and is based on causal logic. For the third condition, we assume that no event can last beyond its impact. For example, a fire cannot finish later than its impact, vegetation change. On the contrary, the impact of vegetation change lasts much longer than the event of the fire. This condition may not be valid for all domains, but for the domain of vegetation modeling this is a basic assumption. Given these three conditions, only a subset of the temporal interval relations is considered plausible and six temporal event-impact relations are defined: *d_before*, *d_meets*, *d_overlaps*, *d_finished-by*, *d_starts*, or *d_coincident* (where *d* indicates a durative event, see Figure 1).

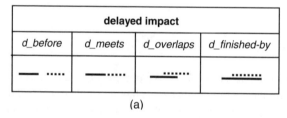

Fig. 1. When events (continuous line) and impacts (dashed line) are both modeled as intervals, possible event-impact relations include (a) delayed impacts: *d_before*, *d_meets*, *d_overlaps*, *d_finished-by*, and (b) immediate impacts: *d_starts*, and *d_coincident*

Four of these relations involve the start of the impact occurring after the start of the event. We refer to these four relations more generally as *delayed impacts*. This corresponds to a case where the event starts and finishes before the impact starts (*d_before*), the event finishes at the same time the impact starts (*d_meets*), the event starts earlier than the impact but finishes after the impact has started (*d_overlaps*), or the event starts earlier than the impact and finishes with the impact (*d_finished-by*) (Figure 1a).

The remaining two cases, *d_starts*, where the event and the impact start at the same time but the event finishes earlier than the impact, and *d_coincident*, with the event and the impact starting and finishing at the same time, are classified as *immediate impacts* (Figure 1b). The following Table 1 illustrates these event-impact relations with examples from the vegetation domain.

Table 1. Examples from the vegetation domain for possible event-impact relations

Relation	Event	Impact	Comment
d_before	fertilizing	grassland growth	Grassland growth starts at a time after fertilizing, since the nutrients need some time to initiate the growth.
d_meets	fire	regrowth of forest	The regrowth of a forest area starts immediately once the fire is extinguished.
d_overlaps	pollution	trees dying off	Trees die off at a time after the soil or air has been polluted since the pollutants need some time to reach the trees' systems. The tree dying goes on even after the pollution stopped as the trees are already weakened.
d_finished_by	grazing of heathland	growth of heath vegetation	Sheep grazing is used as a management tool to clear young trees overgrowing heathland. The regrowth of heath vegetation starts after a certain amount of young trees have been grazed. As soon as grazing stops, trees regrow and suppress the growth of heath vegetation.
d_starts	fire	destruction of old growth forest	At the time a fire starts, old trees in a forest die off and continue dying even after the fire has stopped.
d_coincident	storm	uprooting of trees	Trees start to get uprooted in a forest when a storm begins. Once the storm stops, no more trees fall over.

The categorization of relations between events and impacts modeled as intervals has been examined in the work of [7] where the terms *cause* and *effect* are equivalent to our use of event and impact. The first two of the conditions described above are similarly employed in this earlier work for distinguishing a set of causal temporal relations. In [7], depending on the starting time, the effect can be *delayed* with respect to the starting time of the cause or the effect can start *simultaneously*. A further categorization of delayed effects is provided by distinguishing *threshold delays*, where the change can only deliver its effect by reaching a certain level over a certain time period, from *diffusion delays*, where the cause and effect are spatially not co-located and the cause will need time to reach its effect. Effects that start simultaneous to their cause are further subcategorized depending if their duration is shorter, equally long or longer to their underlying cause [7]. In this work, we consider a further refinement based on the condition that an event has to finish either before or at the same time as the impact finishes.

3.1.2 Events as Instants and Impacts as Intervals

In this paper, the focus is on modeling punctual events and their relations with impacts that are durative and modeled as intervals. Although events are commonly modeled as intervals, in this work abstracting the event to an instant while treating the impact as an interval allows us to highlight additional semantics associated with impacts. Events in a nature reserve or national park, like a forest fire or clearing activity, typically take place in a very short amount of time (hours or days) as compared with the resulting vegetation changes that often happen or last over long time periods (months or years). Compared to Allen's thirteen possible interval-interval relations, temporal point-interval relations are reduced to only five possibilities: *before, starts, during, finishes* and *after* [17]. Based on the conditions that the event has to start before or at the same time as its impact, only two event-impact relations are possible when the event is punctual: *p_before* or *p_starts* (with *p* denoting a punctual event).

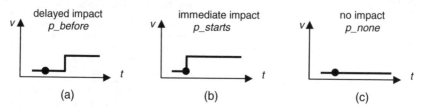

Fig. 2. Impact categories based on when the impact occurs relative to punctual events: (a) delayed impact: *p_before*, (b) immediate impact: *p_starts*, and (c) no impact: *p_none*. Continuous lines show an impact as the change of an object's state over time *t* with respect to an attribute value *v*. A point symbol is used for the event.

As with the case of interval events, *p_before* is considered a *delayed impact* (Figure 2a) and *p_starts* is considered an *immediate impact* (Figure 2b). An example of a delayed impact occurs when the event of seeding a forest (modeled as a punctual event) is followed by a period of no value change for the attribute *vegetation_height*. After some delay the value will start to increase. An example of an immediate impact is when planting trees in a forest area results in the instant change of the value of the attribute *tree_density*. Note that both these examples involve an increase in the attribute value. Of course, there are other cases such as the event of clearing trees in a forest area where the attribute value, in this case *tree_density*, decreases in response to an event.

Not all events have an impact on objects. There can also be the case *p_none* where an event has *no impact* implying no change to any attribute value of the object (Figure 2c). For example, a particular storm event may have no impact on any of the attributes of a meadow.

3.1.3 Classification of Impacts

Further refining the classification of impact types, attributes can be distinguished as: *qualitative attributes* for impacts corresponding to qualitative change and *quantitative attributes* for impacts corresponding to quantitative change. The first category includes nominal and ordinal attributes. Nominal attributes refer to an attribute domain that consists of simple labels that cannot be ordered, for example, vegetation types. Ordinal data is similarly qualitative in the sense that it reflects a rank order although without a measurable variable attached to the ranking, for example low, medium or high vitality of the vegetation. Interval and ratio attributes, on the other hand, are quantitative attributes that are measured on a scale. Ratio variables are measured with and interval variables without respect to a fixed point. Attributes such as *vegetation_height* and *tree_density* are ratio attributes since they are measured on a scale and can be related to a fixed point, in this case zero.

Based on this understanding of attributes, impacts can be distinguished further into two basic types: *abrupt impacts* and *gradual impacts*. We assume that attribute values are measured over a discrete set of time points T. Depending on the sampling rate of the application domain, time points could be measured, for example, in days, months or years. An impact function I_A is a function from T to the set of attribute values A, i.e.,

$$I_A: T \to A, t \mapsto I_A(t).$$

The first step to analyze the impact of an event is to identify a threshold value ε that is used to define discernable impacts. A *discernable impact* is defined as an attribute value change that is greater than or equal to the threshold value ε. A threshold value $\delta > \varepsilon$ is introduced to distinguish a discernable impact from an abrupt impact. These threshold values ε and δ are domain specific and can either be given by domain experts or be computed via data mining.

An *abrupt impact* involves an immediate change (increase or decrease) of an object's attribute value, that is greater than or equal to the threshold value δ. For quantitative attributes values, this would involve a jump, or sudden drop, with one state holding over an interval and then another state holding over an immediately following interval (cf the definition of discrete change in [8]). This can be formalized as follows:

$I_A|_{T'}$ is abruptly increasing iff for the finest division of the interval T', i.e., $T' = [t_0, t_1,..., t_n]$ there is an $l \in \{1, ..., n\}$ such that

$$\frac{I_A(t_l) - I_A(t_{l-1})}{t_l - t_{l-1}} \geq \delta$$

The finest division of the interval T' depends on the sampling rate that could even vary within the same sampling set. For example, if the change of an attribute value is sampled every third year, the impact is only abrupt if the yearly average change rate is greater or equal to the threshold value δ.

A *gradual impact* is reflected in the progressive change of an object from one state to another. Progressive change proceeds without significant jumps from one state to another. For quantitative variables there will be states of intermediate attribute values that follow a trend of increase or decrease.

$I_A|_{T'}$ is gradually increasing iff $I_A|_{T'}$ is not abrupt and for all $t, t' \in T'$ with $t < t'$ holds

$$I_A(t') - I_A(t) > 0 \quad \wedge \quad \exists t_1 t_2 \in T', t_1 < t_2 : I_A(t_2) - I_A(t_1) \geq \varepsilon$$

For gradually increasing impacts, the attribute value change of all possible intervals has to be positive and the overall change has to be greater than or equal to the threshold ε. Gradually decreasing and abruptly decreasing impacts are formalized respectively.

No impact is defined as a stable state of the attribute value, where attribute value changes are smaller than the threshold value ε:

$I_A|_{T'}$ is no impact iff for all $t, t' \in T'$:

$$|I_A(t) - I_A(t')| < \varepsilon$$

In the case of ordinal data, an abrupt impact would involve an immediate change of ranking, while a gradual impact involves movement through a series of rankings, for example, from *very low* to *low* over *medium* to *high* and finally *very high*. For an ordered set of non-numerical attribute values $C = \{c_1, c_2, ..., c_m\}$ the formalization is as follows:

$I_A|_{T'}$ is abruptly increasing iff for the finest division $[t_0, t_1, ..., t_n]$ of the interval T', there is an $l \in \{1, ..., n\}$ and $i, j \in \{1, ..., m\}$ such that

$$I_A(t_l) = c_i \text{ and } I_A(t_{l-1}) = c_j \text{ and } i - j > 1$$

$I_A|_{T'}$ is gradually increasing iff $I_A|_{T'}$ is not abruptly increasing and for all $t, t' \in T'$ with $t < t'$ holds

$$I_A(t_l) = c_i \text{ and } I_A(t_{l-1}) = c_j \text{ and } i - j = 1$$

$I_A|_{T'}$ is no impact iff for all $t \in T'$ there is an index i such that

$$I_A(t) = c_i$$

We assume here that a gradual increase for ordinal attributes is characterized by the change of an attribute value from one ranking to the next, while an abrupt increase is characterized by the change of an attribute value of at least two rankings. However, the number of rankings can vary depending on the application domain.

For nominal attributes an abrupt impact would involve a change to another class (e.g., from forest to heathland). A gradual impact for nominal variables cannot be smooth or progressive but may involve a set of intermediate classes. For example, a progressive change of the vegetation type from *heathland* to *forest* is characterized by an increasing number and growth of trees and can be described by intermediate forms like *woody heathland* and *heathy woodland*.

The formalization of these impact types shows that the mapping of changes to impact types depends crucially on two factors: the determination of the threshold values ε and δ and the sampling rate which determines the time points of T and sets the granularity. We assume here that the granularity is held constant in each application domain as otherwise the distinction between abrupt and gradual impacts can vary, e.g., at one granularity an impact may be modeled as abrupt while at a more refined granularity the impact may be treated as gradual.

On this account, we get two basic types of impacts involving nominal attributes: *abrupt* and *gradual* and four basic types of impacts for attributes that exhibit an increase or decrease in their value (ordinal, interval or ratio): *abrupt increase*, *abrupt decrease*, *gradual increase*, and *gradual decrease*. All of these impacts types can happen as either immediate or delayed impacts with respect to the occurrence of an event.

The categorization of impacts into abrupt or gradual (and increase or decrease) can be used as evidence for detecting causal relations between impacts and events.

Depending on the application domain, these qualitative descriptors of impacts can be used to establish rules that are applied to determine relations between the type of impact and possible events that might have caused the changes. An example is the rule that abrupt decrease in vegetation height followed by a gradual increase can only be caused by a clearing or fire event.

3.1.4 Modeling Evolving Impacts

The impact types described in the previous section can be considered as primitives, which can be further combined to describe the subsequent evolution of objects in a geospatial domain and capture semantics of change. These combinations of changes are referred to as *evolving impacts*. For example, a volcanic eruption can result in a sudden impact on the elevation of an area when the stream of lava changes the elevation of a surface by several meters causing a long-term and possibly permanent change to the value of elevation in the affected area. These changes can be modeled as an *abruptly increasing* impact on the attribute *elevation,* followed by a time interval where no further change to *elevation* occurs (Figure 3a). An *abruptly increasing* impact followed by a *gradually decreasing* impact (Figure 3b) has a sudden and immediate increasing change on the attribute value of an object that then subsequently decreases over time. For example, after a flooding event, the value of the attribute *soil_moisture* increases suddenly, but then decreases slowly with the drying of the soil. A *gradually increasing* impact where there is a progressive upward change of an attribute followed by no further change, occurs, for example, after planting trees. In this case, the attribute value of *vegetation_height* increases gradually with the growth of the trees, but once the trees have reached their maximum height they do not grow anymore and the value of the attribute *vegetation_height* is maintained (Figure 3c). A *gradually increasing* impact on an object followed by a *gradually decreasing* impact (Figure 3d), for example, occurs after the fertilizing of a meadow where the value of the attribute *vegetation_growth* increases gradually, but then subsequently decreases gradually after the nutrition is exhausted.

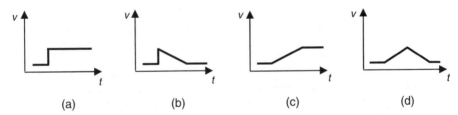

Fig. 3. Examples of combinations of impact types: (a) *abrupt increase < no change*, (b) *abrupt increase < gradual decrease*, (c) *gradual increase < no change*, and (d) *gradual increase < gradual decrease* (< refers to *followed_by*)

The following Table 2 shows pictograms of all possible pairwise combinations of the four basic impact types for attributes that are measured as ordinal, interval or ratio values: *abrupt increase, abrupt decrease, gradual increase,* and *gradual decrease*. Included are also combinations with no impact, where the attribute value does not change from its original value immediately after the event (delayed impact) or does not change further after one of the four impact types occurred (lasting impact).

Table 2. All possible pairwise combinations of the four basic impact types: *abrupt increase* (abr-inc), *abrupt decrease* (abr-dec), *gradual increase* (grad-inc), *gradual decrease* (grad-dec), and *no impact* (no). Impact types that occur first are listed in the left column and subsequent impact types in the upper row. Points represent the event instant.

first \ subsequent	abr-inc ⌐	abr-dec ⌐	grad-inc ╱	grad-dec ╲	no —
abr-inc ⌐					
abr-dec ⌐					
grad-inc ╱					
grad-dec ╲					
no —					

Combinations of impact types that are depicted in the diagonal axis include one basic impact type followed by the same impact type, which is equivalent to just the basic impact type. These cells are grayed out in Table 2. For example, *gradual increase* followed by *gradual increase* stays *gradual increase*. If we exclude the basic impact types of the diagonal (gray cells in Table 2), we obtain twenty possible combinations of plausible impact types (white cells in Table 2).

Two of these combinations involve a sudden change of attributes that is reversed immediately. For example, an *abrupt increase* followed at once by an *abrupt decrease* of the attribute value or vice versa would result in a sudden spike or a dip that could lead to no net change at the end because this combination reverses the impact immediately. The four plausible impact combinations in the bottom row of Table 2 depict delayed impacts, where no impact occurs immediately after the event but some time later. The right column includes four impact combinations where the attribute value first changes but then stays on that changed level without varying any further (lasting impacts).

The combinations grad-inc/grad-dec and grad-dec/grad-inc refer to a *gradual increase* followed by a *gradual decrease* and vice versa. These combinations are quite common, for example, in the domain of vegetation change where a fire can result in a gradual increase in the *number_of_species*, since the burned ground provides a good basis for invading species, followed by a gradual decrease in the value of *number_of_species* after the original vegetation becomes dense again and suppresses other species.

As a next step, matrices have to be developed for attributes of the respective application domain (see an example for the vegetation domain in section 4) relating impact types to event types with respect to the change of an attribute value. Attribute changes corresponding to the behavior of one of the shown pictograms in Table 2 help to identify events that could have possibly triggered the change, if the event is unknown.

3.2 Modeling Event-Impact Relations Spatially

To model the temporal event-impact relations introduced in section 3.1 spatially, impact maps are generated for each attribute that changes over time. From a cognitive point of view impact maps are useful since they abstract relevant information about attribute change avoiding unnecessary detail (e.g. long plant lists) and provide a qualitative description of change for an area of interest. An impact map is generated for a chosen time interval and attribute, for example, to show the change of the attribute *vegetation_height* for the time period from 1980 to 1990 (see Figure 4a). The map shows the evolution of impacts over the chosen time period for each area by generating impact types depending on the temporal characteristics of the attribute change.

Such an impact map is generated for each attribute that has been measured in the chosen area and time frame, in our example *vegetation_coverage* (Figure 4b) and *vegetation_type* (Figure 4c). Note that the attribute *vegetation_type* is nominal and the observed nominal changes are displayed in the map, e.g. change from forest to heathland. By intersecting all available impact maps in an area over the chosen time frame we obtain a composite map of qualitative changes (Figure 4d), i.e., the evolution of impact types, linked to a composite table of all attributes for all intersected areas (Figure 4e).

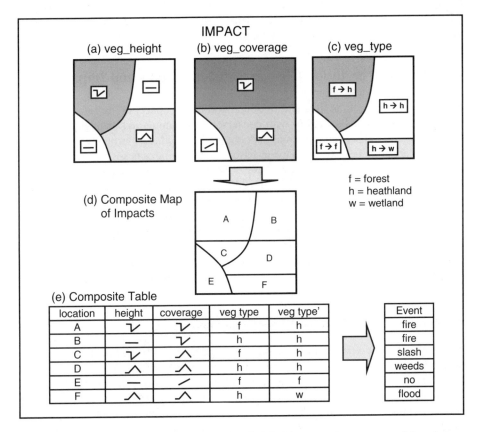

Fig. 4. Impact maps of the attributes *vegetation_height* (a), *vegetation_coverage* (b), and *vegetation_type* (c) are intersected to a composite map of impacts (d). A composite table (e) shows the evolution of impacts for each attribute in each intersected area and is used to derive the event that most likely underlies these impacts.

Event-impact rules provided by domain experts that state which events typically result in which evolution of impact types are used to derive events that most likely have happened in an area according to the observed impact type. These are typical examples for such event-impact rules, specified for a forest area:

E = fire   ↔ H (⇂) ∧ C (⇂) ∧ T (f,h)
E = slashing ↔ H (⇂) ∧ C (⌒) ∧ T (f,h)
E = flood   ↔ H (⌒) ∧ C (⌒) ∧ T (f,w)

(E = event, H = *vegetation_height*, C = *vegetation_coverage*, T = *vegetation_type*, f = *forest*, h = *heathland*, w = *wetland*)

The first rule of these denotes that if a fire event happens in a forest area then the attribute values of *vegetation_height* and *vegetation_coverage* both show an abrupt decrease followed by a gradual decrease and the attribute *vegetation_type* changes from forest to heathland. The next section shows how the concept of event-impact relations, evolving impacts and impact maps is applied to vegetation monitoring data in a nature reserve near Melbourne, Australia.

4 Modeling Vegetation Change Based on Events and Impacts

Modeling events, such as fires and flooding, and their impacts in the form of vegetation change is especially helpful with conservation planning and management of nature reserves (e.g. [22]) and national parks (e.g. [23]). Other environmental modeling tasks would also benefit from an event-based model of impacts such as forestry [20], soil erosion studies [21], land-use changes [16] or wildfire modeling [26,27]. The following section shows how the developed conceptual model for events and impacts is applied to vegetation and event data for a nature reserve near Melbourne, Australia, the Royal Botanic Gardens of Cranbourne. In such nature reserves often events happen and their impacts on the vegetation have to be estimated or alternatively, vegetation changes occur without knowing the underlying events. We illustrate a methodology for deriving events from vegetation changes (impacts) that have happened. A prototype implemented in ArcMap/ArcObjects allows users to query impacts and events.

4.1 Inference Rules for Impact-Event Relations

The Royal Botanic Garden in Cranbourne comprises roughly 360 hectares of native heathlands, wetlands and forest. The park management keeps track of events that occur in the nature reserve including wild and controlled fires, clearing, slashing (cutting trees to approximately 1m height of the trunk), and sand mining (prior to its reserve status). Vegetation changes have been monitored by ecologists through field work and additionally derived from aerial photographs over the last decades. Attributes used for indicating vegetation changes are *vegetation_type*, for example, heathland or swamp scrub, *number_of_species*, *vegetation_cover* and for some areas, *vegetation_height*. The available monitoring data consists mainly of plant lists including the coverage of each species. These lists can be sorted into ecological plant groups, for example, forest, heath, pioneer, and weed species. The sampling rate for the vegetation monitoring is yearly in most areas. To determine the type of impact (abrupt, gradual, no impact) from the observed vegetation change, the two thresholds ε and δ are defined for each (quantitative) attribute:

number_of_species:	$\varepsilon = 3$ species	$\delta = 10$ species
vegetation_cover	$\varepsilon = 5\ \%$	$\delta = 20\ \%$
vegetation_height	$\varepsilon = 0.1$ m	$\delta = 1$ m

The threshold ε specifies if we have a discernable change in the attribute value, for example a change of the *species_number* by two species is not considered a discernable change. The thresholds ε and δ are used to define gradual change: an increase or decrease of an attribute value that lies between ε and δ is considered gradual, for example, an increase of *vegetation_cover* between 5% and 20%. Changes that meet or exceed the threshold δ and occur within one sampling step, in this case, one year, are considered to be abrupt, for example, a change of the attribute value *vegetation_height* by 1 m or more in one year.

For attribute values that are measured quantitatively, for example, *number_of_species* or *vegetation_cover*, the typical evolution of impacts as well as the impact types associated with events are known by domain experts. Based on ecological experience, event-impact rules are then defined that state which impact type is typically associated with which kind of event (see section 3.2). A matrix relating the response in the *vegetation_cover* of ecological plant groups to events is used to show, for example, the evolution of impacts on actual plant species (Figure 5).

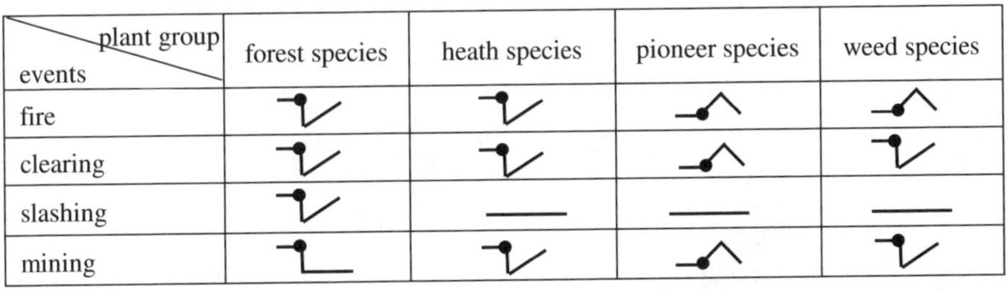

events \ plant group	forest species	heath species	pioneer species	weed species
fire	⌐\/	⌐\/	‾/\	‾/\
clearing	⌐\/	⌐\/	‾/\	⌐\/
slashing	⌐\/	—	—	—
mining	⌐_	⌐\/	‾/\	⌐\/

Fig. 5. Event-impact matrix, showing impact types occurring as a result of events with respect to the change of the attribute *vegetation_cover* of ecological plant groups like forest, heath, pioneer, and weed species

Weed species, for example, *Lantana* and *Plantain*, exhibit a *gradual increase* followed by a *gradual decrease* regarding their coverage when faced with a fire event since they spread on burnt soil but are not very shade tolerant after being overgrown by other species. With a clearing event these species show an *abrupt decrease* followed by a *gradual increase* since they are first cleared away and then begin to spread again gradually. Pioneer species, for example, *Birch* and *Fireweed*, are known for invading disturbed areas easily including poor quality soils with few nutrients, but are then later replaced by other species. This is the reason why a slashing event has *no impact* on these species since cutting trees to 1 m height does not provide disturbed soil for them to invade. However, pioneer species find optimal conditions after a mining event when the soil of an area is disturbed profoundly. These species exhibit a *gradual increase* followed by a *gradual decrease* when they become overgrown by other species.

Such a matrix is generated for each quantitative attribute, in this case another matrix was created for *vegetation_height*. Moreover, inference rules are generated for nominal and ordinal attributes, for example change of *vegetation_type*:

Vegetation Type Change:	Possible Events:
forest → heathland	fire, clearing
heathland → wetland	flooding
heathland → forest	no event

4.2 A Prototype Impact Analyzer

A prototype *Impact Analyzer* is implemented using ESRI's ArcMap™9.1 and the COM-based development technology ArcObjects to demonstrate the application of event-impact modeling in the Royal Botanic Gardens of Cranbourne (see Figure 6). The user selects an area from an impact map that is stored as an ArcMap feature shapefile for which she intends to analyze impacts and events and opens the window of the *Impact Analyzer*.

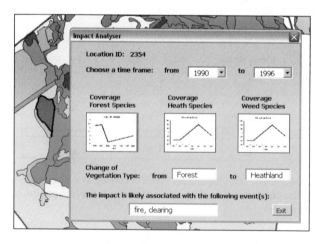

Fig. 6. Prototype for analyzing impacts and events. The user selects an object (polygon) in an impact map and chooses a time frame to display the vegetation changes over the chosen years. The evolving impact types relating to *vegetation_coverage* of forest, heath and weed species are shown. Two possible events (fire, clearing) are proposed that underlie these noted impacts.

The prototype allows the user to choose a time frame (e.g. from 1990 to 1996) and displays how the coverage of forest, heath and weed species has varied over the chosen time interval. Attribute changes are shown as graphs that correspond to the qualitative classification of evolving impact types. A sudden decrease, for example, of the attribute value *coverage* for forest species from 80% to 10% over one year between 1990 and 1991 is classified as an abrupt decrease. This sudden drop is followed by a gradual increase of the value by 10% per year for six years until a total of 60% in the year 1996. For forest species, we get the evolving impact type of abrupt decrease followed by gradual increase. The coverage of heath species and weed species shows a different impact evolvement: gradual increase followed by gradual decrease.

The *Impact Analyzer* uses the event-impact matrices that have been created for this application domain (see, for example, Figure 5) as well as the impact-event rules for the nominal attribute *vegetation_type*, and displays the result of the query by proposing all possible event types that coincide with the results from the matrices and rules. This prototype serves as a decision support system to provide users with one or more event types as a possible explanation for the detected change in forest, heath and weed species coverage. A domain expert may still be required to confirm which event type is the most likely explanation.

5 Conclusion and Future Work

In this work, a new qualitative classification of impacts on geospatial objects is developed based on investigating temporal event-impact relations. Six possible event-impact relations are identified with events and impacts modeled as intervals (*d_before*, *d_meets*, *d_overlaps*, *d_finished-by*, *d_starts*, and *d_coincident*). In order to model long-term impacts like vegetation changes that result from short-term events, such as a fire, these relations are reduced further by focusing on instantaneous events having durative impacts. The result is two possible event-impact relations: *p_before* (delayed impact) and *p_starts* (immediate impact).

A subsequent classification of impacts reveals two basic impact types based on changes to nominal attribute values: *abrupt* and *gradual* and four basic impact types for changes to attributes that exhibit an increase or decrease in their value (ordinal, interval or ratio): *abrupt increase, abrupt decrease, gradual increase,* and *gradual decrease*. All of these types can happen as either immediate or delayed impacts with respect to the occurrence of an event. These basic impact types are formalized using an attribute-based impact function to abstract these qualitative impact types from quantitative change data. Since impacts may evolve over time, the basic impact types were combined pairwise including combinations with no (further) impact to model delayed and lasting impacts. The results are 20 possible combinations for evolving impacts. These combinations are displayed as pictograms and used in qualitative impact maps that show the extent of an impact type. Event-impact rules are given for the vegetation domain to show how likely events, such as, fire or slashing, can be derived from a given set of evolving impacts regarding different attributes, for example, *vegetation_height* and *vegetation_coverage*.

The model of event-impact relations is applied to the vegetation monitoring and event data of a nature reserve near Melbourne, Australia. Based on this model, we developed the prototype *Impact Analyzer* using ArcMap/ArcObjects that enables the user to query impacts and events on the vegetation data of the Royal Botanic Gardens of Cranbourne. With the *Impact Analyzer* the user is able to derive possible underlying events if only data about the change to vegetation, i.e. impacts, is known and not the actual events that happened. For example, if ecologists discover an *abrupt decrease* followed by a *gradual increase* in the coverage of forest species but no impact on heath and weed species, it can be inferred from the event-impact analysis that the likely prior event is slashing. Such results can potentially be used for predicting future changes after an event has happened or to model a system's vulnerability to different events and so define spatially distributed risk levels for nature reserves under different scenarios. Our approach aims to support users in understanding changes that have happened and how these changes are related to events. This knowledge can help users in analyzing events and their impacts in their application domain and support decisions for planning and management.

Future work includes the refinement of event-impact rules and inference matrices in order to identify unique events from impacts. The temporal event-impact relations will also be extended to model interval-interval rather than point-interval relations to allow modeling of durative events that might be important in other application domains. These future extensions of the model will provide more support for event-based modeling of dynamic domains as well as a better understanding of event-impact relations within different domains.

Acknowledgements

Kathleen Stewart Hornsby's research is supported by the National Geospatial-Intelligence Agency under research grants HM1582-05-1-2039 and NMA201-00-1-2009. Inga Mau's work is funded by a Melbourne Research Scholarship (MRS) of the University of Melbourne. Special thanks go to Terry Coates of the Royal Botanic Gardens Cranbourne and the Royal Botanic Gardens Board Victoria for providing the vegetation data.

References

1. Allen, E., Edwards, G., Bedard, Y.: Qualitative Causal Modeling in Temporal GIS. In: Kuhn, W., Frank, A.U. (eds.) COSIT 1995. LNCS, vol. 988, pp. 397–412. Springer, Heidelberg (1995)
2. Allen, J.F.: Maintaining Knowledge about Temporal Intervals. Communications of the ACM 26, 832–843 (1983)
3. Beller, A.: Spatial/Temporal Events in a GIS. In: Proceedings of the GIS/LIS 91 Conference, Atlanta, Georgia, vol. 2, pp. 766–775 (1991)
4. Chen, J., Jiang, J.: An Event-Based Approach to Spatio-Temporal Data Modeling in Land Subdivision Systems. GeoInformatica 4, 387–402 (2000)
5. Claramunt, C., Thériault, M.: Managing Time in GIS: An Event-oriented Approach. In: Clifford, J., Tuzhilin, A. (eds.) Recent Advances in Temporal Databases: Proceedings of the International Workshop on Temporal Databases, pp. 23–41. Springer, Heidelberg (1995)
6. Cole, S., Hornsby, K.: Modeling Noteworthy Events in a Geospatial Domain. In: Rodríguez, M.A., Cruz, I., Levashkin, S., Egenhofer, M.J. (eds.) GeoS 2005. LNCS, vol. 3799, pp. 78–92. Springer, Heidelberg (2005)

7. El-Geresy, B.A., Abdelmoty, A.I., Jones, C.B.: Spatio-Temporal Geographic Information Systems: A Causal Perspective. In: Manolopoulos, Y., Návrat, P. (eds.) ADBIS 2002. LNCS, vol. 2435, pp. 191–203. Springer, Heidelberg (2002)
8. Galton, A.: Qualitative Spatial Change. Oxford University Press, Oxford (2000)
9. Galton, A.: Fields and Objects in Space, Time, and Space-Time. Spatial Cognition and Computation 4, 39–68 (2004)
10. Galton, A., Worboys, M.: Processes and Events in Dynamic Geo-Networks. In: Rodríguez, M.A., Cruz, I., Levashkin, S., Egenhofer, M.J. (eds.) GeoS 2005. LNCS, vol. 3799, pp. 45–59. Springer, Heidelberg (2005)
11. Grenon, P., Smith, B.: SNAP and SPAN: Towards Dynamic Spatial Ontology. Spatial Cognition and Computation 4(1), 69–104 (2004)
12. Halls, P.J., Polack, F.A.C., O'Keefe, S.E.M.: A New Approach to the Spatial Analysis of Temporal Change Using Todes and Neural Nets. Cybergeo (The European Journal of Geography) Article 139 (2000)
13. Hornsby, K., Egenhofer, M.: Qualitative Representation of Change. In: Frank, A.U., Hirtle, S. (eds.) COSIT 1997. LNCS, vol. 1329, pp. 15–33. Springer, Heidelberg (1997)
14. Hornsby, K., Egenhofer, M.: Identity-Based Change: A Foundation for Spatio-Temporal Knowledge Representation. International Journal of Geographical Information Science 14, 207–224 (2000)
15. Kowalski, R.A., Sergot, M.J.: A Logic-Based Calculus of Events. New Generation Computing 4, 67–95 (1986)
16. Marceau, D.J., Guindon, L., Bruel, M., Marois, C.: Building Temporal Topology in a GIS Database to Study the Land-Use Changes in a Rural-Urban Environment. The Professional Geographer 53, 546–558 (2001)
17. Ohlbach, H.J.: Relations between Fuzzy Time Intervals. In: Proceedings of Temporal Representation and Reasoning, 2004. TIME 2004. 11th International Symposium, pp. 44–51 (2004)
18. Peuquet, D.J.: Making Space for Time: Issues in Space-Time Representation. Geoinformatica 5, 11–32 (2001)
19. Peuquet, D.J., Duan, N.: An Event-Based Spatio-Temporal Data Model for Geographic Information Systems. International Journal of Geographical Information Systems 9, 7–24 (1995)
20. Rasinmäki, J.: Modelling Spatio-temporal Environmental Data. Environmental Modelling & Software 18(10), 877–886 (2003)
21. Renschler, C.S., Diekkrüger, B., Mannaerts, C.: Regional Soil Erosion Risk Evaluation Using an Event-Based GIS-Model-Scenario Technique. In: Laflen, J., Tian, J., Huang, C. (eds.) Soil Erosion & Dryland Farming, pp. 365–379. CRC Press, USA (2000)
22. Twumasi, Y.A.: The Use of GIS and Remote Sensing Techniques as Tools for Managing Nature Reserves: The Case of Kakum National Park in Ghana. In: IEEE International Geoscience and Remote Sensing Symposium, Sydney, Australia, July 9-13, pp. 3227–3229. University of New South Wales (2001)
23. White, J.D., Ryan, K.C., Key, C.C.: Running SW Remote Sensing of Forest Fire Severity and Vegetation Recovery. International Journal of Wildland Fire 6(3), 125–136 (1996)
24. Worboys, M.F.: Modelling Changes and Events in Dynamic Spatial Systems with Reference to Socio-economic Units. In: Frank, A.U., Raper, J., Cheylan, J.-P. (eds.) Life and Motion of Socio-Economic Units, ESF GISDATA series, vol. 8, pp. 129–138. Taylor and Francis, London (2001)

25. Worboys, M.F., Hornsby, K.: From Objects to Events. In: Egenhofer, M.J., Freksa, C., Miller, H.J. (eds.) GIScience 2004. LNCS, vol. 3234, pp. 327–343. Springer, Heidelberg (2004)
26. Yuan, M.: Use of Knowledge Acquisition to Build Wildfire Representation in Geographical Information Systems. International Journal of Geographic Information Science 11(8), 723–745 (1997)
27. Yuan, M.: Representing Complex Geographic Phenomena with both Object- and Field-like Properties. Cartography and Geographic Information Science 28(2), 83–96 (2001)

Preferred Mental Models: How and Why They Are So Important in Human Reasoning with Spatial Relations

Marco Ragni[1], Thomas Fangmeier[2], Lara Webber[2], and Markus Knauff[2]

[1] Department of Computer Science, Georges-Köhler-Allee
79110 Freiburg, Germany
ragni@informatik.uni-freiburg.de
[2] Center for Cognitive Science, Friedrichstraße 50
79098 Freiburg, Germany

Abstract. According to the mental models theory, humans reason by constructing, inspecting, and validating mental models of the state of affairs described in the premises. We present a formal framework describing all three phases and testing new predictions about the construction principle humans normally use and about the deduction process itself – the model variation phase. Finally, empirical findings in support of these principles are reported.

1 Introduction

Everyday spatial reasoning is strongly connected to the extensive use of spatial relations which locate one object with respect to others. This covers a whole range of situations: from finding where Munich is in relation to Vienna, to deciding which shelf goes where when constructing a cabinet from assembly instructions. In the present paper we restrict the focus of investigation to relational reasoning problems in the spatial domain. However, the results are also applicable to the temporal domain as well. A typical example is:

The hammer is to the right of the pliers.
The screwdriver is to the left of the pliers.
The wrench is in front of the screwdriver.
The saw is in front of the pliers.

Which relation holds between the wrench and the saw?

The four sentences given in the example above are called *premises*, each premise consists of the *terms* (tools), and the *relation* between these tools (i.e. right of). A premise like "The hammer is to the right of the pliers" consists of a 'reference object' (RO), in this case the pliers, and a 'to be located object' (LO), the hammer. The question refers to a possible *conclusion* that follows from the premises. The correct answer for this question is that the wrench is to the left of the saw.

There are basically two main cognitive, as well as mathematical, approaches about how humans solve these problems: *syntactic* and *semantic* theories. For example, Rips (1994) suggested that humans solve these problems by applying formal transitivity rules to the premises, whereas the *mental model theory (MMT)*, proposed by Johnson-Laird and Byrne (1991), suggests that people draw conclusions by constructing and inspecting a spatial array that represents the state of affairs described in the premises (see also Johnson-Laird, 2001). The latter seems to be the more reasonable theoretical account in terms of empirical and neuronal evidence (Goodwin & Johnson-Laird, 2005; Fangmeier, Knauff, Ruff, & Sloutsky, 2006).

A limitation of the MMT so far is that this theory does not explain the difficulty humans have with complex relations. The MMT explains the complexity of reasoning tasks in terms of the number of models which need to be investigated, but neglects the complexity of the construction of the models. This is, we believe, where the *relational complexity theory (RCT)* introduced by Halford (1993) comes into play. We believe that this approach can help to address the complexity of model generation, and the complexity of processing the differences between distinct models. Different models representing a spatial description can be measured in terms of cognitive economicity (Halford, 1993; Goodwin & Johnson-Laird, 2005). However, a weakness of the RCT is that it does not explain how the reasoning process itself works. In other words both theories are limited to some extent, and more recent research suggests that a combination of these theories is a more fruitful way of explaining how humans reason deductively (Goodwin & Johnson-Laird, 2005; Ragni, Fangmeier, Webber, & Knauff, 2006). In the following we will describe the theory of preferred mental models (PMMT) for human spatial reasoning in more detail. PMMT is a combination, and to some degree an extension, of MMT and RCT.

In this paper we summarize the current state of our project in which we try to integrate the MMT and RCT to explain human spatial reasoning and why some problems are more difficult to solve than others. In section 2 we review some basic concepts of the Preferred Mental Model Theory and in section 3 we present a generalized computational model for human reasoning with spatial relations based on this theory. This generalization implies a cognitive measure, which is introduced in section 4. Section 5 presents an analysis of the most important reasoning phase – the model variation phase. In section 6, we present empirical data in support of our computational model. Finally, section 7 discusses the results presented in the paper and gives a short overview of some questions that are left open.

2 The Theory of Preferred Mental Models

First of all, we have to define what a (mental) *model* is. A model is defined in the usual logical sense as a structure in which the premises are true. Psychologically, a model is an internal representation of objects and relations in spatial working memory that matches the state of affairs given in the premises of the reasoning task. From the *representational view*, the model could account for *metrical* or *relational* information. The former is the more constraining (i.e. stronger) and usually identified with visual mental images. The latter is less constraining (i.e. weaker) and typically identified with spatial representations (Schlieder & Berendt, 1998). Schlieder has developed a computational model for spatial reasoning with intervals with continuous structures. However, his linearization principle cannot be applied in our case, since it only explains the economic arrangement of intervals in the model construction phase, and not in the other phases. Following the principle of representational parsimony, our account is based on relational information alone. Thus, models are spatial representations that are more abstract than visual images. They avoid excessive visual detail to bring out salient information for inferences (Knauff & Johnson-Laird, 2002).

2.1 The Deduction Process

The semantic theory of MMT is based on the mathematical definition of deduction, i.e. a propositional statement φ is a consequence of a set of premises ψ, if in each model of ψ, φ is true. In other words, there is no counter-example, i.e. a model where ψ holds but not φ. In mathematical logic the way models are generated is not of importance so in this sense the MMT is more concrete.

The MMT describes the reasoning process in three distinct phases (Johnson-Laird & Byrne, 1991). In the *comprehension phase*, reasoners construct a mental model that reflects the information from the premises. If new information is encountered during the reading of the premises it is immediately used in the construction of the model. During the *description phase*, this model is inspected to find new information that is not explicitly given in the premises. Finally, in the *validation phase* alternative models are searched that refute this putative conclusion.

However, some questions remain open with regards to how people deal with multi-model problems. For example, which model is constructed first, and does this model construction adhere to certain principles? Why do reasoners neglect some models? All these questions are not answered by the classical mental model theory.

2.2 Preferred Mental Models

The preferred mental model theory (Knauff, Rauh, & Schlieder, 1995; Knauff, Rauh, Schlieder, & Strube, 1998; Ragni, Knauff, & Nebel, 2005; Ragni, et al., 2006) is based on a combination of MMT and RCT, since the MMT is used to explain the deduction process while the RCT is used to explain human preferences in constructing certain models. The first two phases are identical to the phases identified in the MMT (see above), although we talk about the comprehension phase in terms of construction and the description phase in terms of model inspection. The main difference in the theories is that Knauff et al. (1998) argue that the empirical data does not support the notion that people do construct several models. Instead, people only construct one model and then vary this model by a process of local transformation. This model, which is called the preferred mental model (PMM) is easier to construct and to maintain in working memory compared with all other possible models (Knauff, et al., 1998; Ragni, et al., 2006). In the model variation phase this PMM is *varied* to find alternative interpretations of the premises (e.g. Rauh, Hagen, Knauff, Kuß, Schlieder, & Strube, 2005).

3 The SRM Model

In (Ragni et al., 2005), we proposed a computational model for spatial reasoning by means of mental models. Our SRM model (Spatial Reasoning by Models) maps spatial working memory to a two-dimensional array and uses a spatial focus that places objects in the array, manipulates the position of objects, and inspects the array to find spatial relations that are not given in the premises. In the following we summarize how the three inference steps work.

3.1 Model Construction Phase

We assume that in the construction phase humans follow the principle of economicity (Manktelow, 1999), i.e. a model is constructed incrementally from its premises. Such a model construction process saves working memory capacities because each bit of information is immediately processed and integrated into the model (Johnson-Laird & Byrne, 1991). Therefore, the information from the premises does not have to be stored. Vandierendonck, Dierckx, and De Vooght (2004), introduce the idea of a focus performing this manipulation of the objects and additionally, as in case of indeterminacy of the premises, annotations to objects are written. We outline the SRM-model using the following example:

> A is to the left of B.
> B is to the left of C.
> B is to the left of D.
> D is to the left of E.

The focus inserts in a spatial array the first object A, it then moves to the right of A and inserts the object B, then it moves to the right of B and inserts object C. This description is so far deterministic, since it leads to one only model.

> A   B   C

Then the SRM processes the third premise "B is to the left of D". The focus moves back to the already inserted object B and moves one step to the right to insert object D, it finds the cell occupied (by object C) so it moves to the right of C inserts object D, and makes the annotation (rB) on D to indicate indeterminacy. This gives us the first model

> A   B   C D_{rB}

and this model is constructed according to the insertion strategy we call *first free fit (fff)*. This means that an object is inserted at the first free position. Alternatively, object D could be placed into the cell on the right of B. If this cell is occupied by another object (object C), this is then shifted to the next cell. This strategy is called *first fit (ff)* principle, and it gives us the second possible model.

> A   B D_{rB} C

In other words the *ff* principle always inserts the object at the next position that fulfills the spatial relation specified in the premise. The next premise "D is to the left of E" is processed. Since in the model variation phase all possible models must be producible, we need to annotate the newly inserted object if this is related to an already annotated object (in this case the position of E depends on the position of D, and D itself is dependent on object B). We get now three possible models, which are constructed according to the different strategies:

PMM	MMM	FMM
A B C D$_{rB}$ E$_{rD}$	A B D$_{rB}$ C E$_{rD}$	A B D$_{rB}$ E$_{rD}$ C

These three models have not been constructed according to the same construction principles but the first model (which is called the preferred mental model, PMM) by the *fff*-principle, the second by a combination of the *ff* and *fff*-principle (a mixed mental model, MMM) and the third by the *ff*-principle (called FMM). By assuming a consequent insertion principle the second model should be ignored in constructing models. The *fff*-principle can be explained using the SRM model (Ragni et al., 2005) as the more economic strategy. Such preferences lead to the PMMT which is a combination of MMT and RCT (Ragni et al., 2006). The *ff*-strategy could be explained linguistically, i.e. humans tend to interpret a relation like "to the right" as "directly to the right". However empirical findings (which will be presented later) show that the PMM is the first constructed model in nearly all cases. For this reason we continue our investigations by taking this model:

A   B   C   D$_{rB}$ E$_{rD}$

3.2 Model Inspection Phase

Having given this model we can derive by inspecting it new facts, like C is to the right of A, D is to the right of A, E is to the right of C and so on. Following the above mentioned deduction principle a conclusion like E is to the right of C is only true if and only if it is true in all models. This brings us to the model variation phase.

What happens if a given conclusion must be verified? Assume that the SRM should validate the conclusion "E is to the right of C". In this case, the focus moves to the C (RO) and then scans the array to the left to find the E (LO). Since the conclusion is true in the model at hand the SRM generates the output "valid conclusion".

3.3 Model Variation Phase

We come now to the core of the deduction phase – the phase where the putative conclusion is checked in all models consistent with the premises. However, we do not assume that humans check all models (due to limitations of time and memory capacity), so which models are checked? And is there a common principle?

We definitely assume that there is no iteration of the first two phases in which arbitrary alternative models are generated and inspected in turn (Johnson-Laird & Byrne, 1991). Instead, the SRM model follows the idea that humans tend to preserve an initially built model as much as possible. For these reasons the SRM changes the model at hand by changing the positions of two neighbored objects. This principle is called the principle of continuous changes. Such changes are precise in a generalized neighborhood graph (Ragni & Wölfl, 2005). As a consequence of this principle it is possible to explain, why humans sometimes find counter-examples and why sometimes they overlook them – if the counter-example is very distant from the PMM, then – we assume – the process of variation is terminated before the counter-example is found (due to the amount of variation). The minimal changes follow the principle of "conceptual neighborhood" which we have empirically determined in recent studies (Rauh et al., 2005). The principle says that alternative models are generated by local transformations, i.e. moving one object in the model. To find the next alternative model, the SRM model starts from the RO of the conclusion and first checks if the next objects have annotations with respect to the LO. As already mentioned, this annotation basically stores the relation that must hold between RO and LO. If this is the case, then the SRM model starts to change the position of the objects as long as the constraint from the annotation is satisfied. This takes it stepwise to alternative models but also has the consequence that models which are difficult to reach are thus more likely to be neglected than models which are only minor revisions of the PMM.

$$A \quad B \quad C \quad D_{rB} \quad E_{rD}$$

We present now the model variation phase for the above given model and we want to validate or falsify the conclusion E is to the right of C. Obviously, in order to check if the relation *right of* between E and C holds in all models, we have to investigate the annotations of E and C, since an annotation on another term does not have an influence (in one-dimensional problems) on the position of the terms E and C. The essence of the algorithm is that the focus searches for the RO C in the model. Then the focus checks if the RO C has an annotation. In our case the LO E has an annotation, therefore the model variation starts with this object. To check if E is right of C means, we have to move E in the contrary direction as far as possible. Since it has an annotation (relating E and D), and since D is the direct left neighbor of E, E cannot be moved. So in this case, we move D to the left and get the following model:

$$A \quad B \quad D_{rB} \quad C \quad E_{rD}$$

The model inspection shows that E is in this model still to the right of C. In the next step, we can move E to the left and get the counter-example:

$$A \quad B \quad D_{rB} \quad E_{rD} \quad C$$

4 Complexity of Spatial Reasoning with Mental Models

Our aim is to explain human difficulties with relational deduction tasks. We claim that complexity in spatial reasoning stems from two different sources, namely the relational complexity of the premises or conclusion (*description complexity*), and the complexity to deduce, from a given set of premises, a conclusion (*deduction complexity*). The *description complexity* is best described via the RCT as it provides the best explanation for the difficulty humans have with decomposing complex relations, where for *deduction complexity* the main concept is an abstract "unit" that stands for the number of operations in the array and the number of relations that must be handled. Therefore, the relational complexity is able to explain the difficulty of complex relational terms (in both the construction and inspection phase) or the difficulty in understanding and decomposing complex concepts, where the deduction complexity explains the difficulty in representing and reasoning with such problems. For instance a ternary relation such as "A is in-between B and C" can be decomposed (in the spatial array) into binary relations "B is to the right of A" and "B is to the left of C" or vice versa. The reason for the decomposition is due to the tendency of individuals to reduce the representational complexity as much as they can (Halford, Wilson, & Phillips, 1998).

The deduction complexity is the complexity to validate or falsify from a given set of premises a putative conclusion. Since the deduction process consists mainly of model construction, inspection, and variation, and since these processes are done by the focus, the difficulty of tasks clearly depends on the operations of the focus. We abstract from different costs for different operations of this machine and use a *uniform complexity measure*. Nonetheless, we can show that the empirical differences in reasoning difficulty can be captured by this measure (Ragni et al., 2005; Ragni et al., 2006). We show how this measure works on our example:

>   A is to the left of B.
>   B is to the left of C.
>   B is to the left of D.
>   D is to the left of E.

The focus inserts in a spatial array the first object A, (cost 1 unit) then moves to the right of A (cost 1 unit) and inserts the object B (1 unit), then moves to the right of B (1 unit) and inserts object C (1 unit) and so on.

This complexity measure allows us to state rather precise predictions on the processing time of specific spatial reasoning problems (Ragni, et al., 2005). The measure can differentiate between problems with different term and premise orders (Ragni et al., 2005) and, thus, can also account for the "figural effect" in relational reasoning (Johnson-Laird & Bara, 1984; Knauff et al., 1998). This complexity measure makes several predictions about the construction of PMMs, the difficulty of more complex relations, and the strategies used in model variation to revise the model at hand. Obviously the explanation of these effects provides insight into how humans reason. By analyzing which models are ignored, we would know why humans err. In the next section we present empirical data in support of the SRM.

5 An Analysis of the Model Variation Phase

The SRM-model assumes a continuous change of models, and since time and working memory is limited, we assume that humans investigate only a small number of models. If the counter-example is over a certain distance from the first generated mental model, the preferred mental model, than the counter-example is ignored. We will conceptualize this continuous change through the generalized neighborhood graph (Ragni & Wölfl, 2005). The concept of neighborhood graph was introduced by Freksa (1991) for conceptualizing the continuous change of one base relation of spatial calculi into another. The generalized neighborhood graph is based on the notion of continuity but generalizes the classical neighborhood graph (Freksa, 1991) to conceptualize the similarity of models or in our case neighbored models.

>   A is to the left of B.
>   B is to the left of C.
>   B is to the left of D.
>   D is to the left of E.

Leads to 3 valid models: (PMM) A B C **D E**
                                A B **D** C **E**
                                A B **D E** C

(PMM) (MMM) (FMM)

Fig. 1. The Generalized Neighborhood Graph (GNG)

The advantage of having a generalized neighborhood graph is that we can determine the *transformation distance* between models by an assumed continuous transformation (see Fig. 1). In other words, it takes "longer" to transform the PMM into the model on the right (which is constructed according to the *ff*-principle, FMM) than it does to transform it into the model in the middle (which is constructed according a mixed strategy, MMM). We call this distance *the transformation distance*. From a formal perspective we cannot identify the transformation distance in human reasoning but we will investigate this question empirically in Experiment 3. The same question is of interest, if we analyze invalid models. If we change in a valid model two neighbored objects in such a way that it contradicts the premise, or two very distant objects like the first and the last object – which model is more likely to be rejected by humans?

6 Experimental Investigations with Individuals

We investigated some important questions about the *construction* process and the *variation* process with indeterminate relational reasoning problems. First, we examined the construction phase and questioned which of the previous introduced principles first fit (*ff*) or free first fit (*fff*) are more likely to be used when participants construct possible models from an indeterminate problem. Second we addressed the question of whether the level of the complexity affects the PMM, the corresponding accuracy and reaction times during the verification. The third question deals with the variation phase during validation. We investigated how and when people vary the principles in a verification task when they are offered a non-preferred mental model (¬PMM).

Preferred Mental Models: How and Why They Are So Important in Human Reasoning   183

Table 1. Different problems [a] – [f]: The left column contains the premise structure of the different problems, the middle column contains the possible models and the last column contains additional comments for each possible model. For the question dealing with the *fff*-principle we used all depicted problems [a] to [f], for the question dealing with the relational complexity we used just problems [a] and [b]. The problems [a] and [b] are indeterminate two-dimensional with two possible models; the problems [c] and [d] are one-dimensional with three or six possible models, and the problems [e] and [f] are determinate one- and two-dimensional with only one possible model. The models indicated in bold letters are the hypothesized PMM for each problem constructed in line with the *fff*-principle. The mirrored two-dimensional problems [a] and [b] are not shown.

	Problem	*Possible models*					*Comment*
[a]	A is to the left of B. C is to the right of A. D is behind C. E is behind A.	(1)	**E** **A**	**B**	**D** **C**		*fff* (PMM) U shape
		(2)	E A	D C	B		*ff* (¬PMM) U shape
[b]	A is to the left of B. C is to the right of A. D is behind C. E is behind A.	(1)	**E** **A**	**B**	**C** **D**		*fff* (PMM) Z shape
		(2)	E A	C D	B		*ff* (¬PMM) Z shape
[c]	B is to the right of A. C is to the right of B. D is to the right of B. E is to the right of C.	(1)	**A**	**B**	**C**	**D**	**E**	*fff* (PMM)
		(2)	A	B	D	C	E	*ff* (¬PMM)
		(3)	A	B	C	E	D	-- (¬PMM)
[d]	B is to the right of A. C is to the right of B. D is to the right of B. E is to the right of B.	(1)	**A**	**B**	**C**	**D**	**E**	*fff* (PMM)
		(2)	A	B	C	E	D	--- (¬PMM)
		(3)	A	B	D	E	C	--- (¬PMM)
		(4)	A	B	D	C	E	--- (¬PMM)
		(5)	A	B	E	C	D	--- (¬PMM)
		(6)	A	B	E	D	C	*ff* (¬PMM)
[e]	A is to the left of B. C is to the right of B. D is behind C. E is behind A.		E A	B	D C			determinate
[f]	B is to the right of A. C is to the right of B. D is to the right of C. E is to the right of D.		A	B	C	D	E	determinate

6.1 Does the fff-Principle Reflect How Humans Construct Preferred Mental Models?

The *first* question we investigated was if participants adopt the *fff*-principle when constructing a PMM. Twenty participants were shown different one- and two-dimensional problems with one or more possible models (see Table 1). The five terms were randomly replaced with the name of a fruit (lemon, orange, kiwi, peach, mango, and apple). The premises were presented to the participants on a computer screen. Each premise was presented sequentially (in a self-paced manner), and remained on the screen until the presentation of the fourth (and final) premise. After the button press all the premises were then removed and the participants were asked to draw the model on a sheet of paper. They were free to draw more than one model if they noticed that this was possible (but only 10% of the participants did!). However, they were neither instructed to draw more than one model, nor told that in some tasks more than one model was possible. We found that 78% of participants drew the PMM.

These results support the assumption that participants prefer to construct models in line with the hypothesized *fff*-principle which leads to the PMM. Even though there are more than two models possible, the majority of the participants drew only one: the PMM.

6.2 Does Relational Complexity Interact with the Construction of PMM?

The *second* question examined whether relational complexity affects the PMM. Participants were shown two-dimensional problems with two possible models (see Table 1 problems [a] and [b]). The procedure for the presentation of the premises followed the same format as in the prior experiment. After deletion of the premises a conclusion (set of relations) was presented on the screen. We asked the participants about the relations between the different terms in the model constructed from the four premises.

The complexity level differs in three steps: binary, ternary, and quaternary (see Table 2). In binary relations we asked only if a term is "near to" another term. In ternary relations we asked about two relations with three terms and in quaternary relations we asked about two relations between four terms. The comparison "near to" means that one term is in direct contact with another, regardless of the dimension (horizontal, vertical, diagonal). If one imagines a grid with one term in the center then this means another term that is "near to" can be in one of eight possible positions. All participants were given training before the experiment commenced to ensure that they had correctly understood the problems and the relations between the terms.

Table 2. An example for a two-dimensional problem with two possible models and offered relations that vary the complexity: binary, ternary and quaternary

Possible models	Complexity level	Offered relations
PMM E D A B C	Binary	Is C near to B and B near to C?
	Ternary	Is C as near to B as B is near to D?
¬PMM E D A C B	Quaternary	Is A as near to D as C is near to E

In comparison to the binary relations the errors increased the higher the complexity (see Fig. 2), which also resulted in an increase in the corresponding reaction times.

Participants rejected the offered relations although they are logically correct. These results indicate that the processing efforts for higher complexity level (ternary and quaternary) led to more errors. But do higher complexity levels affect the PMM, the ¬PMM or both?

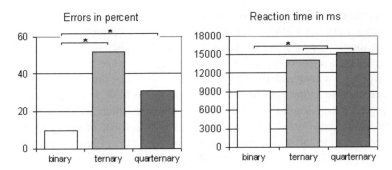

Fig. 2. Errors and reaction times for the logically valid offered relations with different complexity steps (binary, ternary and quaternary). The asterisks denote that the difference was significant $\alpha \leq 0.05$.

To deal with this question we used additional sets of relations (see Table 3). The logically valid relations are introduced in Table 2, the other relations hold only for the PMM, or hold only for the ¬PMM. We were especially interested in the proportions of acceptance and denial of relations that hold only in one of these possible models (PMM or ¬PMM).

Table 3. An example of a two-dimensional problem with two possible models and offered relations. However, only the first offered relations [1] are valid in the logically sense, [2] holds only in the PMM and [3] holds only in the ¬PMM. All offered relations in the table have a ternary complexity. The other complexity levels are also used but not shown. Note that a cell that is in contact to another is "near to". If you imagine a grid with a term in the center than all eight possible positions around are "near to".

Possible models	Offered relations (ternary complexity)	Logically valid	Holds in the PMM	Holds in the ¬PMM
PMM E D A B C	[1] Is C as near to B as B is near to D?	Yes	Yes	Yes
	[2] Is E near to B as B near to C?	No	Yes	No
¬PMM E D A C B	[3] Is E near to C as C near to B?	No	No	Yes

The most important finding was that we did not find the same rejection rate for the two offered relations that hold only in one model. If the participants are able to construct both models then they should be able to find the inconsistencies in the offered relations that hold only in one of the possible models. The offered relation should then be rejected. Interestingly this was not the observed behavior of the participants, since it was more difficult to accept relations that only hold in the ¬PMM as compared to those that hold only in the PMM.

The question remains that if the participants react in a consistent manner, is this because they have the PMM in mind more often and thus neglect the ¬PMM. If so then they should not find inconsistencies for offered relations that hold only in the PMM, but should reject offered relations that hold only in the ¬PMM, and vice versa. For this reason we analyzed all answer pairs with the same premises but with offered relations that hold either only in PMM or only in ¬PMM. If the participants have PMM in mind they should say "yes" to the relation which is only correct for the preferred model and "no" for the one that was only correct for the ¬PMM. For this analysis we excluded all combinations in which each of the offered relation pairs were both answered with "yes" or "no". Overall we found 77% answer pairs for the PMM and only 23% answers for the ¬PMM whereas the reaction times for the preferred and non-preferred conclusions did not differ.

Several outcomes resulted from these findings. Higher complexity in the offered relations led to higher processing efforts. Furthermore, it seems to be easier to reject an offered relation which holds only in the ¬PMM than to reject one that holds only in the PMM. In addition we found again a clear preference for the PMM when we analyzed the consistency of answer pairs in which the relation holds only in one possible model.

6.3 How Is the PMM Varied to Find Alternative Models?

The *third* question investigated the variation of constructed models during the final step of the inference process. We were interested in the accuracy and reaction times for different offered possible valid and invalid models. Note that we use the term "valid" not in the logical sense but as in a "valid" model is consistent with the problem (premises) and an "invalid" model is inconsistent with the problem (premises). Participants were shown indeterminate problems (see Table 4).

Each premise consists of two terms of fruits (apple, pear, peach, kiwi, and mango) and a relation ("left of" or "above"). Here we, again, summarize only the main findings, for a full report see Ragni, Fangmeier, & Knauff (in prep.).

For each problem one of six models was offered, three valid and three invalid. The valid models can only be constructed using the *fff*-principle (PMM), the *ff*-principle, or a mixture of both principles. An additional variation was the premise order of the first and second premise. The three invalid models were constructed by varying the terms in the PMM in one of three ways: a *near* exchange of the second and third term, a *middle* exchange of the second and fourth term and a *far* exchange of the first and last term.

Table 4. Indeterminate 5-term series problems with 4 premises. At half of the problems the first and second premise was exchanged (I-II and III-IV). Each premise consists of two terms (fruits kiwi, mango, pear, apple, peach [in German: Kiwi, Mango, Birne, Apfel, Pfirsich]) and a relation ("left of" or "above"). One of the six models was offered in one problem. The models [1] to [3] are possible – valid – models ("valid" means consistent with the problem); they are constructed corresponding to [1] the *fff*-principle (PMM), [2] a mixture of the *fff*- and the *ff*-principles, and [3] the *ff*-principle. The models [4] to [6] are invalid models ("invalid" means inconsistent with the problem) with an exchange in [4] the second an third term (near exchange), [5] in the second and fourth term (middle exchange) and [6] in the first and last term (far exchange).

	Problem	Offered model
I	1. A is to the left of B. 2. B is to the left of C. 3. B is to the left of D. 4. D is to the left of E.	[1] **A B C D E** [2] **A B D C E** [3] **A B D E C**
II	1. B is to the left of C. 2. A is to the left of B. 3. B is to the left of D. 4. D is to the left of E.	[4] **A C B D E** [5] **A D C B E** [6] **E B C D A**
III	1. A is above B. 2. B is above C. 3. B is above D. 4. D is above E.	[1] [2] [3] [4] [5] [6] **A A A A A E** **B B B C D B** **C D D B C C** **D C E D B D** **E E C E E A**
IV	1. B is above C. 2. A is above B. 3. B is above D. 4. D is above E.	

The participants read the four premises self-paced in a sequential order. Each premise was displayed in the middle of the screen and disappeared before the next was shown. The offered models consisted of either a horizontal or vertical enumeration of the terms (written out fruit names). The participants were asked to decide whether the offered model is a possible model of the previous premises.

As in the previous results we found the highest correct answers (92%) if the participants validated models which were congruent with the *fff*-principle and the reaction times were the fastest ($M = 3797$ ms, $SD = 1640$ ms). In comparison to the earlier problems we presented problems with more than two alternative models which could be constructed by one of the two principles, but also by a mixture of both (mix-principle). From the construction point of view the *fff*- or the *ff*-principle makes much more sense with regard to simplicity and using a single strategy in comparison to the mix-principle. However, we found that the *fff*-principle was better than the *mix*-principle and the *mix*-principle was in turn better than the *ff*-principle (see Figure 3). The same pattern was obtained in the reaction times. Note, that the participants' verification of the offered models constructed by the *ff*-principle was at chance level.

For the invalid models we found an increase in the error rate from far to near exchange and the reaction times were also faster for far and middle. As in the results before the accuracy for constructed models in line with the *fff*-principle were the highest and the reaction time the fastest. Once again these indicate that it is much more likely that the participants construct a PMM (*fff*-principle), since the accuracy was nearly perfect and they reacted faster if they were asked to validate a PMM.

Interestingly we found a pattern for the other two offered models that were from the construction point of view most improbable. We hypothesized that the *mix*-principle needs a strategy change within a single problem. However, from the validation point of view it is very comprehensible. If we assume that the participants constructed the PMM (following the *fff*-principle), then it is easier to exchange only the fourth term with the third (you get the model that is constructed after the *mix*-principle). The variation of the PMM to validate the model which is constructed in accordance to the *ff*-principle needs more effort, since two terms have to be exchanged and more relations (and rules) have to taken into account.

For the invalid models we found patterns that would be expected if the participants had the PMM in mind, since we constructed the invalid models starting from the PMM. The nearer the exchanged terms were the harder it was to reject the invalid model.

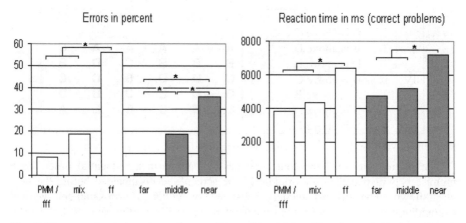

Fig. 3. Errors and reaction times for possible valid models (*fff*-, mix- and *ff*-principle) and invalid models (far, middle and near exchange of terms). Bar charts with white surfaces were results from possible valid models, whereas bar charts with grey surfaces were invalid models. The asterisks denote that the difference was significant $\alpha \leq 0.05$.

7 General Discussion

We have presented a formal theory and tested new (and hitherto untested) predictions made by our formal SRM-model. The predictions we made regarding insertion principles, the model variation phase, and the search for counter-examples, and all theoretical predictions have been corroborated by the experiments, and therefore indirectly support the formal SRM-model. The purpose of the SRM-model is to have a computational theory for relational reasoning. Nonetheless, the implied complexity measure of the SRM is powerful enough to be a base for the introduction of a cognitive complexity theory. This cognitive complexity theory is built on two complexity measures; one the *description complexity*, the complexity involved in decomposing complex relations in smaller and easier relations, which we have shown in (Ragni et al., 2006) corresponds to, or is an adoption of, Relational Complexity introduced by Halford and colleagues. The other complexity measure is based on the deduction core of the SRM, and is called *model complexity*.

Using the generalized neighborhood graph (Ragni & Wölfl, 2005) we were not only able to explain the concept of continuous transformations (used by humans) in the model variation phase, but also to introduce the important parameter that we have called *transformation distance*. This parameter is not only important to explain individual performance (one can think about the working memory limitations which can have an influence about the number of models inspected), but also to explain, why humans make logical flaws, the distance from a PMM to a counter-example is too high and therefore the counter-example is never found. However, more empirical data is needed in order to define the limits of human transformation distance.

The main aim of the experimental investigations with individuals was to identify the strategies humans use when reasoning about spatial relations. We corroborated the *fff*-principle, for both one- and two-dimensional problems. Participants tended to draw the PMM based on this principle. Furthermore, we revealed a number of additional results: First, for the model *inspection phase*, our findings suggest that ternary and quaternary relations are more difficult: both resulted in more errors and longer latencies. This result extends the findings of Goodwin and Johnson-Laird, (2005) who found people had difficulty in constructing a mental model when the premises consisted of more complex relations. Their findings can also be modeled in the SRM framework. Second, if the participant has the PMM in mind then it is easier to reject a putative conclusion which holds only for the ¬PMM, than it is to reject the conclusion which was only valid for the PMM.

Once a model is constructed (usually the PMM) we found that participants employ different strategies to verify an offered model (a ¬PMM) during the *variation phase*. If the participants have to vary a model they generally start with the PMM. The "shorter" the transformation distance between models the easier it is to reach a counter-example. The "greater" the transformation distance the more unlikely it is that a counter-example will be found. This is strong support for the PMMT and the corresponding computational model.

Acknowledgments

This research was supported by grants to MK from the DFG (German National Research Foundation) in the Transregional Collaborative Research Center, SFB/TR 8 project. The authors would like to thank, Bernhard Nebel, Geoff Goodwin and Phil Johnson-Laird for comments and suggestions.

References

Fangmeier, T., Knauff, M., Ruff, C.C., Sloutsky, V.: FMRI evidence for a three-stage model of deductive reasoning. Journal of Cognitive Neuroscience 18, 320–334 (2006)

Freksa, C.: Qualitative Spatial Reasoning. In: Mark, D., Frank, A. (eds.) Cognitive and Linguistic Aspects of Geographic Space, pp. 361–372. Kluwer, London (1991)

Goodwin, G.P., Johnson-Laird, P.N.: Reasoning about relations. Psychological Review 112, 468–493 (2005)

Halford, G.S.: Children's understanding: The development of mental models, NJ: Erlbaum. Hillsdale (1993)

Halford, G.S., Wilson, W.H., Phillips, S.: Processing capacity defined by relational complexity: implications for comparative, developmental, and cognitive psychology. Behavioral and Brain Sciences 21, 803–831 (1998)

Johnson-Laird, P.N.: Mental models and deduction. Trends in Cognitive Sciences 5, 434–442 (2001)

Johnson-Laird, P.N., Bara, B.G.: Syllogistic inference. Cognition 16, 1–61 (1984)

Johnson-Laird, P.N., Byrne, R.M.J.: Deduction, Hove, UK: Erlbaum (1991)

Knauff, M.: The cognitive adequacy of Allen's interval calculus for qualitative spatial representation and reasoning. Spatial Cognition and Computation 1, 261–290 (1999)

Knauff, M., Johnson-Laird, P.N.: Visual imagery can impede reasoning. Memory & Cognition 30, 363–371 (2002)

Knauff, M., Rauh, R., Schlieder, C.: Preferred mental models in qualitative spatial reasoning: A cognitive assessment of Allen's calculus. In: Proceedings of the Seventeenth Annual Conference of the Cognitive Science Society, Mahwah, NJ, pp. 200–205. Erlbaum (1995)

Knauff, M., Rauh, R., Schlieder, C., Strube, G.: Continuity effect and figural bias in spatial relational inference. In: Proceedings of the Twentieth Annual Conference of the Cognitive Science Society, Mahwah, NJ, pp. 573–578. Erlbaum (1998)

Manktelow, K.I.: Reasoning and Thinking. Psychology Press, Hove (UK) (1999)

Ragni, M., Fangmeier, T., Knauff, M. (in prep.) A Computational Level Theory of Spatial Reasoning.

Ragni, M., Fangmeier, T., Webber, L., Knauff, M.: Complexity in Spatial Reasoning. In: Proceedings of the 28th Annual Cognitive Science Conference, Lawrence Erlbaum Associates, Mahwah, NJ (2006)

Ragni, M., Knauff, M., Nebel, B.: A Computational Model for Spatial Reasoning with Mental Models. In: Bara, B.G., Barsalou, L., Bucciarelli, M. (eds.) Proceedings of the 27th Annual Cognitive Science Conference, pp. 1064–1070. Lawrence Erlbaum Associates, Mahwah, NJ (2005)

Ragni, M., Wölfl, S.: On Generalized Neighborhood Graphs. In: Furbach, U. (ed.) KI 2005. LNCS (LNAI), vol. 3698, Springer, Heidelberg (2005)

Rauh, R., Hagen, C., Knauff, M., Kuß, T., Schlieder, C., Strube, G.: Preferred and alternative mental models in spatial reasoning. Spatial Cognition and Computation 5, 239–269 (2005)

Rips, L.J.: The psychology of proof: Deductive reasoning in human thinking. The MIT Press, Cambridge, MA, US (1994)

Schlieder, C., Berendt, B.: Mental model construction in spatial reasoning: A comparison of two computational theories. In: Schmid, U., Krems, J.F., Wysotzki, F. (eds.) Mind modelling: A cognitive science approach to reasoning, pp. 133–162. Pabst Science Publishers, Lengerich (1998)

Vandierendonck, A., Dierckx, V., Vooght, G.D.: Mental model construction in linear reasoning: Evidence for the construction of initial annotated models. The Quarterly Journal of Experimental Psychology 57A, 1369–1391 (2004)

The Spatial and the Visual in Mental Spatial Reasoning: An Ill-Posed Distinction

Holger Schultheis, Sven Bertel, Thomas Barkowsky, and Inessa Seifert

SFB/TR 8 Spatial Cognition, Universität Bremen, Germany
{schulth,bertel,barkowsky,seifert}@sfbtr8.uni-bremen.de

Abstract. It is an ongoing and controversial debate in cognitive science which aspects of knowledge humans process visually and which ones they process spatially. Similarly, artificial intelligence (AI) and cognitive science research, in building computational cognitive systems, tended to use strictly spatial or strictly visual representations. The resulting systems, however, were suboptimal both with respect to computational efficiency and cognitive plausibility. In this paper, we propose that the problems in both research strands stem from a misconception of the visual and the spatial in mental spatial knowledge processing. Instead of viewing the visual and the spatial as two clearly separable categories, they should be conceptualized as the extremes of a continuous dimension of representation. Regarding psychology, a continuous dimension avoids the need to exclusively assign processes and representations to either one of the categories and, thus, facilitates a more unambiguous rating of processes and representations. Regarding AI and cognitive science, the concept of a continuous spatial / visual dimension provides the possibility of representation structures which can vary continuously along the spatial / visual dimension. As a first step in exploiting these potential advantages of the proposed conception we (a) introduce criteria allowing for a non-dichotomic judgment of processes and representations and (b) present an approach towards representation structures that can flexibly vary along the spatial / visual dimension.

1 Introduction

In this contribution we address issues of modeling visual and spatial aspects in mental reasoning with spatial information. More specifically, we focus on information on how a certain number of objects can be related to one another in space, with regard to topology, distance, and directional relations.

Whether the corresponding mental representations of such information can be more adequately described by means of non-modal structures such as propositions (Jahn, 2003) or one-, two- and three-dimensional arrays (e.g., Ragni, Knauff, & Nebel, 2005; Glasgow & Papadias, 1992), or through diagrams (Chandrasekaran, Kurup, Banerjee, Josephson, & Winkler, 2004) or images (Bertel, Barkowsky, Engel, & Freksa, 2006; Kosslyn, 1994) is the topic of much ongoing research. While the group of non-modal structures is frequently linked to reasoning with spatial mental models (Johnson-Laird, 1983), diagrams and images are usually employed to capture mental representations evoked in mental imagery (Kosslyn, 1994). In terms of working memory processes, spatial mental models are often described as mainly relying on central executive, non-modal functions (Gillhooly, Logie, Wetherick, & Wynn, 1993), though not exclusively (De Vooght & Vandierendonck, 1998), while mental imagery has been suggested to heavily tap visuo-spatial subsystems (Kosslyn & Thompson, 2003; Ishai & Sagi, 1997; Kosslyn & Sussman, 1995).

Consequently, for spatial reasoning, tasks are usually modeled as *either* employing spatial or visual mental representations (i.e., being based on spatial mental models or on images), but rarely both. This is somewhat in-line with double dissociations found to exist between visual and spatial short-term memory (Klauer & Zhao, 2004). Also, existing comparative approaches of image- and model-based reasoning with same or comparable tasks are aimed at dissociating the two conditions (e.g. Knauff & Johnson-Laird, 2002, for relational reasoning) rather than establishing if, how, and when spatial and visual processing co-occurs.

On a neuropsychological level, evidence has long been collected for a functional and anatomical separation of two major pathways in higher-level visual processing (Ungerleider & Mishkin, 1982; Haxby et al., 1991). Originating occipitally, a ventral pathway runs to the inferior temporal lobe and processes object properties such as color and shape (thus labeled "what" pathway), while a dorsal pathway projects to posterior parietal areas and processes spatial attributes and movements ("where" pathway).

Inter-individual differences have been reported for the use of spatial and visual representations, for example, for abstract problems (Hegarty & Kozhevnikov, 1999) or during wayfinding (Aginsky, Harris, Rensink, & Beusmans, 1997). Such differences led to the postulation of different individual cognitive styles, such as "verbalizers", "spatializers", or "visualizers". More recently, object and spatial subgroups have been proposed for the visualizers (Kozhevnikov, Kosslyn, & Shephard, 2005; Kozhevnikov, Hegarty, & Mayer, 2002), leading to further differentiations.

Given all this evidence for *different types* of mental representations, it hardly comes as a surprise that even for spatial reasoning tasks mental representations have been often modelled as either visual or spatial and that, frequently, properties of either format have been investigated independently of the other format. Consequently, a main research focus has been on dissociating between rather than on integrating different representation formats in spatial reasoning. However, while the dissociation between the visual and the verbal may be adequate for many types of mental reasoning, we argue that it led and still leads to misconceptions about the nature of representations in *spatial* knowledge processing. Instead of viewing the visual and the spatial as two clearly separable categories, for spatial reasoning, they should be conceptualized as the extremes of a continuous dimension of representation with various types of representations in-between.

While it has been shown that for simple relational problems "visual" strategies are often slower than "spatial" ones (Knauff & Johnson-Laird, 2002), visual representations (i.e., mental images) can be suspected to have advantages over spatial mental models in situations where much spatial information needs to be integrated or high degrees of visual detail play role (cf. Kosslyn & Thompson, 2003). Despite strong computational advantages of spatial mental models in terms of parsimony, there exists no elegant computational approach today for flexibly and dynamically integrating topological, distance, and directional knowledge into a spatial (but non-visual) representation format. In the current contribution, we suggest that a continuum of visual / spatial representations may best serve to model the different aspects of mental reasoning with spatial information.

In the following, we first focus on existing conceptions of describing spatial mental knowledge processing and identify two core problems. We then argue for an integrative view of visual / spatial aspects in mental spatial knowledge processing that may overcome the strict separation between the two modalities. We propose scalable representation structures as a modeling conception for describing mental spatial knowledge processing that comprises both spatial and visual aspects. Finally, we briefly discuss a number of open issues to be addressed in future research.

2 Existing Conceptions and Problems

Mirroring the above stated distinction between the visual and the spatial, research in psychology as well as in AI and cognitive science have mainly treated the spatial and the visual as clearly separable aspects of tasks. In doing so, psychological research has mainly concentrated on the question of which (properties of) entities are–or even must be–represented or processed either visually or spatially (cf. Kosslyn & Thompson, 2003) and the question of which individual traits of a person might indicate her capability to represent or process entities visually or spatially (cf. Kozhevnikov, Hegarty, & Mayer, 2002).

A major focus of AI and cognitive science on the other hand has been on the conceptualization of different types of representation structures for building artificial cognitive systems (cf. Glasgow & Papadias, 1992) or for modeling natural cognitive systems like humans (cf. Barkowsky, 2002). In particular, the representation structures proposed so far were usually aimed at being either visual representations or spatial representations.

Thus, much research in psychology and AI / cognitive science has–at least implicitly–assumed a strict distinction between visual and spatial aspects in mental spatial knowledge processing. On closer inspection, however, there are at least two problems associated with such a stance, namely (1) that there currently exists no consensus as to what the defining characteristics of visual or spatial mental representations really are, and (2) that current artificial cognitive systems and models of mental spatial knowledge processing either employ strictly spatial or strictly visual representations. These problems will be briefly outlined in the following two sections.

2.1 Problem 1 – Spatial and Visual: Where's the Difference?

A crucial prerequisite for categorizing entities with respect to their visual or spatial nature are clear criteria for judging entities as being visual or spatial. Similarly, to distinguish people regarding their capability to represent or process entities visually or spatially requires tests which selectively tap this capability. To construct such tests, again criteria are needed to decide which test items are instrumental in determining the relevant ability. Such criteria, however, currently do not seem to exist.

On the one hand there is lack of clarity regarding what prevalent tests like the mental rotation test (Vandenberg & Kuse, 1978) and the Minnesota Paper Form Board

(Likert & Quasha, 1941) actually measure. The former, for instance, was intended to measure the ability to visualize and mentally manipulate spatial arrangements such as three-dimensional block figures. In accord with this conception, the test has subsequently been employed in several studies (e.g., Charlot, Tzourio, Zilbovicius, Mazoyer, & Denis, 1992; Mellet, Tzourio, Denis, & Mazoyer, 1995) to identify the imagery (i.e., visualization) abilities of study participants. Yet, other researchers (e.g., Kosslyn & Thompson, 2003) have claimed that the mental rotation test does only measure *spatial* reasoning abilities, but not visualization capabilities. Remarkably, none of these works mentions clear criteria on which grounds the test is thought to measure visual or spatial abilities. Rather, both the researchers utilizing the test to measure visualization and the researchers utilizing the test to measure spatial abilities seem to assume it being obvious that the test measures what they utilize it for.

In addition, there is a more general disaccord with respect to the question of which (properties of) entities are spatial and which ones are visual. Levine, Warach, and Farah (1985), for example, assume that an analog clock face is something visual, whereas Kosslyn and Thompson (2003) judge clock faces and the position of the corresponding clock hands as being spatial. A similar confusion exists with respect to the shape of objects. In some work (cf. Kosslyn & Thompson, 2003), shape is assumed to be a prime example for a visual property. Other researchers have, however, argued that shape can be and is represented and processed spatially (e.g. Leeuwenberg, 2004). Like with the employment of tests, criteria-based justifications for one or the other stance are hard to find in the literature.

These examples illustrate that unambiguous, objective criteria for judging entities as being visual or spatial currently do not seem to exist. Consequently, categorizing entities and persons according to the visual / spatial dichotomy has given rise to a number of different categorizations and to some confusion regarding the characteristics of the visual and the spatial. Such conceptual disarray poses a serious problem for investigating and realizing the processes and representations in spatial knowledge processing in natural and artificial cognitive systems.

2.2 Problem 2 – Integrating Spatial Representations

Numerous ways of representing and processing spatial knowledge have been devised in AI both for explanatory purposes in cognitive models and for technical purposes in diagrammatic reasoning systems. The most widely used 2-dimensional structure is the regular rectangular raster array (e.g., Glasgow & Papadias, 1992; Khenkhar, 1991; Ragni, Knauff, & Nebel, 2005). This type of representation structure supports, for instance, translation operations, neighborhood-based processes like region growing operations, or topological relations. To more adequately model operations like rotation, scaling, or symmetry detection, Funt (1980) used a circular representation structure consisting of individual processors arranged in the form of concentric rings. Other types of representation which have been proposed are, for example, vector images (Barkowsky, 2002), bitmap images (Kosslyn, 1980) or less complex qualitative or metrical linear ordering structures (e.g., Schlieder & Berendt, 1998; Chang, Shi, & Yan, 1987).

Despite this diversity, two aspects are characteristic for nearly all of the approaches in AI and cognitive science. First, the representation structures devised so far often are applicable only to specific types of spatial knowledge. Put differently, such representation structures are only appropriate to represent certain types of spatial knowledge, for example just knowledge about orientation or just knowledge about topology. As a consequence, most of the existing structures do not allow representing spatial situations which require integrating several types of spatial knowledge. Notable exceptions from this rule are representation structures like bitmap or vector images which have been termed diagrammatic representations: by depicting the spatial situation in a 2D plane all aspects essential to spatial knowledge processing (i.e., distance, orientation, topology, shape) can be represented in an integrated way. Consider, for example, the following situation: Larissa returns from a holiday and tells her friend that she has just been to a beautiful place called San Giovanni a Piro. As Larissa's friend has never been to San Giovanni a Piro, she asks Larissa about the location of the place. Larissa's description could sound as follows: San Giovanni a Piro is a small village in the south of Italy on the Tyrrhenian coast, not far from Naples but not as far as the tip of the boot of Italy.

There are several kinds of spatial knowledge involved here and they have to be considered in combination to determine the location of San Giovanni a Piro (SGP). The topological relations are: in (Italy, SGP) and touches (SGP, Tyrrhenian coast). A cardinal direction relation is south (SGP, central Italy), distance information includes not far from (SGP, Naples) and not as far from as (SGP, Naples, tip of boot of Italy). Last, there is shape information involved, such as shape of (tip of boot of Italy).

Second, like in psychological research, AI and cognitive science work does seem to underlie–at least implicitly–the assumption that each representation structure is necessarily either spatial or visual. Therefore, computational systems are either restricted to the processing of specific types of spatial knowledge (where integration of different knowledge types is not required), or they comprise several representation structures some of which are spatial and some of which are visual. A common approach in the latter type of systems is to utilize the (specific) spatial representation structures wherever this is sufficient and to employ the visual representations for situations with greater complexity or to benefit from their higher specificity (e.g., Chandrasekaran et al., 2004).

However, utilizing distinct spatial and visual representations of the same situation concurrently entails to either store all information about a spatial situation redundantly or to newly and fully construct a visual representation from the knowledge already stored in the spatial representations whenever needed. In both cases the system would be suboptimal with respect to computational efficiency. Moreover, it seems implausible that such a computational system is a good model of human cognition, because this would be in disaccord with the idea that information processing in the human mind is organized according to optimize the information processing demands (cf. Collins & Quillian, 1969).

Regarding cognitve plausibility one might argue that both entertaining multiple representations and redundantly storing information are fundamental properties of cognition in some contexts and, accordingly, the procedure employed in current computational systems is not necessarily cognitively implausible. Such argument, however, neglects the fact that evidence for multiple representations in spatial cognition usually suggests that these representations differ considerably with respect to their content (e.g., Brockmole & Wang, 2002; Carlson, 1999). In other words, if multiple representations store information redundantly they do so only partially. On the contrary, the usual computational approaches would suggest that the human cognitive system would maintain multiple representations such that some representations do not hold any information not already contained in some other representation. Such an assumption seems to depart too far from cognitive efficiency too be accepted without explicit empirical support.

Thus, in such "hybrid" approaches the problem of computational inefficiency and–in the case of cognitive science–also of cognitive implausibility arises.

2.3 Lessons Learned

The arguments in the preceding two sections illustrate two crucial problems associated with the spatial vs. visual distinction: first, there is currently no consensus as to what the defining characteristics of visual or spatial mental representations really are. As a consequence, psychological research reports ambiguous results with respect to (a) which entities are processed visually and which ones are processed spatially, and (b) which individual traits indicate an inclination to process information visually or spatially. Second, current artificial cognitive systems and models of mental spatial knowledge processing either employ strictly spatial or strictly visual representations. This seems, on the one hand, computationally inefficient and, on the other hand, cognitively implausible.

In the following, we propose an approach to tackle both of these problems. The basis for our approach is the observation that both problems can be traced back to the tendency existing in psychology as well as AI and cognitive science of trying to strictly separate the spatial from the visual. We claim that such a strict distinction is (a) not possible and (b) not reasonable. In our view the spatial and the visual are not two separable categories, but should better be viewed as two extremes of a continuous dimension. Taking such a position allows addressing the above mentioned issues.

First of all, assuming a continuous visual / spatial dimension explains why it was not possible until now to achieve a clear and unequivocal categorization of (properties of) entities and a person's processing mode as being either spatial or visual: some entities, for instance the mental representations of the clock faces and clock hands discussed above–which have been claimed to comprise visual as well as spatial aspects–, could be assumed to be lying somewhere well in-between the two extremes of the visual / spatial dimension and therefore cannot be exclusively assigned to either of the two categories. Accordingly, a more appropriate characterization of entities on the visual / spatial dimension would be a comparative one which allows identifying whether an entity is more spatial or more visual than one or several others.

Second, regarding AI systems and cognitive models, the concept of a continuous dimension suggests to employ not only representation structures which are either spatial or visual; rather, representation structures should be employed which lie in-between the two extremes. Such "intermediate" structures integrate spatial and visual aspects and potentially allow making a smoother transition between strictly spatial and strictly visual representation structures. To illustrate this consider the following example: A computational system starts working on a spatial reasoning task for which it employs a spatial representation structure R_{S1} (e.g., containing topological knowledge). Assume further that at some point during reasoning the integration of another representation structure R_{S2} (e.g., containing directional knowledge) becomes necessary to represent the situation. One option would be to now create a strictly visual representation R_V (i.e., a picture) and to copy all information contained in R_{S1} and R_{S2} to R_V. Format changes would be required for all content copied (e.g., from a specific spatial format to a specific visual format). A different option that utilizes the postulated visual / spatial continuum of representations would involve slight changes to R_{S1} and R_{S2} and the association of their respective content; it could lead to the formation of an integrated spatial representation R_I that also permits to represent the new situation, however, at lower computational costs than R_V would require because only little recoding and copying would be needed in its creation. In our view, integrated spatial representations potentially provide for lean (i.e., economic) and cognitively plausible representation structures.

Summing up, the problems associated with the spatial / visual distinction in psychology, AI, and cognitive science can be avoided by assuming a continuum underlying the spatial / visual dimension. However, such an assumed continuum raises several new questions like, for example, as to what the criteria for determining the position of an entity on the continuum really are or how one can realize intermediate representation structures in a computational system. In the next section, we will give first answers to these and related questions thus evolving a new conception of the spatial / visual distinction as well as introducing first ideas towards a new type of representation structures for spatial reasoning in computational systems.

3 Beyond a Strict Separation of Spatial and Visual

As explained in the previous section a strict separation of spatial and visual which implies two categories with clear cut boundaries entails a number of serious problems. One way to avoid these problems is to assume a continuous dimension with spatial and visual as extremes. Such bipolar conception is, however, only useful if there are clear criteria based on which one can–at least in an ordinal way–place (properties of) entities and / or representation structures unambiguously on this dimension. Such criteria will be detailed in the following.

3.1 Criteria for the Spatial/Visual Dimension

On the basis of the properties of cases where visual representation structures have been employed in computational systems or of situations in which humans have been assumed to use visual mental systems we propose four criteria for the placement of spatial representations on the spatial / visual dimension. In other words, in most or all circumstances (a) which necessitated the use of visual representation structures in computational systems or (b) in which psychological studies detected the employment of visual processing, the following four criteria were more strongly pronounced than when no visual processing took place:

- a high number of types of spatial relations included in the representation,
- a high number of different spatial relations included in the representation,

- a high degree of specificity of the relationship (i.e. the degree of completeness of the set of relations) of each pair of included entities, and
- a high degree of exemplarity of entities and relations included in the representations.

Each of the criteria is proportionally related to the visual side of the continuum, that is, for example, the higher the specificity the closer is the corresponding entity or representation supposed to be to the visual endpoint of the spatial / visual dimension.

In the rest of this section the four criteria will be detailed in turn.

Number of Types of Relations. This criterion refers to the types of spatial relations involved in the mental processing of a spatial reasoning tasks. As argued above the spatial representation structures usually employed in computational systems are knowledge type specific, that is, are intended to and only permit to represent one kind of spatial knowledge like, for instance, orientation knowledge. As soon as several types of knowledge have to be or are represented in an integrated way, normally diagrammatic representation structures are used. Accordingly, we assume that processes and / or representation structures are the more visual the more types of relations are involved. Essentially, this means that more integrated and thus more complex processes and representations are viewed as being more visual.

Number of Relations. Due to their integrated nature, visual processes and visual representations tend to comprise not only more types of relations, but also deal with a higher number of relations than spatial processing and spatial representations, respectively. That is, in an integrated, visual representation typically more spatial relations are–if only implicitly–specified between the represented entities. Therefore, our second criterion to judge an entity's / a representation's position along the spatial visual dimension is the number of relations involved.

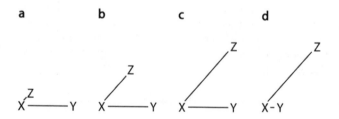

Fig. 1. Four possible interpretations of the situation X is west of Y and Z is northeast of X

Specificity. Knowledge about spatial situations may–and often does–allow for several interpretations. Consider, for example, the following (spatial) facts a person might know about the three entities X, Y, and Z: X is west of Y and Z is northeast of X. Although these two facts may be quite helpful for the task the person tries to accomplish the precise position of the three entities to each other is not unambiguously given. Knowing just the two above stated facts, the orientation relation between Y and Z could be Z is west of Y, Z is northwest of Y, Z is north of Y, or Z is northeast of Y (see Fig. 1a, b, c, and d, respectively). In other words the two given facts do not fully specify the spatial situation between the three entities with respect to their mutual orientation. However, depicting the two facts in the two-dimensional plane as in Fig. 1 ineluctably specifies the relation between Z and Y. Thus, one defining characteristic of visual processes and representations is their higher specificity when compared to more ambiguous representations (cf. Stenning & Oberlander, 1995). Consequently, we chose specificity as a further criterion for locating processes and representations on the spatial / visual dimension.

Exemplarity. In psychology, certain aspects of entities like color, shape, or texture have been assumed to entail more visual processing (e.g., Knauff & Johnson-Laird, 2002; Levine, Warach, & Farah, 1985). Likewise, visual representation structures used in spatial knowledge processing in AI and cognitive science are usually more appropriate for representing such details as specific shapes and relations than spatial representation structures are (cf. Barkowsky, 2002).

At the same time, such details will be so much more likely to be processed or represented visually the more the reasoning process involves concrete exemplars instead of prototypes / categories of objects. If, for example, one needs to reason about the internal layout of a familiar office building (say, for navigating in it) it is more probable that the reasoning process will involve color, shape, and texture information about that building than if one reasons only about the internal structure of office buildings in general.

As a result of these two observations, we propose that the extent to which concrete exemplars (instead of prototypes / categories) are part of the reasoning process forms one criterion for judging processes and representations with respect to their location on the spatial / visual dimension (cf. categorical and coordinate pathways, Kosslyn, 1994).

To illustrate the criteria and how they might be applied, consider the two representations given in Fig. 2a and 2b, respectively, of the same spatial situation comprising a church, a pond, and a tree. The representation in Fig. 2a just encodes two spatial relations between the three objects, namely that the tree is left of the church and that the pond is in front of the church. In contrast, the representation in Fig. 2b represents a number of additional spatial relations between the three objects, for example the information that the pond is in front of the tree, the distance information that both the tree and the pond are close to the church, and the additional directional information that the church is right of the tree and behind the pond. Accordingly, the second representation does not only encode more spatial relations between the objects, but also more types of spatial relations. Furthermore, whereas the first representation does not specify the precise position of the three objects (apart from the fact that the tree is left of and the pond is in front of the church), the second one does. Finally, the second representation can be seen as encoding concrete exemplars of the categories tree, church and pond, since the church, for instance, is rather small and warped (which can be assumed not to be the properties of a prototypical church). Thus, compared to the representation in Fig. 2a the representation in Fig. 2b has more types of spatial relations, more spatial relations, a higher specificity, and a higher exemplarity and, therefore, will be judged to be more visual than the first representation.

This example shows that the criteria proposed both nicely allow locating representations on the visual-spatial dimension and lead to judgments which are in accord with previous categorizations in which the representation of Fig. 2b would have been more visual than the representation in Fig. 2a. Consequently, the first problem arising in assuming a continuous spatial / visual dimension, namely devising consistent criteria which allow continuously judging processing / representations, has been satisfactorily engaged. Yet, the second problem, that is, what kind of representation structures might occupy intermediate position on the spatial / visual dimension has not been discussed until now. In the following section, aspects pertaining to this second problem will be presented.

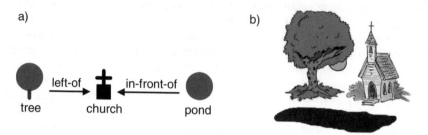

Fig. 2. Two representations of spatial relations between a church, a pond, and a tree

3.2 Scalable Representation Structures

As argued above, assuming a spatial / visual continuum does not only avoid category assignment problems, but also potentially permits to realize computationally more efficient and cognitively more plausible computational systems. A prerequisite to the latter, however, is the availability of representation structures which are located between the extreme end points of the dimension, that is, representation structures which are neither exclusively spatial nor exclusively visual. Ideally, these representation structures should be such that they allow incrementally building a more and more complex / visual representation of spatial situations, based on the current task demands. Due to their incremental, demand driven refinability we term such representation structures *scalable representation structures*.

Although existing representation structures for computational cognitive systems may vary in the amount of information they represent (i.e., in their complexity), they usually do so confined to one type of spatial knowledge, that is, are not scalable across knowledge types. In fact, scalable representation structures in the sense just mentioned currently do not exist. Due to the advantages scalable representation structures might yield, this seems to be a serious lack of current research and, as a consequence, we took first steps in devising such structures.

Before further describing our approach, however, one fundamental issue needs to be addressed: are scalable representation structures to be conceived of (a) as one single representation structure incrementally being upgraded and growing more and more in complexity or (b) as a system of separate distributed specific representation structures which are more and more tightly combined on demand? In the next section these two possibilities will be discussed.

Starting from the results of this discussion we will present a more detailed account of the requirements for scalable representation structures as well as first ideas towards their realization in the subsequent section before finally discussing open issues with respect to our approach and their potential merit for psychological research.

One for All or All for One? The defining characteristics of scalable representation structures are that they (a) are incrementally refinable on demand and (b) can occupy intermediate locations on the spatial / visual dimension. From these characteristics alone, however, it is not clear how scalable representation structures should be conceived of regarding their organization. On the one hand, scalable representation structures could be viewed as being *monolithic*, that is, being constituted by a single representation structure which is–at least potentially–able to integrate all (types of) spatial relations at once. Regarding the aim of building a computational model of human mental spatial knowledge processing such a stance would, furthermore, posit that human cognition relies on such a monolithic representation structure. On the other hand, scalable representation structures could also be conceived of as being *composed*, that is, consisting of several distinct smaller (and more specific) representation structures. Since the representation structures to be devised are meant to improve existing and / or to facilitate the building of new computational cognitive systems, we chose to base our decision regarding the two possibilities of conceptualizing scalable representation structures on already existing evidence from psychological as well as AI /cognitive science research.

Regarding psychology, it seems to be a common assumption that spatial knowledge about the environment is stored in the form of several "...smaller chunks, each of which is encoded by a separate representation." (Brockmole & Wang, 2002). And indeed there are several studies indicating that humans maintain different representations for spatial reasoning tasks with respect to (a) the content of these representations (i.e., different hierarchically representations; see, for instance, Hirtle & Jonides, 1985; McNamara, 1986), (b) whether object information (what) or location information (where) is represented (e.g., Hegarty & Kozhevnikov, 1999; Klauer & Zhao, 2004), and (c) whether relations are represented egocentrically or allocentrically (Easton & Sholl, 1995; Rieser, 1989). Thus, there seems to be ample psychological evidence supporting composed scalable representation structures.

This evidence is further corroborated by the AI / cognitive science literature. Numerous different types of representation structures both for diagrammatic reasoning systems (e.g., Chandrasekaran, Kurup, Banerjee, Josephson, & Winkler, 2004) and cognitive models (e.g., Ragni et al., 2005) have been proposed so far. Notably, it turned out that the proposed structures individually only support a certain (small) number of representational requirements. Put differently, until now AI / cognitive science researchers have not been able to devise a single, monolithic representation structure for spatial reasoning. In contrast, it has been claimed (cf. Sloman, 1985) that several special purpose representation structures are more sensible than one general purpose representation.

In accord with these converging evidence from the two fields of research we assume that scalable representation structures should be conceptualized a being composed of several specialized representation structures. A more detailed account of how scalable representation structures may be constructed from a number of distinct, dedicated representation structures will be given in the next section.

Integration Approaches. As indicated by the above section there seems to be ample evidence and common agreement that spatial information is represented distributed across several separate representation structures. There is, however, considerable less work regarding the question how these separate representation structures could be combined or integrated such that spatial reasoning across the full range of possible spatial relations could be realized, since previous research efforts have mainly concentrated on specific aspects of spatial reasoning in isolation from each other solely considering, for example, cardinal directions (e.g., Frank, 1995) or topological relations (Egenhofer, 1994). One notable exception to this rule is the work of Sharma (1996), who develops a formalism to combine topological and direction relation for spatial inferences in Geographic Information Systems (GIS). Similarly, the TPCC calculus (Moratz, Nebel, & Freksa, 2003) combines directional and distance knowledge in a formal framework.

Although these are important approaches regarding the integration of several knowledge types in spatial reasoning, there are at least three shortcomings associated with these approaches with respect to scalable representation structures: first, though, these approaches combine some of the separate types of spatial knowledge they normally do not encompass all types (e.g., Sharma, 1996, neglects distance information). Second, the approaches are meant to be solutions to some technical problem. Consequently, their main focus is on technical aspects and, in particular, they do not try to be and are not cognitively plausible. Third, the reasoning formalisms proposed do not include a specification of the structures in which the combined knowledge types are represented, that is, the approaches do not assume / specify any particular representation structures. To build a computational model of mental spatial knowledge processing, however, (a) representation structures need to be devised, (b) all types of knowledge have to be integrated, and (c) both former points have to be realized in a cognitively plausible way. Despite several cognitive models of spatial reasoning such a cognitive account of scalable representation structures currently does not seem to exist. As with the more technical approaches, previous cognitive models of spatial reasoning seem to have concentrated merely on single types of knowledge (see, e.g., Ragni et al., 2005; Goodwin & Johnson-Laird, 2005).

Accordingly, a computational account of integrated mental spatial knowledge processing currently does not exist. Since this seems to be a serious lack, we are currently working on developing such an account. As the first step in that development, we chose to concentrate on scalable representation structure, that is, how separate knowledge types might be represented and integrated in a cognitively adequate way.

Like explicated in the previous section, it seems most plausible to conceive scalable representation structures as arising from the interplay of distributed, separate representation structures which are combined on demand. Yet, such a conception entails the problem of coordinating the individual structures, that is, the individual knowledge represented in these structures. Coordination is necessary, because the different types of knowledge are dependent on one another. If, for instance, some entity X is known to be a proper part of some other entity Y, and Y is known to be north of a third entity Z, the direction relation between X and Z can not be arbitrary. Thus, if (the combination of) the separate representation structures should form a coherent representation of some spatial situation the consistency of the knowledge stored in the individual representation structure needs to be ensured. Any conception of scalable representation structures therefore needs to include some mechanism(s) which realize(s) this mutual adjustment. Before one can devise such mechanisms the mutual dependencies of the different knowledge types must be known. Accordingly, our first step towards developing scalable representation structures was to identify these dependencies by building composition tables across different knowledge types.

Table 1. Composition table giving the possible topological relations between Y and Z (shown in the corresponding table cells) given both a topological relation between Y and X ($Y R_1 X$) and a distance relation between X and Z ($X R_2 Z$); TOP is the set of all possible topological relations; see text for details

$X R_2 Z$ / $Y R_1 X$	cl	md	fr
equal	TOP	disjoint	disjoint
disjoint	TOP	TOP	TOP
tangent	TOP	$TOP\backslash${equal, in t/nt}	$TOP\backslash${equal, in t/nt}
overlaps	TOP	$TOP\backslash${equal, in t/nt}	$TOP\backslash${equal, in t/nt}
in/t	TOP	disjoint	disjoint
in/nt	TOP	disjoint	disjoint
contains/t	TOP	$TOP\backslash${equal, in t/nt}	$TOP\backslash${equal, in t/nt}
contains/nt	TOP	$TOP\backslash${equal, in t/nt}	$TOP\backslash${equal, in t/nt}

Composition tables in general state all relations between two entities Y and Z which are possible in the light of given relations between X and Y as well as X and Z. As discussed above and shown in Fig. 1, for example, given X south-west of Z and X west of Y possible direction relations[1] between Y and Z are Z west of Y, Z north-west of Y, Z north of Y, and Z north-east of Y. A complete composition table not only states the possible relations for one pair of relations that hold between Y and X and X and Z, but for all potential pairs of relations. Furthermore, like in this example, composition tables usually are concerned with only one type of knowledge, that is, the given relations as well as the determined possible (given one pair) relations belong to the same type of spatial knowledge (direction in the presented example). Yet, to identify the mutual dependencies between different types of knowledge such composition tables are not sufficient. Instead, one needs to take into account which relations of a certain knowledge type K_a are possible in the light of two relations R_1, R_2 where R_1 is of type K_a and R_2 is of another knowledge type. Table 1 presents an example of such a composition table across knowledge types. Shown in the table cells are those topological relations which could hold between Y and Z given the topological relation indicated by the relation in the corresponding row of the first column and the distance relation given in the corresponding column of the first row. For example, if Y and X are equal and X and Z have medium (third column) or far (fourth column) distance to each other Y and Z have to be disjoint.

---

[1] Using cardinal directions and assuming an 8-sector model with the directions north, north-west, west, south-west, south, south-east, east, and north-east.

Analogously to the construction of Table 1, we constructed cross composition tables for all six possible pairings of types of spatial knowledge, that is, distance / direction, direction / distance, distance / topology, topology / distance, topology / direction, and direction / topology. For each of these pairings we distinguished which of the two paired knowledge types was inferred (i.e., which knowledge type was inside the table cells). Thus, overall, 12 composition tables were created. To allow constructing complete, unique, and correct composition tables without the construction process getting out of hand we had to take several additional assumptions. Although these assumptions restrict our approach to a subset of all possible spatial situations, such a procedure seemed justified as a very first step towards scalable representation structures.

The assumptions we took were:

1. Possible distance relations between entities are close (cl), medium (md), and far (fr).
2. Possible direction relations are north (N), east (E), south (S), and west (W).
3. Possible topological relations are equal, disjoint, tangent, overlaps, in/tangential (in/t), in/not tangential (in/nt), contains/tangential (contains/t), and contains/not tangential (contains/nt) which are defined as in the RCC-8 calculus (Cohn et al., 1997).
4. If the distance between two entities is medium or far they have to be disjoint.
5. Two entities can only be close to each other if no other entity lies between them.
6. Entities are two-dimensional convex regions.

Building on these assumptions the 12 cross composition tables allowed identifying the mutual dependencies of the knowledge types which are shown in Table 2. One dependency, for example, is the fact that the direction relation between two entities X, Y is imposed on the entities contained by X / Y (see first line in the direction / topology row).

In developing scalable representation structures the crucial issue is to take these dependencies into account, that is, to devise mechanisms that ensure that the rules stated in Table 2 are not violated[2]. But how can such mechanisms be realized, given the assumption that the overall spatial representation is the combination of several distinct partial representations? We assume that the basic cohesion between the distinct substructures arises from links connecting the entities which are part of the individual representations. More precisely, those entities in the separate representations that stand for the same object of the represented situation are mutually and completely linked (i.e., every entity in one substructure links to the corresponding entity in all other substructures). Structurally, these links are all that is necessary to realize the functionality of scalable representation structures. First, if during mental spatial knowledge processing the need for considering an additional type of spatial knowledge arises, an appropriate representation (for that knowledge type) can be created and then linked to the already existing representation structures. In this way the overall representation structure can be easily extended on demand and, thus, is truly

---

[2] This is not to say that composition tables or processes utilizing them would be part of mental spatial knowledge processing. The composition tables are only meant to constitute a list of requirements which results the to be developed cognitively plausible mechanisms of integrated mental spatial knowledge processing should tend to yield (see also the Open Issues section below).

scalable. Furthermore, according to the rules in Table 2, the links allow realizing the consistency between the individual representation structures. If some new information regarding some entities comes to the knowledge of the cognitive system, that is, this knowledge is incorporated in one of the substructures, corresponding entities in the other substructures can be identified via the links and possible changes to these other structures (due to the dependencies stated above) can be applied. In our current conceptualization of such adjustment procedures we assume sets of representation pair specific processes which realize the adjustment. Put differently, for each pair of knowledge types like, for instance, topology and direction, there exists a set of processes specific to these particular knowledge types which ensure the consistency between the individual representation structures of these types.

Table 2. Mutual dependencies between the different knowledge types as identified by the cross composition tables. X, Y and Z are distinct entities.

Knowledge Types	Dependencies
topology / distance	**If** X equal Y **Then** every Z with distance d to X / Y has distance d to Y / X
	If X has distance d to Y and Y is inside Z **Then** X's distance to Z cannot be bigger than d
	If X has distance d to Y and Y contains Z **Then** X's distance to Z cannot be smaller than d
	If X has distance md or fr to Y and Y contains Z **Then** X and Z are disjoint
	If X has distance md or fr to Y and Y is in, overlaps or touches Z **Then** X cannot contain Z
direction / distance	**If** X is close to Y and Y is *dir* of Z, where *dir* is a direction relation **Then** X is dir_{inv} of Z cannot hold, where dir_{inv} is the inverse of *dir*
direction / topology	**If** X inside Y and Y is *dir* of Z **Then** X is *dir* of Z holds
	If X equal Y and Y is *dir* of Z **Then** X equal Z cannot hold.
	If X is *dir* of Y and Y inside Z **Then** X cannot contain Z

Open Issues. As stated above, the current conceptions are but a first step towards scalable representation structures, though, in our view, it is a promising one. There are several issues which need further clarification. For example, the mutual dependencies identified so far are based solely on cross composition tables of the form K_a, $K_b => K_a$ / K_b, where K_a and K_b are different knowledge types. Sharma (1996) has termed this type of cross composition *heterogeneous*, and identified two other forms of cross composition which he termed *mixed* (K_b, $K_b => K_a$) and *integrated* (K_{a1}, K_{b1}; K_{a2}, K_{b2} => K_{a3}, K_{b3}). To give a complete account of dependencies between representations of different knowledge types, it seems necessary to also respect mixed and integrated forms of cross composition.

Apart from these more technical concerns working on the development of scalable representation structures, we identified additional psychological issues: So far, we have described our ideas towards scalable representation structures. To develop a satisfactory cognitive model of spatial reasoning, which is capable of integrating different kinds of spatial knowledge, we need to consider the following questions, which only empirical research can answer: (A) how well are different kinds of spatial knowledge integrated by humans; is the integration symmetric or asymmetric, regarding the ordering of the knowledge fragments to be integrated? (B) Consider, for example, a spatial reasoning task which involves two spatial relations of different types. Are all dependencies between knowledge types described equally respected by humans? Does a resulting inference depend on the order of the given premises? Are some of the dependencies preferred or omitted during mental reasoning? (C) If, for example, one gets to know an orientation relation between two entities, will mental knowledge representation "automatically" entail an inference of distance information (e.g., as it would be the case in diagrammatic reasoning)? How tightly are the different knowledge types coupled? (D) How fast is an integration of additional spatial knowledge with an existing mental representation? What if the knowledge is of a type that, so far, was not included? Considering the limitation of the working memory, how many different representation structures can be successfully integrated? (E) In case of changes in already instantiated knowledge structures, new information has to be updated and communicated among the coupled representation structures. How fast is such propagation of additional knowledge? And again, due to the limitations of memory, do all of the integrated knowledge structures remain consistent, or how many integrated knowledge structures can be kept consistent at a time?

Our cognitive modeling approach aims at specifying scalable representation structures capable of integration of different kinds of spatial knowledge. In doing so, the spatial relations and inference rules have to be made explicit. However, further empirical research is needed to shed light on the visual and the spatial in spatial reasoning and to provide important information on concrete properties of the proposed scalable representation structures.

4 Conclusion

This contribution has proposed that instead of conceptualizing mental representations in spatial reasoning as either exclusively spatial or visual in format, different representation formats located on a continuum between these two extremes may be more adequate. Four criteria have been identified that are suggested to be positively proportional to visual characteristics of a representation: a high number of types of spatial relations, a high number of different relations, a high degree of specificity of relationships between entities, and a high degree of exemplarity of entities and relations. We have put forward the notion of scalable representation structures whose position on the visual / spatial continuum changes with the integration of additional spatial knowledge. While a first approach that exploits inherent interdependencies between different types of spatial knowledge for mental spatial reasoning is suggested and discussed, a number of open issues remain for further research before adequate scalable spatial representations can be introduced for computational cognitive modeling. It was the aim of our paper to initiate a debate about properties of potential, integrated (i.e., cross-type) representation structures in mental spatial reasoning. Attempts to dissociating between mental faculties often lead to an increase in detailed knowledge; however, trying to understand the mechanisms behind integrating different faculties in reasoning is equally important.

Acknowledgements

In this paper work done in the project R1-[ImageSpace] of the Transregional Collaborative Research Center SFB/TR 8 Spatial Cognition is presented. Funding by the German Research Foundation (DFG) is gratefully acknowledged. We also thank two anonymous reviewers for constructive comments.

References

Aginsky, V., Harris, C., Rensink, R.: Two strategies for learning a route in a driving simulator. Journal of Environmental Psychology 17, 317–331 (1997)

Barkowsky, T.: Mental representation and processing of geographic knowledge - A computational approach. Springer, Heidelberg (2002)

Bertel, S., Barkowsky, T., Engel, D.: Computational modeling of reasoning with mental images: basic requirements. In: Fum, D., Missier, d.F., Stocco, A. (eds.) Proceedings of ICCM 2006, Trieste, pp. 50–55. Edizioni Goliardiche, Trieste (2006)

Brockmole, J.R., Wang, R.F.: Switching between environmental representations in memory. Cognition 83, 295–316 (2002)

Carlson, L.A.: Selecting a reference frame. Spatial Cognition and Computation 1(4), 365–379 (1999)

Chandrasekaran, B., Kurup, U., Banerjee, B., Josephson, J.R.: An Architecture for Problem Solving with Diagrams. In: Blackwell, A., Marriott, K., Shimojima, A. (eds.) Proceedings of Diagrams 2004, pp. 151–165. Springer, Heidelberg (2004)

Chang, S.K., Shi, Q.Y., Yan, C.W.: Iconic indexing by 2-D string. IEEE Transactions on Pattern Analysis and Machine Intelligence 9(3), 413–428 (1987)

Charlot, V., Tzourio, N., Zilbovicius, M., Mazoyer, B., Denis, M.: Different mental imagery abilities result in different regional cerebral blood flow activation patterns during cognitive tests. Neuropsychologia 30, 565–580 (1992)

Cohn, A.G., Bennett, B., Gooday, J.M., Gotts, N.: RCC: a calculus for region based qualitative spatial reasoning. GeoInformatica 1, 275–316 (1997)

Collins, A.M., Quillian, M.R.: Retrieval time from semantic memory. Journal of Verbal Learning and Verbal Behavior 8, 240–247 (1969)

De Vooght, G., Vandierendonck, A.: Spatial Mental Models in Linear Reasoning. Kognitionswissenschaft 7(1), 5–10 (1998)

Easton, R.D., Sholl, M.J.: Object-array structure, frames of reference, and retrieval of spatial knowledge. Journal of Experimental Psychology: Learning, Memory, and Cognition 21(2), 483–500 (1995)

Egenhofer, M.J.: Deriving the composition of binary topological relations. Journal of Visual Languages and Computing 5, 133–149 (1994)

Frank, A.: Qualitative spatial reasoning: Cardinal directions as an example. International Journal of Geographical Information Systems (1995)

Funt, B.: Problem-solving with diagrammatic representations. Artificial Intelligence 13, 201–230 (1980)

Gilhooly, K.J., Logie, R.H., Wetherick, N.E., Wynn, V.: Working memory and strategies in syllogistic reasoning tasks. Mem Cognit 21, 115–124 (1993)

Glasgow, J., Papadias, D.: Computational imagery. Cognitive Science 16, 355–394 (1992)

Goodwin, G.P., Johnson-Laird, P.N.: Reasoning About Relations. Psychological Review 112(2), 468–493 (2005)

Haxby, J.V., Grady, C.L., Horwitz, B., Ungerleider, L.G., Mishkin, M., Carson, R.E., Herscovitch, P., Schapiro, M.B., Rapoport, S.I.: Dissociation of object and spatial visual processing pathways in human extrastriate cortex. Proceedings of the National Academy of Sciences 88, 1621–1625 (1991)

Hegarty, M., Kozhevnikov, M.: Types of visual-spatial representations and mathematical problem solving. Journal of Educational Psychology 91(4), 684–689 (1999)

Hirtle, S.C., Jonides, J.: Evidence of hierarchies in cognitive maps. Memory & Cognition 13, 208–217 (1985)

Ishai, A., Sagi, D.: Visual imagery facilitates visual perception: Psychophysical evidence. Journal of Cognitive Neuroscience 9(4), 476–489 (1997)

Jahn, G.: Hybrid representation of spatial descriptions. International workshop Spatial and Visual Components in Mental Reasoning about Large-Scale Spaces (01-02 September 2003), Bad Zwischenahn, Germany (2003)

Johnson-Laird, P.N.: Mental models. Harvard University Press, Cambridge, MA (1983)

Khenkhar, M.: Object-oriented representation of depictions on the basis of cell matrices. In: Herzog, O., Rollinger, C.-R. (eds.) Text Understanding in LILOG. LNCS, vol. 546, pp. 645–656. Springer, Heidelberg (1991)

Klauer, K.C.: Double dissociations in visual and spatial short-term memory. Journal of Experimental Psychology: General 133(3), 355–381 (2004)

Knauff, M., Johnson-Laird, P.N.: Visual imagery can impede reasoning. Memory & Cognition 30(3), 363–371 (2002)

Kosslyn, S.M.: Image and Mind. Harvard University Press, Cambridge, MA (1980)

Kosslyn, S.M.: Image and brain - The resolution of the imagery debate. MIT Press, Cambridge, MA (1994)

Kosslyn, S.M., Sussman, A.L.: Roles of imagery in perception: Or, there is no such thing as immaculate perception. In: Gazzaniga, M.S. (ed.) The cognitive neurosciences, pp. 1035–1042. MIT Press, Cambridge, MA (1995)

Kosslyn, S.M., Thompson, W.L.: When is early visual cortex activated during visual mental imagery? Psychological Bulletin 129(5), 723–746 (2003)

Kozhevnikov, M., Hegarty, M., Mayer, R.E.: Revisiting the visualizer-verbalizer dimension: Evidence for two types of visualizers. Cognition & Instruction 20, 47–78 (2002)

Kozhevnikov, M., Kosslyn, S., Shepard, J.: Spatial versus object visualizers: A new characterization of visual cognitive style. Memory & Cognition (in press)

Leeuwenberg, E.: Structural information theory and visual form. In: Kaernbach, C., Schröger, E., Müller, H. (eds.) Psychophysics beyond sensation, pp. 481–505. Lawrence Erlbaum, Mahwah, NJ (2004)

Levine, D.N., Warach, J., Farah, M.: Two visual systems in mental imagery: Dissociation of "what" and "where" in imagery disorders due to bilateral posterior cerebral lesions. Neurology 35, 1010–1018 (1985)

Likert, A., Quasha, W.H.: Revised Minnesota Paper Form Board Test (Series AA), New York. The Psychological Corporation (1941)

McNamara, T.P.: Mental representations of spatial judgments. Cognitive Psychology 18, 87–121 (1986)

Mellet, E., Tzourio, N., Denis, M., Mazoyer, B.: A positron emission tomography study of visual and mental spatial exploration. Journal of Cognitive Neuroscience 16, 6504–6512 (1995)

Moratz, R., Nebel, B., Freksa, C.: Qualitative spatial reasoning about relative position: The tradeoff between strong formal properties and successful reasoning about route graphs. In: Freksa, C., Brauer, W., Habel, C., Wender, K.F. (eds.) Spatial Cognition III, pp. 385–400. Springer, Heidelberg (2003)

Ragni, M., Knauff, M.: A Computational Model for Spatial Reasoning with Mental Models. In: Bara, B.G., Barsalou, L., Bucciarelli, M. (eds.) Proceedings of the 27th Annual Cognitive Science Conference, p. 1797. LEA (2005)

Rieser, J.J.: Access to knowledge of spatial structure at novel points of observation. Journal of Experimental Psychology: Learning, Memory, and Cognition 15(6), 1157–1165 (1989)

Schlieder, C., Berendt, B.: Mental model construction in spatial reasoning: A comparison of two computational theories. In: Schmid, U., Krems, J.F., Wysotzki, F. (eds.) Mind modelling: A cognitive science approach to reasoning, learning and discovery, pp. 133–162. Pabst Science Publishers, Lengerich (1998)

Sharma, J.: Integrated Spatial Reasoning in Geographic Information Systems: Combining Topology and Direction. Ph. D. Thesis, University of Maine (1996)

Sloman, A.: Why we need many knowledge representation formalisms. In: Bramer, M. (ed.) Proceedings BCS Expert Systems Conf. 1984, Cambridge University Press, Cambridge (1985)

Stenning, K., Oberlander, J.: A cognitive theory of graphical and linguistic reasoning: logic and implementation. Cognitive Science 19, 97–140 (1995)

Ungerleider, L.G., Mishkin, M.: Two cortical visual systems. In: Ingle, D.J., Goodale, M.A., Mansfield, R.J.W. (eds.) Analysis of visual behavior, pp. 549–586. MIT Press, Cambridge, MA (1982)

Vandenberg, S.G., Kuse, A.R.: Mental Rotations, a group test of three-dimensional spatial visualization. Perception and Motor Skills 47, 599–604 (1978)

Grounded Perceptual Schemas: Developmental Acquisition of Spatial Concepts

Amitabha Mukerjee and Mausoom Sarkar

Department of Computer Science and Engineering
Indian Institute of Technology, Kanpur 208 016, India
{amit,mausoom}@cse.iitk.ac.in

Abstract. Hand-engineered definitions of spatial categories are increasingly seen as brittle and spatial concepts in human interactions may need to learn these in terms of perceptually grounded "image schemas". Here, we present a developmental approach for the acquisition of grounded spatial schemas in a perceptual agent. We assume a capability for dynamic visual attention, and perceptual notions of wholeness and proximity. We first learn perceptual-object to linguistic-name mappings from simple 2D multi-agent visual streams co-occurring with word-separated utterances. Mutual information based statistical measures are seen to be sufficient to identify nominal participants in a simple discourse, based on a synthetic model of dynamic visual attention. Next, we use this knowledge of nominals to ground the semantics of spatial relations in language. We show that a notion of proximity between perceptual objects is sufficient to obtain a pre-verbal notion of graded spatial poses. Once linguistic data is superimposed on this, simple associative structures lead to distinctions such as "in" or "out". Finally we also show how this can lead to a model of actions, where verbs are learned along with the associated argument structures.

1 Spatial Learning

Models of spatial categories are rarely discrete. Most of the time they involve gradations, and one assesses the *degree* to which object A is "on", "in", or "in front of" B. Approaches toward the formalization of space, such as those focusing on topological or qualitative models [4], or potential fields [21] have either focused on the distinctions between these categories in terms of discrete sets, or used continuum fields with programmer-selected parameter sets. In any case, such models are difficult to define in any reliable manner. More importantly, these models are very brittle, and often fail under contextual variation.

In contrast, the human infant acquires these words from a grounded context, and forms perceptual schemas which reflect underlying concepts independent of the word used. These schemas can be simplified into discrete propositional structures, but the perceptual schema is always available for disambiguating conflicts arising out of the discrete categorization. Such models, often called *Image Schema* in Cognitive Linguistics [15] or *Perceptual Schema* in Experimental Psychology [18], involve abstractions on low-level features extracted from sensorimotor modalities. It is widely believed in cognitive science that the process of category formation operates on these inputs to define the structure of our conceptual space.

In this work, we propose a technique for the grounded acquisition of perceptual schemas as a mechanism for defining spatial primitives. We assume a developmental perspective involving pre-verbal cognition - i.e. the early stages of cognition in an infant, where categorial abstractions are formed from perceptual inputs [18]. Subsequently, combined with the ability to chunk linguistic structures, this leads to the associations of words with these prior categories. Here we extend this model to learn simple spatial prepositions in their most elementary contexts.

1.1 Capabilities of the Learner

The model assumes that the learner has the ability to segment the scene based on coherently moving blobs [27], and that it has a measure of perceptual proximity. In addition, the learner has the ability to assign degrees of perceptual salience to different objects, which is simulated in this work by a computational model of dynamic visual attention. In the first part of this work, we present a mechanism for learning object-label mappings based on matching a sparse word-separated text with the visual data, to learn nominals such as "square" or "circle" (Section 3). This follows an emerging consensus that language learners can acquire some nominals from language usage based on correlation alone, ([1]. Unlike other models of noun learning that look at single objects [26], or use words [28], our visual input involves multiple agents and the text involves unrestrained user commentary.

Given a knowledge of grounded object labels, we then embark on the attempt to develop a spatial model for distinguishing the simple prepositions "in" and "out". We assume that the learner has a perceptual ability to identify the most proximal object, which leads to a discretization of the visual space into "voronoi regions". A number of continuum features could have been used for learning the spatial primitives, but we have chosen to use the *Stolen voronoi area* [7] (section 4). We use a topographical neural network (kohonen map) to model the resulting spatial relations. Finally, we show how this model can be used to extend to the learning of spatial actions such as "run", "hit", or "chase", and to learn their argument structures (section 5).

Given the ambitious nature of this exercise, we limit ourselves to an extremely simple visual input, but the linguistic commentary obtained on this is unrestricted spoken (transcribed) text. We use a 2D video derived from the social psychology work of Heider and Simmel [9]. The co-occurring text were collected as part of an experiment on how users segment events into hierarchical subtasks [19][1] In this task, users were asked to segment the actions in the scene and also to describe the action in an unconstrained narrative. Consequently, the linguistic input has the wide variety expected in multiple articulations for the same scene (see Table 1 below) .

---

[1] We are grateful to Bridgette Martin and Barbara Tversky for sharing this video which was prepared by Bridgette Martin Hard and her colleagues, and also for the transcriptions of the co-articulated text which was collected by her.

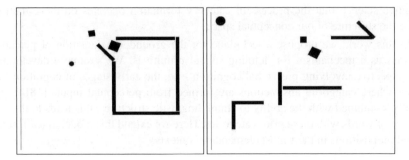

Fig. 1. *Input Videos.* Chase sequence (derived from [9] and Hide and Seek sequence - both created by Bridgette Martin [19]). The same three agents, "big square", "small square" and "circle" participate in two activities in two different spaces.

Table 1. *Description of the Events by Subjects.* Differing statements by two subjects in the [*Chase* video]

Start Frame	End Frame	Subject One	Subject Two
617	635	the little square hit the big square	they're hitting each other
805	848	the big square hit the little square	and they keep hitting each other
852	1100	the big square hit the little square again; the little circle moves to the door; the big square threatens the little circle	now the circle is blocking the entrance for the big square; now the circle is inside the square
1145	1202	the big square goes inside the box; (and) the door closes	another square went inside the big square

1.2 Spatial Features

In the spatial domain, there have been several attempts at defining spatial relations involving continuum measures defined over different geometric features on object pairs. Many of these measures involve point attributes such as potential fields, but our interest here is more on area measures since perceptually the objects constitute an area. Among the methods proposed here we may mention force histograms [20] [24], which define directional relationships between arbitrary shaped objects. These can be trained (using an ANN) to learn the semantics of "right ,above, left and below". On Functional properties of objects are the focus of the Lockwood and Forbus model (*SpaceCase*) [16] which gives a functional account of prepositions that can account for psychological data [8] that reflect how in/on judgments are affected by geometry of the ground, animacy of the ground, animacy of the figure and function of the ground. In this work however, we are trying to simulate a far more primitive learner, and while prior concepts such as animacy may be available, we are trying to acquire the very basic semantics of space, and we focus on measures that provide simple shape based measures. Here we find some measures based on direct area overlap [29], or on voronoi influence zones [7, 6]. The former requires a rectangular zone model heuristic and is somewhat sensitive to the size of the object, whereas the latter approach uses *Stolen Voronoi Area* which depends only on a notion of proximal object. Here we adopt the latter approach in the following experiments, and our feature set is defined in terms of how much of the zone of influence is lost by the insertion of the located object. The distinction between closed (bounded) zones and open (unbounded) ones constitute a key perceptual signature of containment.

Earlier work on preposition grounding includes Regier [25] who uses static and moving object scenes with single word labels. He uses a complex neural network inspired from neuropsychological and cognitive evidence. In the later part of this paper we try to ground few spatial relations*in, out* in simplistic situations from a video and its narration. In our work, we attempt to look at just two prepositions related to the containment concept which is among the earliest concepts available to an infant (six months, [3]). This relation relates to the prepositions "in" and "out", which are among the earliest spatial terms learned [2]. However, widespread polysemy in usage of "in" and "out" make it difficult to pinpoint their meanings.

The learned image schema for relative spatial poses is in the nature of a topographical neural net, also known as a kohonen map or a Self-Organizing Map (SOM). This is an unsupervised clustering method and requires no labels or other priors. Once the clustering is obtained, it can be mapped to linguistic data with very few exemplars.

Two video sequences were used in the experiments. Both show two squares and a circle moving in a 2D space "Fig.1". The first sequence (*Chase*) shows a room with a door, and a large square that chases the other two. The second (*Hide & Seek*) shows a game of hide & seek.

2 Synthetic Models of Visual Attention

Here, the learner is presented with only the visual stream and is not in the presence of the speaker, attention is mediated by visual saliency alone, and not by cues received from the speaker's gaze, under an assumption which we call the Perceptual Theory of Mind, following the Theory of Mind model [1].

Computational models of Visual Attention involve bottom-up and top-down processes. While top-down processes vary depending on task requirements, bottom-up aspects are more stable and have been encoded for static images [12] based on parallel extraction of intensity, colour and orientation contrast feature maps. Colour and intensity contrast maps are obtained as feature pyramids (maps at different scales), along with center-surround maps (multi-scale difference of feature maps). The center-surround feature processing is similar to the difference of gaussian convolved images (DOGs), present in the LGN or V1 regions of the mammalian cortex. For orientation specific processing, gabor filters are used with different frequencies and at different scales to generate the orientation specific feature map.

The main difficulty of using the Itti/Koch model here is that it does not handle dynamic scenes, i.e. the attention processes do not model scene motion. This model is extended here to model dynamic scenes based on motion saliency. Motion saliency is computed from the optical flow, and a confidence map is introduced to record the uncertainty accumulating at scene locations not visited for some time. A small foveal bias is introduced to mediate in favour of proximal fixations as opposed to large saccadic motions. The saliency map is the sum of the feature maps and confidence maps, mediated by the foveal bias, and a Winner-Take-All (WTA) isolates the most conspicuous location for the next fixation. The overall architecture is highlighted in "Fig.2".

2.1 Perceptual Theory of Mind

The Theory of Mind hypothesis [1] holds that the learner has a model for several aspect of the speaker's mind, at various levels from a sensitivity to the object being attending to, to belief structures (e.g. children under three are found to be incapable of entertaining false beliefs). In this work, we focus at the lowest end of this spectrum, and focus on what we call the *Perceptual Theory of Mind*. While much of the Theory of Mind work has focused on gaze following based on cues from the speaker's eyes or her gaze direction, the Perceptual Theory of Mind makes a much weaker claim: in the absence of direct cues from a speaker, it assumes that the speaker would have attended to those parts of the scene that the learner also finds salient. This is probably a valid assumption for children from the age of six months onwards [13], although the mechanisms for perceptual salience are themselves being developed at this stage. In our work, we do not specify a particular development status for our learning agent, but assume this model to infer that the scene objects being attended to by the agent were also salient for the speaker at the moment of utterance.

Language Acquisition experiments tend to cast doubts on the efficacy of a purely associationist model of learning words, and it is true that a large percentage of our vocabulary is not learned using multimodal inputs but from reading. But even here, a case can be made that a small core of grounded words constitute the key to learning other words. Indeed, proponents of a semantic approach to language often categorize concepts in terms of a small set of semantic primes, on which other words must be built [11]. This work presents some evidence that for some of the earliest words, multimodal associations mediated by attentional processes provide strong and reliable cues for learning nominals and their properties, verbs and their argument and event structures.

3 Object-Name Learning

Grounded semantics for nominals appearing in the text are associated with the corresponding image object token based on the following steps:

1. *Tracking and Recognition.* Object recognition is based on shape.
2. *Synthetic Gaze Prediction* Estimates the entities that are being attended to during a particular utterance, under the perceptual theory of mind assumption.
3. *Associative Learning.* The words in the commentary are associated with their perceptual correlates.

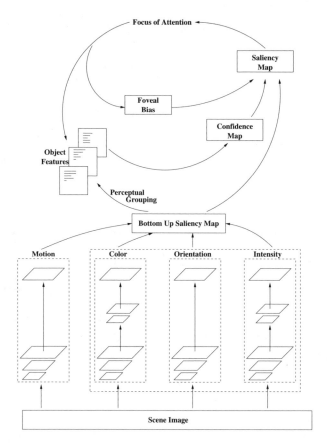

Fig. 2. *Bottom-Up Dynamic Visual Attention Model.* The feature maps for static images (colour, intensity and orientation) are extended with a motion saliency map (based on optical flow). In addition a confidence map records which sites have not been visited for a longer time. Winner-Take-All determines the next fixation.

3.1 Tracking and Shape-Based Recognition

Spelke [27], among others has demonstrated that infants perceive objects as connected blobs with coherent motion. In our videos, these blobs are rather simple, and these are extracted and tracked for the duration of the video. A global list of all the objects and their pose and orientation is used for computation of motion primitives.

Shape recognition is required to obtain shape universals for different object tokens with the same perceptual signature. This is important in both correlating objects in the same scene (squares) and also across different perceptual input situations, as in combining word associations across the two separate videos. Shape matching for 2D objects is implemented based on a histogram of the tangent direction at each point along the boundary, a simple scale, rotation and transformation invariant metric that is cognitively plausible. The normalized (scale invariant) histograms were circularly shifted and compared using statistical divergence function as in Roy [26] to determine the closest transformation between two shapes, which serves as our model of shape similarity.

$$d_v(X, Y) = \sum_i \frac{(x_i - y_i)^2}{(x_i + y_i)}$$

where $X = \bigcup_i x_i$ and $Y = \bigcup_i y_i$ are two histograms and x_i, y_i are the values of a histogram.

The synthetic model of visual attention for dynamic scenes was used to predict the gaze for the two videos. The predicted gaze for the two videos are shown in "Fig.3".

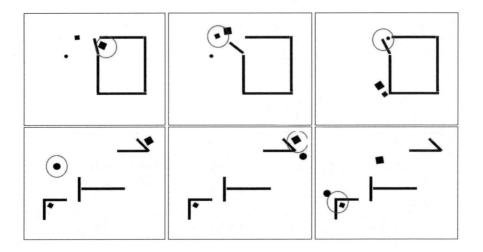

Fig. 3. *Focus of attention on the video.* Red circles represent the focus of attention; Oscillating attention between multiple objects makes multiple participants more likely.

3.2 Association of Meanings with Words

The attended objects are now associated with the temporally correlated words using one of two probability measures. At this point, we also assume that the learner has been exposed to other linguistic fragments before this, so that words like "the" and "is" are known to be more general than this discourse context, and are not applied to this situation. (In the BNC, "the" occurs 1500 times more frequently than "square", say). Using perceptual equivalence relations based on shape, we associate the objects with words from the multiple narratives using the probability measures outlined below.

Two measures for associating the words with the objects are used.

1. Mutual Information Measure.
2. Joint Probability Measure.

Mutual information Measure. After temporally correlating words with objects the association is defined as the product of mutual information of word w_i and object o_j with their joint probability.

$$A = \Pr(w_i, o_j) \log \frac{\Pr(w_i, o_j)}{\Pr(w_i)\Pr(o_j)}$$

If W and O ($W = \bigcup_i w_i$ and $O = \bigcup_i o_i$) are two random variables then their Mutual Information $I(W, O)$ would be

$$I(W, O) = \sum_i \sum_j \Pr(w, o_j) \log \frac{\Pr(w_i, o_j)}{\Pr(w_i)\Pr(o_j)}$$

where $\Pr(w_i, o_j) \log \frac{\Pr(w_i, o_j)}{\Pr(w_i)\Pr(o_j)}$ is the contribution of each word object pair.

Joint-Probability Measure. This measure is the product of the conditional probability of object o_j given word w_i with their joint probability.

$$A = \Pr(w_i, o_j) \Pr(o_j|w_i)$$

The conditional probability will give high values for the words that occurred specifically for a given object, while joint probability will highlight the number of word object pairs. Hence this measure should bring out the stronger labels associations.

Both the measures showed object names being highly associated with the corresponding object. Results are shown for the chase video and both videos together "Fig.4 and Fig.5" respectively.

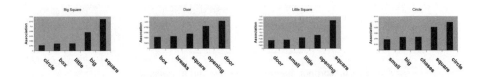

Fig. 4. *Early Learner: Noun Learning using First Encounter.* Association using Joint Probability of words with Big Square, Door, Little Square and Circle in Chase video. Most objects except "little square" are well characterized.

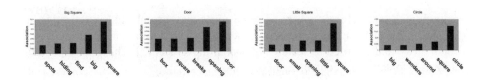

Fig. 5. *Noun Learning using Multiple Encounters.* Associating words using Joint Probability - remembering shapes across two encounters (both videos).

4 Spatial Prepositions of Containment: "in" and "out"

Grounded learning of spatial modifiers involves defining concept classes to represent spatial distinctions. Clearly these would need to be graded (or continuum or fuzzy). As explained earlier, we choose voronoi model of space [7, 6] as this requires very little of the learner other than the capacity to identify the most proximal region, and it can also give a graded measure of membership.

A Voronoi representation of space divides the space into tessellations (voronoi regions) where each region contains a site - traditionally, this is a point, but it may also be a line or an area; here we use the line segments bounding an area as our sites. All points in the voronoi region are closer to its site than to any other site. Voronoi diagrams of points and line segments are shown in fig 6. Voronoi diagram for any arbitrary shape can be determined by defining it as a voronoi digram of points and line segments. In our model for line segments, we distinguish (as do most algorithms) the two end points from the body of the line; this results in two internal boundaries for the voronoi diagram - the separator between this end-point and the interior of the line - and are useful in discriminating the nature of the intersection between zones, and are retained.

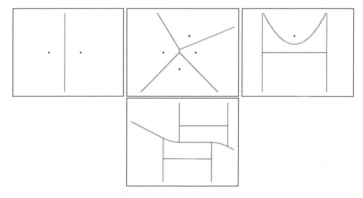

Fig. 6. Voronoi diagrams of points and line segments

The voronoi model of a space changes as agents move in it. This dynamic nature results in qualitative changes as the boundary between two sites or objects (reflecting adjacency of those objects) shifts to some other objects. For the *empty* space, each part of the boundary has its zone of influence. Some zones may be bounded, reflecting proximity with other objects, where as some zones may be unbounded, reflecting an open nature. As an agent enters this space, the areas of these zones are reduced (stolen voronoi area), and are used as features of spatial movement [7]. This relation is ultimately used to define the containership concept.

4.1 Learning Containment

Containment is one of the earliest concepts in our repertoire, yet it offers tremendous complexity. In adult usage, containment is affected by function which results in a wide variety of ramifications [5]. However, to an early learner, the prototypical interpretation of containment and container emerges based on abstracting on percepts, possibly earlier than six months [3, 2].

The voronoi model used here as the learning feature captures the influence of one object over another. We use the ratio of stolen area from bounded and unbounded voronoi regions as a binary feature. The ratio is calculated between the voronoi area consumed by the introduction of the new object and the initial voronoi area of the reference object, and distances between features are determined in terms of an euclidean metric to construct the kohonen map (Fig. 7).

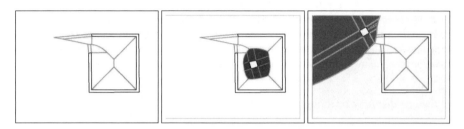

Fig. 7. (a) Initial object voronoi (b) and (c) Dark area shows the stolen voronoi area

4.2 Self Organising Map

A self organising map is an unsupervised learning method based on associative memory, also known as the kohonen map [14]. A grid of neurons is defined, each with a k dimensional weight vector during the learning process so that proximal regions represent "proximal" patterns in the input data. From an initial assignment of random weights, each data point finds its best matching unit/neuron (BMU) and the weights of neurons in the neighbourhood of BMU are modified towards the weight vector of the input, using the equation

$$W(t+1) = W(t) + \theta(t)\alpha(t)(D(t) - W(t))$$

where $W(t)$ is the weight vector of the neuron, $D(t)$ id the input vector, $\theta(t)$ is the neighbourhood function which constraints the amount of influence on the neighbours, and $\alpha(t)$ is the learning rate.

We use a simple kohonen map with a gaussian neighbourhood function and a linearly decreasing learning rate. The grid size of 30x40 nodes represents the clustering space. Visualization of the SOM was done using a unified distance matrix(U-matrix), which shows the distance between neighbouring units weight vectors as a 3-d landscape. It is calculated as the sum of the magnitudes of weight vectors of all immediate neighbours. The 3d landscape is visualised by mapping the U-matrix height into the RGB space. The uniform patches of color represent zones of uniform height representing some cluster.

Two maps were trained based on input data of different spatial poses. The first map shows clear colour (height) gradations for poses inside and outside the space. Among

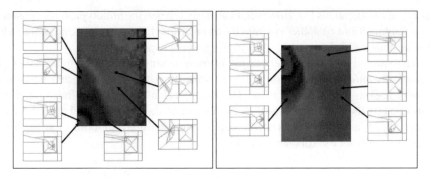

Fig. 8. *Visualizations of Perceptual Schemata.* The Self-Organizing Map acts as our perceptual schema. Given an input, it can identify what cluster it belongs to and with what confidence. In the opposite direction, given a conceptual description it can generate a maximum likelihood prototype pose. (a) map for the complete in-out cluster, and (b) trained only on "in" examples.

the regions that may be thought of as "in", the points at the center are furthest from the out region, and the points near the door lie close to the boundary, which itself shows a clear gradation in colour 8(a). Similarly, the out regions are also graded by distance from the center. In a second experiment, we clustered the zones that correspond mainly to the "in" region (corresponding to situations where the feature ratio of bounded-area-stolen is non-zero). Here we find finer distinctions emerging that may be taken to read "corner" or "at-the-edge" or "center". 8(b).

4.3 Mapping to Prepositions "in" and "out"

The top level clusters were then matched with words occurring in the the user's commentaries on the video, using the same mutual information measure as used for nouns. Initially, we expected this to fail on such sparse data, since many of the uses of "in" or "out" varied widely owing to the wide-ranging polysemy in the meanings of these very basic words. Also there were only 41 instances of "in" and 28 of "out". For example, there are no instances, among the 28 examples of "out" in the data, of a sentence like "The circle is out of the box" - there are some instances of "outside" but we treat it as a separate word. Here are some examples of "out" from the user (spoken) commentaries on the videos:

1. `little block comes out.`
2. `circle is out in the open.`
3. `square is moving around the box trying to get out.`
4. `the big square moved out of the corner.`
5. `he's trying to push his way out of the square.`
6. `and they all spread out again.`
7. `the door shuts the square out of the box.`

Sentence 1, with the verb "come" is a prototypical interpretation of "out" as the motion goal-specifier. Sentence 2 has a much more specific meaning, but it spatially it is consistent. Sentences 3, 4 and 5 have similar meanings, but the contexts make them poor exemplars. Sentences 6 would have been similar, but it has enough of the "out" in it to correlate well. The verb-preposition complex "spread out" (7) reflects a completely different usage. In each instance, the specific meaning carried by these prepositions are considerably modified by the regulating verb. Similar instances of polysemy with "in" are preponderant in the data.

However, we were ourselves surprised to find that for both of these prepositions, there are enough correlations so that the spatial cluster for the prototypical "in" and "out", emerge with the highest mutual correlation in both cases with the corresponding words in our simple statistical measure. We feel that this may be because even in the metaphorical and other conventionalized uses of this preposition, it still retains some of the basic sense, so that the correlations come out. We observe that "out" is in a tie with "of" which seems to indicate the construct "out of". The particle "of" occurs almost twice as frequently as "off" in general text, so it is likely to drop off in the specificity measure if we were to include other (non-spatial) contexts. Note that since it is assumed that nouns/object labels are already known, these were removed from the word set, but then these labels are far less frequent and would are unlikely to constitute an obstacle. Thus, given that a pre-verbal system (or infant) has a notion of what constitute in /out (containment) then it is very straightforward for her to learn a rudimentary grounded schema for "in" or "out" from a handful of examples.

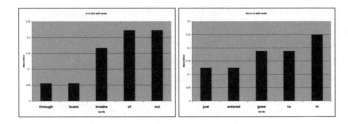

Fig. 9. *Preposition Learning.* Matches between clusters and words in text a) category in-out; b) category in.

5 Learning Action Labels

The last part of this work demonstrates the learning of action names. Since actions are pre-linguistically categorized based on the number of participants in the action, this also results in discovering the valence of the verb. We use different k-ary feature sets depending on the number of participants, and actions, which require a temporal component, are learned using a recurrent neural network (SRN) [22]. Owing to paucity of data for learning actions additional movies were created for training.

5.1 Features for Spatio-temporal Analysis

The feature set determines the dimensionality of the space in which the word meaning is grounded. The valence of the predicate is a crucial input for this information, it is seen that a *monadic* feature set is sufficient for actions like run (intransitive), whereas a *dyadic* feature set is needed in actions like chase (transitive).

There is considerable evidence that infants have pre-linguistic perceptual notions for Path concepts such as source, trajectory etc, and also other notions such as Up-Down, Containment, Force, Part-Whole and Link [17]. Some of these aspects have also been implemented in computational systems [23]. In this work, we assume that the pre-linguistic visual system has the capability to abstract the following:

(a) Shape classes (square vs circle, on a high contrast image)
(b) Motion characteristics for individual agents (monadic features)
(c) Motion characteristics for pairs agents (dyadic features)

Specifically, we define the following abstract features:

- *Monadic Features:*
    1. Velocity- vx and vy in respective direction
    2. Acceleration- ax and ay in respective direction
    3. θ- Angular displacement of the object
    4. $d\theta$- Change in displacement of the object.
    5. ω - Angular velocity of the object.
    6. α - Angular acceleration of the object.
- *Dyadic Features*
    1. Proximity- It is inverse of the boundary-to-boundary distance between two objects.
    2. Relative Velocity between the two objects
    3. Relative Acceleration between the two objects
    4. Measure of Parallelism between the direction of motion of the two objects i.e. $\cos(\hat{v}_a - \hat{v}_b)$
    5. Leader A - Measure of leadership (in motion) of A w.r.t B i.e $\cos(\hat{v}_a - \theta_{ba})$
    6. Leader B - Measure of leadership (in motion) of B w.r.t A i.e. $\cos(\hat{v}_b - \theta_{ab})$
    7. Chamfer- Measure of chamfering between the two objects i.e. $\frac{\sin(\hat{v}_a - \theta_{ba}) + \sin(\hat{v}_b - \theta_{ab})}{2}$

For example, an action such as "X chases Y" may associate *Proximity* (indicative of spatial clustering between X and Y), *Parallelism*, high *Leader X* and low *Relative Velocity*. The dyadic parameter *Chamfer* feature reflects if two objects are moving together or are moving one ahead of the other this feature, when low, may indicate a follow or chase action; when close to 90, it is indicative of a move-together action.

5.2 Verb Learning Results

Assuming that each video sequence is of l duration during which a verb is reported for τ time units and not reported for $l - \tau$ units. The Detection Accuracy, or true positives, is computed as a percentage of positive classifications out of τ. The False Negatives, again, are computed as a percentage of τ. However, the false positives as well as focus mismatches are computed as a percentage of $l - \tau$. Results are presented for three learning scenarios:

Timescale = 2530 frames (81 seconds)

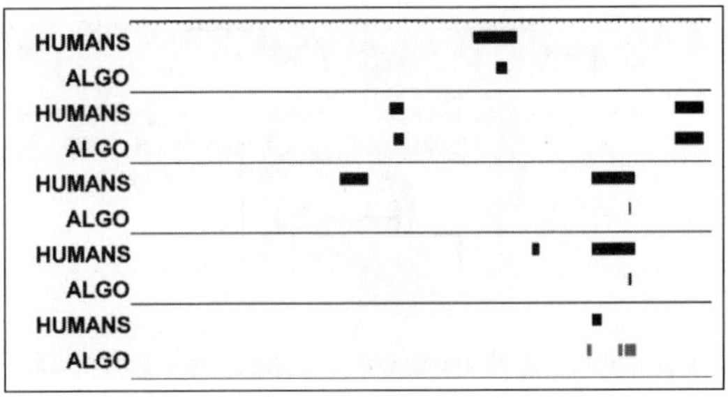

Fig. 10. *Comparing machine and human classifications*, for *verb*: *chase*. Each row is a different pairwise combination of agents.

Timescale = 2530 frames (81 seconds)

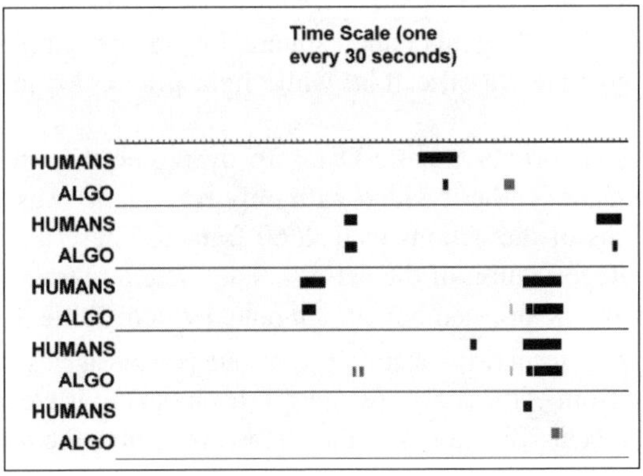

Fig. 11. *Comparing outputs - trained on synthetic data.* One-Verb-One-Network for *chase* video on Synthetic Data.

Table 2. *Interval results.* One-Verb-One-Network, Unsupervised.

Verb	TP	FP	FN	FM
hit	3	3	1	1
chase	6	0	3	4
come Closer	6	20	7	24
move Away	8	3	0	14
spins	22	0	1	9
moves	5	1	2	7

where TP=True Positive, FP=False Positive, FN=False Negative and FM=Focus Mismatch.

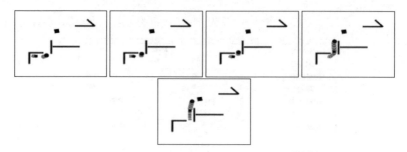

Fig. 12. Example of classification of **chase** event, one for every 1 Second (30 frames), between the circle and the small square

1. *One-Verb-One-Network:Human Subject Tags*. Here Different SRNs are trained for each action cluster; Table 2 presents the results with unsupervised text and image correlations based on the statements of human subjects. "Fig.12" shows some of the video fragments which are classified by the learning system as chase.

   "Fig.10, Fig.11 and Fig.13" present results along a time line, each row reflects a different combination of agents (small square, big square, circle). Dark gray color indicates false positive classification, while light gray color indicates focus mismatches.

2. One-Verb-One-Network: Synthetic Data: To overcome the very severely limited data in the videos, a synthetic video with only two agents was created, executing canonical versions of the actions over 2660 frames. These may be thought of as pre-linguistic categorization of the actions. The system is trained on the synthetic data, and tested on the original human-annotated video; Table 3 shows the results.

   Between the two scenarios - learning from one pair of agents in a human-tagged data vs learning from elaborately created synthetic data, it is clear that having more data improves the classification accuracy. However, since the training set has only

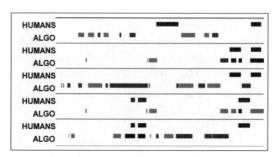

Fig. 13. A comparison of system and human descriptions of *come closer* verb over a time line. Note the large extent of focus mismatch (light grey).

Table 3. Percentage results for One-Verb-One-Network Synthetic Data

Verb	TP	FP	FN	FM
chase	52.59%	0.21%	47.09%	1.24%
come Closer	53.61%	11.29%	45.52%	20.41%
move Away	65.30%	12.07%	33.37%	17.15%

where TP=True Positive, FP=False Positive, FN=False Negative and FM=Focus Mismatch

two agents and has no focus mismatch, the extent of this problem is actually higher when trained on synthetic data.

3. All-Verbs-One-Network: Synthetic Data

    Here we train a single recurrent network to distinguish among all the verbs, and use the synthetic video for training. Only dyadic features are used. There is one input neuron for each feature, one output neuron for each verb and the hidden layer has the same number of nodes as the input layer. An additional output neuron indicates no output.

    The accuracy of events detected improves considerably if we constrain the sequence based on the participating agents as well as the verb. The Interval accuracy is computed as a ratio of correctly classified intervals to the total intervals for each pair of objects. Table 4 lists the performance of the system in this mode.

Table 4. Frame-by-frame Interval Accuracy results for All-Verbs-One-Network

Video	Total Frames	Objects involved	Frames Classified	Correct Classification	Interval Accuracy %
Hide & Seek	2530	SS, C	2492	2380	94.07
		SS, BS	2329	2156	85.22
		BS, C	2530	2028	80.16
Chase	2530	SS, C	2334	2285	90.30
		SS, BS	2329	2156	85.22
		BS, C	2384	2207	87.23

SS = Small Square, BS = Big Square, C = Circle. Percentage accuracy computed over total number of frames and not on number of classified frames.

6 Conclusion

We have developed (from an extremely simple visual stimulus) a coherent approach to acquiring spatial primitives and also action primitives that build on spatial structures. In particular, we have made no prior assumptions on knowledge about the agents or domain of action, the only abilities inherent in the learner is that it has a model of visual attention, and that it can identify perceptually proximal objects. Based on this, it forms pre-verbal conceptual schema for frequently occurring clusters. These are subsequently mapped into labels for objects (nouns), spatial poses (prepositions) and actions (verbs). In addition, participants in the preposition and verb structures are identified in terms of the grounded semantics of the relational schema.

The image schema is stored in a topographical neural net, which can then determine quickly if a given novel input belongs to some class or not. For the action learning, the image schema must encode temporal data and this is achieved through an Recurrence Neural Net (SRN).

A crucial aspect of the current work is that the set of features used in the preposition and verb learning are presumed to be available to the learner, and are not learned per se. While one may argue that some of these notions (such as size or motion trajectory information) may be preferred as a part of the innate perceptual apparatus, we strongly suspect that some of the relative motion parameters may actually be learnable given enough training data, and we hope to explore this with the additional data.

In terms of spatial language, this work makes no attempt to learn the syntactic structure of the statements per se, and effects such as tense, case, gender etc are not modeled at all. An important extension would be to learn more grammatical structures as in the work of [10]. In our case, once the verbal heads of various phrases are known, and with some knowledge of closed-class words, it would be possible to identify some of the roles played by grammatical elements in different constructions appearing in the narrative.

This work raises several questions that may prove to be important. First, is this experiment scalable to visual data more complex than toy 2D videos? The answer to this appears to be yes, as we shall soon be reporting. However, as visual complexity increases, it requires more and more skill in tracking and object consolidation to keep track of the motions. A second and possibly more important question would refer to the conceptual ramifications, even in space, of the simple concepts that one runs. One can consider for instance, a "road going to Bremen." How does one tune an action of go as a motion to the fixed concept carried here, what Talmy has called *fictive* motion? Even further afield, one can "go" to an internet site, or my thoughts may "go" to you. Clearly these transfers define new senses, but they are possibly built up on the primitive grounded notion that was first captured in our sensorimotor mode. The larger issues thrown up by these constitute what may be called metaphorical extension, and perhaps the type of grounded models learned through multimodal interaction is precisely the kind of in such this may indeed constitute one of the more important challenges for reasoning as a whole in the coming years.

Acknowledgments

We thank Bridgette Martin Hard and Barbara Tversky for initial discussions regarding the idea behind this paper. The videos and the text were provided by Bridgette Martin Hard following initial discussions at the Spatial Cognition summer institute 2003 at Bad

Zwischenahn. We also thank Vivek Singh for his visual attention model, and Dana for his help in developing it. V. Shreenivas and Achla Raina helped in the verb acquisition work.

References

1. Bloom, P.: How Children Learn the Meanings of Words. MIT Press, Cambridge, MA (2000)
2. Bowerman, M.: Learning how to structure space for language: A crosslinguistic perspective. In: Bloom, P., Peterson, M.A., Nadel, L., Garrett, M.F. (eds.) Language and Space, June 1999, MIT Press, Cambridge, MA (1999)
3. Casasola, M., Cohen, L.B., Chiarello, E.: Six-month-old infants categorization of containment spatial relations. Child Development 74, 679–693 (2003)
4. Cohn, A.G., Hazarika, S.: Qualitative spatial representation and reasoning: An overview. Fundamenta Informaticae 46(1/2), 1–29 (2001)
5. Coventry, K.R.: Function, geometry, and spatial prepositions: Three experiments. Spatial Cognition and Computation 1, 145–154 (1999)
6. Edwards, G., Ligozat, G., Gryl, A., Fraczac, L., Moulin, B., Gold, C.M.: A voronoi-based pivot representation of spatial concepts and its application to route descriptions expressed in natural language. In: Proceedings, 7th International Symposium on Spatial Data handling, Delft, The Netherlands, pp. 7B1–7B15 (1996)
7. Edwards, G., Moulin, B.: Towards the simulation of spatial mental images using the voronoi model. In: IJCAI-95. Workshop on the Representation and Processing of Spatial Expressions, Montreal, Canada (August 19, 1995), pp. 63–74 (1995)
8. Feist, M.I., Gentner, D.: Factors involved in the user of in and on. In: Proceedings of the Twenty-fifth Annual Meeting of the Cognitive Science Society (2003)
9. Heider, F., Simmel, M.: An experimental study of apparent behavior. American Journal of Psychology 57, 243–259 (1944)
10. Ford, D.P.: Learning grammatical constructions in a miniature language from narrated video events. cognitive science (2003)
11. Goddard, C., Wierzbicka, A. (eds.): Meaning and Universal Grammar: Theory and Empirical Findings, Amsterdam. John Benjamins [Studies in Language Companion Series, 60], vol. 1 (2002)
12. Itti, L.: Models of Bottom-Up and Top-Down Visual Attention. PhD thesis, California Institute of Technology, Pasadena, California (January 2000)
13. Flavell, J.H.: Theory-of-mind development: Retrospect and prospect. Merrill-Palmer Quarterly 50, 274–329 (2004)
14. Kohonen, T.: The self-organizing map. Proceedings of the IEEE 78(9), 1464–1480 (1990)
15. Langacker, R.: Foundations of Cognitive Grammar, vol. 1. Stanford University Press, Stanford, CA (1991)
16. Lockwood, K., Forbus, K., Usher, J.: Spacecase: A model of spatial preposition use. In: Proceedings of the 27th Annual Conference of the Cognitive Science Society, Stressa, Italy (2005)
17. Jean, M., Mandler, J.M.: How to build a baby ii: Conceptual primitives. Psychological Review 99, 587–604 (1992)
18. Mandler, J.M.: Foundations of Mind. Oxford University Press, New York (2004)
19. Martin, B., Tversky, B.: Segmenting ambiguous events. In: Proceedings of the 25th annual meeting of the Cognitive Science Society Crucial for our Data-Collection chapter (2003)

20. Matsakis, P., Wendling, L.: A new way to represent the relative position between area objects. IEEE Trans. Pattern Analysis and Machine Intelligence 21(7), 634–643 (1999)
21. Mukerjee, A.: Neat vs scruffy: A review of computational models for spatial expressions. Representation and processing of spatial expressions, 1–37 (1998)
22. Mukerjee, A., Vaghela, P.B., Shreeniwas, V.: Pre-linguistic verb acquisition from repeated language exposure for visual events. In: Proceedings International Conference on Natural Language Processing (ICON-2004) (19-22, 2004)
23. Oates, T., Cohen, P.R., Atkin, M.S., Beal, C.R.: Building a baby. In: Proceedings of the 18th annual conference of the Cognitive Science Society, pp. 518–522 (1996)
24. Matsakis, P., Bondugula, R., Keller, J.: Force histograms and neural networks for humanbased spatial relationship generalization. In: Proceedings of Int. Conf. on Neural Networks and Computational Intelligence (2004)
25. Regier, T.: A model of the human capacity for categorizing spatial relationships. Cognitive Linguistics, 63–88 (1995)
26. Roy, D.: Integration of speech and vision using mutual information. In: International Conference on Acoustics, Speech and Signal Processing(ICASSP 2000) (2000)
27. Spelke, E.S.: Principles of object perception. Principles of object perception. Cognitive Science 14, 29–56 (1990)
28. Steels, L.: Language learning and language contact. In: Workshop Notes of the ECML/MLnet Workshop on Empirical Learning of Natural Language Processing Tasks, ECML 1997, pp. 11–24 (1997)
29. Vorwerg, C., Socher, G., Fuhr, T., Sagerer, G., Rickheit, G.: Projective relations for 3d space: Computational model, application, and psychological evaluation. In: AAAI/IAAI, pp. 159–164 (1997)

Modeling Human Spatial Memory Within a Symbolic Architecture of Cognition

Carsten Winkelholz[1] and Christopher M. Schlick[2]

[1] Research Establishment for Applied Science (FGAN),
Neuenahrer Strasse 20, 53343 Wachtberg, Germany
winkelholz@fgan.de
[2] Institute of Industrial Engineering and Ergonomics, RWTH Aachen University,
Bergdriesch 27, 52062 Aachen, Germany
c.schlick@iaw.rwth-aachen.de

Abstract. This paper presents a study on the integration of spatial cognition into a symbolic theory. The concepts of encoding object-locations in local allocentric reference systems and noisy representations of locations have been integrated into the ACT-R architecture of cognition. The intrinsic reference axis of the local reference systems automatically result from the sequence of attended locations. The first part of the paper describes experiments we performed to test hypotheses on the usage of local allocentric reference systems in the context of object-location memory in graphical layout structures. The second part describes in more detail the theory and its integration into ACT-R. Based on the theory a model has been developed for the task in the experiments. The parameters for the noise in the representation of locations and the parameters for the recall of symbolic memory chunks were set to values in the magnitude quoted in literature. The model satisfyingly reproduces the data from user studies with 30 subjects.

1 Introduction

Symbolic theories of cognition are appealing for studies in the field of human-computer interaction. Symbolic theories allow the expression of the cognition process in relation to the task and the visual elements of the interface. One main issue for cognitive modeling in this field is the integration of visual information. Ehret [5] and Anderson et al. [2] describe symbolic models that reproduce learning curves for the location of information on a display. In these examples the underlying mechanism for the learning of locations is the same as for the learning of facts. After some practice the location of specific objects, such as menu buttons, can be retrieved without a time consuming random visual search and encoding of labels. In both studies the locations of visual objects are represented in absolute screen coordinates, and no noise in the representations of the scalar values has been assumed. Accordingly, the success of a retrieval only depends on the number of repetitions of the location of an item. However, the location of an object can only be identified within a frame of reference. In experimental psychology it is well accepted to divide the frames of references into the following two categories: an egocentric reference system, which specifies the location of an object with respect to the observer, and an environmental (allocentric) reference system, which specifies the location of an object with respect to elements and features of the environment. A good review of the experimental evidence on the usage of these different reference systems in human spatial memory is given by McNamara [9]. This aspect of human spatial memory implicates that the structure of a graphical layout might affect the performance of object-location memory. Object-location memory in the context of graphical layout-structures has already been investigated in the field of information visualization by Tavanti & Lind [13] and Cockburn [4]. These studies showed that different kinds of displays influence performance in object-location retrieval from memory. In both studies the memory task was to associate alphanumerical letters to object-locations. Cockburn suspected that a horizontal oriented layout facilitates the formation of effective letter mnemonics, whereas Tavanti & Lind speculated that a more 'natural' appearance of a visualization enhances object-location memory. Both studies did not consider spatial relations of objects to each other within the structure as a factor. Therefore, we performed our own experiments in which the structure of the object-configurations was varied. In the first part this paper reports the design and results of these experiments. A detailed analysis of the user traces within this experiment suggests that users choose the last two attended locations as a reference axis to which they encode the currently attended location. The second part of the paper describes how we integrated this fact in combination with the concept of noisy location dimensions into the visual module of the symbolic ACT-R [2] architecture of cognition. Based on this extended visual module a symbolic model for the memorizing-task has been formulated. The parameters for noise in the location representation and activation decay of memory chunks have been set to fit the data and are compared to values quoted in literature.

2 Experiments

The task of our experiments was to memorize a randomly created sequence of highlighted objects from different structures. The number of correct repeated sequences is used as a measure of performance. This kind of memory task allows an effective analysis of the errors made by subjects. Two experiments were performed. The first experiment investigated the factor horizontal vs. vertical oriented layout structure and the factor of the existence vs. non-existence of symmetric features in the layout structure. The second experiment focused on the investigation of noise in the encoding of spatial object-to-object relations. In the following the procedure of the experiment is only sketched. More details can be found in [18]. Thirty volunteer subjects (only male, average age 35) were recruited from the staff of our institute to perform both experiments. All subjects had normal or corrected-to-normal vision. Three sets of different structures were created. Each structure consisted of red spheres of equal size. The layout structures were presented against a black background on a 21" VGA monitor with a resolution of 1280x1024 pixels. The monitor was in front of the subjects within 2 feet. Subjects were asked to respond by clicking with a mouse.

2.1 Experiment 1

The first experiment aimed at showing whether the performance of recalling object-locations is still improved in the horizontal oriented structures, even if no alphanumerical letters are used as retrieval keys.

Materials. Fig. 1 shows the three structures that were used in the first experiment. Each structure consists of 25 spherical items. The first structure represents a 2D display of a tree-structure, like it is used in most common graphical user interfaces. The second structure is horizontal oriented and exhibits some symmetrical features. The third structure is equivalent to the first one except that it is rotated counterclockwise by 90 degrees.

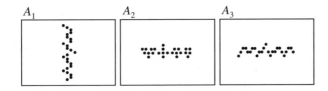

Fig. 1. Set of object configurations used in Experiment 1

Design and Procedure. In each encoding retrieval trial, the subject was presented with one structure. After an acoustical signal, the computer started to highlight objects of one randomly created sequence. Only one object of the sequence was immediately highlighted. The sequences were five items long. The highlighted object differed from the unhighlighted objects by color (blue instead of red), increased size and a cross that appeared within its circle shape. The end of a sequence was indicated by a second acoustical signal. Subjects were instructed to repeat the highlighted objects in correct order, by clicking them with the mouse. After five objects had been clicked, another acoustical signal rang out. After a short break the next sequence was presented to the subject. All sequences were created randomly with the property that no object is highlighted twice in succession. New random sequences were created for each subject in order to avoid the event that an easy sequence was created by chance for any structure (e.g., all objects of a sequence are only in one row). By creating random sequences for each subject the factor of the sequence itself is balanced among subjects. In order to examine a specific factor in detail the establishment of a sequence for all test subjects is of interest. This was done in parts in the second experiment reported below.

Results. The number of correct and erroneous repeated sequences for each structure is shown in Table 1.

Table 1. Contingency table (2×3) of correct and erroneous sequences in set A of Experiment 1

	A_1	A_2	A_3
Correct seqs	46	61	63
Erroneous seqs	74	59	57

The effect of structure approaches significance (2×3 contingency table $p = 0.056$, $\chi^2 = 5.77$). When comparing the numbers of correct repeated sequences between each pair of structures with a one-sided analysis of the corresponding 2×2 contingency tables, the exact Fisher test yields that performance in the horizontal oriented structures is significantly higher ($p < 0.05$), whereas the symmetric features in the structure did not show any significant effect.

The most important result of Experiment 1 is that it shows that the horizontal oriented structures do improve performance, even if no alphanumerical letters are used as retrieval keys.

2.2 Experiment 2

The second experiment aimed at showing how the usage of local frame of references in human spatial memory, as they are discussed by McNamara [9], affects the performance of object-location encoding/retrieval in dependence on different graphical layout structures.

Materials. The structures used in the second experiment are shown in Fig. 2. They are divided into two subsets because the limited pool of subjects did not allow the testing of all permutations needed to prevent order effects. The structures

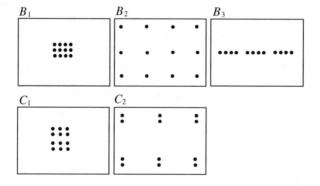

Fig. 2. Set of object configurations used in Experiment 2

in set B and C were created to test some factors assumed to play an important role in the process of object-location encoding/retrieval in structures. The motivation to choose these structures is founded in the assumptions and expectations from before the experiments were performed. Mainly the following factors were expected to contribute to the overall performance:

1. Hierarchical features,
2. Noise in the allocentric location-representation,
3. Noise in the egocentric location-representation,
4. Higher availability from memory of locations represented in allocentric frames of references if objects are in spatial vicinity.

The last factor seems plausible because the effort to assess spatial object-to-object relations is smaller if objects are close together; possibly no eye movement is needed. This last factor would give the narrow matrix B_1 an advantage over the wide matrix B_2 in respect to performance of object-location encoding/retrieval. The other factors listed above may also contribute. The noise in the location-representation is more grievous in the linear structure than in the matrices since there is only one dimension that contributes information. In the case of the matrices though, direction also contributes. Table 2 shows which structure profits by which factor compared to another structure in its set. A + sign in one cell means that the structure of the row takes an advantage over the structure in the column in respect to the factor of the table; a − sign indicates the opposite. The factor of hierarchical features is balanced within each set, so this factor is not included in the tables. (For this purpose the linear structure B_3 has been separated into three groups with four objects). To estimate the overall performance, the tendencies shown in the tables must be quantified. Furthermore, not every factor might contribute equally to the overall performance. Without any computational model as described in Sect. 3 and 4 only speculation can occur about these questions. However, in the setup used in the experiment,

Table 2. The factors that the structures profit from in the structures of B and C (FOR - frame of reference)

	Less noise in allocentric FOR			Less noise in egocentric FOR			Higher availability of allocentric FOR		
	B_1	B_2	B_3	B_1	B_2	B_3	B_1	B_2	B_3
B_1		0	++		− −	−		++	+
B_2	0		++	++		+	− −		−
B_3	− −	− −		+	−		−	+	

	C_1	C_2		C_1	C_2		C_1	C_2
C_1		+			−			+
C_2	−			+			−	

it can be assumed that the differences in the noise of the egocentric location-representation are nearly negligible because the changes in the average visual angles between the different objects in the scene are small compared to the human field of view. This is in contrast to an allocentric location-representation, where the angles take values on the whole range. The effect of noise in the allocentric location-representation in the structure B_1 and B_2 is expected to have an equal effect because all relative distances are equal. It was expected that the effect of decrease in performance in the linear structure would be very distinct. Structure C_1 and C_2 differ only by the distances between the six pairs of objects; the distances between the two objects within a pair are equal. The hypothesis for this structure is that for transitions within a sequence between objects of two far distant pairs it will become more difficult for the subject to encode the location of the object within a pair because the directions do not significantly differ. One predefined sequence was used to show this effect. User traces can be used for the parameterization of stochastic models. The regularities found by the algorithms can be analyzed and interpreted [16].

Design and Procedure. The experimental design was similar to Experiment 1. This time the sequences were six items long. With one exception all sequences were created randomly for each subject. One sequence for the structures of set C was predefined. As mentioned above, this was done so experimental tracing data could be effectively analyzed. The sequence was predefined for the structures C_1 and C_2, respectively. The predefined sequence is shown in Fig. 3 on the left. Its usage within the experiments was such that the probability that subjects remembered the sequence from a previous presentation was low. Furthermore, this effect had been balanced between the structures C_1 and C_2.

Results – Performance. The numbers of correct repeated sequences are shown in the contingency Table 3. The performance in the linear structure is significantly lower than in the structures of the matrices (exact Fisher-test $p < 0.001$). Although the number of correct sequences in the narrow structure is slightly higher than in the wide matrix, this difference is not significant. For C_1 and C_2 in Table 3 the number of correct and erroneous sequences from the randomly created sequences and the predefined sequence are combined. According to this contingency table, performance in C_2 is significantly lower than in C_1 (exact Fisher-test $p < 0.05$).

Table 3. Contingency table (2×3) of correct and erroneous repeated sequences in set B and C

	B_1	B_2	B_3		C_1	C_2
Correct seqs	38	34	16	Correct seqs	35	25
Erroneous seqs	22	26	44	Erroneous seqs	25	35

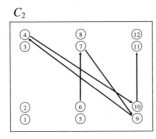

context → next symbol	interpretation
b-6-(7-9) → 4,3	The spatial vicinity, caused subject to mistake 3 for 4
b-(6-7) → 9,10	The spatial vicinity, caused subject to mistake 9 for 10
b-6-(7-10) → 3 b-6-7-10-(3) →10	If subject mistakes 10 for 9, he also mistakes 3 for 4.
b-6-7-9-(4) →10,11 b-6-7-9-(4-11) →unif.	Omission of object 10

Fig. 3. Contexts of erroneous behavior found by the parameterization of a stochastic model. Left: The structure with symbols assigned to the objects and the predefined test sequence. Right: Table with contexts and possible interpretation.

Analysis of errors. A look at the errors subjects made in their answers provides more insight into the underlying cognitive processes. To analyze the answer sequences for the predefined sequence in set C we used a modified algorithm for variable length Markov chains (VLMC) [14,3] to parameterize a stochastic model by the answer sequences. Roughly speaking, this algorithm can be seen as a filter for subsequences (called contexts) from the data that contain predictive information. We modified this algorithm in such a way that only contexts that contain significant predictive information in a more statistical sense are included into the model [17]. The conventional algorithms do not appropriately consider that sample sizes for different contexts vary in the data. To apply this algorithm to the answer sequences the objects in the structure must be assigned to symbols. The contexts of erroneous behavior found by this method in the answer sequences of the structures C_1 and C_2 are shown in Fig. 3. In the first column of the table the contexts found by the algorithm are shown in parentheses followed by an arrow and the most probable object occurring next in the answer sequences, if this context is given. E.g. $(7, 10) \to 3$ means: If subjects had clicked on object 7 followed by object 10, the most probable object they will click next is object 3. Multiple symbols/numbers listed on the right side of an arrow are ordered by their probabilities, with the first in the list being the object that is most probable of being next. On the right of an arrow possible next symbols are listed, as long as their frequencies for the given context meet one of the two conditions. First, the frequency is significantly higher than for the symbols with lower frequencies. Second, the frequency does not differ significantly from the frequency of the symbol with the next higher probability. On the left side of the arrow, also the most probable sequence that lead to the context is given, where the symbol b stands for the beginning of the answer sequence. In structure C_2 with the more distant pairs there are more contexts concerned with the confusion of the objects within the pairs of the upper left and lower right corners, whereas for structure C_1 there are more contexts concerning the omission of an object. The most notable context for structure C_2 is $(7, 10) \to 3$. The angle between the line from 7 to 10 and the line from 10 to 3 is similar to the angle between the lines 7 to 9 and 9 to 4. Therefore, this context indicates that subjects used the relative change in the direction of two transitions as a reminder.

2.3 Conclusions

The results of these two experiments make the following suggestions with regard to a computational model: first, as Wang et al. [15] suggested, the model should encode spatial object-to-object relations between the previously and currently attended objects as memory chunks. Second, the relation between three objects should also be encoded. As will be discussed in the next section, this can be interpreted since the visual system uses the connection line between two objects as an allocentric reference axis. Third, the results from the comparison of the horizontal and vertical oriented structures in the first experiment suggest that noise in the distance dimensions of spatial memory is distorted towards a higher accuracy in the horizontal direction. Fourth, subjects need not necessarily gaze at objects they are attending in order to assess their locations. During all experiments eye-movement data was collected. Because of a failure of the tracking system the recorded data was very noisy. However, the eye-movement data revealed that in structure B_2 subjects tended to fixate on the middle of the screen. Obviously, in B_2 it is sufficient to fixate on a location in the middle of the screen to asses most of the spatial object-to-object relations. According to theories of visual attention, moving attention is possible without moving fixation. Therefore, the effort to repeat transitions of the sequences in structure B_1 and B_2 is similar, yet different in structure C_2; here, subjects needed to move fixation to resolve which object within a pair had been highlighted.

3 Theory

3.1 Rules for the Cognitive Process

One very popular architecture for cognitive modeling is ACT-R [2]. ACT-R consists of several modules that are controlled by a central production system. These modules are the visual module, memory module, manual module for controlling the hands, and the goal module for keeping track of current goals and intentions. The central production system interacts with these modules by the symbolic content of their specific buffers. The process is described by a set of rules. In each step of the cognition process one rule that matches the pattern of information in the module buffers is selected. The rule is able to make requests to modules, e.g., to retrieve some information from memory. The retrieved information is loaded into the buffer of the memory and in the next step a new rule applies to the new content of the buffers. The minimal time for one cycle to be completed is 50 ms. The time consumption of a request and the outcome are determined by the design of the module. The working principles of the modules are determined by sub-symbolic mechanisms which may incorporate complex formulas. The most elaborated module of ACT-R is the memory module. The memory is assumed to be a collection of symbolic entities $d^{(i)}$ called chunks. The probability of a successful retrieval and the retrieval time are determined by the activation of the memory chunks, which is calculated by a formula taking into account the time spans in which a chunk has already been retrieved, the strength of its association with the current goal and the similarity of the attributes in a request to the values in the attributes of a memory chunk. This central equation of ACT-R is given by:

$$a_i(t) = b_i(t) + \sum_j w_j s_{ji} + \sum_k u_k m_{ki}. \qquad (1)$$

The base activation $b_i(t)$ decays logarithmically over time and increases each time the memory chunk is retrieved. The parameters s_{ji} reflect the frequency of how often chunk $d^{(i)}$ has been retrieved if the symbolic value of attribute ν_{jg} of the goal was identical to the current value. The parameter m_{ki} is the similarity parameter, and we think could best be interpreted as the log-probability $m_{ki} = \ln(P(\nu_{kx} = \nu_{ki}))$ that the value in attribute ν_{kx} in the request x is identical to the value in the attribute ν_{ki} of the chunk $d^{(i)}$. The parameters w_j and u_k are weighting factors reflecting the importance of a single attribute. To decide during a simulation which chunk will be retrieved from memory, noise is added to (1), and the random variable

$$A_i = a_i(t) + X + Y \qquad (2)$$

is considered. The random variables X and Y are independent normal distributed with a mean of zero and a variance σ_X, σ_Y. The value of the first random variable X is added when the chunk has been created. And the second one Y is added when $a_i(t)$ is reevaluated. The memory chunk with the highest activation will be retrieved. If the activation of no memory chunk exceeds a threshold τ_a a failure will be retrieved. Because of the decaying of the base activation a memory chunk will be forgotten unless it is not frequently retrieved. The time needed for a successful recall also depends on the activation by the relation $t \propto e^{-A}$. The higher the activation, the faster a memory chunk can be recalled.

The visual model has recently been added to the theory. The current design of the visual module is specialized in reading and finding objects with specific features on the screen. Visual attention is guided by the commands activated by the selected rule. An object location is represented by its coordinates in pixels on the computer screen. In the following, we will present our approach to adding the concept of noisy spatial relations to the visual module.

3.2 Locations and Reference Systems

As mentioned above, the visual module of ACT-R encodes object-locations in the reference-system of the screen, which is equivalent to creating all spatial object relations to one edge of the screen. The recently proposed extension called ACT-R/S [6] focuses on an egocentric frame of reference. However, according to Mou & McNamara [10] humans also use reference systems concerning the intrinsic axis of the object configuration. E.g., two salient objects create an axis that is used to specify the location of other objects. The most natural way to integrate this into the concept of attention of the visual module is to consider the last two attended objects as an axis of reference. This is an extension to the proposal of Johnson et al. [8,15] considering only the previously attended object in creating object-to-object relations. This means that the distance is represented in a pure environmental reference system, and the direction in respect to the egocentric perceived up-vector, which is defined by the orientation of the retina in space. In this sense, we call this reference system "semi-allocentric" because the encoded information is not independent from the position and orientation of the observer. However, creating object-location memory chunks in this "semi-allocentric" reference system is less effort for the visual system because it only needs to keep track of two objects, whereas in the case of the pure allocentric reference system, three objects are needed. This point will be discussed in more detail below in the context of visual indices. We considered all three different reference systems that are summarized in Fig. 4. The introduction of object-relations based on three objects is important for three reasons. First, it fits well with the concept of intrinsic axes in the object configuration as reported by Mou & McNamara. Second, the concept of angles is essential to most cognitive operations in geometrical tasks. Third, it is the simplest percept for spatial memory chunks that allows reconstructing object-locations, even if the whole configuration is rotated.

3.3 Noise

The variances in pointing errors of recalled object-locations require the dimensions stored in the memory chunks to be noisy. A clear definition of the reference systems is needed to integrate noise into the stored dimensions. Huttenlocher et al. [7] showed that the distribution of pointing errors supports the assumption that subjects imagine object-locations on a plane relative to a center in polar coordinates. We generalized this to use spherical coordinates in respect to an extension of the visual module in three dimensions. This also has some interesting implications on the representation of locations on a screen though, as will be discussed in the following. Spherical coordinates are a system of curvilinear coordinates that are natural for describing positions on a sphere or spheroid. Generally, θ is defined to be the azimuthal angle in the xy-plane from the x-axis, ϕ to be the polar angle from the z-axis and r to be distance (radius) from a point to the origin. In the case of the allocentric reference system this means that if the three points v_{-2}, v_{-1}, v_0 were attended and v_0 has to be represented in a local allocentric reference system, the point v_{-1} defines the origin, the polar axis is given by (v_{-1}, v_{-2}), and the local spherical y-axis points orthogonal into the screen. For the semi-allocentric reference system, v_{-1} is again the origin, but the polar axis is parallel to the vertical axis of the screen and the x-axis is parallel to its horizontal axis. The viewpoint of the subject is the origin in the case of the egocentric reference system. In the typical scenario of a user interacting with symbols on the screen the differences in the angles and distances between symbols represented in the egocentric system are very small compared to the differences if represented in an allocentric or semi-allocentric reference system. Therefore, if the same magnitude of noise is assumed in all reference systems, memory chunks represented in the egocentric reference system would be far more inaccurate compared to object-locations represented in the other two reference systems, and as a result, can nearly be neglected. The next question is, if θ, ϕ, and r should be considered as single, independent memory chunks. Because it is impossible to imagine a distance without a direction and an angle without corresponding lines, it is reasonable to combine distance and angle as one percept in one memory chunk. This does not mean that the dimensions cannot be separated later. E.g., it should be possible to extract the r-dimension as a distance and apply it to a different direction, as originally perceived. This kind of transformation corresponds to a mental rotation, but these are probably post processing activities of the cognitive system.

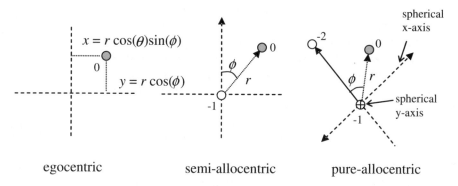

Fig. 4. Three different frames of reference, that can be defined according to how many attended object locations are considered. The objects are attended in the order (v_{-2}, v_{-1}, v_0).

For modeling the noise it is assumed that the dimensions of a location are buffered in the neural system and the representation is noisy. A spatial location is represented by $d(r, \theta, \phi, \theta', \phi')$, where r, θ, ϕ are the spherical coordinates of the pure allocentric reference system and θ', ϕ' additionally hold the polar and azimuth angles in the semi-allocentric reference system. These values are summarized into one symbol representing this relation as one percept. The angles of the semi-allocentric reference system are integrated into the symbol according to the assumption that we are not able to imagine any angle between two directions without the two directions themselves. In this sense, the symbol of a location encoded into a general allocentric reference system contains both the semi-allocentric and the pure-allocentric reference systems. However, we do distinguish a symbol that encodes only the dimension in the semi-allocentric reference system to reflect that the visual system may only focus on a single direction. Finally, the values of the dimensions are interpreted as the mean values of a noisy neural representation.

$$P(r_x, \theta_x, \phi_x, \theta'_x, \phi'_x | D = d(r, \theta, \phi, \theta', \phi')) = \qquad (3)$$
$$f(r, r_x) f(\theta, \theta_x) f(\phi, \phi_x) f(\theta', \theta'_x) f(\phi', \phi'_x)$$

Where $f(a, a_0)$ is a logistic distribution with mean a_0 and variance σ_a, the noise in the r-dimension is biased. We used the logistic distribution for computational reasons. The bias depends on the vertical or horizontal orientation of distance to be estimated. The noise σ_r is relative to r, which means that it is scale invariant. As the final noise in the r-dimension we use:

$$\sigma_r(\phi', r) = (f_{\sigma_r} + (1 - f_{\sigma_r}) \cos^2(\phi')) \sigma_r r \qquad (4)$$

If during a simulation the cognitive system requests an object-location based on a symbolic entity $d(r, \theta, \phi, \theta', \phi')$, the values of the dimensions are set to random values $\tilde{d}$ according to the noise given by (3). After the noise has been added to the location request it is decided if the values are latched on possible features in the display. Therefore, the object-locations $x^{(i)}(r, \theta, \phi, \theta', \phi')$ of all features in question are calculated in the current local reference system corresponding to the reference system in the request. The probability $P_{x^{(i)}}$, that visual attention is caught by feature $x^{(i)}$, and the probability P_0 that it is not, are given by:

$$P_{x^{(i)}} = \frac{P(\tilde{d}^{(x)} | x^{(i)})}{V^{-1} + \sum_i P(\tilde{d}^{(x)} | x^{(i)})} \qquad P_0 = \frac{V^{-1}}{V^{-1} + \sum_i P(\tilde{d}^{(x)} | x^{(i)})} \qquad (5)$$

These equations express the posterior probability $P(x^{(i)} | d^{(x)})$ that if a noisy location $d^{(x)}$ from the neural representation in memory is given the location results from the feature $x^{(i)}$. The parameter V^{-1} describes the weight of a noisy background. The likelihood functions $P(d^{(x)} | x^{(i)})$ are the truncated logistic distribution as if the feature $x^{(i)}$ would have been the stimulus and are similar to (3). The process of the projection of a noisy neural location representation from memory onto a new percept is illustrated in Fig. 5. The most rational choice would be the feature with the maximum probability according to (5). However, we assume this decision to be noisy as well, so the visual system maps the request on the features with a probability given by (5). A similar mapping scheme can be applied to map a perception to already existing memory chunks. Whenever a location has been encoded into the symbolic entity d it will be stored in memory and the retrieval is determined by the memory module and equation (1). This noise model has two interesting properties. First, because the truncated logistic distribution is asymmetric, the expected report of an object-location is biased away from the reference axis. This is the same effect as reported at categorical boundaries by Huttenlocher et al. [7]. Second, for object-locations on a flat screen the values of θ are discrete $\theta = \{\pi/2, 0, -\pi/2\}$ and encode whether the object-location in question is on the left side or the right side, or aligned, when facing into the direction of the reference axis. This is consistent with the assumption of interpreting the reference axis as a categorical boundary, where θ encodes the category. Thus, the categorical boundaries of Huttenlocher et al. are simply projected egocentric reference axes.

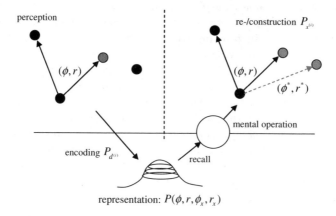

Fig. 5. Perception, representation and re-/construction of a location

3.4 Visual Indices

It is evident that humans browsing a graphical structure encode environmental characteristics of object-locations, say the spatial relations to objects nearby. The crucial point in encoding such environmental features is that after some objects in the environment have been attended, attention needs to return to a specific location previously attended. If this return depends on such noisy operations as those so far described, the cognitive system will hardly return to a specific reference point. At this point the concept of visual indexing, or FINST - FINger INSTantiation [11], is needed. According to this theory, the cognitive system has *"access to places in the visual field at which some visual features are located, without assuming an explicit encoding of the location within some coordinate system, nor an encoding of the feature type"*. Experiments show that the number of FINSTs in the visual system is limited to four or five. To implement environmental scan patterns, FINSTs need to enable the visual system to access previously attended locations without or at least with minimal noise. In the visual module of ACT-R the concept of FINSTs is currently only used to determine if a location has already been attended, but it gives no direct access to such an indexed location. In our simulations we gave the cognitive system direct access to an indexed location by determining the visual index through the sequential position in the chain of attended locations. This index can be used to return (or avoid returning) attention to a particular location in the chain of attended locations.

3.5 Competitive Chunking

As a human subject learns object-locations in a graphical structure he/she becomes familiar with the structure after some time. This means that he/she recognizes environmental features faster and is able to link environmental features more efficiently to object-locations. This implicit learning is similar to the effect that subjects are able to learn sequences of letters more efficiently if the sequences contain well-known words or syllables. Servan-Schreiber & Anderson [12] discussed this effect in the context of a symbolic theory as a competitive chunking (CC) mechanism. According to the theory of CC, a memory chunk for the sequence can be compressed by replacing elements of the sequence by references to subchunks having a high activation, and therefore can be retrieved quickly and reliably from memory. Memory chunks need to be declared in ACT-R in advance. Subsequently, the mechanism of emerging subchunks is not part of ACT-R. To investigate such a mechanism in the context of object-location memory we extended the formulae 1 for the activation of chunks in memory by a term a_i^{CC}, calculating correlations across attributes of chunks in memory, which results in virtual sub-chunks supporting their container chunks. We derived this term heuristically and it is given by:

$$a_i^{CC} = a_i + c_{CC} \sum_{m=1}^{n_s} \sum_{n=1}^{n_s} \sum_{k=1}^{N} I_{mni} K_{mnik} \ln(1 + e^{c_d b_k}) \qquad (6)$$

The index k runs through the chunks of the same kind; the index m and n through the slots of the chunk type. The parameter K_{mnik} compares the similarity of the slot values and is given by:

$$K_{mnik} = \begin{cases} P(\nu_{mk} = \nu_{mi}) P(\nu_{nk} = \nu_{ni}) \text{, if } P(\nu_{mk} = \nu_{mi}) P(\nu_{nk} = \nu_{ni}) > c_\tau \\ 0 \qquad\qquad\qquad\qquad\qquad\qquad\qquad\qquad\qquad\text{, otherwise} \end{cases} \qquad (7)$$

Hence, K_{mnik} is the probability that both values are equal. The parameters $P(\nu_{ab} = \nu_{cd}) = p_{abcd}$ express the probability that the values ν_{ab} and ν_{cd} both result from the same source. For scalar values they can be calculated by a Bayesian approach. To limit the contributions, K_{mnik} is cut by a threshold c_τ. Thus, roughly speaking, the sum of the K_{mnik} over the slot pairs is a measure of how many equal slot values chunks $d^{(i)}$ and $d^{(k)}$ share. If only K_{mnik} is used as a factor for the CC, slots also contribute, whose values are equal over all chunks. This means that they do not carry any information. Therefore, we introduced the factor I_{mni} that estimates how much normalized information the knowledge of the value $V_m = \nu_{mi}$ in attribute m of memory chunk $d^{(i)}$ contains about the values V_n in attribute n of the other chunks.

$$I_{mni} = 1 - \frac{H(V_n|V_m = \nu_{mi})}{H(V_n)} \qquad (8)$$

If ν_{mi} contains no information about V_n, I_{mni} is zero. If V_m is fully determined by the knowledge of ν_{mi} then I_{mni} is 1. If the slots only contained clearly distinguishable symbolic values, the entropies in (8) could be calculated by their frequencies. In the case of spatial memory chunks though, the similarities have to be taken into account. With the abbreviation $P(\nu_{ab} = \nu_{cd}) = p_{abcd}$ the entropies can be estimated by

$$H(V_n) = -\frac{1}{N} \sum_{k=1}^{N} \ln \frac{\sum_{k'}^{N} p_{nknk'}}{N} \qquad (9)$$

and

$$H(V_n|V_m = \nu_{mi}) = -\frac{1}{\sum_{k'}^{N} p_{mimk'}} \sum_{k=1}^{N} p_{mimk} \ln \frac{\sum_{k'}^{N} p_{nknk'} p_{mimk'}}{\sum_{k'}^{N} p_{mimk'}} \qquad (10)$$

In the limit $P(\nu_{nk} = \nu_{nk'}) = \delta(\nu_{nk}, \nu_{nk'})$ of clearly distinguishable slot values the equations (9) and (10) are identical to a formula estimating the probabilities of the entropies for the information by the frequencies of the slot values. Furthermore, the contribution of each chunk is weighted by a factor according to its basis activation b_k with a lower bound to zero and approximating b_k for large activations. Due to the additional term (6) in the activation equation, virtual subchunks emerge through the clustering of attribute values, which support their container chunks (Fig. 6).

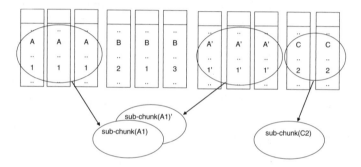

Fig. 6. Virtual sub-chunks from correlations in attributes across parent chunks. The rectangles symbolize the parent chunks with their attributes.

4 Simulation

4.1 The Model

The model we developed is based on the theory sketched in the previous section. It is similar to a conventional model of list memory [1] and it describes the encoding and retrieval stage of the memorizing task. The previously highlighted objects up to the current location are mentally rehearsed during the encoding of the sequentially presented object-locations. This rehearsal serves to boost the activation of the corresponding memory chunks so they can be recalled reliably later on. Different to common models of list memory, the model encodes environmental-features of the object-locations (say, spatial relations to objects in the vicinity) during the rehearsal or checks if one of the objects retrieved to the environmental feature matches the environment of the currently attended object. If the environment does not match, the reference system is restored through the visual indices, and a new guess is made excluding the denied object-location. The environmental features are encoded in competing chunks with a symbolic tag to the corresponding object-location and spatial relations to objects in its neighborhood. To check an environmental feature is time consuming because it must be retrieved from memory. Therefore, the cognitive system needs to find a tradeoff between the loss of time in the rehearsal and the reliability of the location of a rehearsed object. In ACT-R it is possible to let different rules compete. Which rule will be selected is determined stochastically proportional to parameters reflecting the probability that the selection of each rule has induced a feeling of success in the past. These parameters can be learned during the simulation. In our simulation one rule for skipping and one rule for actually performing the validation of a location by an environmental feature compete. The answer stage is equal to the rehearsal stage, except that environmental features are not encoded anymore and are validated for every location because time pressure is no longer present. Overall, the model contains 142 rules. This unexpected high number of rules results from the time pressure set on the task. At any possible stage the model needs to check if a new object is highlighted, which leads to a lot of exceptions. The ACT-R parameters for retrieval of memory chunks were set to the defaults reported in literature [2]. The variance $\sigma_{(\theta,\phi)}$ of the noise for the angular dimension was set to 0.06 radians and to 0.08 for the r-dimension. The skewing factor f_r in (4) of the noise in the r-dimension was chosen to be 0.8. The parameter for the background noise was set to $V = 2.e3$. The same parameters and rules were used for all simulations and graphical structures.

4.2 Validation of the Model

We simulated the experiments with 30 subjects ten times and compared the mean values to the data. The model satisfyingly reproduces the overall performance (Fig. 7) of the subjects ($R^2 = 0.83$), though there is one disturbing discrepancy: the model exhibits a different performance for structure A_2 and A_3, yet a statistical test reports no difference for the data. The structures A_2 and A_3 are both horizontally oriented, but A_2 is more regular than A_3 and contains some symmetrical features. Therefore, we expected that in A_2 subjects become familiar with the structure earlier by the CC mechanism. On the other hand, the environmental features are more distinct in the less symmetric structures and therefore provide more information for the validation of an object in the sequence. As reported in Sect. 2.1, the analysis of the learning curves of the subjects demonstrated these effects as a tendency, but could not be significantly demonstrated. In the overall performance these two effects should have been balanced to result in the observed error rate. Obviously, this is not the case. We primarily integrated the encoding and validation of environmental characteristics into the model in order to investigate the effect of becoming familiar with the structure. The model, however, did not show these effects either (see Fig. 8a). In the end it seems that the observed data could have been explained better without considering the validation of objects by their environmental characteristics. To examine the contribution of the encoding of environmental characteristics and CC we made simulation runs without CC and without the encoding of environmental features. The results are displayed in Fig. 7 and show that the encoding

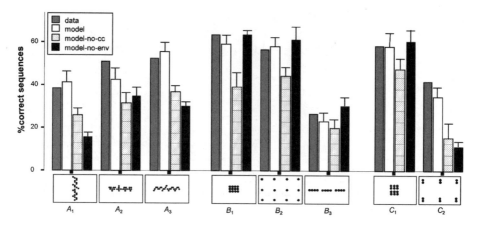

Fig. 7. Performance of different models compared to the data (model: model with encoding of environmental features and competitive chunking [CC] mechanism enabled, model-no-cc: model with encoding of environmental features, but without the CC. model-no-env: model without encoding environmental features).

of environmental features does not have the same effect for all structures. Interestingly, the performance of the model not encoding environmental features is similar to the complete model for the structures B_1, B_2, and C_1. Performance is even lower for these structures if environmental features are encoded without CC. However, for the structures A_1, A_3, and C_2 the model seems to significantly benefit from the encoding of environmental features. This effect can be explained by the observation that during the encoding of environmental features the model loses a lot of time, especially if activation of memory chunks for environmental features are low. In this case the sequences cannot be rehearsed so often to boost activation. If spatial relations of preceding objects in the sequence are as distinct as in B_1, B_2, and C_1, the advantage of encoding environmental features clearly does not compensate the disadvantage of time loss. Only if CC is assumed is the activation of environmental features in memory high enough to make the encoding of environmental features paying. For structures A_1, A_3, and C_2 the cognitive model benefits from encoding environmental features, even if no CC supports their activation. For the unsymmetrical structures A_1 and A_3 the environmental features are more distinct and therefore more effective for the validation of object-locations. Altogether, the attempt to fit the model without encoding of environmental features to the data required lowering the variances in the angular dimensions to 0.01 radians, which is only approximately 15% of the variance assumed for the model with encoding of environmental features, and therefore much lower than reported in literature [7]. Accordingly, the consideration of strategies to validate object-locations are mandatory. Unfortunately, the model encoding environmental features became very complex, and the new integrated mechanism, such as the CC, should be further evaluated. Regardless, the complete model reproduces some other effects that may result from encoding and

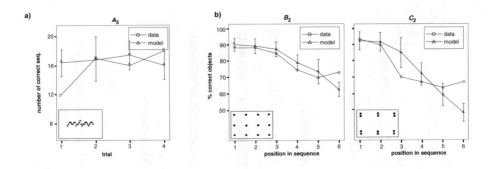

Fig. 8. Performance of subjects and model. Left: Overall performance. Right: Dependency of the performance from the position in the sequence.

validating environmental features. This is displayed in Fig. 8b. Here the number of correct repeated objects is plotted over the position of the object in the sequence for two structures. Both curves show a plateau at the beginning up to the third object. In the model this plateau results from the environmental features. Because one object of the sequence is mainly learned by the spatial relation to its predecessor, each correct reconstructed object of the sequence strongly depends on a successful reconstruction of its predecessor. As a result, the cognitive system focuses on the first objects in the rehearsal to boost the activation of the corresponding memory chunks for a reliable recall. An analysis of the model shows that the stronger decline after the second and third object in the curve of the model results from an unreliable recall of environmental features rather than a failed recall of the primary spatial relation. The curves of the subjects show an additional plateau at the end of the sequences. The error rate at the end of the sequence is lower than for the whole sequence. This means that some objects of the sequence have been reconstructed based only on the environmental characteristics here. This hypothesis is supported by the observation that this plateau is more distinct for structure C_2. The model only validates an object by the environmental features and repeats applying the spatial relation as long as the validation fails. It contains no rules for a strategy for skipping one object of the sequence and trying to identify the next object only by its environmental features. Therefore, the model only reproduces the plateau at the beginning. Nevertheless, when gathering the measuring points of all curves for all structures together the model achieves a correlation to the data with $R^2 = 0.92$. Hence, the performance of the model is generally satisfying. Additionally, the parameters for the variances in the location representation are in the magnitude reported by Huttenlocher et al. [7]. Huttenlocher et al. reported $\sigma_\phi = 0.17$ (the model: 0.06) and $\sigma_r = 0.025$ (the model: 0.08). The higher variance in the angular dimension reported by Huttenlocher et al. may be explained by a missing reference point in their experimental design. Subjects had to infer the reference point as the center of a circle which increased uncertainty. Subjects might encode the distance as a distance to opposite reference points on the border of the circle. This might decrease the overall measured variance in the distance of the answers. In the model, the parameter for the variance refers to one single representation. If strategies of subjects are assumed using multiple representation, the overall variance in the answers is decreased as well.

5 Conclusion

This paper described extensions to the visual-module of the ACT-R theory which enable the development of very detailed models for the visual working memory. The concepts were derived from well-known effects in experimental psychology. Overall, the modeling gave us a deeper insight into the mechanisms and bottlenecks of encoding object-locations. One challenge in modeling the memorizing task was the limited number of FINSTs. The number of FINSTs limits the complexity of environmental features that can be encoded. This is interesting with respect to visual working memory in three dimensions. Encoding of an object-location in a real allocentric local reference system in three dimensions needs at least three object-locations to define a reference plane. This reduces the number of free FINSTs in an encoding task. This might explain why spatial reasoning in three dimensions is more difficult for most people than spatial reasoning in two dimensions. In future work we will extend the concepts described in this paper to three dimensions. The future work will elaborate the chosen Bayesian approach and will try to develop an integrated model of human cognition when interacting with graph-based structures. Furthermore, we are aiming at additional validation studies based on very simple geometrical tasks without time pressure.

References

1. Anderson, J.R., Bothell, D., Lebiere, C., Matessa, M.: An integrated theory of list memory. Journal of Memory and Language 38, 341–380 (1998)
2. Anderson, J.R., Bothell, D., Byrne, M.D., Douglass, S., Lebiere, C., Qin, Y.: An integrated theory of the mind. Psychological Review 111(4), 1036–1060 (2004)
3. Bühlmann, P., Wyner, A.J.: Varaible Length Markov Chains. Annals of Statistics 27(1), 480–513 (1999)
4. Cockburn, A.: Evaluating spatial memory in two and three dimensions. International Journal of Human-Computer Studies 61(3), 359–373 (2004)
5. Ehret, B.D.: Learning where to look: Location learning in graphical user interfaces. CHI Letters 4(1), 211–218 (2002)
6. Harrison, A.M., Shunn, C.D.: ACT-R/S: Look Ma, No "cognitive-map"! In: International Conference on Cognitive Modeling, pp. 129–134 (2003)
7. Huttenlocher, J., Hedges, L.V., Duncan, S.: Categories and Particulars: Prototype Effects in Estimating Spatial Location. Psychological Review 98(3), 352–376 (1991)
8. Johnson, T.R., Wang, H., Zhang, J., Wang, Y.: A Model of Spatio-Temporal Coding of Memory for Multidimensional Stimuli. In: Proceedings of the Twenty-Fourth Annual Conference of the Cognitive Science Society, pp. 506–511. Lawrence Erlbaum Associates, Mahweh, NJ (2002)

9. McNamara, T.P.: How are the locations of objects in the environment represented in memory? In: Freksa, C., Brauer, W., Habel, C., Wender, K. (eds.) Spatial cognition III: Routes and navigation, human memory and learning, spatial representation and spatial reasoning, pp. 174–191. Springer, Heidelberg (2003)
10. Mou, W., McNamara, T.P.: Intrinsic frames of reference in spatial memory. Journal of Experimental Psychology: Learning, Memory, and Cognition 28, 162–170 (2002)
11. Pylyshyn, Z.W.: The role of location indexes in spatial perception: A sketch of the FINST spatial-index model. Cognition 32, 65–97 (1989)
12. Servan-Schreiber, E., Anderson, J.R.: Chunking as a mechanism of implicit learning. Journal of Experimental Psychology: Learning, Memory, and Cognition 16, 592–608 (1990)
13. Tavanti, M., Lind, M.: 2D vs. 3D, Implications on Spatial Memory. In: Proceedings of the IEEE Symposium on Information Visualization, pp. 139–145 (2001)
14. Ron, D., Singer, Y., Tishby, N.: The Power of Amnesia: Learning Probabilistic Automata with Variable Length. Machine Learning 25(2/3), 117–149 (1996)
15. Wang, H., Johnson, T.R., Zhang, J., Wang, Y.: A study of object-location memory. In: Gray, W., Schunn, C. (eds.) Proceedings of the Twenty-Fourth Annual Conference of the Cognitive Science Society, pp. 920–925. Lawrence Erlbaum Associates, Mahweh NJ (2002)
16. Winkelholz, C., Schlick, C., Motz, F.: Validating Cognitive Human Models for Multiple Target Search Tasks with Variable Length Markov Chains. SAE-Paper 2003-01-2219. In: Proceedings of the 6th SAE Digital Human Modeling Conference Montreal, Canada, (2003)
17. Winkelholz, C., Schlick, C.: Statistical Variable Length Markov Chains for the Parameterization of Stocastic User-Models from Sparse Data. In: Proceedings of IEEE International Conference on Systems, Man and Cybernetics (2004)
18. Winkelholz, C., Schlick, C., Brütting, M.: The effect of structure on object-location memory. In: Proceedings of the Twenty-Sixth Annual Conference of the Cognitive Science Society, pp. 1458–1463. Lawrence Erlbaum Associates, Mahweh, NJ (2004)

Updating in Models of Spatial Memory

Björn Rump and Timothy P. McNamara

Department of Psychology, Vanderbilt University, 111 21st Ave South
Nashville, TN 37203
bjoern.rump@vanderbilt.edu

Abstract. This chapter discusses a new model of spatial memory and updating. The model includes an egocentric subsystem that computes and represents transient self-to-object spatial relations and an environmental subsystem that forms enduring representations of environments using intrinsic reference systems. Updating occurs in both subsystems, but only the egocentric subsystem readily provides object locations relative to any adopted orientation. In the absence of visual support, updating in the egocentric subsystem is limited, and object locations may have to be retrieved from the orientation dependent environmental subsystem. The model is evaluated in light of the results of numerous studies from the areas of spatial memory and spatial updating and contrasted with two alternative models. Furthermore, results are presented that suggest that interobject spatial relations are preferentially represented when they are aligned with intrinsic reference directions in the environmental subsystem.

1 Introduction

Effective navigation abilities are crucial for the survival of almost every living mobile species. They are essential, for instance, for finding the way back to a previously discovered source of food or water, for safely returning home after a sudden change of weather, and for not getting lost in a complex environment like a cave. For most modern humans, effective navigation skills have become less critical for daily survival, but many common activities are nevertheless still characterized by the need to navigate successfully between places.

Successful navigation relies on two components. First, the organism needs to be able to represent the locations of significant objects or landmarks in the environment. Second, the organism needs to be able to stay oriented with respect to these represented elements. As the organism moves, the spatial relations between the organism and the elements in the environment constantly change. In order to stay oriented, spatial updating processes need to be invoked to compensate for these changes.

This differentiation into representation and updating can also be found in research on human spatial cognition. One prominent line of research focuses on the nature of spatial representations (e.g., the reference systems employed). This research typically involves paradigms in which participants first learn a layout of objects and then are tested on their memories of the layout while at a remote location. Another line of research is primarily concerned with people's spatial updating abilities, that is, their ability to keep track of the locations of objects in their environment as they move within that environment under various (cue) conditions.

This chapter begins by reviewing significant findings in those two research areas. Then a model of spatial memory and updating is presented and evaluated in light of these findings. Two other existing models of spatial memory and updating are discussed in terms of their relationship to our model and in terms of their ability to account for the available findings from the research on spatial memory and updating. Finally, the results of recent experiments are discussed that provide insights into how interobject spatial relations are represented in spatial memories.

2 Spatial Updating

A typical study in the area of spatial updating was presented by Rieser, Guth, and Hill (1986). Participants studied the locations of five objects on the floor and then pointed to individual objects while blindfolded to establish baseline performance. In one condition, participants were then guided to a novel position[1] and asked to point to the objects again. In another condition, participants were presented with a novel position and asked to point to the objects after moving to the novel position in imagination. When participants had physically moved to the novel position, pointing responses were only slightly slower and less accurate than during baseline pointing. In contrast, when participants had moved to the novel position in imagination, considerable costs in terms of both speed and accuracy were present relative to baseline pointing performance.

Results such as these are commonly interpreted as implying that updating involves demanding transformation processes and that these computations are greatly facilitated in the presence of physical locomotion (e.g., Farrell & Robertson, 1998). In the absence of physical locomotion (i.e., when adopting a novel position merely in imagination), it is hypothesized that these computations have to be carried out unfacilitated, which in turn produces considerable costs in terms of latency and accuracy.

A number of studies have been conducted to identify which of the cues that are normally associated with physical locomotion are sufficient for efficient spatial updating. It has been frequently shown that simulated optic flow by itself is not sufficient for efficient spatial updating (e.g., Chance, Gaunet, Beall, & Loomis, 1998; Klatzky, Loomis, Beall, Chance, & Golledge, 1998; Péruch, May, & Wartenberg, 1997; but see also Kearns, Warren, Duchon, & Tarr, 2002). A prerequisite for efficient updating seems to be that the person's physical position changes. Whether this position change is accomplished through passive transport, which primarily provides vestibular cues, or through active movement, which provides additional proprioceptive and efferent cues, does not matter in many circumstances (e.g., Wang & Simons, 1999; Wraga, Creem-Regehr, & Proffitt, 2004). There is, however, evidence that those additional cues become beneficial when the movement trajectory is more complex (e.g., Sholl, 1989; Yardley & Higgins, 1998).

---

[1] For convenience, the term position will be used to refer to the combination of a particular location and a particular orientation in space.

In principle, any movement can be decomposed into individual rotations (i.e., changes in orientation) and translations (i.e., changes in location). Early evidence suggested that while updating based on imagined rotations is difficult (with the difficulty increasing as a function of rotation angle), updating object locations after imagined translations could be done without additional costs (e.g., Presson & Montello, 1994; Rieser, 1989). Further investigations suggested that updating after imagined translations is not cost-free (e.g., Easton & Sholl, 1995; May & Wartenberg, 1995) but still considerably easier than updating after imagined rotations, even when the two movement types are equated in terms of how much they change the person-to-object spatial relations (May, 2004).

That updating is not only benefited in the presence of physical locomotion, but takes place automatically, was demonstrated by Farrell and Robertson (1998). Participants in their study were required to physically rotate to a novel orientation, but point to objects as if they were still facing the initial orientation. Performance was as poor in this *ignore rotation* condition as in the imagined rotation condition, indicating that participants were unable to voluntarily refrain from updating. This finding has been replicated repeatedly using similar paradigms, providing strong support to the notion that updating takes place automatically in the presence of physical locomotion (Farrell & Thomson, 1998; May & Klatzky, 2000; Waller, Montello, Richardson, & Hegarty, 2002; Wang & Simons, 1999).

As discussed previously, the advantage of physical updating over imagined updating is commonly interpreted as indicating that updating involves complex transformations of object coordinates and that these transformations are highly facilitated in the presence of physical locomotion (e.g., Farrell & Robertson, 1998). An alternate account posits that the difficulty of updating after imagined movements does not stem from a lack of such facilitation, but that it instead results from interference that is caused by a conflict between the awareness of one's physical position in an environment and the discrepant position one has to adopt in imagination (e.g., May, 1996).

These competing accounts have been tested by May (1996, 2004). May (1996) showed that disorientation actually improved performance on pointing to objects relative to imagined rotations, suggesting that updating in imagination is impaired by interference (see also Waller, Montello, Richardson, & Hegarty, 2002). In a more elaborate formulation of the nature of interference, May (2004) proposed that interference arises from conflicts between incompatible object location codes at the sensorimotor level, which are specified relative to the physical position, and object location codes at the cognitive level, which are specified relative to the imagined position. This hypothesis is supported by his finding that for both imagined rotations and translations, performance degrades monotonically as a function of the amount that the direction of the to-be-retrieved object relative to the imagined position differs from the direction of the to-be-retrieved object relative to the physical position ("object direction disparity"). To account for the finding that imagined rotations are more difficult than imagined translations (e.g., Rieser, 1989), even when object direction disparity is controlled for (May, 2004), a second source of interference is proposed that only applies to imagined rotations. This second source of interference, referred to as head-direction disparity, reflects conflicts that arise from having to specify an object direction relative to an imagined orientation that is different from one's physical heading.

The interference account is further supported by May's (2004) finding that providing participants with additional time between the presentation of the to-be-imagined position and the presentation of the target object improved overall performance, but did not reduce the difficulty differences between different imagined rotations or translations. This finding contradicts the notion that moving to a new position in imagination is difficult because demanding transformation processes need to be carried out without the facilitative effect of physical locomotion. Additional time after the presentation of the to-be-imagined position should have allowed participants to start, or (for longer delays) even finish, adopting the to-be-imagined position, which would have reduced or even eliminated the difficulty differences between different imagined rotations or translations. The interference account, however, has no problems accounting for this finding because the postulated interference does not come into effect until the target object is retrieved. Therefore, the additional time could improve general task performance, but it would not affect the relative difficulty of different imagined rotations or translations.

In summary, the observed difficulty of imagined updating does not seem to stem from having to carry out demanding transformation processes unfacilitated. Instead, it seems to be the consequence of conflicts that are created by the need to imagine a position that is different from one's actual position in the environment.

3 Spatial Memory

In attempts to investigate the nature of long-term spatial representations, one property that has received much of the attention of researchers is the orientation dependence of spatial memories. The question is whether spatial representations can be accessed equally well from all possible orientations, or if not, which orientations are preferred in the representation. Early evidence suggested that memories of small scale spaces are orientation dependent, whereas representations of room-sized or larger spaces are not (e.g., Presson, DeLange, & Hazelrigg, 1989; Presson & Hazelrigg, 1984). Subsequent attempts to find orientation independent performance for room-sized spaces have not been uniformly successful. Using methods that very closely mimicked those used by Presson and colleagues, orientation independent performance was either not observed (Roskos-Ewoldsen, McNamara, Shelton, & Carr, 1998) or observed only under very specific experimental circumstances (Sholl & Nolin, 1997).

Today, a large body of evidence indicates that spatial memories are orientation dependent (e.g., Diwadkar & McNamara, 1997; McNamara, Rump, & Werner, 2003; Shelton & McNamara, 1997; Valiquette, McNamara, & Smith, 2003; Waller, Montello, Richardson, & Hegarty, 2002; Werner & Schmidt, 1999; Yamamoto & Shelton, 2005). Preferred orientations in spatial memory can be used to make inferences about the reference system used to represent the information. The assumption is that spatial relations can be directly retrieved for those orientations that are aligned with the utilized reference system, whereas spatial relations have to be inferred for orientations that are not aligned with the employed reference system, which is assumed to produce costs in terms of speed and accuracy (Klatzky, 1998).

Various ways of classifying spatial reference systems have been proposed (e.g., Levinson, 1996), but for the purpose of studying human spatial cognition, a distinction between egocentric and allocentric reference systems has proven useful. In egocentric reference systems, locations are specified with respect to one's own body or part of one's body (e.g., in retina, arm, or trunk coordinates). In allocentric reference systems, spatial relations are specified independently of the observer. An allocentric reference system can be defined, for instance, by an individual object, by a layout of objects, by cardinal directions, or by a geometrical feature of the environment (e.g., Hartley, Trinkler, & Burgess, 2004).

Typically, the orientations that are better accessible are those that were egocentrically experienced during learning (e.g., Shelton & McNamara, 1997), which suggests that spatial memories are stored in terms of static egocentric representations. Several recent studies, however, have shown that preferred orientations can be determined by properties of the surrounding environment (e.g., McNamara, Rump, & Werner, 2003; Shelton & McNamara, 2001a). Furthermore, orientations that were never actually experienced can become preferred if they are made sufficiently salient through instructions (e.g., Mou & McNamara, 2002).

Results such as these are in clear contrast to the hypothesis that spatial memories are specified with respect to an egocentric reference system. Instead, they suggest that spatial memories are specified with respect to allocentric reference frames and that saliency determines which particular reference directions are selected.

4 A Model of Spatial Memory and Updating

Mou, McNamara, Valiquette, and Rump (2004) presented a model of spatial memory and updating that consists of two subsystems. The *environmental subsystem* is concerned with representing spatial environments in an enduring manner. Interobject spatial relations are represented independent of the observer using allocentric reference frames. This subsystem is largely influenced by the research described in the previous section. The *egocentric subsystem* represents object locations in the form of transient self-to-object relations. The main purpose of this subsystem is to guide interaction with the immediate surroundings. This model was inspired in part by theories of the relations between visually-guided action and visual perception (e.g., Creem & Proffitt, 2001; Milner & Goodale, 1995; Rossetti, 1998; Rossetti, Pisella, & Pelisson, 2000) and shares its basic architecture with the recently proposed model by Waller and Hodgson (2006).

The process of forming a representation in the environmental subsystem can be described as follows (McNamara, 2003). When a person is learning the layout of a novel environment, the spatial structure of the environment must be interpreted in terms of a spatial reference system. It is suggested that this process is analogous to determining the "top" of a figure, in that a conceptual "north" is assigned to the layout, which in turn determines its perceived orientation (e.g., Rock, 1973). The layout is interpreted in terms of an intrinsic reference system, that is, an allocentric reference system that is defined by the layout of to-be-represented objects (e.g., Tversky, 1981). A given collection of objects will have an infinite number of possible intrinsic reference axes, but because of the particular arrangement of objects and because of spatial and non-spatial properties of the objects, some axes will be more salient than others and will therefore be more likely to be selected as reference axes (e.g., Steck, Mochnatzki, & Mallot, 2003; Vorwerg, 2003). Features of the surrounding environment, such as the walls of a room, can also make some intrinsic axes more salient than others. Typically, egocentric experience is the dominant cue for selecting reference axes, but intrinsic axes that do not correspond to the study view can be preferred if they are made sufficiently salient (e.g., Mou & McNamara, 2002). Once a person has represented a spatial layout in terms of a particular intrinsic reference system, he or she usually continues to interpret the layout through this reference system, even when seeing it from a novel position. If, however, the novel perspective is aligned with salient axes in the environment and the layout was represented using a reference system that was not aligned with these axes, then the layout can be reinterpreted in terms of a new reference system (e.g., Shelton & McNamara, 2001a).

The egocentric subsystem does not form enduring representations. Instead, object locations are represented at the sensory-perceptual level in terms of self-to-object relations. These self-to-object relations are transient and decay relatively rapidly in the absence of perceptual support. Through deliberate rehearsal, they can be maintained even in the absence of perceptual support, but it is proposed that such a process is limited to a small number of self-to-object relations. The primary role of this system is to guide locomotion through and interaction with the immediate surroundings when vision is available. The representational format of self-to-object relations is well suited for these purposes as it readily provides the endpoints for goal directed action without further computations.

When moving through a familiar environment, updating takes place in both subsystems, but in different ways. In the egocentric subsystem, the transient self-to-object relations are updated as long as perceptual support is available. This updating is efficient and requires minimal attentional control. Once perceptual support is diminished (e.g., when walking in the dark or with eyes closed), the self-to-object relations start to decay and updating becomes more effortful and restricted to a small number of objects. Updating in this subsystem allows, for instance, for evading obstacles and keeping track of surrounding objects to interact with them, but it does not prevent a person from getting lost. In order to stay oriented, a person must know his or her position with respect to familiar objects in the environment, and this knowledge must be represented over the long term (e.g., Loomis & Beall, 1998). This is where updating in the environmental system comes into play. Updating in the environmental subsystem consists of keeping track of one's location and orientation with respect to the intrinsic reference system that is used to represent the locations of the objects in the environment. For this purpose, the body of the observer is treated like any other object in the environment. It is important to note that only the position of the observer is updated but not the intrinsic reference system itself. Because interobject spatial relations are specified with respect to the intrinsic reference system and not with respect to the position of the observer and because the reference system itself is not updated, spatial relations can only be directly retrieved when the person's orientation corresponds to one of the reference directions in the representation. When a person's orientation does not correspond to one of the reference directions, relative object location must be inferred, even though the person's position is properly updated

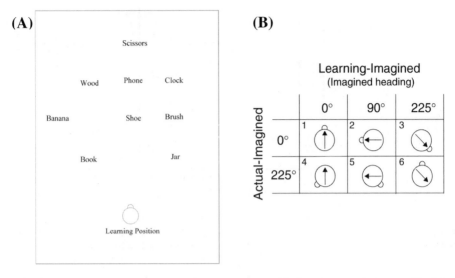

Fig. 1. (A) Layout of objects used by Mou at al. (2004). **(B)** Experimental design used by Mou et al. (2004). Head-nose icons indicate actual headings; arrows indicate imagined headings. Headings and differences between headings are measured counterclockwise.

in the environmental system and the person is therefore oriented with respect to the environment.

Mou et al. (2004) conducted a series of experiments to determine if the reference system used to represent the spatial structure of a layout is indeed not updated during physical locomotion. In other words, they tested if spatial relations can be retrieved more efficiently when the imagined heading equals the learning heading, as studies on spatial memory involving remote testing have shown, even when participants are tested within the studied environment and allowed to physically adopt the tested imagined heading. In addition to manipulating the disparity between the imagined heading at the time of test and the learning heading, a second variable was varied: the disparity between the actual heading at the time of test and the imagined heading at the time of test. As previously described, studies on spatial updating have shown that performance is substantially impaired when the imagined heading does not match the actual heading (e.g., Presson & Montello, 1994; Rieser, 1989).

In one experiment (Experiment 2), participants studied the layout depicted in Figure 1A from a single viewpoint and then walked to the center of the layout (i.e., the shoe) to be tested. The task was to point to objects relative to either the imagined heading that corresponded to the learning heading (imagined heading = 0°) or to an imagined heading that was 90° or 225° disparate (in counter-clockwise direction) from the learning heading. The disparity between the actual heading and the imagined heading was manipulated between participants. In one group, participants always physically adopted the imagined heading on which they were tested (A-I=0° group). This condition resembled the physical updating conditions of previous experiments (e.g., Rieser, 1989). In a second group, participants always turned to a facing

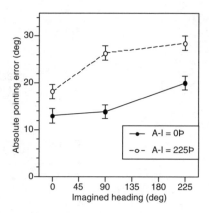

Fig. 2. Angular error in pointing judgments in Mou et al.'s (2004) second experiment as a function of actual-imagined (A-I) heading and imagined heading. Error bars are confidence intervals corresponding to ± 1 standard error of the mean, as estimated from the ANOVA.

direction that was 225° disparate from the tested imagined heading (A-I=225° group). Figure 1B depicts the experimental design for a better understanding of the resulting actual and imagined headings.

Participants started each trial facing the 0° direction and then turned with eyes open to the appropriate facing direction, from which they were allowed to review the layout again. Once they had closed their eyes and indicated that they were ready, the actual trial was presented to them (e.g., "Imagine you are facing the banana. Point to the book."). The delay between the closing of eyes and the presentation of the target object was on the order of ten seconds. After responding, participants always turned back to the 0° facing direction.

The most important result was that, as depicted in Figure 2, participants in the A-I=0° group were able to report object directions both faster and more accurately from the imagined heading of 0° than from the imagined heading of 225°. In other words, even though these participants were actually facing the direction they were asked to imagine facing and had turned to this direction with open eyes, they were not able to recover the spatial structure of the layout as efficiently from the heading of 225° as from the 0° learning heading. Performance was as good on the imagined heading of 90° as on the imagined heading of 0°, indicating that participants had represented the layout in terms of two orthogonal reference axes (e.g., Mou & McNamara, 2002). Participants in the A-I=225° group responded overall considerably slower and less accurately than participants in the A-I=0° group, which replicated the previously found costs for making spatial judgments from an imagined heading that is different from one's actual body orientation. As participants in the A-I=0° group, these participants also exhibited superior performance on the imagined heading of 0° in comparison to the imagined heading of 225°. They, however, did not show an advantage for the 90° imagined heading for reasons that are not entirely clear.

The inferior performance for the imagined heading of 225° relative to the imagined heading of 0° in both groups indicates that participants relied on an orientation dependent representation for making the spatial judgments and not on efficiently updated self-to-object relations. Moreover, the reference system underlying the representation was evidently not updated as participants physically turned. However, as evidenced by the poorer performance in the A-I=225° group, participants did update their own orientation with respect to the environment as they turned and were therefore aware that their actual orientation was not the same as the imagined heading. In the context of the model, this updating corresponds to the updating of the position of the observer with respect to the reference system used to represent the structure of the layout in the environmental system. It is important to note that the inferior performance for the imagined heading of 225° cannot simply be attributed to a noisy updating process that accumulates error with increasing turning angle. In the A-I=225° group, participants did not have to physically turn for the imagined heading of 225° but they had to turn for the other two imagined headings (see again Figure 1B). Yet, performance was worst on this imagined heading.

In another experiment (Experiment 1) in this series, participants were required to make judgments of relative direction (e.g., "Imagine you are standing at the wood, facing the jar. Point to the clock.") instead of egocentric pointing judgments (e.g., "Imagine you are facing the phone. Point to the brush."). This task therefore required imagined translation in addition to imagined rotation. Results of this experiment were very similar to those of the just-described one, the main difference being that the effect of imagined heading was even larger. Our interpretation of this difference is that participants might have been able to rely on residual self-to-object relations in the egocentric subsystem for the egocentric pointing judgments in the first experiment but not for the judgments of relative direction in the second experiment, as the latter require the recovery of object-to-object spatial relations.

The key finding of these experiments was that participants were not able to update efficiently even after physically turning with open eyes, which suggests that spatial updating is considerably more limited than previously thought. If reference systems underlying long-term representations are typically not updated and updating in the proposed egocentric system is quite limited, how can this be reconciled with the numerous findings of efficient updating after physical locomotion that were discussed in earlier parts of this chapter? Some of those studies used layouts that were considerably less complex than the ten object layout that Mou et al. used. Presson and Montello (1994), for instance, used only three objects, Wraga et al. (2004) used four objects, and Waller et al. (2002) used four-point paths. The number of to-be-updated self-to-object relations might have been small enough in these studies to fit within the capacity limits of the proposed egocentric subsystem. A similar explanation could apply to Rieser's (1989) study. Here, a layout of nine objects was used, but participants were presented with the target object as they were turned to the new orientation. This could have allowed them to only keep track of that single object as they turned and thereby stay within the capacity limitations of the egocentric subsystem. In addition, the delay between locomotion and testing was shorter than the ten seconds in Mou et al.'s (2004) experiments, so that self-to-object relations would have been probed at an earlier stage in the decay process.

It is also possible that the efficiency of physical updating has been overestimated in previous studies because many of them have focused primarily on contrasting physical updating with imagined updating (e.g., Rieser, Guth, & Hill, 1986; Wraga, Creem-Regehr, & Proffitt, 2004). The evident superiority of physical updating over imagined updating could have led authors to conclude that efficient updating occurs after physical locomotion but not after imagined locomotion. As discussed in the second section of this chapter, there is evidence that the high costs associated with imagined updating can be attributed to a large extent if not completely to conflicts that arise from having to adopt a position in imagination that is different from one's actual position (e.g., May, 1996). A close inspection of the results of previous updating studies that used more complex layouts reveals that performance was indeed often worse when participants responded after physical locomotion than when they responded relative to the original learning position (e.g., Farrell & Thomson, 1998; Rieser, Guth, & Hill, 1986; Wang, 2004; Wraga, Creem-Regehr, & Proffitt, 2004).

These costs of physical updating were considerably smaller than the large effects of imagined locomotion and did not reach significance. In most of these studies, only 10-15 participants were tested, presumably because of the large size of the effects of imagined locomotion. It is likely that the small sample sizes have prevented the statistical detection of existing physical updating costs. Significant costs for physical updating were found by Wraga (2003) and May (1996) using less restricted sample sizes. In May's study, as in Mou et al.'s (2004), updating costs were not determined by the amount of physical turning but instead by whether or not a given orientation was misaligned with the dominant geometric structure of the environment (e.g., walls of the surrounding room, salient axes in the layout). This supports the notion that the observed costs were not merely the consequence of a noisy updating process, but instead caused by participants relying on an enduring representation that was specified with respect to a reference system that was intrinsic to the represented environment.

In light of these considerations, the results of numerous experiments on spatial updating are compatible with the proposed model of an egocentric subsystem that is capable of providing updated self-to-object relations but is quite limited in terms of its capacity and maintenance time, and an environmental subsystem that contains enduring representations that are specified using intrinsic reference frames, which are typically not updated as a person moves. More systematic research, though, is needed to identify the exact limitations of egocentric updating. Particularly, parameters like the number of objects and the temporal aspects of testing need to be systematically varied within the same experimental methodology. A recent study by Waller and Hodgson (2006) suggested that the amount of locomotion can also be a relevant factor. Their participants appeared to have relied on egocentric updating for rotation angles smaller than 135° and on enduring representations for larger angles.

One aspect of the model that has not been considered so far is how the increased difficulty of imagined updating in comparison to physical updating (e.g., Rieser, 1989) is explained. It is proposed that in order to indicate object locations that are not available in the egocentric subsystem, the required spatial relations are retrieved or inferred from the environmental subsystem and then mapped onto an egocentric frame of reference for the purpose of making the pointing judgments. When the imagined position does not equal the actual position, the retrieved spatial relation also needs to be aligned with the egocentric reference frame, which is proposed to involve effortful processing.

Conceptually, this explanation in terms of demanding alignment processes is quite different from an interference-based explanation like the one proposed by May (2004), but it makes very similar predictions for many situations. For instance, under conditions of remote testing or after disorientation, the actual position with respect to the retrieved environment is no longer specified, so that effortful alignment processes are no longer required. Similarly, it can also explain why additional time after the presentation of the to-be-imagined position does not reduce the relative difficulty of different imagined movements, because the proposed alignment process could not be initiated until the target object is retrieved. Problematic for the alignment notion, however, is the finding that the difficulty of imagined translations is determined by the disparity of the target object's direction relative to the imagined and the actual position.[2] The alignment hypothesis predicts that if performance were affected by imagined translations, it should be affected by the amount the spatial relation has to be translated in order to become aligned with the egocentric reference frame, but not by the object direction disparity of the target object.

One way to reconcile this finding with the model is by assuming that the retrieved and egocentrically aligned spatial relation is interfered with by residual self-to-object relations in the egocentric subsystem. This explanation would only involve one source of interference. May's (2004) second source of interference, head-direction disparity, would not be necessary because its effects are predicted by the alignment process.

5 Other Models

Our model is not the first to attempt to provide a unifying framework for spatial memory and spatial updating. Perhaps the most prominent ones are Wang and Spelke's model (Wang & Spelke, 2000, 2002; see also Wang, 1999) and Sholl's model (Easton & Sholl, 1995; Sholl, 2001; Sholl & Nolin, 1997). Our model shares a number of general similarities with these models but differs in important details.

Wang and Spelke's (2000, 2002) model of spatial memory and updating comprises three subsystems. The first subsystem represents the locations of objects in the surrounding environment in terms of dynamically updated self-to-object relations. This dynamic egocentric subsystem is in principle very similar to our egocentric subsystem. However, the self-to-object spatial relations stored in this subsystem, even though not enduring, are far less transient than in our model. The subsystem as a whole also plays a more important role in that it is considered the primary system for guiding human navigation and wayfinding. The other two subsystems are thought to complement this primary system. The second subsystem is similar to our environmental subsystem in that it holds enduring allocentric representations. In contrast to our model, though, this subsystem represents the geometric shape of the environment (e.g., walls of a room or other geometrical features of the environment) but not the locations of individual objects or landmarks. The primary role of this system is to support reorientation in case the dynamic egocentric system breaks down. The third subsystem stores the appearance of familiar landmarks and places in the form of view-dependent representations and serves primarily for place recognition. Like the dynamic egocentric system, this system represents object locations with respect to an egocentric reference frame, but the views stored in this system are static and enduring, and are therefore not updated.

---

[2] This does not apply to imagined rotations, because for imagined rotations, rotation angle and object direction disparity are identical.

Wang and Spelke's model strongly emphasizes the role of dynamically updated self-to-object relations. While in our model the egocentric system is primarily concerned with guiding locomotion through and interaction with the immediate surroundings when vision is available, their dynamic egocentric system plays a central role even in large-scale navigation and is overall characterized as more potent. Representations in their dynamic egocentric system are maintained in the absence of perceptual support as long as the person stays oriented, and even though the need for a limit on the number of simultaneously maintained and updated self-to-object relations is acknowledged, the system is not characterized as limited to a small number of objects. Although more systematic investigations are needed to identify the exact limitations, egocentric updating seems to be considerably more limited than postulated by Wang and Spelke.

At present, the model does not contain any specifics about the way in which knowledge is accessed from an imagined position that does not correspond to one's physical position in the environment. Therefore it does not provide explanations of the effects of imagined rotations and translations or why reporting object directions relative to an imagined novel position is improved after a person is disoriented.

Enduring representations of object locations are stored in collections of familiar views according to Wang and Spelke. This aspect of the model correctly predicts orientation dependent performance for situations in which object locations are not available in the dynamic egocentric system (e.g., when retrieving remote environments or when pointing to local objects after disorientation), but it also predicts that the preferred orientations should be the ones that correspond to familiar views. This prediction has been contradicted by the results of several experiments on spatial memory that have shown that properties of the environment can determine which orientations are preferred (e.g., McNamara, Rump, & Werner, 2003; Shelton & McNamara, 2001a) and that preferred orientations can be orientations that were never actually seen (e.g., Mou & McNamara, 2002; Shelton & McNamara, 2001b; see also Waller, 2006). Equally problematic for the notion of stored views are the findings of May (1996) and Mou et al. (2004) that when participants were not able to update efficiently, performance was only impaired for orientations that were misaligned with the dominant geometric structure of the environment (e.g., walls of the surrounding room, salient axes in the layout), but not for other novel orientations. In light of these findings, the notion that enduring memories of object locations exist solely in collections of familiar views is not tenable.

Sholl's model (Easton & Sholl, 1995; Sholl, 2001; Sholl & Nolin, 1997) contains an allocentric object-to-object subsystem and an egocentric self-reference subsystem. The object-to-object system holds long-term spatial representations, which can be conceptualized as networks of nodes representing the individual objects. The spatial relations between objects are specified through vectors that connect proximate objects with each other. These vectors are only defined locally (i.e., with respect to neighboring objects) so that a representation in this subsystem does not possess a global reference direction. Because spatial relations are not specified relative to a global reference direction, a coordinate system needs to be superposed onto the vector space in order to access the stored information. This process will be explained shortly.

The self-reference system codes object locations in terms of self-to-object spatial relations using the front-back, left-right, and up-down body axes (e.g., Franklin &

Tversky, 1990). It operates at both the sensorimotor level and at the representational level. At the sensorimotor level, it provides a framework against which all spatially directed motor activity is organized. The self-to-object relations are maintained even in the temporary absence of visual support and are updated automatically as a person moves, but they are not maintained over the long term. At the representational level, the role of the self-reference system is to mediate the retrieval of spatial information from the object-to-object system. By superposing the self-reference system onto a vector space in the object-to-object system, object locations can be retrieved as self-to-object relations through the self-reference system.

Sholl's model posits that the representational self-reference system can be superposed onto a vector space at any location and in any orientation, so that spatial relations can be accessed equally well from all possible imagined positions. This, however, is only true for the retrieval of remote environments. When a person is oriented within a familiar environment, the position of the representational self-reference system in the vector space for that environment is fixed to the current physical position, because under these circumstances, the representational self-reference system is attached to the sensorimotor self-reference system, which in turn is tightly coupled to the physical body. When moving in a known environment (i.e., one that has been represented in the long term object-to-object system) two forms of updating occur simultaneously. First, the temporary self-to-object relations that are stored in the sensorimotor self-reference system are updated accordingly. Second, the engagement of the sensorimotor self-reference system with the surrounding environment determines the position of the representational self-reference system in the vector space that represents the objects in the current environment. Therefore, as the person moves in the environment, the position of the representational self-reference system is properly updated so that object locations represented in the object-to-object system are readily available relative to the current physical position of a person.

Sholl's sensorimotor self-reference system is similar to our egocentric subsystem in that its main purpose is to provide a framework for spatially directed motor activity. But as in Wang and Spelke's dynamic egocentric subsystem, representations in this subsystem are not transient and updating in the absence of perceptual support is not limited to a small number of objects. A major difference between Sholl's model and both Wang and Spelke's and our model is that in her model, updating with respect to the enduring object-to-object system readily provides object locations relative to any physical position that a person adopts. However, performance for updating after physical locomotion is not always equally good for all considered physical orientations (e.g., May, 1996; Mou et al., 2004). Similarly, Sholl's model predicts that when retrieving remote environments, performance should be orientation independent, because representations in the object-to-object system are not specified with respect to global reference directions. A large number of studies, however, have shown orientation dependent performance for remote testing even when environments were learned from multiple views (e.g., McNamara, Rump, & Werner, 2003; Shelton & McNamara, 2001a; Werner & Schmidt, 1999), which makes the claim of orientation independent representation perhaps the most problematic aspect of the model.[3]

Imagined updating, according to Sholl's model, requires detaching the representational self-reference system from the sensorimotor one and moving it to the appropriate position in the object-to-object vector space. It is postulated that the representational self-reference system is moved through processes of mental rotation and translation. These transformation processes are assumed to be demanding and thereby responsible for the difficulty of imagined updating. Disorientation is proposed to result in the disengagement of the sensorimotor self-reference system from the environment, which in turn causes the representational system to detach from the sensorimotor one. With the position of the representational self-reference system no longer fixed in the vector space, demanding transformation processes are no longer required and the representational self-reference system can be superposed onto the vector space at any desired position. This process allows Sholl's model to account for the finding that reporting object directions relative to an imagined position is improved after disorientation (e.g., May, 1996).

Problematic for the model, however, is the finding that additional time after the presentation of the to-be-imagined position does not reduce the relative difficulty of different imagined movements (May, 2004). According to the model, the additional time could be used to detach the representational self-reference system and move it to the appropriate position in the vector space, which would reduce or even eliminate the difficulty differences between different imagined movements. Also problematic is the finding that the difficulty of imagined translations is affected by the disparity of the target object's direction relative to the imagined and the actual position. According to the model, moving the representational self-reference system is an analog process, so the difficulty should be determined by the to-be-translated distance and not by the target object direction disparity.

It may be possible to make Sholl's model compatible with these findings by replacing its explanation of the difficulty of imagined updating with an interference based explanation. This modification could, for instance, be conceptualized as interference between the sensorimotor and the representational self-reference system and between the conflicting object locations within the two systems.

6 How Are Interobject Spatial Relations Represented in Memory?

According to our model, the particular reference directions used to specify interobject spatial relations is a central property of long-term spatial representations. When accessing a representation from an imagined heading that corresponds to an employed reference direction, the stored spatial relations can be directly retrieved. For other imagined headings, spatial relations need to be inferred, which produces measurable performance costs. A question that has remained unanswered is how spatial relations are actually represented in memory. Our conjecture is that a spatial representation in

---

[3] More recent versions of the model (e.g., Holmes & Sholl, 2005) suggest that representation in the orientation independent object-to-object system might be limited to very familiar environments, but orientation dependence has even been observed for environments that participants frequently navigated for a period of at least one year (Werner & Schmidt, 1999).

the environmental subsystem can be conceived as a network of nodes representing individual objects that are interconnected by vectors. Each vector represents the distance and direction between two objects, where the direction is specified relative to an employed reference direction (e.g., Vorwerg, 2003). Due to combinatorial complexity, the spatial relations between all possible pairs of objects are not represented equally well. Some are represented with high fidelity and others are represented with low fidelity or are simply not represented at all.

Recent experiments in our lab investigated whether the fidelity with which the relation between two objects is represented is affected by whether the two objects fall along a reference direction. In other words, the goal was to first assess which reference directions were apparently selected by participants, using their performance on various imagined headings, and then in a second step, to test whether participants were better at pointing to a target object when the direction of the target object relative to the imagined position was the same direction as one of the apparent reference directions.

In one experiment, participants studied a configuration of nine objects from one of two viewpoints (0° or 180°) that were aligned with a square mat on which the objects were placed and the walls of the surrounding rectangular room. Participants were blindfolded before entering the room and were then disoriented and led on a convoluted path to the appropriate viewing position. After learning, participants made judgments of relative directions in a remote room. The trials were constructed using eight equally spaced directions (defined relative to the room) for both imagined heading and allocentric pointing direction (the direction of the target object relative to the imagined position).

As illustrated in Figure 3A, performance as a function of imagined heading followed a clear sawtooth pattern in both groups. Pointing judgments were more accurate for imagined headings that were aligned with the edges of the mat and the

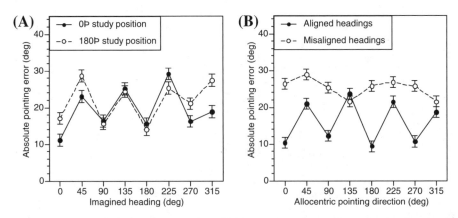

Fig. 3. Angular error in judgments of relative direction as a function of **(A)** imagined heading and study position and **(B)** allocentric pointing direction and imagined heading collapsed across study position. Aligned directions are 0°, 90°, 180°, & 270°; misaligned directions are 45°, 135°, 225°, & 315°. Error bars are confidence intervals corresponding to ± 1 standard error of the mean, as estimated from the ANOVA.

walls of the room than for other imagined headings, suggesting that participants in both groups had represented the layout using those four "aligned" directions as reference directions. The same four aligned directions also showed a benefit for allocentric pointing direction, but this effect was not independent of imagined heading. As Figure 3B shows, pointing to a target object in an aligned direction only improved performance when participants were imagining an aligned heading. For misaligned imagined headings, no such effect was present.

In a second experiment, participants learned the same layout but were neither disoriented nor led on a convoluted path to the appropriate study position. These variations of study conditions were conducted as part of a series of experiments that tried to identify the factors that determine the number of reference directions selected. In this second experiment, the reference directions selected, as indicated by performance on the various imagined headings, was not uniform across participants. Whereas some participants had apparently represented the layout relative to all four aligned directions, others had represented the layout only relative to the 0° and the 180° directions (see Figure 4A). The former participants showed effects of allocentric pointing direction that were comparable to the effects found in the first experiment. Their performance was improved for all four aligned allocentric pointing directions, but only within aligned imagined headings. The latter participants, in contrast, only showed a benefit for the allocentric pointing directions of 0° and 180°, and this effect was only present for the imagined headings of 0° and 180° (see Figure 4B).

The results of both of these experiments paint a clear picture. Directions that showed a benefit in imagined heading, which directions by conjecture were used as reference directions, were also benefited in allocentric pointing, but only when participants imagined facing one of them at the same time. These results support the notion that the spatial relation between two objects is preferentially represented when

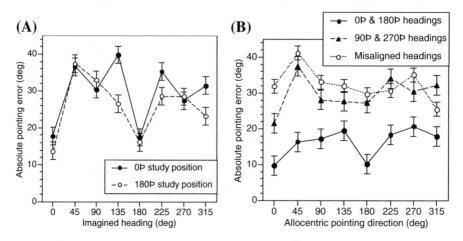

Fig. 4. Results from participants who were classified as having only used 0° and 180° as reference directions. Angular error in judgments of relative direction as a function of **(A)** imagined heading and study position and **(B)** allocentric pointing direction and imagined heading collapsed across study position. Misaligned directions are 45°, 135°, 225°, & 315°. Error bars are confidence intervals corresponding to ± 1 standard error of the mean, as estimated from the ANOVA.

the two objects fall along an employed reference direction. When making spatial judgments from imagined headings that correspond to a reference direction, pointing to objects that lie in the direction of a reference direction is easy, because their relative location is represented with high fidelity or probability. Pointing from those headings to objects that do not lie in the direction of a reference direction is more difficult, because these relative object locations are only represented with lower fidelity or are not represented at all. The spatial information needed to make such pointing judgments must be inferred from relations that are represented.

For judging spatial relations relative to an imagined heading that does not correspond to a reference direction, our model in its current form only states that spatial relations need to be inferred, but does not provide specifics on how such an inference process would operate. The described effects of allocentric pointing direction, however, constrain the possibilities for such a process. One obvious possibility, that the required spatial relation is first retrieved relative to a reference direction and then rotated to match the imagined heading, can be ruled out. Such a process would incorrectly predict a benefit for allocentric pointing directions that correspond to reference directions even from imagined headings that do not correspond to reference directions, because the first step of this process would be sensitive to differences in the fidelity with which interobject spatial relations are represented.

Narrowing down the possibilities for the inference process further will be an important goal of future research. This research should allow for a better understanding of how relative directions are determined from imagined headings that do not correspond to employed reference directions.

7 Summary and Outlook

The goal of this chapter was to summarize a new model of spatial memory and updating. The model proposes an environmental subsystem that is concerned with representing the spatial layout of familiar environments over the long-term, and an egocentric subsystem that holds transient self-to-object spatial relations that are mainly used for guiding locomotion through and interaction with the local environment when vision is available.

Representations in the environmental subsystem are specified with respect to a single or a small number of reference directions that are intrinsic to the represented environment. When imagining a heading that is aligned with a reference direction, the stored spatial relations can be directly retrieved; for other imagined headings, the required spatial relations need to be inferred. The selection of particular reference directions also results in the preferential representation of those interobject spatial relations that are aligned with one of the employed reference directions.

Updating in the environmental subsystem consists of updating the position of the observer with respect to the intrinsic reference system that underlies the representation of the current environment, but it does not involve updating the reference system itself. Because the reference system itself is not updated, the stored spatial relations can only be directly retrieved when the person's orientation corresponds to one of the reference directions. When a person's orientation does not correspond to one of the reference directions, relative object location must be inferred, even though the person's position is properly updated in the environmental subsystem and the person is therefore oriented with respect to the environment.

Updating in the egocentric subsystem, in contrast, readily provides object locations relative to the current position, regardless of the person's orientation. This system, however, operates best when vision is available. In the absence of visual support, maintaining and updating the self-to-object relations in this subsystem becomes very limited.

This notion of one system that provides updated self-to-object relations but is limited in its capacity and maintenance time, and another system that holds enduring representations that are orientation dependent, is in good agreement with studies showing no updating costs for simpler layouts (e.g., Presson & Montello, 1994; Wraga, Creem-Regehr, & Proffitt, 2004) and studies showing updating costs for more complex layouts (e.g., May, 1996). Further investigations, however, are required to identify the precise limitations of egocentric updating.

Unclear at this point, is also which factors determine the number of reference directions selected. In all of our experiments, in which participants showed improved performance on more than one imagined heading, the benefited headings were orthogonal or opposite to each other, which suggests that multiple reference directions can only be selected when they are orthogonal or opposite to each other. It seems that in order for multiple reference directions to actually be selected, the environment needs to provide a very salient axis or a pair of very salient orthogonal axes. How salient an axis is perceived to be, however, seems to be not only determined by the properties of the to-be-represented environment but also heavily influenced by the learning experiences. These experiences include, for instance, which directions are actually experienced or emphasized through instructional manipulations, and apparently even whether a person was guided to a study position while disoriented or not, as illustrated by the experiments discussed in the previous section.

Finally, more work is needed to better understand the processes involved in making spatial judgments relative to imagined headings that do not correspond to employed reference directions. This research should investigate the processes that detect mismatches between imagined headings and represented reference directions, as well as the subsequent inferential processes invoked to determine the required relative directions.

References

Chance, S.S., Gaunet, F., Beall, A.C., Loomis, J.M.: Locomotion mode affects the updating of objects encountered during travel: The contribution of vestibular and proprioceptive inputs to path integration. Presence: Teleoperators and Virtual Environments 7(2), 168–178 (1998)

Creem, S.H., Proffitt, D.R.: Defining the cortical visual systems: "What", "where", and "how". Acta Psychologica 107(1-3), 43–68 (2001)

Diwadkar, V.A., McNamara, T.P.: Viewpoint dependence in scene recognition. Psychological Science 8(4), 302–307 (1997)

Easton, R.D., Sholl, M.J.: Object-array structure, frames of reference, and retrieval of spatial knowledge. Journal of Experimental Psychology: Learning, Memory, and Cognition 21(2), 483–500 (1995)

Farrell, M.J., Robertson, I.H.: Mental rotation and automatic updating of body-centered spatial relationships. Journal of Experimental Psychology: Learning, Memory, and Cognition 24(1), 227–233 (1998)

Farrell, M.J., Thomson, J.A.: Automatic spatial updating during locomotion without vision. Quarterly Journal of Experimental Psychology: Human Experimental Psychology 51(3), 637–654 (1998)

Franklin, N., Tversky, B.: Searching imagined environments. Journal of Experimental Psychology: General 119(1), 63–76 (1990)

Hartley, T., Trinkler, I., Burgess, N.: Geometric determinants of human spatial memory. Cognition 94(1), 39–75 (2004)

Holmes, M.C., Sholl, M.J.: Allocentric coding of object-to-object relations in overlearned and novel environments. Journal of Experimental Psychology: Learning, Memory, and Cognition 31(5), 1069–1087 (2005)

Kearns, M.J., Warren, W.H., Duchon, A.P., Tarr, M.: Path integration from optic flow and body senses in a homing task. Perception 31(3), 349–374 (2002)

Klatzky, R.L.: Allocentric and egocentric spatial representations: Definitions, distinctions, and interconnections. In: Freksa, C., Habel, C., Wender, K.F. (eds.) Spatial cognition: An interdisciplinary approach to representing spatial knowledge, pp. 1–17. Springer, Heidelberg (1998)

Klatzky, R.L., Loomis, J.M., Beall, A.C., Chance, S.S., Golledge, R.G.: Spatial updating of self-position and orientation during real, imagined, and virtual locomotion. Psychological Science 9(4), 293–298 (1998)

Levinson, S.C.: Frames of reference and Molyneux's question: Crosslinguistic evidence. In: Bloom, P., Peterson, M.A., Nadel, L., Garret, M.F. (eds.) Language and Space, pp. 77–1017. MIT, Cambridge (1996)

Loomis, J.M., Beall, A.C.: Visually controlled locomotion: Its dependence on optic flow, three-dimensional space perception, and cognition. Ecological Psychology 10(3-4), 271–285 (1998)

May, M.: Cognitive and embodied modes of spatial imagery. Psychologische Beiträge 38(3-4), 418–434 (1996)

May, M.: Imaginal perspective switches in remembered environments: Transformation versus interference accounts. Cognitive Psychology 48(2), 163–206 (2004)

May, M., Klatzky, R.L.: Path integration while ignoring irrelevant movement. Journal of Experimental Psychology: Learning, Memory, and Cognition 26(1), 169–186 (2000)

May, M., Wartenberg, F.: Rotationen und Translationen in Umräumen: Modelle und Experimente. Kognitionswissenschaft 4, 142–153 (1995)

McNamara, T.P.: How are the locations of objects in the environment represented in memory? In: Freksa, C., Brauer, W., Habel, C., Wender, K.F. (eds.) Spatial Cognition III. LNCS (LNAI), vol. 2685, pp. 174–191. Springer, Heidelberg (2003)

McNamara, T.P., Rump, B., Werner, S.: Egocentric and geocentric frames of reference in memory of large-scale space. Psychonomic Bulletin and Review 10(3), 589–595 (2003)

Milner, A.D., Goodale, M.A.: The visual brain in action. Oxford University Press, Oxford, England (1995)

Mou, W., McNamara, T.P.: Intrinsic frames of reference in spatial memory. Journal of Experimental Psychology: Learning, Memory, and Cognition 28(1), 162–170 (2002)

Mou, W., McNamara, T.P., Valiquette, C.M., Rump, B.: Allocentric and egocentric updating of spatial memories. Journal of Experimental Psychology: Learning, Memory, and Cognition 30(1), 142–157 (2004)

Péruch, P., May, M., Wartenberg, F.: Homing in virtual environments: Effects of field of view and path layout. Perception 26(3), 301–311 (1997)

Presson, C.C., DeLange, N., Hazelrigg, M.D.: specificity in spatial memory: What makes a path different from a map of the path? Journal of Experimental Psychology: Learning, Memory, and Cognition 15(5), 887–897 (1989)

Presson, C.C., Hazelrigg, M D: Building spatial representations through primary and secondary learning. Journal of Experimental Psychology: Learning, Memory, and Cognition 10(4), 716–722 (1984)

Presson, C.C., Montello, D.R.: Updating after rotational and translational body movements: Coordinate structure of perspective space. Perception 23(12), 1447–1455 (1994)

Rieser, J.J.: Access to knowledge of spatial structure at novel points of observation. Journal of Experimental Psychology: Learning, Memory, and Cognition 15(6), 1157–1165 (1989)

Rieser, J.J., Guth, D.A., Hill, E.W.: Sensitivity to perspective structure while walking without vision. Perception 15(2), 173–188 (1986)

Rock, I.: Orientation and form. Academic Press, New York (1973)

Roskos-Ewoldsen, B., McNamara, T.P., Shelton, A.L., Carr, W.: Mental representations of large and small spatial layouts are orientation dependent. Journal of Experimental Psychology: Learning, Memory, and Cognition 24(1), 215–226 (1998)

Rossetti, Y.: Implicit short-lived motor representations of space in brain damaged and healthy subjects. Consciousness and Cognition 7(3), 520–558 (1998)

Rossetti, Y., Pisella, L., Pelisson, D.: New insights on eye blindness and hand sight: Temporal constraints of visuo-motor networks. Visual Cognition 7(6), 785–808 (2000)

Shelton, A.L., McNamara, T.P.: Multiple views of spatial memory. Psychonomic Bulletin and Review 4(1), 102–106 (1997)

Shelton, A.L., McNamara, T.P.: Systems of spatial reference in human memory. Cognitive Psychology 43(4), 274–310 (2001a)

Shelton, A.L., McNamara, T.P.: Visual memories from nonvisual experiences. Psychological Science 12(4), 343–347 (2001b)

Sholl, M.J.: The relation between horizontality and rod-and-frame and vestibular navigational performance. Journal of Experimental Psychology: Learning, Memory, and Cognition 15(1), 110–125 (1989)

Sholl, M.J.: The role of a self-reference system in spatial navigation. In: Montello, D.R. (ed.) COSIT 2001. LNCS, vol. 2205, pp. 217–232. Springer, Heidelberg (2001)

Sholl, M.J., Nolin, T.L.: Orientation specificity in representations of place. Journal of Experimental Psychology: Learning, Memory, and Cognition 23(6), 1494–1507 (1997)

Steck, S.D., Mochnatzki, H.F., Mallot, H.A.: The role of geographical slant in virtual environment navigation. In: Freksa, C., Brauer, W., Habel, C., Wender, K.F. (eds.) Spatial Cognition III. LNCS (LNAI), vol. 2685, pp. 62–76. Springer, Heidelberg (2003)

Tversky, B.: Distortions in memory for maps. Cognitive Psychology 13(3), 407–433 (1981)

Valiquette, C.M., McNamara, T.P., Smith, K.: Locomotion, incidental learning, and the selection of spatial reference systems. Memory & Cognition 31(3), 479–489 (2003)

Vorwerg, C.: Use of reference directions in spatial encoding. In: Freksa, C., Brauer, W., Habel, C., Wender, K.F. (eds.) Spatial Cognition III. LNCS (LNAI), vol. 2685, pp. 321–347. Springer, Heidelberg (2003)

Waller, D.: Egocentric and nonegocentric coding in memory for spatial layout: Evidence from scene recognition. Memory and Cognition 34(3), 491–504 (2006)

Waller, D., Hodgson, E.: Transient and enduring spatial representations under disorientation and self-rotation. Journal of Experimental Psychology: Learning, Memory, and Cognition 32(4), 867–882 (2006)

Waller, D., Montello, D.R., Richardson, A.E., Hegarty, M.: Orientation specificity and spatial updating of memories for layouts. Journal of Experimental Psychology: Learning, Memory, and Cognition 28(6), 1051–1063 (2002)

Wang, R.F.: Representing a stable environment by egocentric updating and invariant representations. Spatial Cognition and Computation 1(4), 431–445 (1999)

Wang, R.F.: Between reality and imagination: When is spatial updating automatic? Perception and Psychophysics 66(1), 68–76 (2004)

Wang, R.F., Simons, D.J.: Active and passive scene recognition across views. Cognition 70(2), 191–210 (1999)

Wang, R.F., Spelke, E S :. Updating egocentric representations in human navigation. Cognition 77(3), 215–250 (2000)

Wang, R.F., Spelke, E.S.: Human spatial representation: Insights from animals. Trends in Cognitive Sciences 6(9), 376–382 (2002)

Werner, S., Schmidt, K.: Environmental reference systems for large-scale spaces. Spatial Cognition and Computation 1(4), 447–473 (1999)

Wraga, M.: Thinking outside the body: An advantage for spatial updating during imagined versus physical self-rotation. Journal of Experimental Psychology: Learning, Memory, and Cognition 29(5), 993–1005 (2003)

Wraga, M., Creem-Regehr, S.H., Proffitt, D.R.: Spatial updating of virtual displays during self- and display rotation. Memory & Cognition 32(3), 399–415 (2004)

Yamamoto, N., Shelton, A.L.: Visual and proprioceptive representations in spatial memory. Memory & Cognition 33(1), 140–150 (2005)

Yardley, L., Higgins, M.: Spatial updating during rotation: The role of vestibular information and mental activity. Journal of Vestibular Research: Equilibrium and Orientation 8(6), 435–442 (1998)

Sensorimotor Interference When Reasoning About Described Environments*

Marios N. Avraamides and Melina-Nicole Kyranidou

Department of Psychology, University of Cyprus, P.O. Box 20537,
1678 Nicosia, Cyprus
mariosav@ucy.ac.cy, se02km9@ucy.ac.cy

Abstract. The influence of sensorimotor interference was examined in two experiments that compared pointing with iconic arrows and verbal responding in a task that entailed locating target-objects from imagined perspectives. Participants studied text narratives describing objects at locations around them in a remote environment and then responded to targets from memory. Results revealed only minor differences between the two response modes suggesting that bodily cues do not exert severe detrimental interference on spatial reasoning from imagined perspective when non-immediate described environments are used. The implications of the findings are discussed.

Keywords: Spatial reasoning, alignment effect, sensorimotor interference.

1 Introduction

1.1 Alignment Effects in Memory and Perception

When confronted with an unfamiliar spatial environment such as when visiting a foreign city for the first time, we often rely on cartographic maps to make navigational decisions. In order to use such maps effectively we spontaneously rotate them into alignment with our perceived viewpoint in the environment (Werner, 2002). In cases when we cannot rotate a map – as in the case of a wall-mounted "You-Are-Here" map – the efficiency of using the map is severely impaired. This phenomenon is well-documented in psychological research and exists even when reasoning about a map that is no longer in view (e.g., Levine, Jankovic, & Palij, 1982; Presson & Hazelrigg, 1984).

In the past, researchers (e.g., Presson & Hazelrigg, 1984) have used the term *aligned trials* to refer to those trials of the experimental design in which a participant is called to make a spatial judgment (e.g., point to a target-object) from a physical perspective that is parallel to the orientation adopted at the time of viewing a map or studying a spatial layout. In perception, aligned trials are thus equivalent to deciding which way to turn when holding a map aligned with one's perceived heading in the environment (e.g., facing north in the environment and holding a north-up map in front of you). Trials in which the participant is physically oriented toward a direction

---

* The presented work is part of an undergraduate thesis by Melina-Nicole Kyranidou.

that is not parallel to the initial study viewpoint have been typically referred to as *misaligned trials*. Reasoning for misaligned trials is thus similar to the reasoning needed to decide how to turn in the environment when the orientation of the map-reader differs from the orientation of the map.

A typical finding from experiments is that people can reason more efficiently when doing so under aligned than misaligned conditions. This effect has been referred to in the literature as the *alignment effect* and is defined as the performance difference between aligned and misaligned trials (e.g., Presson & Hazelrigg, 1984). Used in the context of map reading the term alignment effect refers to the discrepancy between an observer's physical orientation and the initial studied orientation which, as early studies of spatial memory argued, was stored in memory.

However, the term alignment effect, as used by many other researchers (e.g., Roskos-Ewoldsen, McNamara, Shelton, & Carr, 1998), has a somewhat different meaning. A great number of spatial memory experiments adopt the following paradigm. Participants first study a spatial scene and then make spatial judgments from memory about the location of objects in the scene. In many of these experiments testing takes place after participants are guided to a remote location (e.g., Mou & McNamara, 2002). A vast number of experiments on spatial memory converged on a single finding: spatial judgments are faster and/or more accurate when participants adopt imagined perspectives that are aligned with the initial study viewpoint. This finding has also been called an alignment effect. In this context, however, the effect refers to the performance difference between trials in which an imagined perspective, and not the physical perspective of the participant, is aligned or misaligned with the study viewpoint. Recent research has provided evidence that the two types of alignment effects are in fact independent (Mou, McNamara, Valiquette, & Rump, 2004).

The theories of spatial memory that have been developed over the years have been designed around explanations for the latter type of alignment effect (we will use the term *memory alignment effect* for the sake of simplicity). For example, Shelton & McNamara's (1997) visual-spatial snapshot model proposed that spatial memories are defined egocentrically; that is, people maintain memories in the orientations defined by the viewpoints they experience. Easton & Sholl's (1995) model proposes that two systems are used in spatial memory. An object-to-object system codes locations in an allocentric manner while a self-reference system is used to fix the orientation of the object-to-object system. Both these models predict superior performance for trials in which the orientation of the imagined perspective matches that of the study view. A recent theory by McNamara (2003) suggests that spatial memories are not egocentric. Instead, a reference frame intrinsic to the scene is used to encode the layout but this reference frame is stored in memory in a preferred orientation. In a series of experiments McNamara and his colleagues (e.g., Mou & McNamara, 2002) provide convincing evidence that cues, other than egocentric experience (e.g., the internal structure of the layout, instructions etc.), can be used to establish the preferred direction that is stored in memory. Whether the preferred direction that is used to orient a spatial memory is viewing experience or else, theories of spatial memory assume that the memory alignment effect represents a performance advantage for when an imagined perspective matches the direction used to for storage.

If the memory alignment effect is caused by the organizational characteristics of spatial memory, it remains to be seen what exactly causes the other alignment effect caused by the angular discrepancy between the physical perspective of the observer and the study view. More relevant to this effect (we will call this effect *sensorimotor alignment effect*) is the literature on spatial updating.

1.2 Spatial Updating and Sensorimotor Interference

Spatial updating refers to the process of automatically computing the egocentric locations of objects of our immediate environment as we change our position and/or orientation in it. Studies in spatial updating typically require participants to study a spatial scene, move – either physically or by imagination – to a novel standpoint or orientation in the layout, and then localize target-objects by pointing to them (e.g., Presson & Montello, 1994; Rieser, Guth, & Hill, 1986). Often, spatial performance is compared to that of a control condition in which participants point to targets from their initial standpoint/orientation.

The typical result from spatial updating studies is that, compared to pointing to targets from the learning standpoint, pointing performance from imaginal standpoints is inferior. This is especially the case when the imagined perspective adopted at the novel standpoint is misaligned with the physical perspective at the learning standpoint. For example, Rieser (1989) has shown that latencies for pointing to targets increased as a function of the angular disparity of the imaginal standpoint from the initial learning standpoint.

The inferior pointing performance from imaginal standpoints in spatial updating studies has been typically attributed to the lack of vestibular and proprioceptive information during imagined movement; these types of information are believed to be important factors allowing people to automatically update their spatial representations during actual physical movement (Klatzky, Loomis, Beall, Chance, & Golledge, 1998; Rieser, 1989). The finding that performance does not suffer when the novel standpoint/orientation is adopted by physical instead of imagined movement corroborates the central role of vestibular and proprioceptive information in spatial updating.

Recently, May (2004) has proposed a novel hypothesis, termed *the sensorimotor interference hypothesis*, to provide another account for the inferior spatial performance associated with responding from imagined standpoints. According to May, in order to point to the location of an object from an imagined standpoint, a person has to overcome conflicts that are caused by the discrepant physical (sensorimotor) and the imagined egocentric (i.e., relative to the imagined position) locations of objects. May argues that sensorimotor codes specifying the objects' locations relative to the actual position and orientation of the person are automatically activated. Therefore, an observer needs to suppress this sensorimotor information in order to select the appropriate response. Suggestive of these conflicts is May's finding that reaction times for pointing toward targets from imagined standpoints vary as a function of *object-direction disparity* (i.e., the angular difference between the sensorimotor and the imaginal response vectors).

In addition to object-direction disparity, May proposes a second source of interference. He argues that even when the appropriate response vector is chosen, in order to be executed by pointing, it needs to be specified from a reference frame centered on and oriented with the person's body. This type of interference, termed *head-direction disparity*, explains why performance from imagined standpoints suffers dramatically when adopting the imagined standpoint entails performing mental rotations instead of translations only.

The strong coupling of pointing with the body is evidenced in studies that employ disorientation. In one study, May (1996) had people point to target-objects of a spatial layout under 3 conditions. In one condition, named embodied repositioning, participants were allowed to turn their bodies into the requested facing direction before pointing to a target. In another condition, termed cognitive repositioning, participants remained at their initial standpoint and simply imagined adopting the requested facing direction. Finally, the disoriented repositioning condition was similar to cognitive repositioning with the exception that participants were rotated to the left and right for 5 seconds after having studied the layout. Results revealed superior pointing performance for embodied repositioning. More importantly, however, performance was worse in the cognitive repositioning than in the disoriented repositioning condition. Similar results are provided by Waller, Montello, Richardson, & Hegarty (2002). This experiment showed an alignment effect (i.e., inferior performance for trials in which the imagined heading was misaligned with the observer's actual orientation) which was diminished in a disorienting condition. The results by May (1996) and Waller et al. (2002) suggest that performance in such perspective-taking tasks can be improved, and the alignment effect can be reduced, if the influence of discrepant bodily cues is diminished through disorientation. These findings seem to indicate that at least part of the alignment effect is due to sensorimotor interference (object-direction disparity and head-direction disparity) that is present when responding from imagined perspectives.

1.3 Sensorimotor Interference and Manual Responding

The majority of studies using perspective-taking tasks either in the context of spatial perception/memory or spatial updating have employed manual responses such as pointing with the arm or a pointer (e.g., Avraamides, Loomis, Klatzky, & Golledge, 2004; Rieser, Guth, & Hill, 1986), turning the body toward the direction of the target (Avraamides, Klatzky, Loomis, & Golledge, 2004; Klatzky et al., 1998), walking to targets (Loomis, Lippa, Klatzky, & Golledge, 2002; Riecke, van Veen, & Bülthoff, 2002) or selecting iconic arrows (De Vega & Rodrigo, 2001). This raises the possibility that processes specific to the response mode itself are in part responsible for the alignment effect. Such responses seem so strongly anchored to the reference frame of the human body that using them from imagined perspectives might be problematic.

Indeed, a study by Avraamides et al. (2004) revealed that the large systematic pointing errors people make when performing a triangle-completion task without physically moving are eliminated when pointing is replaced by a verbal response (describing angles in degrees). It seems then that language might be a more flexible response medium than pointing when it comes to responding from imagined perspectives. The fact that in everyday life we typically use pointing from our own standpoint and only seldom from imagined standpoints suggests that is possible that pointing has become through practice so strongly coupled with the orientation of our body that it is difficult to use from a different orientation.

A recent study by Avraamides and Ioannidou (2005) has contrasted a pointing-like response mode with verbal responding in a perceptual perspective-taking task. The task was designed to involve the reasoning required to use maps that are aligned or misaligned with the observer's viewing perspective. Participants, viewed displays of a rectangular table portraying 6 people sitting around it in various arrangements. With the display perceptually available at all times, they were then instructed to adopt the perspective of one of the characters, and locate from that perspective a second character (e.g., "imagine you are person x, locate person y"). Two response-mode conditions were included in the experiment. In the labeling condition, participants responded by selecting the verbal label describing the relative position of the target. In the pointing response condition, they responded by selecting an iconic arrow pointing toward the desired direction. Results revealed faster and more accurate performance with verbal responding than pointing with arrows. More importantly, the alignment effect was less pronounced in the labeling condition. Accuracy for pointing diminished as misalignment (i.e., as the angular deviation between the imagined perspective and the participant's physical perspective/orientation of the display) increased. In contrast, accuracy in the verbal responding condition was equal for all levels of imagined perspective. In terms of latency, response times increased with greater misalignment of the imagined perspective but the increase was substantially steeper in the pointing response mode.

Avraamides and Ioannidou (2005) explained their findings in terms of sensorimotor interference. They argued that using responses that are strongly anchored to the reference frame of the human body poses unnatural demands on people. That is, to respond people first need to ignore their physical orientation in order to determine the correct response vector but then need to use their body to execute the response. This is, in fact, the head-direction disparity account proposed by May (2004).

1.4 Sensorimotor Interference and Reasoning About Remote Environments

Although there is increasing evidence that sensorimotor interference can explain, at least part of, the sensorimotor alignment effect, there is still not enough research on how bodily cues affect performance in spatial memory experiments. As already mentioned, studies in spatial memory often move participants to a different room for testing purposes (e.g., Mou & McNamara, 2002) and sometimes even had each participant face a different direction at the time of test (Avraamides & Kelly, 2005). It seems that researchers have taken measures against that possibility of confounding their results with unwanted sensorimotor effects, nevertheless, it is not yet known whether such effects are present when reasoning about remote environments.

A very recent study by May (in press) attempted to explore this. In this study participants first learned a spatial layout in one room and then were tested in localizing objects from imagined perspectives either in the same room or a different room. The critical question was whether testing at a remote location would spatially decontextualize the participant and therefore reduce the conflicts between the actual and imagined perspectives and hence attenuate interference costs. The results from the two experiments conducted were mixed. In the first experiment which used a single-trial design (i.e., participants switched imagined perspective on a trial-by-trial basis) there was no significant difference between the two room conditions. Whether tested in the same room or at a different one, participants were more accurate and faster to respond in aligned than misaligned trials. In fact, performance decreased with the angular deviation between imagined perspective and the physical perspective of the participant. In the second experiment which used blocked-trial testing (i.e., participants adopted a perspective and located from it multiple objects) an attenuation of interference costs for the remote location testing was evidenced. Alignment effects were present in the remote room testing as well but to a much lesser degree.

The goal of the present work is to examine whether sensorimotor interference, head-direction disparity in particular, is present when reasoning about remote spatial environments. Our methodology is, however, very different from that used by May. First, because in everyday life we often reason about environments that we encode through language (e.g., when comprehending driving instructions), we have chosen to focus on spatial scenes that are conveyed through language. Second, we have followed the design of Avraamides and Ioannidou (2005) in which the response mode was manipulated. The rationale is that if head-direction disparity is an issue when reasoning about remote environments, the alignment effect will be greater when responding in a body-defined manner.

The general methodology of our experiments is as follows. Participants first read narratives that describe a remote location (e.g., a museum, a hotel lobby etc). Then they are requested to respond to statements of the form "face x, find y", where x and y are objects described in the narrative. In one condition, hereafter the pointing condition, participants respond by selecting among iconic arrows. In another condition, hereafter the labeling condition, they respond by choosing among verbal labels. In order to determine the role of head-direction disparity we contrast performance in these two conditions. Following the logic of Avraamides and Ioannidou (2005) we will attribute any differences in the magnitude of the alignment effect between the two response modes to the presence of different levels of head-direction disparity.

2 Experiment 1

The first experiment is designed to examine whether sensorimotor interference is present when responding with a pointing-like manner in a task that entails reasoning about a remote environment encoded in memory through language. With remote environments there are no object-direction disparity conflicts, as no automatic activation of sensorimotor codes can take place.

If heading-direction disparity poses a problem for pointing then similar results that those of Avraamides and Ioannidou (2005) should be expected. That is, participants should be substantially slower and make more errors in pointing than labeling and this should be the case when responding takes place from misaligned perspectives. On the other hand, if there is no heading-direction disparity influence when reasoning about remote environments, similar alignment effects between the two response modes should be expected.

Based on the premises of the various spatial memory theories a memory alignment effect is predicted with both response modes. This alignment effect would indicate the use of a preferred direction for orientating spatial memories upon reading the narratives. The critical issue here is whether the alignment effect would be equivalent in the two response modes or not. A greater alignment effect for pointing would suggest the presence of additional interference costs specific to this particular response mode.

2.1 Method

Participants
Twenty-six students attending an introductory Psychology course at the University of Cyprus served as participants in the experiment in exchange for course credit or small monetary compensation.

Materials
Four narratives adopted from Franklin and Tversky (1990) and Bryant and Wright (1999) – also used in Avraamides (2003) – were translated in Greek and used in the present experiment. Each narrative described a naturalistic scene in which four critical objects were located around the participant occupying the front, back, left, and right direction as specified by an egocentric reference fame. For each participant, two of the narratives were randomly assigned to the pointing response condition and two in the labeling condition. Half of the participants were presented with the pointing narratives first while the other half began with the labeling narratives. In both response conditions participants responded by pressing keys on the keyboard. In the labeling condition the keys used were "y", "u", "i", and "o". The keys were marked with the initials of the greek translations of the terms "front", "back", "left", and "right". For the pointing condition iconic arrows were placed on the arrow keys of the numeric keypad in the appropriate egocentric arrangement[1]. Each narrative along with the response trials that followed were presented on desktop computers running the E-Prime software package.

Design
A 2 x 3 x 2 within-subjects factorial design, with response mode (pointing vs. labeling) and imagined-perspective (0°, 90°, 180°)[2] as the within-subjects factors and order of tasks (pointing first vs. labeling first), was adopted.

2.2 Procedure

Prior to the experimental trials for each response mode, two types of practice trials were administered. First, participants were asked to imagine being at specific

---

[1] A number of studens (whose data are not included in the analyses reported here) participated in a condition in which the iconic arrows were placed in a linear arrangement on the same keys used in the labeling condition. No different results than those reported here were found.
[2] No differences between left and right were found in any of the analyses. Therefore we have combined the two in the 180° imagined perspective.

locations on campus or the nearby area and locate familiar landmarks. Before the pointing blocks, participants responded by extending their arm to point to the landmarks from their imagined position. Before the labeling blocks, they responded by describing the relative to their imagined position location of the landmarks. This block of trials was completed when the experimenter judged that the participant was competent in responding from an imagined position. Then, participants carried out a number of practice trials on the computer which aimed at familiarizing them with using the keys to select the appropriate response from the set of alternatives. For the labeling condition, arrows were presented on the screen as probes and participants had to respond as fast as possible by pressing the key with the verbal label describing the pointing direction of the arrow. Similarly, prior to the pointing condition, participants pressed they key marked with the arrow that pointed the same way as the probe. Trials for this familiarization phase were followed by 2 blocks of experimental trials in the corresponding response mode. In each block a narrative was presented and participants were given unlimited time to study it, visualize themselves in it, and remember the objects around them. After two filler sentences were presented, 16 localization trials followed. For each trial, the sentence "Face x, Find y" (where x and y were objects from the narrative) was presented on the screen. At this point, participants had to imagine facing object x and locate object y from the imagined perspective. The 16 trials for each block were created by presenting all possible combinations of the four objects as referents for the imagined heading and as targets. Trials were presented randomly for each participant. A total of 64 trials (32 in each response mode) were administered to each participant. In the labeling condition participants indicated their response by pressing the key marked with the initial of the appropriate verbal term. For the pointing-mode blocks, responses were indicated by pressing the key marked with the arrow pointing to the appropriate direction.

2.3 Results

Familiarization Phase
The data from the practice trials were analyzed to examine whether differences between the two response modes existed before the experimental trials. In both response modes participants responded to the orientation of a presented arrow-probe using the same 4 keys. The keys were marked with the initial of verbal labels or arrows depending on the response mode to be used in the experimental trials that followed. Because the familiarization phase involved no misaligned perspectives we expected that performance would be either equal or superior for pointing. Pointing from one's actual perspective is a well-practiced task and does not require mapping labels to the various regions of space. Accuracy was very high (95%), therefore, latencies became the primary focus. A repeated-measures analysis of variance (ANOVA) with response mode and order of conditions as factors was performed. The only significant result was a main effect for response mode, $F(1,22)=22,04$, MSE=389586, $p<.001$. Participants were faster in the pointing (1134 ms) than in the labeling response mode (1980 ms).

Experimental Phase
Accuracy data and latencies for correct responses were analyzed using a repeated-measures ANOVA with response mode, imagined perspective, and block as within-subject factors and order as between-subject factors. Data from two participants were discarded from all analyses because of very low accuracy (<30%).

Accuracy: The average accuracy was 83.8% for pointing and 85.2% for labeling; this difference was not statistically significant, p=.12. An alignment effect was, however, present. Accuracy was higher for the 0° perspective (89.8%) than for the 180° and the 90° perspectives (81.5% and 82% respectively), $F(2,44)=6.82$, MSE=.03, p<.01. More importantly though, a significant response mode x imagined perspective interaction was obtained, $F(2,44)=4.85$, MSE=.02, p<.01. Simple main effect analyses revealed a significant effect for imagined perspective in the pointing response mode, $F(2,46)=8.10$, MSE=.02, p<.001. As seen in Figure 1, performance for the aligned perspective (0° perspective) was more accurate than performance for the 180° and 0° perspective, p's<.01. Performance for the 180° and the 90° perspectives did not differ significantly, p=.48. In the labeling condition, although performance was somewhat higher for the 0° perspective, the simple main effect analysis showed that the alignment effect was not statistically significant, p=.36. To further compare the two response-mode conditions for an alignment effect, a new variable was created by subtracting for each subject the accuracy for the 90° perspective from that for the 0° perspective[3]. A paired-samples

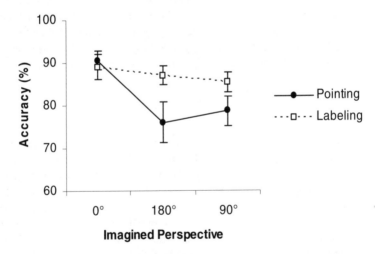

Fig. 1. Average Accuracy for each imagined heading as a function of response mode, Exp.1

---

[3] Although trials for the 180° perspective can be regarded as misaligned trials, they represent a special case as the imagined perspective is the exact opposite of the learning/physical perspective. Because it is highly possible that participants use special strategies for this imagined perspective (which may be different for the two response modes), we excluded those data from this analysis only.

t-test revealed that the alignment effect was greater for pointing than labeling, t(23)=2.8, p<.01. As seen if Figure 1, this was the case because accuracy for the 90° imagined perspective was lower in the pointing than the labeling condition.

A marginally significant response mode x task order interaction was also obtained, F(1,22)=3.45, MSE=.08, p=.07. As seen in Figure 2, participants who performed the pointing task first were significantly more accurate in the labeling response mode (89.6%) than the pointing-mode (78%). A pair-wise t-test showed that this difference was statistically significant, p<.05. In contrast, participants who performed labeling first were equally accurate in the two tasks (85.6% and 84.7% respectively for pointing and labeling.

Latency: Data in which latency exceeded the mean latency of the response mode x block cell of each participant by 2.5 standard deviations were omitted from the analyses.

Latency was longer for the pointing (4626 ms) than labeling (4435 ms) but the difference was not statistically significant, p=.56. As with accuracy, a significant response mode x order interaction was obtained in the latency analysis, F(1,22)=7.59, p<.05.

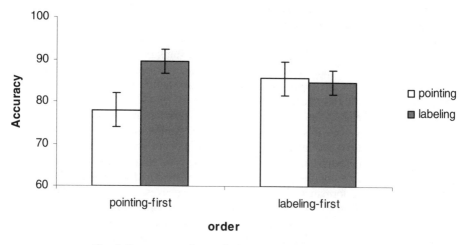

Fig. 2. Response mode x order interaction for accuracy data

As shown in Figure 3, the pointing response mode was slower when completed before than after labeling, t(22)=2.12, p<.05. In contrast, performance in the labeling task was not affected by order, p=.81. Furthermore, latency became shorter in the second block of each response-mode condition, F(1,22)=19.56, MSE= 4522245. This effect did not interact with any other variables.

Although the effect of imagined perspective was not significant (p=.14), participants were faster when responding from the 0° perspective (4204 ms), than the 180° (4510 ms) and the 90° (4776 ms) perspectives. The interaction between imagined perspective and response mode was also non-significant, p=.90.

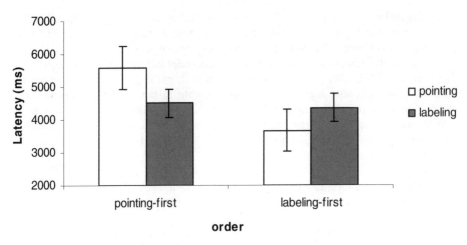

Fig. 3. Response Mode x Order interaction for latency data

2.4 Discussion

In line with theories of spatial memory an alignment effect was present when responding by pointing. Participants in that condition performed better at locating targets from the aligned 0° perspective than both the 90° and the 180° perspectives. The alignment effect in the labeling condition was smaller. Although mean accuracy varied in the predicted manner, the effect of imagined perspective was not significant. Mean latencies for both response conditions were as predicted, even though the effect fell a bit short from significance.

The main difference in the accuracy data for two response modes was that, when the imagined perspective was misaligned with the actual perspective of the participants and the study viewpoint (i.e., 90° and 180° perspectives) participants made substantially more errors when pointing than when labeling. This finding is compatible with the sensorimotor interference hypothesis and the previous results from Avraamides and Ioannidou (2005). Inferior pointing performance from imagined perspectives can be attributed to the presence of head-direction disparity. When having to point, participants had to first compute the appropriate response vector from the imagined perspective but then transformed it to their actual perspective in order to select and iconic arrow. This transformation process might have added an extra cost to responding in this condition.

Although our accuracy results are compatible with the premises of the sensorimotor hypothesis, they were complicated by the presence of a response mode x order interaction. Our analyses indicated that pointing was severely affected, both in terms of accuracy and latency, by the order in which the two tasks were completed. In contrast, labeling was relatively unaffected by order. Although, higher-order interactions involving task order were not significant, it is possible that the order moderated the effects of other variables. What is particularly striking is the very weak alignment effect in the labeling condition whereas other studies have reported alignment effects in almost identical settings (Avraamides, 2003). Therefore, before drawing definite conclusions from this experiment, we replicated Experiment 1 using a between-subjects design and a greater number of participants.

3 Experiment 2

Experiment 2 followed the logic of experiment 1. In order to avoid the order effect, we have modified the experiment so that two separate groups of people would participate in each response mode. Each condition included all 4 narratives; with 4 blocks of trials in each response mode we were also able to examine the effect of practice. One possibility regarding practice is that interference costs could be present initially but disappear after some task exposure.

The main objective was again to compare the magnitude of the alignment effect for the two response modes. Experiment 1 indicated that the alignment effect was more evident for pointing than labeling due to inferior performance from misaligned trials.

3.1 Method

3.1.1 Participants
Sixty-eight students from an introductory Psychology course at the University of Cyprus participated in the experiment in exchange of course credit. Thirty-four participants were randomly assigned to each of the two response-mode conditions.

3.1.2 Materials
Experiment 2 incorporated the same 4 narratives that were used in Experiment 1. Each participant was presented with all 4 narratives and responded by selecting iconic arrows or verbal labels depending on the response-mode condition assigned to.

3.1.3 Design
Experiment 2 followed a 3 x 4 x 2 mixed-factorial design with imagined perspective (0°, 90°, and 180°) and practice (1st block, 2nd block, 3rd block, and 4th block) as the within-subjects factors and response mode (pointing vs. labeling) as the between-subjects factor.

3.2 Procedure

The procedure was identical to Experiment 1 with the exception that each participant completed all 4 narratives (and the preceding practice and familiarization trials) with either pointing or labeling.

3.3 Results

Familiarization Phase
An independent-samples t-test on accuracy data for the familiarization phase revealed that accuracy was superior for the pointing response mode (95.8%) than the labeling response mode (81.6%), $t(63)=2.54$, $p<.05$. Furthermore, latency was substantially shorter in the pointing (1081 ms) than the labeling condition (2144 ms), $t(63)=2.00$, $p<.01$.

Experimental Phase
Accuracy data and latencies for correct responses were analyzed using a repeated-measures ANOVA with imagined perspective, and practice (defined as block number) as

within-subject factors and response mode as a between-subjects factor. Data from three participants were discarded from all analyses because of very low accuracy (<25%).

Accuracy: The average accuracy was 84.4% and did not differ between the two response modes, p=.92. As in Experiment 1, a significant alignment effect was observed. Accuracy was higher for the 0° (89.6%) than the 180° and 90° imagined perspectives (81% and 82.5% respectively), F(2,126)=19.73, MSE=.03, p<.001. A significant response mode x imagined perspective interaction was also obtained, F(6,378)=2.29, MSE=.02, p<.05.

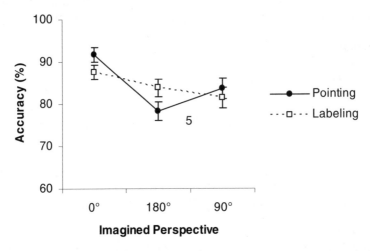

Fig. 4. Average Accuracy for each imagined heading as a function of response mode, Exp.1

Simple main effect analyses revealed a significant effect for imagined perspective in the pointing response mode, F(2,60)=28.85, MSE=.005, p<.001. As seen in Figure 4, performance for the 0° perspective was more accurate than performance for the 180° perspective which was in turn more accurate than performance for the 90° perspective, p's<.01. A simple main effect for imagined perspective was also present in the labeling condition, F(2,66)=3.64, MSE=.009, p<.05. Performance for the 0° perspective was more accurate than performance for the 180° and the 90°, p's<.05. However, performance did not differ between the 180° and the 90° perspectives, p=.3.

As in Experiment 1, a new variable was created by subtracting for each particpant the accuracy for the 90° perspective from that for the 0° perspective in order to compare the two response-mode conditions for an alignment effect. An independent-samples t-test revealed that the alignment effect was equal for pointing and labeling, t(63)=2.8, p<.52.

The analyses also revealed that accuracy increased with the practice, F(3,189)=13.02, MSE=.045, p<.001. Furthermore, the effect of practice interacted with the effect of imagined perspective, F(6,378)=2.29, MSE=.023, p<.05. The interaction was due to the fact that the alignment effect was greater in the first block of trials and was reduced later. Although the alignment effect in the first block of trials tended to be higher for pointing (15.7%) than labeling (9.6%), the 3-way interaction did not approach statistical significance, p=.49.

Latency: Trials in which latency exceeded the mean latency bor the block x imagined perspective cell of each participant by 2.5 standard deviations were omitted from all analyses.

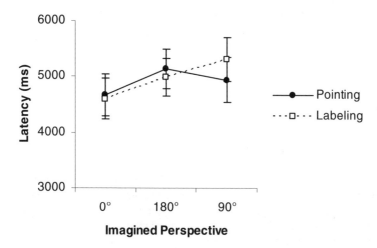

Fig. 5. Latecny as a function of imagined perspective and response mode

A repeated-measures analysis revealed that latency decreased with practice, $F(3,180)=29.57$, MSE=90999824, $p<.001$. Practice did not interact with any other variables. Therefore, in order to reduce the number of missing data due to incorrect responses and hence the elimination of participants from the analysis, we averaged latencies over blocks and carried out a repeated-measures ANOVA with imagined perspective as the within-subjects factor and response-mode condition as the between-subjects factor. The only significant effect was a main effect for imagined perpective, $F(1,63)=4.85$, MSE=1234492, $p<.05$. Within-subject contrasts revealed a significantly faster performance for the 0° perspective (4635 ms) than for either the 180° (5062 ms) or the 90° (5119 ms) perspectives. Figure 5 shows the average latencies for each imagined perspective in the two response modes.

As seen in Figure 5, the alignment effect was greater for labeling than pointing. An independent-sample t-test revealed, however, that the 435 ms difference between the two response modes was not statistically significant, p=.22.

3.4 Discussion

In contrast to Experiment 1, the results of Experiment 2 indicated a clear alignment effect for both response modes in accuracy and latency. More importantly, the magnitude of the alignment effect was equal in the two response modes.

The differences between the two response modes were only minor when compared to those of Avraamides and Ioannidou (2005). Accuracy for the aligned 0° imagined perspective was higher in pointing than labeling indicating a facilitatory effect for the alignment of the physical reference frame or study viewpoint with the imagined perspective to be adopted. Also, accuracy for the 90° perspective was more accurate in the labeling condition but as indicated by the latency data this could have been a speed-accuracy trade off; latency was significantly shorter for that perspective in pointing than labeling.

An additional goal of Experiment 2 was to determine the effect of practice and particularly whether practice interacted with other important variables. First, both our accuracy and latency analyses confirmed the presence of a practice effect; participants were both more accurate and faster as blocks increased. More importantly, however, in the accuracy analysis, the effect of practice interacted with the alignment effect. A greater alignment effect was present in the first block of trials and there was a tendency for a bigger effect in pointing than labeling. Thus, it seems that some conflicts were initially present for pointing but soon went away as a result of practice.

4 General Discussion

The two experiments presented here were designed to examine the presence of an alignment effect in a spatial reasoning task involving remote environments held in memory and compare the magnitude of the effect in two different response-mode conditions.

First, a memory alignment effect was predicted with both response modes. According to various theories of spatial memory (e.g., Easton & Sholl, 1995; Shelton & McNamara, 1997; McNamara, 2003) performance is expected to be superior when an imagined perspective is aligned with the direction used in the organization of memory. In the present experiment, participants studied narratives that described spatial scenes from a particular viewpoint, which is similar to observing a spatial layout from a single stationary standpoint. Based on spatial memory accounts, we therefore predicted a performance advantage for when participants imagined adopting the study view at test. Results from Experiment 1, at least for pointing, were in line with this prediction. More clearly, Experiment 2 supported our expectation for both pointing and labeling.

An important question sought after in the present study was whether the alignment effect would be equal for the two response modes or not. Greater alignment effect for pointing would indicate that in addition to the memory alignment effect we expected, a sensorimotor alignment effect was present.

Based on previous research (Avraamides & Ioannidou, 2005; May, in press) we expected that performance would be particularly impaired when having to point from misaligned perspectives. Responding with iconic arrows maintains the characteristics of real pointing (i.e., the directions are body-defined) and it is believed to rely strongly on the human body. Prior research with body-centered response modes has indicated that they are more difficult to use from imagined perspectives presumably due to their strong attachment to the human body (e.g., May, 2004; Wraga, 2004). Nevertheless, our results suggest that the impairment was not as severe as we expected. In Experiment 1 there was lower accuracy for pointing than labeling from imagined responses but no difference was obtained in the latency data. The effect of accuracy suggests that some interference was associated with pointing from imagined perspectives. The results of Experiment 2 did not replicate this finding. In Experiment

2, pointing was associated with a facilitatory effect for aligned trials. This finding replicated the results from the familiarization phase – which included only aligned trials – showing superior performance for pointing than labeling. It is unclear why the pattern of results differed between the two experiments. One possibility is that the effects of order were more complicated than what was captured by the significant task order x response mode interaction. As a result of the within-subject design, participants, for example, could have carried strategies from one task to the other (e.g., performed pointing in a verbally-mediated fashion).

Despite that some differences were present between pointing and labeling, it seems that the differences were rather small when compared to those found in other experiments (Avraamides et al., 2004; Avraamides & Ioannidou, 2005). In particular, compared to the results of Avraamides and Ioannidou who used similar response modes in a perceptual task, even the interference found in the accuracy of Experiment 1 is rather small in comparison.

One possibility as to why the interference effects were not severe in the present experiments has to do with the fact that the environments used in this study were remote. We believe that there is a fundamental difference between reasoning about immediate and remote spatial environments.

In our view, when encoding an immediate spatial layout people do so using a sensorimotor framework (see also De Vega & Rodrirgo, 2001); that is, they encode the layout either using a body-centric reference frame or an allocentric reference frame that interfaces in some way – depending on the spatial memory theory adopted – with the orientation of the physical body (Easton & Sholl, 1995; McNamara, 2001). Responding by pointing also makes use of the reference frame defined by the physical body. In other words, the same reference frame is involved in both the storage and response levels. When the orientations of storage and response do not match, as in the case of responding from misaligned perspectives, conflicts arise as suggested by May (2004). In terms of heading-direction disparity, a response vector has to be computed based on the orientation at the storage level but then executed based on the orientation of the response level. When pointing from a misaligned perspective, since the same sensorimotor framework is used for the both storage and response, but with two different orientations, interference seems inevitable. As showed by studies employing disorientation (May, 1996; Waller et al., 2002) not knowing about your current orientation with respect to the layout reduces these conflicts.

On the other hand, when people maintain mental representations about remote spatial scenes we argue that they do so by using a cognitive/representational framework; that is, the encoded scenes our stored in a memorial framework that is disengaged from their actual body. The response level still uses a body-centric reference frame but, in the case of remote environments, storage and response are governed by different systems. Heading-direction disparity still exists but we suggest that the transformation of the response vector from the storage system to that of the response can take place more easily as the two systems rely on separate frameworks.

In support of this account are the preliminary results of an experiment using spatial scenes that are learned with direct experience in Virtual Reality (Jonathan Kelly, personal communication, April 2006). In this experiment, participants learned a spatial layout from a particular viewpoint and were then tested in a series of perspective-taking trials in the same or a different room. Results revealed both a memory and a sensorimotor alignment effect for participants tested in the same room but only a memory alignment effect for participants tested in a different room.

Our present findings suggest that if a sensorimotor alignment effect is still present for pointing with remote environments, this effect is substantially attenuated. Compared with the results from other studies showing clear interference in both accuracy and latency (e.g., Avraamides & Ioannidou, 2004), the present findings seem to suggest that sensorimotor conflicts are weakened when reasoning takes place about a non-immediate environment. Nevertheless, further research is needed to determine if that is actually the case.

The present study contains many limitations than need to be addressed in future studies before reaching definitive conclusions. For example, only 4 response alternatives that deviated 90 degrees from the orientation of the physical perspective were used in this study. It might be the case that pointing toward canonical directions is easy and interference effects would show up if objects are placed at other locations. The complexity of the pointing response is one of many factors that need to be examined by future work in this field.

Acknowledgments. We are grateful to all the students who participated in these experiments.

References

Avraamides, M.N.: Spatial updating of environments described in texts. Cognitive Psychology 47, 402–431 (2003)

Avraamides, M.N., Ioannidou, L.M.: Locating Targets from Imagined Perspectives: Labeling vs. Pointing. In: Proceedings of the XXVII Annual Meeting of the Cognitive Science Society, Stresa, Italy (July 2005)

Avraamides, M.N., Kelly, J.W.: Imagined perspective-changing within and across novel environments. In: Freksa, C., Knauff, M., Krieg-Brückner, B., et al. (eds.) Spatial Cognition IV. Reasoning, Action, and Interaction. LNCS (LNAI), pp. 245–258. Springer, Heidelberg (2005)

Avraamides, M.N., Klatzky, R.L., Loomis, J.M., Golledge, R.G.: Use of cognitive vs. perceptual heading during imagined locomotion depends on the response mode. Psychological Science 15, 403–408 (2004)

Avraamides, M.N., Loomis, J.M., Klatzky, R.L., Golledge, R.G.: Functional Equivalence of Spatial Representations Derived From Vision and Language: Evidence From Allocentric Judgments. Journal of Experimental Psychology: Learning, Memory, and Cognition 30, 801–814 (2004)

Bryant, D.J., Wright, G.W.: How body asymmetries determine accessibility in Spatial Frameworks. The Quarterly Journal of Experimental Psychology 52A, 487–508 (1999)

De Vega, M., Rodrigo, M.J.: Updating spatial layouts mediated by pointing and labeling under physical and imaginary rotation. European Journal of Cognitive Psychology 13, 369–393 (2001)

Easton, R.D., Sholl, M.J.: Object-array structure, frames of reference, and retrieval of spatial knowledge. Journal of Experimental Psychology: Learning, Memory, & Cognition 21, 483–500 (1995)

Franklin, N., Tversky, B.: Searching imagined environments. Journal of Experimental Psychology: General 119, 63–76 (1990)

Klatzky, R.L., Loomis, J.M., Beall, A.C., Chance, S.S., Golledge, R.G.: Spatial updating of self-position and orientation during real, imagined, and virtual locomotion. Psychological Science 9, 293–298 (1998)

Levine, M., Jankovic, I.N., Palij, M.: Principles of spatial problem solving. Journal of Experimental Psychology 111, 157–175 (1982)

Loomis, J.M., Lippa, Y., Klatzky, R.L., Golledge, R.G.: Journal of Experimental Psychology: Learning, Memory, and Cognition 28, 335–345 (2002)

May, M.: Imaginal repositioning in everyday environments: Effects of testing method and setting. Psychological Research (in press)

May, M.: Imaginal perspective switches in remembered environments: Transformation versus interference accounts. Cognitive Psychology 48, 163–206 (2004)

May, M.: Cognitive and embodied modes of spatial imagery. Psychologische Beiträge 38, 418–434 (1996)

McNamara, T.P.: How are the locations of objects in the environment represented in memory? In: Freksa, C., Brauer, W., Habel, C., Wender, K. (eds.) Spatial cognition III: Routes and navigation, human memory and learning, spatial representation and spatial reasoning, pp. 174–191. Springer, Berlin (2003)

Mou, W., McNamara, T.P., Valiquette, C.M., Rump, B.: Allocentric and egocentric updating of spatial memories. Journal of Experimental Psychology: Learning, Memory, & Cognition 30, 142–157 (2004)

Mou, W., McNamara, T.P.: Intrinsick frames of reference in spatial memory. Journal of Experimental Psychology: Learning, Memory, & Cognition 28, 162–170 (2002)

Presson, C.C., Montello, D.R.: Updating after rotational and translational body movements: Coordinate structure of perspective space. Perception 23, 1447–1455 (1994)

Presson, C.C., Hazelrigg, M.D.: Building spatial representations through primary and secondary learning. Journal of Experimental Psychology: Learning, Memory, and Cognition 10, 716–722 (1984)

Riecke, B.E., van Veen, H.A.H.C., Bülthoff, H.H.: Visual homing is possible without landmarks-A path integration study in virtual reality. Presence: Teleoperators and Virtual Environments 11, 443–473 (2002)

Rieser, J.J., Guth, D.A., Hill, E.W.: Sensitivity to perspective structure while walking without vision. Perception 15, 173–188 (1986)

Rieser, J.J.: Access to knowledge of spatial structure at novel points of observation. Journal of Experimental Psychology: Learning, Memory, & Cognition 15, 1157–1165 (1989)

Roskos-Ewoldsen, B., McNamara, T.P., Shelton, A.L., Carr, W.: Mental representations of large and small spatial layouts are orientation dependent. Journal of Experimental Psychology: Learning, Memory, & Cognition 24, 215–226 (1998)

Shelton, A.L., McNamara, T.P.: Multiple views of spatial memory. Psychonomic Bulletin & Review 4, 102–106 (1997)

Sholl, M.J., Nolin, T.L.: Orientation specificity in representations of place. Journal of Experimental Psychology: Learning, Memory, and Cognition 23, 1494–1507 (1997)

Werner, S.: Cognitive reference systems and their role in designing spatial information displays. Künstliche Intelligenz 4, 10–13 (2002)

Waller, D., Montello, D.R., Richardson, A.E., Hegarty, M.: Orientation specificity and spatial updating of memories for layouts. Journal of Experimental Psychology: Learning, Memory, & Cognition 28, 1051–1063 (2002)

Wraga, M.: Thinking outside the body: An advantage for spatial updating during imagined versus physical self-rotation. Journal of Experimental Psychology: Learning, Memory, & Cognition 29, 993–1005 (2003)

Mechanisms for Human Spatial Competence

Glenn Gunzelmann[1] and Don R. Lyon[2]

[1] Air Force Research Laboratory
[2] L3 Communications at Air Force Research Laboratory
6030 South Kent Street
Mesa, AZ, United States
{glenn.gunzelmann,don.lyon}@mesa.afmc.af.mil

Abstract. Research spanning decades has generated a long list of phenomena associated with human spatial information processing. Additionally, a number of theories have been proposed about the representation, organization and processing of spatial information by humans. This paper presents a broad account of human spatial competence, integrated with the ACT-R cognitive architecture. Using a cognitive architecture grounds the research in a validated theory of human cognition, enhancing the plausibility of the overall account. This work posits a close link of aspects of spatial information processing to vision and motor planning, and integrates theoretical perspectives that have been proposed over the history of research in this area. In addition, the account is supported by evidence from neuropsychological investigations of human spatial ability. The mechanisms provide a means of accounting for a broad range of phenomena described in the experimental literature.

Keywords: Spatial Cognition, Cognitive Architecture, Computational Model, Frame of Reference, Vision, Representation, Mechanism, ACT-R.

1 Introduction

In this paper, we present a broad theoretical architecture for understanding human spatial competence. Human spatial abilities are brought to bear in a variety of contexts, and in a variety of ways. Spatial information processing is utilized for navigation and wayfinding [1], [2], map reading and orientation [3], [4], [5], and spatial transformations like mental rotation [6], [7]. However, spatial abilities are also recruited for syllogistic reasoning tasks [8], problem solving [9], [10], and language processing [11], [12]. This flexibility and diversity requires that an account of human spatial abilities be able to address a range of specific abilities within the context of overall cognitive functioning.

In addition to breadth, an understanding of human spatial competence requires a grasp of the details of the mechanisms involved in encoding, processing, and using spatial knowledge. This includes questions concerning how spatial information is represented, as well as the mechanisms that are available for manipulating those representations [13], [14], [15]. The literature contains many theories that address various aspects of spatial information processing, including representations of environmental information [11],

[16], [17], visuospatial working memory [18], [19], reasoning with spatial mental models [20], [21], mental imagery [14], [22], and navigation [23], [24], [25]. What currently does not exist, however, is an integrated theory that provides an account of human performance across different domain areas.

The theory presented in this paper addresses each of these general areas of human spatial competence, to provide broad coverage on how humans encode, store, and use spatially-based information to perform a variety of tasks in different domains. Because of the scope of the challenge, we have tried to strike a balance between presenting the breadth of the theory, while describing the components in sufficient detail to permit a thorough evaluation. We have grounded the account in the ACT-R cognitive architecture [26], which provides a well-validated theory of overall human information processing. We do this to connect our work to a more general theory of human cognition. This provides us with important constraints on our account and allows us to focus more specifically on mechanisms for spatial information processing, since the existing ACT-R architecture provides validated mechanisms for other critical components of the human cognitive system. Although the mechanisms we propose are not implemented yet, they are specified in enough detail to identify accounts for various phenomena, some of which are described briefly in the remainder of this chapter. To begin, we address several important issues in the realm of spatial competence in the next several subsections. Dealing with critical concepts from the literature at the outset hopefully will clarify our approach and simplify the discussion of other points in the remainder of the paper.

1.1 The Cognitive Map

Tolman's seminal article, "Cognitive Maps in Rats and Men" [27], is generally associated with the origin of modern research into spatial information processing. Since then, the term *cognitive map* has played a central role in theorizing about human spatial abilities. Many theories have been developed that claim humans automatically generate an exocentric cognitive map of the environment based upon experience in a space (c.f. [28], [29], [30]). Proponents of these theories have pointed to the discovery of place cells in the rat [31] and human [32] hippocampus as key evidence for this view. The alternative that is most commonly offered is egocentric encoding of spatial information, where the locations of items in the environment are encoded with respect to the coordinate system defined by the location and orientation of the viewer (e.g., [11], [33]).

Evidence has accumulated on both sides of this debate (e.g., [34], [35], [36], [37]). However, we find the evidence arguing against the exocentric cognitive maps as the default representational format for human spatial representations to be compelling. This is not to say that humans can not or do not sometimes represent space using exocentric reference frames. Rather, our claim is that humans do not *automatically* construct a *cognitive map*[1] of the environment based on visual perception. Instead, we believe that spatial information is encoded in a fragmented manner by default, using

---

[1] We use the term 'cognitive map' to refer to the notion of an internal, exocentric representation of space that is akin to a paper-based map. While the term initially held a much broader connotation, this has been largely lost in current usage.

multiple coordinate systems to represent spatial locations. Initially, early vision utilizes a retinotopic coordinate system, which can be used for guiding and directing visual attention [38]. We propose that the perceptual system generates two enduring, high-level encodings of spatial location from visual input, one based on the egocentric frame of reference (distance & bearing from self), and one based on a frame of reference defined by salient features of the environment (e.g., the boundaries of a room or a prominent landmark). The evidence for these representations comes from functional considerations, described next, and findings from neuropsychological research (Section 4).

Importantly, egocentric and exocentric frames of reference support different functions within the system (e.g., [33]). Encoding location with respect to an egocentric frame of reference facilitates acting on objects in the world ([17], [39], [40]). To interact with an object, it is critical to have knowledge of the relationship between oneself and the object. In addition, this representation of location is a primitive in visual perception, where perceived distance and bearing of an object can be inferred directly from the visual stimulus [33]. In contrast, location information based upon an exocentric frame of reference is important for grounding spatial information in the environment and for computing spatial relations. For these tasks, it is necessary that locational information be represented within a common coordinate system. The egocentric reference frame is not appropriate for such tasks, since any movement or rotation by the viewer produces a change to the coordinate system [33]. Thus, location information based on an exocentric reference frame is needed to link locational information for multiple objects for making spatial judgments. Spatial processes, in conjunction with imagery, can be applied to generate more complex representations for multiple objects from these elements as well (e.g., a cognitive map). However, this is an *effortful* process that inherits the *error* and *bias* that is associated with human visual perception, not an automatic, unconscious process providing an integrated representation of the environment.

1.2 Hierarchical Encoding

There is substantial evidence for a hierarchical component to spatial information processing (e.g., [41], [42], [43]), and any serious theory of human spatial competence needs to account for these findings. In our account, hierarchical phenomena arise as a consequence of the frames of reference used for visual encoding. A frame of reference is used to encode visual information, based upon the contents of the visual experience. To take a famous example from Stevens & Coupe [43], when studying a map of the United States, San Diego will tend to be encoded with respect to the state of California, and Reno will tend to be encoded with respect to the state of Nevada. To compare the relative locations of these two cities, however, requires that they be positioned within the same frame of reference. In this case, it is necessary to shift to the United States as the frame of reference. The relative spatial locations of the two states within the United States will lead to the typical error (i.e., believing that Reno is farther east than San Diego, when it is actually farther west).

We are unable to provide a full discussion of the mechanisms that would support these operations in this paper. However, the key point with regard to hierarchical encoding is that each item encoded by the system is represented within an exocentric reference frame based upon local, salient features of the environment. Hierarchical phenomena arise because that reference frame is then represented as an item in a larger reference frame. Thus, San Diego (the item) is positioned at a particular location within the state of California (the reference frame). However, California occupies a particular location within the United States. Our assumption that spatial comparisons must be carried out within the same reference frame provides the explanation for why various hierarchical phenomena are found in spatial tasks. Mentally re-encoding location relative to a new reference frame takes time and results in increased error and bias.

1.3 The Imagery Debate

Finally, mental imagery has generated a substantial amount of research and theorizing throughout the history of psychology [13], [14], [15], [22], [44], [45]. A major issue under debate has been whether visual mental images are depictive. That is, do mental images have a spatial extent (in the brain) that preserves the spatial properties of the original stimulus? More generally, the question concerns an issue of whether mental images are encoded in a format that is distinct from other kinds of information stored in the brain.

To resolve this issue, we look to the representations and mechanisms in the ACT-R architecture. ACT-R posits a number of processing modules, which are responsible for different aspects of cognition. In the architecture, there is a vision module, which is specialized for processing visual perceptual information. We agree with Kosslyn and others that mental imagery utilizes many of the same cortical areas and neural pathways as vision [22], [46]. Consequently, our theory tightly couples mechanisms for mental imagery with existing architectural mechanisms for visual perception. The result is that vision and mental imagery operate on the same representations, which are different from other information in declarative memory. It is interesting to note, however, that this distinction is based largely on content. All declarative knowledge in ACT-R, including visual chunks, is represented propositionally. Thus, while visual knowledge is distinct, the representation is not necessarily qualitatively different from other knowledge in memory. This speaks to the more detailed issue of whether visual mental images are depictive in a real sense. One reason for propositional representations of visual information in ACT-R is the architecture's relatively abstract and lean representation of visual information. However, it is also the case that propositional representations are more in line with the existing architecture. To the extent possible, we are working within the overall structure of the architecture, until evidence arises that forces us to rethink some of these assumptions. For now, we believe that the representation of visual information currently instantiated in ACT-R provides an adequate foundation that supports the additional representational components and mechanisms we intend to implement.

2 Unified Theories of Cognition and ACT-R

Cognitive architectures, like ACT-R, EPIC, and Soar, instantiate a theory of the human information processing system in its entirety. These unified theories of cognition [47] contain mechanisms to account for various aspects of human cognitive functioning, including problem solving, perception, and motor actions [26], [47], [48]. One of the challenges associated with developing a cognitive architecture is identifying an appropriate set of mechanisms, which are not only capable of producing solutions to a broad range of tasks faced by humans, but which solve those tasks in a psychologically plausible manner. Because of the prevalence of spatial information processing in human cognition and performance, it is critical to incorporate mechanisms for spatial processing in these theories, particularly as cognitive architectures are applied to increasingly complex, spatially rich tasks. In addition, however, it is vital that theories of spatial competence take seriously the constraints imposed by other components of the human cognitive system, many of which have been implemented in cognitive architectures. Human perception and action is constrained in ways that can significantly influence performance on spatial tasks. In addition, human cognitive limitations, like working memory capacity and long-term memory decay moderate how spatial information is processed and remembered. In short, theories of human cognition cannot ignore spatial information processing, just as theories of spatial competence must take into account other perceptual, cognitive, and motor mechanisms.

For the most part, unfortunately, these research communities have remained disconnected. Our intent is to incorporate what is known about human spatial competence into a cognitive architecture to facilitate developing more precise, and psychologically valid, quantitative accounts of human performance on complex, spatially-demanding tasks. Researchers in the area of spatial cognition have developed a variety of theories to account for human performance in different spatial information processing domains (e.g., [19], [20], [22]). These theories capture important capacities and limitations of human spatial ability. However, they are often not implemented. And, when they are, they are typically not implemented as part of a more comprehensive theory of human cognition (e.g., [21], [49], [50]). In the remainder of this paper, we describe our proposal for linking the insights of this research to a sophisticated, yet general, computational theory of the human information processing architecture.

2.1 ACT-R

A full description of the ACT-R architecture is beyond the scope of this chapter. Thus, only a brief sketch is given here. More detailed descriptions can be found elsewhere (e.g., [26], [51]). ACT-R is a cognitive architecture with a set of core mechanisms that has been used to provide accounts of human performance across a broad range of research domains (see [51] for a review). At the highest level, ACT-R is a serial production system where productions (condition-action pairs) are matched against the current state of the system. On each cycle, a single production is selected and executed (fired), which produces a change in the state of the system, and the cycle begins again. The current state in ACT-R is defined by the contents of a set of buffers. Each buffer is associated with a specialized processing module, and serves as the interface between the module and the production system. We mentioned the vision module above, which has a buffer to represent object properties (*what*), and a second buffer to represent location information (*where*). There is also a declarative memory module with a retrieval buffer, which is specialized for storing and processing declarative knowledge (facts and information stored as *chunks*). Each buffer may hold only a single chunk at any given time, and each module can process only a single request at a time. Thus, modules and buffers are serial as well. Parallelism exists in ACT-R through the simultaneous operation of all of the modules. Subsymbolic mechanisms are implemented within the modules and produce a graded quality in cognitive processes. The speed and accuracy of operations are impacted by continuously varying quantities, like activation for declarative knowledge and utility values for productions.

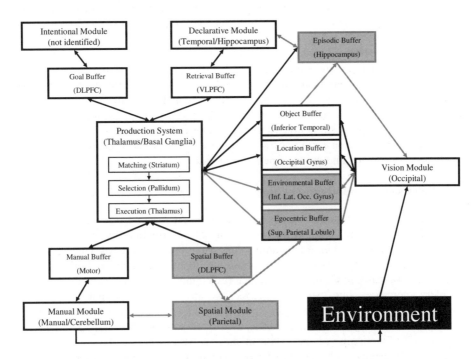

Fig. 1. Schematic illustration of the current ACT-R architecture, with proposed additions included. Structures identified in white represent existing components of the architecture. Grey components represent proposed additions. The *environment* is indicated in black.

The modules generally are driven by requests from the production system. For instance, a production may request a shift of visual attention. The module processes the request and returns the result to the buffer, where it can be accessed by the production system. In the case of shifting attention, the vision module plans and executes the action, and a chunk representing the item being attended is placed into the visual-object buffer. Figure 1 illustrates the major components of the current ACT-R architecture, along with the additions that are proposed in this paper. The current version of ACT-R (ACT-R 6.0) has been designed and implemented to support adding, modifying, or deleting components, out of an appreciation of the limitations of the current architecture and interest in having research explore alternative accounts of cognitive phenomena [26]. This makes ACT-R well-suited for exploring how to account for human spatial competence in the context of a unified theory of cognition. In the next section, we describe our suggested modifications and additions, including how they integrate and interact with the existing architecture.

3 An Architectural View of Spatial Competence

Our account of spatial competence in ACT-R consists of proposals to add a module and several buffers to the architecture, in conjunction with mechanisms to support the kinds of processing performed by humans on spatial information. In general, this proposal is in line with the existing architecture, and with existing practice within the ACT-R community. One exception to this is an explicit proposal for direct communication between modules. Although such connections do not exist in the architecture currently, there is a recognition that they are likely to exist, based both on human performance and neuroanatomy (Anderson, personal communication). We propose a close link between the new spatial module and other modules in the architecture, particularly the vision and motor modules. Overall, we have taken care to ensure that the proposal is consistent, both internally and with ACT-R. Thus, we are confident that the emerging account provides a useful conceptualization of how humans encode, store, and process spatial information.

3.1 Enhanced Visual Representation

The existing representation of visual information in ACT-R is based substantially on the EPIC architecture [48]. It represents visual information by splitting *object* information from *location* information, following the research of Ungerleider & Mishkin [52]. However, these representations are impoverished, due to both historical and technical reasons. Cognitive architectures certainly have not solved the vision problem, nor does our theory. However, we propose to augment the existing representation of visual information, specifically location information, to provide a more psychologically valid representation that is able to support spatial operations.

The basic functioning of the vision module in ACT-R is that the contents of the screen are processed into the *visual icon*, which is a transient representation in a retinotopic frame of reference (actually, locations are based on screen coordinates out of convenience), which is similar to a feature map [53]. Although the ACT-R visual icon is not depictive in the sense that Kosslyn's [22] visual buffer is, we propose that the icon serves similar functions with regard to the construction and use of visual imagery (Section 3.2). Shifts of attention in ACT-R occur when a production includes a request for an attention shift, which specifies constraints on where attention should go. These constraints can be based on *location* (e.g., to the right of where attention is currently), and *features* of the objects displayed (e.g., only blue objects). The constraints are compared to the information available in the icon, and the items that match those constraints are identified. One of the items matching the request is selected (randomly if there are multiple items that match), and attention is shifted to the new location. Once attention "arrives" the visual buffers are populated with chunks representing information about the object. The timing of these operations is based on a vast psychophysical literature.

Egocentric Buffer. The first step in augmenting the representation of location in ACT-R consists of adding a buffer to hold a representation of the egocentric location of the object (the egocentric buffer in Figure 1). Currently, ACT-R's visual-location buffer holds the location of the object using screen-based coordinates. This buffer includes other featural information about the object, including its size, color, and type (e.g., *text* versus *button*), which corresponds roughly to the information available in the visual icon. In practice, this representation is used primarily to support visual search, but it also supports processing of 2D displays, as are commonly used in psychological experiments.

What the existing representation does not support is encoding location in 3D space. Historically, this has not been problematic, since ACT-R (like other cognitive architectures) generally has not been applied to tasks involving complex, 3D environments. This kind of task environment, however, is becoming increasingly common as the applications of work on computational cognitive modeling continue to expand (e.g., [54], [55], [56]). To address this shortcoming, we propose adding an egocentric buffer to hold 3D spatial information. Information encoded in this buffer includes the distance of an object, as well as its bearing, relative to the location of ACT-R in the environment. It also includes an estimate of the absolute size of the object (i.e., not retinal size), as well as the orientation of the object and information about motion (speed and direction). Like existing buffers in the vision module, this information is encoded and updated when visual attention is shifted to a particular object. Note that the existing visual-location buffer remains essential. We believe that visual information represented at the level of features in a retinotopic frame of reference is necessary in the control of visual attention.

Environmental Frame of Reference Buffer. As important as an egocentric encoding of location is for immediate action and processing, it does not provide any information about the location of an object relative to other objects in the world. In Section 1.1, we indicated that representing location information utilizes multiple frames of reference. One of these is based upon the surrounding environment. In virtually any space, there are distinct features that provide a frame of reference for encoding relative locations of objects. This may be a landmark, like the Eiffel Tower in Paris, or geographic feature, like the Pacific Ocean on the California coast in the United States. We propose that the human visual system takes advantage of these features to provide a stable frame of reference for encoding object location. An interesting question exists regarding how a particular reference frame is selected within an environment when multiple options are generally available. We believe that salience plays a key role in this process. However, we suspect that there are large individual differences in this process, which may contribute to performance differences in orientation and navigation tasks [57], [58].

The contents of the environmental buffer provide a basis for calculating spatial relationships among objects. Some proposals suggest that these quantities are computed automatically when visual attention is shifted from one object to another (e.g., [59]). In contrast, we believe that identifying the spatial relationship between two objects is an explicit process of estimation. However, if objects are represented in an exocentric frame of reference, such estimates need not be difficult to compute, and can be determined using immediate visual perception or from memory using mental imagery. So, while you may not have explicitly considered the distance between the phone in your office and the door, you can recall the locations of both items (within the reference frame of your office) from memory and compute that relationship with relative ease when asked. If items are located in different frames of references (e.g., the stove in your kitchen and the computer in your office), additional effort is needed to establish a common frame of reference, but the process will be similar. Mental imagery, described below, supports these operations.

Episodic Buffer. The final modification we propose to the vision module provides a means for consolidating visual experiences into a unitary representation. We accomplish this by proposing an episodic buffer, which links the contents of the visual buffers, and produces an episodic trace of the experience. Our proposal bears significant resemblance to Kosslyn's conceptualization of the role of the hippocampus in representing episodic information [22]. Specifically, we do not propose that all of the information related to a visual experience is represented in this buffer. Rather, we propose that this buffer holds a chunk that encodes pointers to the contents of the other visual buffers (other chunks).

The resulting vision module should operate as follows. When attention is shifted to a new object in the visual field, the vision module updates the visual-object, visual-location, environmental, and egocentric buffers with chunks that represent the information about the object being attended. These processes occur in parallel, through distinct mechanisms in ACT-R, and distinct cortical pathways in the brain (see Section 4). Identifiers for those chunks are specified as slot values in a chunk in the episodic buffer, linking them together in a single episodic representation. All of these chunks are deposited into declarative memory, making the information accessible at later times. In addition, these chunks also are subject to the same activation learning and decay mechanisms as other chunks in memory, meaning that perceptual experience can be forgotten much like other information. These mechanisms already exist in ACT-R. The chunks stored in memory form the basis for mental imagery, which is discussed next.

3.2 Mental Imagery

There is a great deal of evidence suggesting that engaging in mental imagery recruits many of the same areas of the brain as visual perception (c.f., [22]). Based on this literature, we find it appropriate to posit a close link between visual perception and imagery. In fact, our claim, in line with Kosslyn, is that mental imagery does not reflect a distinct and separable component of human cognition. Rather, in this architecture, mental imagery operates through the mechanisms associated with visual perception. Note that in Figure 1, there is no imagery module and no imaginal buffer, in contrast with Anderson et al. [26]. We achieve the functionality associated with mental imagery through the interaction of the vision module with a spatial module that incorporates default features of modules within ACT-R.

Images are generated in this architecture by retrieving episodic perceptual experiences from memory. This process works similarly to retrieving declarative knowledge in ACT-R, such that a request is generated by the central production system to retrieve an episodic chunk from memory, which is placed in the episodic buffer, with pointers to the chunks associated with this memory from the other visual buffers. These chunks can be retrieved based on these references and the information can be propagated to the visual icon, where it is reinstantiated. This process causes attention to be "pulled from" the external environment and focused on this internally-generated visual representation. The result is, essentially, a copy of the original visual experience. However, a variety of errors could occur in this process, including mis-retrievals, which would affect the characteristics of the mental image.

As the preceding description suggests, we posit that mental images are represented at the level of the visual icon in ACT-R. In the current implementation of ACT-R, this is a propositional representation that contains feature-based information about objects, including spatial location, color, and size [51]. Because mental images are effortful to maintain, we posit that the visual icon has a rapid decay rate. It is only by refreshing information that it can be maintained. For items in the visual world, this is effortless, since the electromagnetic radiation impinging on the retina provides constant input regarding visual information in the environment. In the case of mental images, however, attention is required to maintain the image for any significant length of time. As we implement this architecture, we will adapt ACT-R's existing declarative memory decay function for use with this component of the system, with an appropriately higher decay rate.

Of course, mental images can be modified and transformed. The mechanisms available for performing these transformations are described next. We simply note here that the primitive transformations available for manipulating mental images relate to slot values in the chunks created during visual perception. Whereas a variety of transformations are possible, we focus in the next section on *spatial* transformations, which are generated through changes to slots in the chunk in the egocentric buffer.

3.3 A Specialized Module for Processing Spatial (and Magnitude) Information

Spatial information processing is a component of many tasks and cognitive activities. Thus, an account of these abilities in humans must be both general and powerful. The modifications to the vision module described in Section 3.1 are essential to providing a robust representation of object locations in the environment. However, there are no mechanisms in the vision module that directly support spatial transformations, estimations, or calculations. While our discussion centers around processing spatial information obtained through visual perception, we accept that the mechanisms may generalize to other concepts, like volume in auditory perception, brightness in color perception, or intensity in taste, smell, and touch. Since ACT-R has only rudimentary auditory and vocal abilities (and no sense of taste, smell, or touch), these issues, in large part, cannot be addressed in the current architecture. They do, however, offer an interesting direction for future empirical and modeling research.

We propose that mechanisms for processing spatial information are instantiated within a specialized module in ACT-R. There are several key components of this module, which we will examine in turn. The degree to which the different capabilities are actually separable from a cognitive or neuropsychological perspective needs to be carefully evaluated with additional research. In this section, we will attempt to differentiate them while simultaneously providing evidence that they are interdependent. These components of spatial ability include spatial transformations, magnitude estimations, and magnitude computations.

Spatial Transformations. Humans have the ability to maintain and manipulate mental images to perform a number of tasks and activities, from mental rotation [7], to image composition [60], to mental simulation [61], [62]. The ability to transform images and inspect the results is an essential component of spatial reasoning [49]. However, humans are capable of many different kinds of complex spatial image transformations, some of which depend heavily on knowledge and experience and/or are specific to particular object classes (e.g., compressing an accordion-like or spring-like object). Rather than address this broader class of complex transformations, we will initially model a small number of basic, but very frequently used transformations.

Perhaps the best-studied transformation to discuss is mental rotation. Although early research suggested that whole objects might be mentally rotated through intermediate states in a depictive representation [7], subsequent research argues for a more flexible process that can involve focusing on individual object parts, increasing their imagined size if necessary to make a fine discrimination, and changing their visualized position and orientation [6], [22]. Therefore we do not plan to model basic transformations such as size, position and orientation by directly moving the constituent points of an object across the visual icon. Rather, production rules will select a relevant object or object part as the focus of attention, resulting in its position being represented in the egocentric buffer. The production system will also select goal-relevant image transformation processes to alter the relevant slot values of the selected object/part. These transformations include translations, zooming, and mental rotation, which can be achieved by manipulating an image's *distance* and/or *bearing*, *size*, and *orientation* in the egocentric buffer, respectively. The vision module, using direct module-to-module links, recruits the spatial module to perform the operations.

The role of the spatial module in this case is to perform the requested transformations, producing alterations to the representation of the object in the visual icon. In many cases, this will be a complex, iterative process involving several objects or parts at different scales and in different locations. Often the next transformation subgoal will be determined only after the system inspects the results of the previous transformation, so image inspection processes will go hand-in-hand with spatial transformations. In addition, the decay properties of mental images, and perceptual refresh rates of visual stimuli will impact the size of the subgoals and other aspects of these transformation mechanisms. ACT-R already contains processes which control the inspection of simple visual information via the allocation of attention to locations and features represented in the icon.

Although the number of simple transformations that we will model is small, they can be combined using the process just described to create complex manipulations of mental images in the service of spatial cognition. An example of the usefulness of such transformations can be found in an analysis of the performance of expert meteorologists (e.g., [62], [63]). The comments of meteorologists upon viewing a static display of weather patterns give evidence that they are generating weather predictions by imagining a complex series of transformations to the size, location, and orientation of regions affected by various meteorological events.

Magnitude Estimations. Just as the transformation component of the spatial module is closely linked with manipulating mental images, magnitude estimations are associated with encoding information using vision or mental imagery. According to Klatzky [33], estimates of egocentric distance and bearing are primitive values in egocentric frames of reference. In addition, however, humans are able to estimate these relations between arbitrary pairs of objects in the environment, which can be useful for many purposes, including navigation [64]. Of course, there is bias and error associated with these operations, but people are still able to achieve a relatively high degree of accuracy. These processes also appear to be involved in planning and executing motor movements, like reaching to grab a coffee cup on the desk in front of you (e.g., [65]). Estimating magnitudes is a basic function of the spatial module, and is another point at which we posit significant module-to-module communication. In reaching to pick up a coffee cup, for example, detailed spatial information is necessary to plan and execute the appropriate motor actions to grasp the cup. Moreover, people perform these actions precisely, without conscious awareness of the spatial information that is influencing their motor movements [66]. Thus, we believe that these interactions occur through cortical pathways outside the main production cycle in ACT-R.

Under this perspective, the production system of ACT-R is responsible for formulating the high-level action, like "pick up the coffee cup." The motor module is then responsible for determining how to perform that action, which involves interaction with the spatial module to plan the details of the motor movements. Research by Brooks [67] suggests that spatial information processing is required to plan motor movements. It also suggests that there is an overlap between these mechanisms and the mechanisms required to generate, maintain, and inspect mental images, which were described above.

We believe that magnitude estimation is utilized by the vision module as well, when a new item is attended in the environment. When a shift of attention occurs, information about the distance and bearing of the object with respect to ACT-R's location in the environment must be computed. We propose that these operations recruit the spatial module as well. It also may be the case that this component of the spatial module is involved in planning and executing eye movements to bring new items into the focus of attention (e.g., [68]). As noted above, these mechanisms may be applicable more broadly for computing magnitudes other than spatial quantities. However, consideration of those possibilities is beyond the scope of this proposal.

Lastly, the mechanisms of estimation can be engaged by the production system, through the spatial buffer. An explicit attempt to estimate a distance or bearing from one object to another in the environment would be an example of how central cognition may utilize these mechanisms. Such a request would result in a chunk, returned into the spatial buffer, which identifies the objects and the relationship between them. Such explicit requests form the basis of generating a *cognitive map* within this architecture. This set of mechanisms can also compute qualitative estimates of magnitudes, like *close, above, small,* and *far.* Some research has suggested that qualitative (categorical) operations like this are localized in the left hemisphere, while quantitative (continuous) operations (e.g., distance/bearing estimates) are performed in the right hemisphere [69], [70].

Magnitude Computations. In some circumstances, like planning an attention shift or a motor movement, estimates of magnitudes can be useful in isolation. However, some of the most important functions of spatial cognition involve performing computations involving multiple magnitudes. There are a variety of computations that may be performed on magnitude information, including qualitative comparisons (e.g., <, >, =) and quantitative operations (e.g., +, -, /, *). Again, these different types of operations may be performed in different hemispheres in humans (c.f. [69], [70]). This is a sophisticated, and potentially extensive, set of operations to be performed on quantitative information.

We propose that these functions are computed using another set of mechanisms within the spatial module. There is also evidence that these operations are conducted on abstract representations of magnitude, rather than using information embedded in vision (e.g., distance and bearing) or other modality. Neuropsychological research has shown that the angular gyrus, in the posterior inferior parietal lobule, is implicated in the processing of spatial and numerical information (e.g., [71], [72]). We take these findings to suggest that quantitative information of this sort is represented in a common format for performing computations like those mentioned above. Thus, we propose that the outputs of estimation processes are in an abstract, propositional form. Comparisons and computations, then, are performed on this abstract representation. These requests are generated through central cognition, utilizing the spatial buffer mentioned above.

4 Spatial Competence in the Brain

Thus far, the discussion of the architecture for spatial competence has centered around the structure and mechanisms required to support spatial information processing within ACT-R. In this section, we present some information regarding the mapping of those structures and mechanisms to particular brain areas. Neuropsychological evidence concerning spatial abilities in humans is extensive. It has been shown that the parietal lobe is critical in processing spatial information, and a very large number of studies have attributed particular aspects of spatial cognition to particular portions of the parietal lobe and other portions of the cortex (c.f., [73]).

A comprehensive review of the neuropsychological evidence concerning spatial cognition is not presented here. What we do provide is an overview of the mapping of the spatial competence architecture to brain regions without considering the mapping of other components of ACT-R to the brain, which has been addressed elsewhere (e.g., [26]). Along the way, we cite important research that supports our position, but generally do not take time to examine all the perspectives. In addition, area delineations and hypothesized locations should be considered as approximate. There is a great deal of complexity in the human cortex, and we do not wish to suggest that cognitive functions are exclusively localized in the regions we suggest, nor do we believe necessarily that these are the only functions performed by the various locations.

Figure 1 above contains our hypothesized assignment of components of our proposal to brain regions. We have placed the egocentric buffer in the posterior parietal lobe, within the superior parietal lobule. This follows research that has identified a distinction between a ventral *where* or *action* stream and a dorsal *what* stream [52], [66]. We view the egocentric buffer as representing the output of the ventral stream. Next, the environmental buffer, which encodes object location with respect to an exocentric frame of reference, is in the inferior portion of the lateral occipital gyrus. This area and nearby areas (including the parahippocampal cortex) have been associated, variously, with acquiring exocentric spatial information [74], representing the local visible environment [75], perceiving and encoding landmarks [76], [29], encoding 'building stimuli' [74], and encoding of 'large objects' [77]. All of these things can be seen to relate to identifying the location of an object with respect to an exocentric frame of reference based on what is visible in the surrounding environment.

We attribute to the hippocampus the role of encoding episodic information about visual experience (although it is plausible, even likely, that this incorporates other sensory modalities as well). This lines up closely with the description of the function of the hippocampus given by Kosslyn [22]. He states, "...the hippocampus may set up the neural equivalent of 'pointers', linking representations that are stored in different loci..." (p. 223). As noted earlier, others have posited other roles for the hippocampus in spatial cognition, particularly with regard to place cells and the *cognitive map*. Space limitations here prevent us from reviewing and commenting on the evidence relevant to this issue.

Spatial operations take place across the parietal lobe, as noted in Figure 1. However, we posit that the different functions we have identified for the spatial module can be localized to different parts of the parietal lobe. Still, even these more specific references represent substantial abstractions. For instance, the superior parietal lobule has been associated with visuospatial working memory operations [78], [79], [80], and we relate this region to the component of the spatial module that performs spatial transformations. The angular gyrus is active in spatial tasks generally [69], [71] and in tasks requiring calculations, particularly mathematics, more specifically [72], [81]. Thus, we associate the angular gyrus with performing magnitude computations. This conceptualization of the function of this area actually provides a unification across some of the different notions of the role of this portion of the cortex. Finally, proposals for the role of the supramarginal gyrus include directing spatial attention (e.g., [68]), mental imagery (e.g., [82]), and motor preparation [65]. All of these functions fit well with the role attributed to this area in our account, which is performing magnitude estimations. Additionally, these operations all rely on a representation of location (following [52]) to support action (as suggested by [66]). So once again, our theory provides a potential unification for seemingly disparate results.

Mental imagery is captured in Figure 1 in the connections between components of the vision module and the spatial module, which indicate processing links between brain regions. We have not yet associated all of these links with particular pathways in the brain, but there is evidence for at least some of them (c.f., [22]). This proposal for the generation and manipulation of visual mental images lines up well with work by Kosslyn. Certainly, many of the details are missing in the current mapping of components of this account onto the brain, but the emerging view is consistent with what we know about mental imagery and neuropsychology.

Finally, the buffer for the spatial module resides in the frontal cortex. The frontal cortex is associated with high-level planning and goal maintenance activities. Additionally, dorso-lateral prefrontal cortex (DLPFC) shows enhanced activity in performing spatial tasks [83], [84]. We view this activity as stemming from the processing requirements of managing requests for spatial operations and harvesting the results of that processing. This anatomical relationship is similar to the proposed mapping of declarative memory to the brain in ACT-R, where the buffer is in ventro-lateral prefrontal cortex (VLPFC) and the actual storage of declarative information occurs in the temporal lobe and hippocampus [26].

In summary, we have accomplished a tentative mapping of spatial information processing structures and mechanisms to brain areas. The selective review we have presented illustrates that empirical and neuropsychological evidence generally supports the mapping we have developed. There is, as mentioned, a vast psychological literature relating to this topic, and there are many neuropsychological phenomena for which this mapping does not provide an account. As we implement the mechanisms, and validate the performance of the entire system against human empirical and neuropsychological data, we will use key findings in this literature to refine the mechanisms that are implemented to account for the processes that are occurring in the brain when humans perform spatial tasks.

5 Conclusion

We have described a set of mechanisms for human spatial competence. The architecture proposed is consonant with existing empirical and theoretical evidence regarding the capabilities and limitations of human spatial information processing, and is also consistent with current knowledge about the functional neuroanatomy of the brain. In addition, our account is integrated with the ACT-R cognitive architecture, which is a well-validated, quantitative theory of human cognition. As the scope of cognitive architectures expand, and as processing limitations of computer technology are overcome, it is critical that psychologically valid accounts of human spatial competence be implemented in cognitive architectures. Incorporating mechanisms for spatial competence will allow cognitive architectures, like ACT-R, to provide quantitative accounts of human performance in a wider range of task environments. This will be critical for achieving the goals of unified theories of cognition [47].

On the other hand, it is also vital that theories of human spatial competence incorporate mechanisms that account for capacities and limitations in human perceptual, cognitive, and motor performance. In any task, it is the interplay of the entire system that produces the behavior that can be observed. By linking our account to ACT-R, we can leverage the mechanisms of a well-validated theory of the human cognitive architecture. Mechanisms for spatial competence fill in a significant gap in ACT-R's capabilities, just as ACT-R provides detailed mechanisms for memory and performance that link spatial competence to human cognition more broadly. Of course, the proposal we have described in this paper does not address every phenomenon in the literature on human spatial information processing, but it does provide an integrated framework that can be applied widely for understanding the capacities and limitations of human cognition in this area. As the structures and mechanisms are implemented, we will focus on the empirical, theoretical, and neuropsychological details, to ensure that our account is psychologically valid. For example, perhaps the processing mechanisms currently grouped within a single "spatial" module are better conceived as a set of 2, or even 3, separate modules that interact in spatial information processing. This has implications for capacities and processes, and these details will matter when the architecture is utilized to provide quantitative accounts of human performance. This and other issues will be addressed as we move forward. A critical point, however, is that a computational model, implemented within a cognitive architecture, is vital for tackling these issues at this level of detail. Thus, we are enthusiastic and optimistic about the potential for generating a unified, comprehensive account of how humans encode, store, and process spatial information.

Acknowledgements. This work was supported by Grant #02HE01COR from the Air Force Office of Scientific Research (AFOSR). Motivation for the scope of the proposal has come from the Defense Advanced Research Projects Administration (DARPA) Program on Biologically-Inspired Cognitive Architectures (BICA). This research has benefited from discussions with Kevin Gluck, Jerry Ball, Greg Trafton, and other members of the ACT-R community.

References

1. Richardson, A., Montello, D., Hegarty, M.: Spatial Knowledge Acquisition from Maps, and from Navigation in Real and Virtual Environments. Memory and Cognition 27, 741–750 (1999)
2. Cornell, E.H., Heth, C.D., Alberts, D.M.: Place Recognition and Way Finding by Children and Adults. Memory and Cognition 22, 633–643 (1994)
3. Aretz, A., Wickens, C.: The Mental Rotation of Map Displays. Human Performance 5, 303–328 (1992)
4. Barkowsky, T., Freksa, C.: Cognitive Requirements on Making and Interpreting Maps. In: Hirtle, S., Frank, A. (eds.) Spatial Information Theory: A Theoretical Basis for GIS, pp. 347–361. Springer, Heidelberg (1997)
5. Gunzelmann, G., Anderson, J.R., Douglass, S.: Orientation Tasks with Multiple Views of Space: Strategies and Performance. Spatial Cognition and Computation 4, 207–253 (2004)
6. Shepard, R.N., Metzler, J.: Mental Rotation of Three-Dimensional Objects. Science 171, 701–703 (1971)
7. Just, M., Carpenter, P.: Cognitive Coordinate Systems: Accounts of Mental Rotation and Individual Differences in Spatial Ability. Psychological Review 92, 137–172 (1985)
8. Gilhooly, K.J., Logie, R.H., Wynn, V.: Syllogistic Reasoning Tasks, Working Memory and Skill. European Journal of Cognitive Psychology 11, 473–498 (1999)

9. Fincham, J.M., Carter, C.S., Van Veen, V., Stenger, V.A., Anderson, J.R.: Neural Mechanisms of Planning: A Computational Analysis Using Event-Related FMRI. Proceedings of the National Academy of Sciences 99, 3346–3351 (2002)
10. Levine, M., Jankovic, I., Palij, M.: Principles of Spatial Problem Solving. Journal of Experimental Psychology: General 111, 157–175 (1982)
11. Franklin, N., Tversky, B.: Searching Imagined Environments. Journal of Experimental Psychology: General 119, 63–76 (1990)
12. Zwaan, R., Radvansky, G.: Situation Models in Language Comprehension and Memory. Psychological Bulletin 123, 162–185 (1998)
13. Anderson, J.R.: Arguments Concerning the Representations for Mental Imagery. Psychological Review 85, 249–277 (1978)
14. Kosslyn, S.M.: Image and Mind. Harvard University Press, Cambridge, MA (1980)
15. Pylyshyn, Z.W.: Mental Imagery: In Search of a Theory. Behavioral and Brain Sciences 25, 157–238 (2002)
16. Sholl, M.: The Representation and Retrieval of Map and Environment Knowledge. Geographical Systems 2, 177–195 (1995)
17. Tversky, B.: Structures of Mental Spaces: How People Think About Space. Environment And Behavior 35, 66–80 (2003)
18. Baddeley, A.D., Lieberman, K.: Spatial Working Memory. In: Nickerson, R. (ed.) Attention and Performance VIII, pp. 521–539. Erlbaum, Hillsdale, NJ (1980)
19. Logie, R.H.: Visuospatial Working Memory. Lawrence Erlbaum Associates, Hove (1994)
20. Byrne, R.M.J., Johnson-Laird, P.N.: Spatial Reasoning. Journal of Memory and Language 28, 564–575 (1989)
21. Lockwood, K., Forbus, K., Halstead, D.T., Usher, J.: Automatic Categorization of Spatial Prepositions. In: Sun, R., Miyake, N. (eds.) Proceedings of the 28th Annual Meeting of the Cognitive Science Society, pp. 1705–1710. Erlbaum, Mahwah, NJ (2006)
22. Kosslyn, S.M.: Image and Brain. MIT Press, Cambridge, MA (1994)
23. Klatzky, R., Beall, A., Loomis, J., Golledge, R., Philbeck, J.: Human Navigation Ability: Tests of the Encoding-Error Model of Path Integration. Spatial Cognition and Computation 1, 31–65 (1999)
24. Siegel, A., White, S.: The Development of Spatial Representations of Large-Scale Environments. In: Reese, H. (ed.) Advances in Child Development and Behavior, pp. 9–55. Academic Press, New York (1975)
25. Wang, R., Spelke, E.: Updating Egocentric Representations in Human Navigation. Cognition 77, 215–250 (2000)
26. Anderson, J.R., Bothell, D., Byrne, M.D., Douglass, S., Lebiere, C., Qin, Y.: An Integrated Theory of the Mind. Psychological Review 111, 1036–1060 (2004)
27. Tolman, E.C.: Cognitive Maps in Rats and Men. Psychological Review 55, 189–208 (1948)
28. Burgess, N., Jeffery, K.J., O'Keefe, J. (eds.): The Hippocampal and Parietal Foundations of Spatial Cognition. Oxford University Press, Oxford (1999)
29. Mou, W., Mcnamara, T., Valiquette, C., Rump, B.: Allocentric and Egocentric Updating of Spatial Memories. Journal of Experimental Psychology: Learning, Memory, and Cognition 30, 142–157 (2004)
30. Shelton, A., Mcnamara, T.: Systems of Spatial Reference in Human Memory. Cognitive Psychology 43, 274–310 (2001)
31. O'Keefe, J., Nadel, L.: The Hippocampus as a Cognitive Map. Oxford University Press, Oxford (1978)
32. Ekstrom, A., Kahana, M.J., Caplan, J.B., Fields, T.A., Isham, E.A., Newman, E.L., Fried, I.: Cellular Networks Underlying Human Spatial Navigation. Nature 425, 184–187 (2003)

33. Klatzky, R.L.: Allocentric and Egocentric Spatial Representations: Definitions, Distinctions, and Interconnections. In: Freksa, C., Habel, C., Wender, K.F. (eds.) Spatial Cognition. LNCS (LNAI), vol. 1404, pp. 1–17. Springer, Heidelberg (1998)
34. Maguire, E.A., Gadian, D.G., Juohnsrude, I.S., Good, C.D., Ashburner, J., Frackowiak, R.S.J., Firth, C.D.: Navigation-Related Structural Change in the Hippocampi of Taxi Drivers. Proceedings of the National Academy of Sciences 97, 4398–4403 (2000)
35. Redish, A.D., Touretzky, D.S.: Cognitive Maps beyond the Hippocampus. Hippocampus 7, 15–35 (1997)
36. Rolls, E.: Spatial View Cells and the Representation of Place in the Primate Hippocampus. Hippocampus 9, 467–480 (1999)
37. Mcnamara, T., Shelton, A.: Cognitive Maps and the Hippocampus. Trends in Cognitive Sciences 7, 333–335 (2003)
38. Felleman, D.J., Van Essen, D.C.: Distributed Hierarchical Processing in the Primate Cerebral Cortex. Cerebral Cortex 1, 1–47 (1991)
39. Harrison, A., Schunn, C.: ACT-R/S: Look Ma, No "Cognitive-Map"! In: Detje, F., Dorner, D., Schaub, H. (eds.) Proceedings of the Fifth International Conference on Cognitive Modeling, pp. 129–134. Universitats-Verlag, Bamberg, Germany (2003)
40. Previc, F.: The Neuropsychology of 3-D Space. Psychological Bulletin 124, 123–164 (1998)
41. Hirtle, S., Jonides, J.: Evidence of Hierarchies in Cognitive Maps. Memory & Cognition 13, 208–217 (1985)
42. McNamara, T.: Mental Representations of Spatial Relations. Cognitive Psychology 18, 87–121 (1986)
43. Stevens, A., Coupe, P.: Distortions in Judged Spatial Relations. Cognitive Psychology 10, 422–437 (1978)
44. Pylyshyn, Z.: What the Mind's Eye Tells the Mind's Brain: A Critique of Mental Imagery. Psychological Bulletin 80, 1–24 (1973)
45. Titchener, E.B.: Lectures on the Experimental Psychology of the Thought-Processes. Macmillan, New York (1909)
46. Sparing, R., Mottaghy, F.M., Ganis, G., Thompson, W.L., Töpper, R., Kosslyn, S.M., Pascual-Leone, A.: Visual Cortex Excitability Increases during Visual Imagery - A TMS Study in Healthy Human Subjects. Brain Research 938, 92–97 (2002)
47. Newell, A.: Unified Theories of Cognition. Cambridge, MA: Harvard (1990)
48. Kieras, D.E., Meyer, D.E.: An Overview of the EPIC Architecture for Cognition and Performance with Application to Human-Computer Interaction. Human-Computer Interaction 12, 391–438 (1997)
49. Barkowsky, T.: Mental Processing of Geographic Knowledge. In: Montello, D.R. (ed.) Spatial Information Theory - Foundations of Geographic Information Science, pp. 371–386. Springer, Berlin (2001)
50. Gopal, S., Klatzky, R., Smith, T.: Navigator: A Psychologically Based Model of Environmental Learning through Navigation. Journal of Environmental Psychology 9, 309–331 (1989)
51. Anderson, J.R., Lebiere, C.: The Atomic Components of Thought. Erlbaum, Mahwah, NJ (1998)
52. Ungerleider, L.G., Mishkin, M.: Two Cortical Visual Systems. In: Ingle, D.J., Goodale, M.A. (eds.) Mansfield, Analysis of Visual Behavior, MIT Press, Cambridge, MA (1982)
53. Triesman, A.M., Sato, S.: Conjunction Search Revisited. Journal of Experimental Psychology: Human Perception and Performance 16, 459–478 (1990)
54. Best, B.J., Lebiere, C.: Spatial Plans, Communication, and Teamwork in Synthetic MOUT Agents. In: Proceedings of the 12th Conference on Behavior Representation in Modeling and Simulation (2003)

55. Gluck, K.A., Ball, J.T., Krusmark, M.A., Rodgers, S.M., Purtee, M.D.: A Computational Process Model of Basic Aircraft Maneuvering. In: Detje, F., Doerner, D., Schaub, H. (eds.) Proceedings of the Fifth International Conference on Cognitive Modeling, pp. 117–122. Universitats-Verlag Bamberg, Bamberg, Germany (2003)
56. Salvucci, D.D., Macuga, K.L.: Predicting the Effects of Cellular-Phone Dialing on Driver Performance. Cognitive Systems Research 3, 95–102 (2002)
57. Gugerty, L., Brooks, J.: Reference-Frame Misalignment and Cardinal Direction Judgments: Group Differences and Strategies. Journal of Experimental Psychology: Applied 10, 75–88 (2004)
58. Mou, W., Mcnamara, T.: Intrinsic Frames of Reference in Spatial Memory. Journal of Experimental Psychology: Learning, Memory, and Cognition 28, 162–170 (2002)
59. Wang, H., Fan, J., Johnson, T.R.: A Symbolic Model of Human Attentional Networks. Cognitive Systems Research 5, 119–134 (2004)
60. Finke, R.A., Pinker, S., Farah, M.: Reinterpreting Visual Patterns in Mental Imagery. Cognitive Science 13, 51–78 (1989)
61. Hegarty, M.: Mental Animation: Inferring Motion from Static Diagrams of Mechanical Systems. Journal of Experimental Psychology: Learning, Memory and Cognition 18, 1084–1102 (1992)
62. Trafton, J.G., Trickett, S.B., Mintz, F.E.: Connecting Internal and External Representations: Spatial Transformations of Scientific Visualizations. Foundations of Science 10, 89–106 (2005)
63. Bogacz, S., Trafton, J.G.: Understanding Dynamic and Static Displays: Using Images To Reason Dynamically. Cognitive Systems Research 6, 312–319 (2005)
64. Waller, D.: Factors Affecting the Perception of Interobject Distances in Virtual Environments. Presence-Teleoperator and Virtual Environments 8, 657–670 (1999)
65. Decety, J., Kawashima, R., Gulyas, B., Roland, P.E.: Preparation for Reaching: A PET Study of the Participating Structures in the Human Brain. Neuroreport 3, 761–764 (1992)
66. Milner, A.D., Goodale, M.A.: Visual Pathways to Perception and Action. Progress in Brain Research 95, 317–337 (1993)
67. Brooks, L.R.: Spatial and Verbal Components of the Act of Recall. Canadian Journal of Psychology 22, 349–368 (1968)
68. Perry, R.J., Zeki, S.: The Neurology of Saccades and Covert Shifts in Spatial Attention: An Event-Related FMRI Study. Brain 123, 2273–2288 (2000)
69. Baciu, M., Koenig, O., Vernier, M.-P., Bedoin, N., Rubin, C., Segebarth, C.: Categorical and Coordinate Spatial Relations: FMRI Evidence for Hemispheric Specialization. Neuroreport 10, 1373–1378 (1999)
70. Kosslyn, S.M., Sokolov, M.A., Chen, J.C.: The Lateralization of BRIAN: A Computational Theory and Model of Visual Hemispheric Specialization. In: Klahr, D., Kotovsky, K. (eds.) Complex Information Processing Comes of Age: The Impact of Herbert Simon, Erlbaum, Hillsdale, NJ (1989)
71. Ardila, A., Concha, M., Roselli, M.: Angular Gyrus Syndrome Revisited: Acalculia, Finger Agnosia, Right-Left Disorientation and Semantic Aphasia. Aphasiology 14, 753–754 (2000)
72. Dehaene, S., Piazza, M., Pinel, P., Cohen, L.: Three Parietal Circuits for Number Processing. In: Campbell, J.I.D. (ed.) Handbook of Mathematical Cognition, pp. 433–453. Psychology Press, New York, NY (2005)
73. Cabeza, R., Nyberg, L.: Imaging Cognition II: An Empirical Review of 275 PET and FMRI Studies. Journal of Cognitive Neuroscience 12, 1–47 (2000)
74. Aguirre, G.K., Zarahn, E., D'esposito, M.: Neural Components of Topographical Representation. Proceedings of the National Academy of Science 95, 839–846 (1998)

75. Epstein, R., Kanwisher, N.: A Cortical Representation of the Local Visual Environment. Nature 392, 598–601 (1998)
76. Aguirre, G.K., D'esposito, M.: Topographical Disorientation: A Synthesis and Taxonomy. Brain 122, 1613–1628 (1999)
77. Logothetis, N.K.: Vision: A Window on Consciousness, pp. 68–75. Scientific American (November 1999)
78. Jonides, J., Smith, E.E.: Working Memory: A View from Neuroimaging. Cognitive Psychology 33, 5–42 (1997)
79. Smith, E.E., Jonides, J., Koeppe, R.A.: Dissociating Verbal and Spatial Working Memory Using PET. Cerebral Cortex 6, 11–20 (1996)
80. Zago, L., Tzourio-Mazoyer, N.: Distinguishing Visuospatial Working Memory and Complex Mental Calculation Areas within the Parietal Lobes. Neurosci Lett. 331, 45–49 (2002)
81. Duffau, H., Denvil, D., Lopes, M., Gasparini, F., Cohen, L., Capelle, L., Van Effenterre, R.: Intraoperative Mapping of the Cortical Areas Involved in Multiplication and Subtraction: An Electrostimulation Study in a Patient with a Left Parietal Glioma. Journal of Neurology, Neurosurgery, and Psychiatry 73, 733–738 (2002)
82. Knauff, M., Kassubek, J., Mulack, T., Greenlee, M.W.: Cortical Activation Evoked by Visual Mental Imagery as Measured by FMRI. Neuroreport 11, 3957–3962 (2000)
83. Funahashi, S., Bruce, C.J., Goldman-Rakic, P.S.: Mnemonic Coding of Visual Space in the Monkey's Dorsolateral Prefrontal Cortex. Journal of Neurophysiology 61, 331–349 (1989)
84. Wilson, F.A.W., O'Scalaidhe, S.P., Goldman-Rakic, P.S.: Dissociation of Object and Spatial Processing Domains in Primate Prefrontal Cortex. Science 260, 1955–1958 (1993)

Algorithms for Reliable Navigation and Wayfinding

Shazia Haque[1], Lars Kulik[1], and Alexander Klippel[2]

[1] Department of Computer Science and Software Engineering, NICTA Victoria Laboratory
University of Melbourne, Victoria, 3010, Australia
s.haque2@pgrad.unimelb.edu.au
lars@csse.unimelb.edu.au
[2] GeoVISTA Center, Department of Geography
Pennsylvania State University, PA 16802, USA
klippel@psu.edu

Abstract. Wayfinding research has inspired several algorithms that compute the shortest, fastest, or even simplest paths between two locations. Current navigation systems, however, do not take into account the navigational complexity of certain intersections. A short route might involve a number of intersections that are difficult to navigate, because they offer more than one alternative to turn left or right. The navigational complexity of such an intersection may require modified instructions such as *veer right*. This paper, therefore, presents a *reliable path algorithm* that minimizes the number of complex intersections with turn ambiguities between two locations along a route. Our algorithm computes the (shortest) most reliable path, i.e., the one with the least turn ambiguities. Furthermore, we develop a variation of this algorithm that balances travel distance and navigational complexity. Simulation results show that traversing a reliable path leads to less navigational errors, which in turn reduces the average travel distance. A further advantage is that reliable paths require simpler instructions.

1 Introduction

Turn right at the next intersection. Most speakers of the English language are not only able to understand this simple instruction, but will also picture a prototypical intersection at which the above mentioned right turn has to be performed [8,25,12]. This prototypical instantiation of a situation model [27] will work for many intersections, especially in grid-like street networks typical for cities in North America. For historic city centers, as they are often found in European cities such as Paris or Rome, this instruction can easily be ambiguous and we might run into a spatial decision problem.

Knowing of the difficulties of natural language descriptions of complex spatial situations, an up-to-date navigation assistance system would not rely on a simple linguistic description alone but provide additional information about how to proceed. Real world examples can easily be found at Internet route planners providing instructions such as: *After 273 m turn right into Weinberg Strasse.*

The ambiguity of this instruction, however, becomes apparent if a traveler encounters an intersection where two streets equally qualify as 'right'. If a linguistic label such as 'right' applies to more than one turn, we call such a turn *instruction equivalent*. We summarize some important issues why the ambiguity of instruction equivalent turns is hard to resolve even if an instruction supplies a street name:

- Not every street bears a street sign indicating its name.
- Traveling by car leaves the unfamiliar driver often with too little time to find and read the respective street sign.
- Driving by night makes the identification of street signs a futile endeavor.
- The street signs of large intersections with many branches are often too far away making them impossible to read.

One solution is to refine the set of instructions. Instead of using only the set {*left, right, straight*} in combination with a {*street name*}, a refinement could, for example, include the modifiers *bear* and *sharp* leading to the instruction set {*sharp left, left, bear left, straight, bear right, right, sharp right*}. Yet, we have found three further indications in the communicative behavior of route direction givers that show the scrutinies that still persist at intersections with instruction equivalent branches.

1. Direction concepts between the main axes (left, right, straight) are not easily associated with a single linguistic term. While direction changes close to 90 degree left and right are referred to as *left* and *right* with little variation, a plethora of composite linguistic expressions is used in between (i.e. around the diagonals) with often subtle differences in the meaning of the hedge terms [26,2,11,16]. We find expressions such as: *turn slightly right, go right 45 degrees, veer right, sharp right bend* and so forth[1].

2. The conceptualization of directions different from the main axes is not straight forward and the sizes and boundaries of sectors, or more generally, the semantics of corresponding spatial prepositions, are ongoing research questions [9,4].

3. People show conceptual and linguistic difficulties with intersection that deviate from prototypical ones. A comparison of intersections with different navigational complexities revealed the following associated linguistic behaviors (see Figure 1, [14]). The more (structurally and navigationally) complex an intersection is:

- the more verbose are verbal route directions,
- the more varied are the verbalizations,
- the more references are made to the structure,
- the more alternative instructions are offered (redundant information),
- the more modifications are applied to reduce ambiguity.

The approach we therefore present in this article to solve this problem is to calculate the route differently: instead of relying on shortest or fastest algorithm, the navigational complexity of an intersection is taken into account. This complexity is based on how many alternatives are available within the same sector in which the branch to take is located. In other words, the complexity depends on how many *instruction equivalent choices* are available at an intersection. The *unreliability* of a turn is a measure of the probability of a navigator making a mistake while taking that turn which is directly related to the navigational complexity and ambiguity of the decision point/intersection. This means that if, for example, the navigator is told to *turn left* and more than one turn

---

[1] The verbalizations are taken from a data set collected at the University of Santa Barbara, California. They are all instructions for the same turning angle of 140.625 degrees.

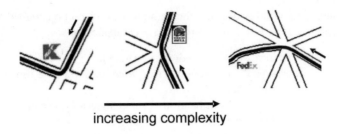

Fig. 1. Increasing navigational complexity of intersections. While the leftmost navigational situation is described as *go all the way down to k-mart and make another RIGHT*, the rightmost navigational situation yields descriptions such as *aand..h.. continue down straight until you co..come to a ..eh .. SIX-intersection.. ROAD.. and youll take thee.. you won.. you will NOT go STRAIGHT, you will go..you will go LEFT on the THIRD (ping).. the third ehm.. intersection. and travel down.. THAT and reach... the destination.*

is present at that intersection that could qualify as a left turn, then the unreliability of that turn increases (depending on the number of similar options available). Accordingly, the unreliability of a route (origin to destination) depends on the unreliability of each turn that the navigator encounters. It has been shown in [23] that drivers are prepared to take suboptimal routes in terms of travel time, if these routes are potentially easier to describe and to follow. Additionally, it is confirmed that successful vehicle navigation systems rely as much on clarity of route instructions as on the length of routes [19].

The remainder of this paper is structured as follows. In Section 2 we introduce our basic algorithm called the *shortest most reliable path algorithm*. Section 3 gives an overview of the simulation environment we used to test the algorithms and to explore the error behavior of a simulated agent. Section 4 details experimental results comparing the classic shortest path algorithm and the shortest most reliable path algorithm. Cognizant of the trade-off between distance and reliability, we introduce a modification of our algorithm to allow for the best balance between travel distance and navigational complexity in Section 5. The new algorithm is compared to the other two in section 6.

2 Shortest Most Reliable Path Algorithm

Our goal is to compute the most reliable route between two locations. Graphs are a standard data structure for representing road and transportation networks. A graph G consists of a set of vertices V and edges $E \subset V \times V$ connecting the vertices. In case of a road network each vertex v represents an intersection and each edge e represents a road. Since the number of roads entering or leaving an intersection is bounded by a small number, we can consider G as a sparse graph which implies that $|E| = \mathcal{O}(|V|)$, where $|E|$ is the number of edges and $|V|$ is the number of vertices in the graph. In a weighted graph a function $w : E \to \mathbb{R}^+$ assigns a weight to each edge $e \in E$. Finding the shortest paths, the paths of least cost between vertices in a weighted graph, is a fundamental network analysis function and a classic problem in computation [6,3].

The following algorithm is based on the simplest path algorithm [7]. The simplest path algorithm aims to minimize the instruction complexity of a route description, i.e., it favors intersections that are easy to describe. It is based on a classification of intersections that assigns a weight to each intersection type [18]. The weight used in the simplest path algorithm represents the instruction complexity to negotiate an intersection. The landmark spider algorithm given in [1] uses the same algorithm as in [7] with the only difference in the weighting function for an intersection, which depends on the distance, saliency, and orientation of a traveler with respect to any landmark present near the intersection. The aim is to generate a *clearest* [1] path in terms of spatial references and landmarks used to describe the route.

The key difference between the shortest and the (most) reliable path algorithm is that a weight w is not only assigned to each edge but also to *each pair of connected edges* in the graph. This approach is inspired by the simplest path algorithm in [7]. The difference, however, is that the weight in [7] reflects the instruction complexity of a decision point, whereas in our approach the weight represents the ambiguity of each turn at an intersection point. The weight r measures how many alternative edges can be considered as instruction equivalent. Since a turn involves two edges, the weight is a function for a pair of connected edges: $r : \varepsilon \to \mathbb{R}^+$, where $\varepsilon = \{((v_i, v_j), (v_j, v_k)) \in E \times E\}$. The weight function r used in the reliable path algorithm characterizes the unreliability measure for a turn represented by an edge pair, i.e., turning from the edge (v_i, v_j) into the edge (v_j, v_k). In addition, the reliable path algorithm takes into account the weight assigned to each edge to compute the shortest most reliable path between the origin and the destination vertex, because there could be many equally reliable paths between the two vertices.

2.1 Computing the Most Reliable Path

The actual shortest most reliable path algorithm, as presented in Algorithm 1, is an adapted version of Dijkstra's shortest path algorithm [6,3]. The input of the algorithm is a graph G that is connected (there is a path from any vertex to any other vertex), simple (there are no edges from a vertex to itself and at most one edge between two different vertices) and directed (each edge is an ordered pair of vertices). The algorithm first initializes all edges connected to the source vertex with an unreliability of zero and distance equal to their respective distance from the source vertex. It also sets the previous connected edge of these edges as NIL. Then, the algorithm iterates through each edge minimizing the cumulative unreliability measure, as well as the cumulative distance from the source (as a second priority). This is achieved in line 8 of Algorithm 1: *competing edges* are all edges that have the same minimum unreliability from the source vertex. Amongst the competing edges the edge which has the minimum distance from the source vertex is selected. The cumulative unreliability as well as cumulative distance from the selected edge to all connected edges is recalculated. The algorithm assigns at each iteration to each edge its previous edge as a predecessor, which is the edge that provides the minimum unreliability and in case of ties the edge that has the shortest distance from the source (see line 15 of Algorithm 1). The predecessor assignment facilitates the reconstruction of the final path. The algorithm iterates until an edge is selected which has the destination vertex d as its terminating vertex. Reconstructing

Algorithm 1. Shortest Most Reliable Path Algorithm

Input : Graph $G = (V, E)$ is a connected, simple and directed graph;
$s \in V$ is the origin vertex; $d \in V$ is the destination vertex;
ε is the set of pairs of (directed) edges that share their "middle" vertex;
$w : E \to \mathbb{R}^+$ is the graph edge weighting function;
$r : \varepsilon \to \mathbb{R}^+$ is the graph turn weighting function;
$w_s : E \to \mathbb{R}^+$ stores the edge weight of the reliable path from s;
$r_s : E \to \mathbb{R}^+$ stores the turn weight of the reliable path from s;
$S = \{\}$ is a set of visited edges; $P = \{\}$ is an ordered set of vertices
Output: The shortest most reliable path $P = \langle s, v_1, \ldots, v_N, d \rangle$

1 $w_s(e) \leftarrow \infty; r_s(e) \leftarrow \infty;$ for all $e \in E$
2 **for** $(s, v_j) \in E$ **do**
3     $r_s(s, v_j) \leftarrow 0$
4     $w_s(s, v_j) \leftarrow w(s, v_j)$
5   previousEdge$(s, v_j) \leftarrow$ NIL
6 **end**
7 **while** $|E \setminus S| > 0$ **do**
8   Find $e \in E \setminus S$ so that amongst competing edges where r_s is minimum, $w_s(e)$ is minimum
9   **if** $e = (v_j, d)$ **then**
10   terminate the loop
11   **end**
12   Add e to S
13   **for** $e' \in E$ **do**
14   **if** $(e, e') \in \varepsilon$ **then**
15   **if** $(r_s(e') > r_s(e) + r(e, e')) \vee ((r_s(e') = r_s(e) + r(e, e')) \wedge (w_s(e') > (w_s(e) + w(e'))))$ **then**
16               $r_s(e') \leftarrow r_s(e) + r(e, e')$
17               $w_s(e') \leftarrow w_s(e) + w(e')$
18   previousEdge$(e') \leftarrow e$
19   **end**
20   **end**
21   **end**
22 **end**
23 Add d to P
24 **while** $e \neq$ NIL **do**
25   Prepend starting vertex of e to P
26     $e \leftarrow$ previousEdge(e)
27 **end**

the shortest most reliable path is then a matter of iterating through the previous edges of d until the previous edge is undefined (NIL), see lines 23–27 of Algorithm 1. As represented in Algorithm 1, the shortest most reliable path algorithm is a *single pair* algorithm. If we exclude lines 9–11 from the algorithm, it iterates until all edges are visited and returns the shortest most reliable paths from the source to all other vertices, i.e., it operates like a *single source* algorithm.

2.2 Computational Complexity

In this section we determine the computational time complexity of the shortest most reliable path algorithm (Algorithm 1). Consider line 8 of Algorithm 1 (which corresponds to the extract minimum operation of Dijkstra's algorithm [6,3]): this operation takes $2|E|$ steps as it has to check all the edges of the graph once, to get the value of minimum unreliability, and again to select the edge with minimum distance amongst those having the same least unreliability. Similarly, in the worst case of a fully connected graph where every pair of vertices is connected by an edge, the FOR loop starting at line 13 would have to compute the unreliability of every edge connected to the selected edge (this operation corresponds to the relax operation in Dijkstra's algorithm [6,3]). In order to compute the unreliability of an edge the algorithm needs to know the orientation of all other edges connected to the selected edge which leads to $2|E|$ steps. Both operations happen $|E|$ times, which leads to a total number of $|E|(2|E|+2|E|)$ steps. However, since geographical networks are sparse graphs with a small limited number of roads at an intersection, the number of steps reduces to approximately $|E|(2|E|)$, which leads to a time complexity of $\mathcal{O}(|E|^2)$.

The shortest path algorithm has a complexity of $\mathcal{O}(|V|^2)$, whereas as the shortest most reliable path algorithm has a complexity of $\mathcal{O}(|V|^4)$ in the case of a fully connected graph because of $|E| = |V|(|V|-1)$. Since most of geographical networks can be considered to be planar graphs, which have a maximum number of $|E| = 3(|V|-2)$ edges, the complexity of the shortest most reliable path algorithm is $\mathcal{O}(|V|^2)$.

The operation of the most reliable simplest path algorithm can be seen as a mapping from the original graph G to a graph $G' = (\widetilde{E}, \varepsilon)$, where $\widetilde{E}$ is the set of edges E ignoring their direction (for details see [7]). The graph G' might not be planar leading to a slower performance by Algorithm 1 as compared to Dijkstra's shortest path algorithm. As compared to the simplest path algorithm, for which the time complexity has been shown to be $\mathcal{O}(|V|^2)$ in [7], the shortest most reliable path algorithm can be slightly slower as it needs to consider the distance as well, whereas the algorithms in [7,1] return the first simplest or clearest paths, respectively.

One way of improving the performance of shortest most reliable path algorithm is to use a binary heap for the operation in line 8, which would lead to $\mathcal{O}(\log |E|)$ steps instead of $\mathcal{O}(|E|)$ leading to an overall time complexity of $\mathcal{O}(|E|\log |E|)$.

3 Simulation and Evaluation of Wayfinding Algorithms

As mentioned above, wayfinding research has lead to a variety of algorithms, each focusing on providing improved navigation assistance for a traveler in its own way. In order to analyze and evaluate these algorithms simulations are necessary. In our case, we were interested in carrying out two classes of simulations: (1) after loading a street network (via a coordinate and adjacency list), execute a particular algorithm for every path present (between two vertices not directly connected) and then analyze the resulting navigational instructions on the basis of distance and unreliability; (2) simulate a human navigator traversing a route (between the same origin and destination), on the basis of instructions provided by different algorithms.

To achieve the above objectives, we developed a simulation environment using Java 2D API. As shown in Figure 2, using this simulation environment we could load a street network graph and select different algorithms to calculate routes. The type of a turn a potential navigator would perceive was based on angles[2] (in degrees) set in the simulation environment. The request for a path could be entered both interactively as well as in batch mode from a file.

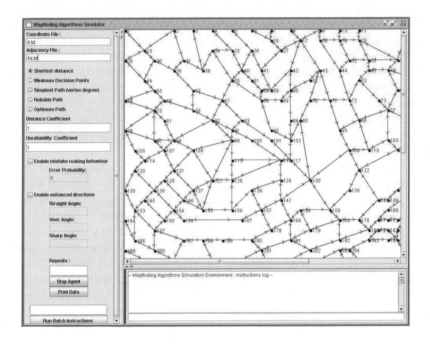

Fig. 2. Wayfinding Algorithms Simulation Environment

The simulated navigator was based on the following assumptions:

- It was error free, in the sense that if it was told to go left it would only choose one of the left turns available (and would not go straight or turn right).
- Depending on the simulation environment user settings it should be able to perceive a turn type based on the angle of the direction change with respect to the current direction of travel.
- The simulated navigator does not make a mistake while executing the first instruction which does not involve a turn.
- In case the simulated navigator made a wrong choice when executing an instruction (correctly) and subsequently was given an instruction which it could not execute,

---

[2] We allowed three types of turns: veer, left/right and sharp depending on angles set for these turns. All turns between $0°$ and the *straight angle* were regarded as straight, between the *straight angle* and the *veer angle* as veer, between the *veer angle* and the *sharp angle* as left/right and between the *sharp angle* and $180°$ as sharp turns by the navigator.

it would realize that it has made a mistake and requery a path to its destination. Similarly, after choosing a wrong turn (while correctly following an instruction) if it ended up at a wrong destination and ran out of instructions, it requeries the path to its destination. A count of requeries reflects the number of times the navigator got 'consciously' lost during the travel.

3.1 Turn Unreliability Weighting Function

The unreliability measure of a turn from edge e into edge e' (both connected by a middle vertex) is given by

$$r(e, e') = \text{count of turns that are instruction equivalent to } e' - 1$$

The type of a turn refers to the linguistic variable used to identify the turn. The unreliability of the entire path R from origin s to destination d is the sum of unreliability measures encountered along the path, given by

$$R(s,d) = \sum_{i=0}^{N-1} r(e_i, e_{i+1})$$

where N is the total number of turns present on the path, $e_0 = (s, v_1)$ and $e_N = (v_N, d)$.

3.2 Comparison of Shortest and Shortest Most Reliable Path Algorithms

Since most of the current automated navigation systems are based on computing the shortest path given a source and destination vertex, we chose to compare the performance of the shortest most reliable path algorithm with that of the shortest path algorithm. This performance evaluation was carried out using the following metrics:

- **Distance traveled:** The distance traveled by the simulated navigator following the path provided by the algorithm.
- **Number of requeries:** This is the number of times the simulated navigator recognized that it is on the wrong path and did a requery from its current location to the destination. This is a measure of the cost that might be incurred by a human navigator sending messages via a mobile device. For our experiment we allowed the navigator up to five requeries per run after which it would finally stop.
- **Stopping distance from destination:** The shortest distance to the destination where the simulated navigator finally stops (after 5 requeries). If the destination was reached successfully this distance would be zero.
- **Total distance:** This is the sum of the distance traveled plus, in case the agent did not reach its destination, the stopping distance from the destination.
- **Missed target travels:** For multiple runs of the simulated navigator, this gives the number of times it failed to reach its destination (after 5 requeries).
- **Actual Unreliability:** This gives the average actual unreliability faced by the simulated navigator while traversing a path (for multiple runs it would give the average value per run per path).

Note that the requery limit and shortest stopping distance to destination metrics actually favor the shortest path algorithm, otherwise considering the unreliability involved on long routes, the simulated navigator could have traveled much more if there were no limit.

4 Experimental Results

Based on the method of path unreliability calculation (see Section 3.1) and using a simple instruction set comprising just 'straight', 'left', and 'right', we evaluated the two algorithms: *shortest path* and *shortest most reliable path*. We tested the algorithms on the street network dataset for the part of the city of Paris comprising more than 120,000 paths (making use of the simulation environment mentioned in Section 3). The results are presented in Figure 3. The bar graphs show for the shortest most reliable path algorithm that most paths had an unreliability of 0 and that the worst unreliability of a path was 4 (for 9 cases). For the shortest path algorithm, most paths had an unreliability between 1 and 10 and the worst unreliability was 15 (10 cases that are barely visible).

Figure 4 details for the same simulation, the ratios of the distances that a navigator would face for individual routes comparing the two algorithms directly. The results show that for more than half of the number of paths the shortest most reliable path algorithm returns a distance 1∼1.4 times (represented as a category distance ratio of 1 in the figure) the distance given by the shortest path algorithm. For around 40000 paths this ratio was between 1.5∼2.4 (represented by a rounded distance ratio of 2 in the figure) and for some rare cases this ratio was significantly higher reaching up to 22.

If the instruction set is more refined, i.e., has more than just 'left', 'right' and 'straight' turns such as 'bear left' or 'turn sharp left', then the calculation of unreliability measure would reflect this finer granularity, i.e., now fewer turns would be ambiguous. This is supported in the current implementation as it only implies specifying more turn types. However, this would come at the cost of higher instruction complexity.

In order to carry out an in depth comparative analysis, using the above mentioned simulation environment, navigator characteristics, and performance metrics, we chose 50 randomly selected paths (origin and destination separated by at least 3 vertices) from the dataset of the part of Paris. We made use of the simple instruction set comprising 'straight', 'left' and 'right' instructions. We ran the simulated navigator 50 times per path with 2 additional conditions for the instruction 'straight', as there is debate whether this direction concept is an axis of a sector [16,12]. It can be assumed, however, that in the absence of competing objects a linguistic expression can be applied more flexible [24] and that in general it can be distinguished between giving an instruction and understanding an instruction. We therefore looked at a second condition for the instruction 'straight' allowing for 12 degrees per side resulting in a 24 degrees sector. The aggregated results for the total 2500 runs are detailed in Table 1. They show the behavior of the simulated agent when it follows the instructions.

As seen from these results, there is a clear advantage for the shortest most reliable path algorithm in terms of missed target travels. Likewise, the stopping distance was closer to the actual destination than in case of the shortest path algorithm. Additionally,

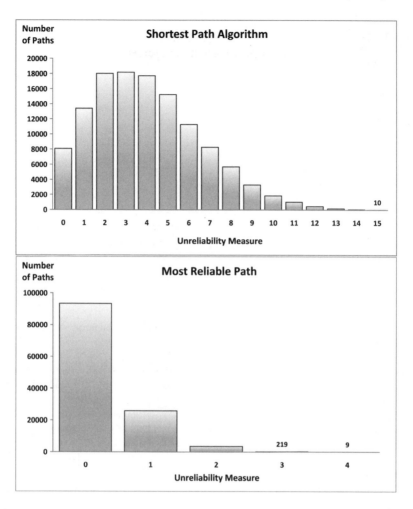

Fig. 3. Comparison of unreliability measures rendered by the two algorithms for the same street network

the number of requeries for the shortest path algorithm was greater by a factor of more than 5. In terms of distance, we see that the total distances traversed by the simulated navigator for the two algorithms are comparable (the distance traversed following the shortest most reliable when the straight angle was set to 0° was only 4.3% greater than the distance of the shortest path; and in case of straight angle set to 12° it was only 6% lesser than the distance of the shortest path). The analysis of individual path runs revealed that for around half the paths (27 for straight angle = 0° and 20 for straight angle = 12°), the distance traversed by the simulated navigator on the reliable path was greater than that traveled while following the shortest route. Amongst these we find for 17 paths in case of the straight angle = 0° and for 13 paths in case of the straight angle = 12° that the reliable distance was more than 10% longer than the shortest distance. A few sample cases are listed in Table 2.

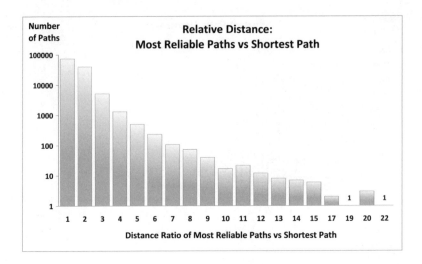

Fig. 4. Number of paths vs rounded ratio of distances

Table 1. Simulation based results for various performance metrics comparing shortest and shortest most reliable path algorithms

Algorithm	Shortest Path	Shortest Most Reliable Path	Shortest Path	Shortest Most Reliable Path
Straight Angle	0°		12°	
Distance Travelled	125719.5	137118.5	106959.5	103114.5
Stopping Distance	6134.1	374.6	2888.5	104.9
Total Distance	131853.6	137493.1	109848	103219.4
Requeries	6773	1291	5209	406
Missed target travel	566	36	319	24
Actual Unreliability	8.3	1.3	5.8	0.5

5 Path Optimization

As seen from the charts in Figures 3 and 4 as well as the simulation results described in Section 4, there are trade-offs meeting the objectives of providing the shortest route and providing a most reliable route to the navigator. The most reliable route has a lower chance of making a mistake due to less ambiguities and reduces the cost associated with requerying the navigation system but it also leads to a greater distance between two locations; computing the shortest route can increase the chance of making mistakes significantly. A ranking of the criteria for human route selection is given in [10]. Amongst those criteria, the overall distance and least time of travel are ranked most highly. Therefore, it is desirable to find a route which optimizes both objectives of shortest distance as well as least unreliability, which in case of an error can result in considerably longer paths.

Table 2. Sample simulation results for four paths comparing Shortest and Shortest Most Reliable Path Algorithm

Algorithm	Unreliability	Requeries	Total distance	Relative distance increase
Shortest	10	226	4246.8	14.7%
Shortest Most Reliable	0	0	4870.5	
Shortest	6	143	2265.78	46.6%
Shortest Most Reliable	1	25	3321	
Shortest	6	134	2237.9	59.0%
Shortest Most Reliable	0	0	3558.5	
Shortest	1	66	1480.6	83.1%
Shortest Most Reliable	0	0	2710.6	

A naïve approach to find a more optimal route would be to consider all paths connecting an origin and a destination and amongst them choose the one which is the most optimal for both criteria. However, considering the worst case of a fully connected graph shows that this strategy is too costly: using the Binomial coefficient $\binom{k}{i}$, which gives the number of generating i unordered outcomes from k possibilities, shows that the maximum number of paths that can exist between an origin and a destination vertex in a connected graph is

$$k!(1 + \sum_{i=1}^{k} 1/i!),$$

where $k = |V| - 2$; $|V|$ is number of vertices present in the graph. The computation of $\mathcal{O}(|V|^k)$ paths is infeasible for practical purposes.

A significantly less costly way of finding an optimal trade-off between the shortest and the most reliable path is to assign an optimum weight to each turn, which depends on both the distance and unreliability involved in taking that turn. This optimum turn weight is given by

$$op(e, e') = \lambda_d w(e') + \lambda_u r(e, e'),$$

where λ_d determines the weight of the impact for the distance of the edge e' and λ_u determines the weight of the unreliability measure when turning from edge e to edge e'. The optimum weight of the entire optimal path OP from the origin s to the destination d is the sum of the optimum weights of the turns encountered on the entire path, i.e.,

$$OP(s, d) = \sum_{i=0}^{N-1} op(e_i, e_{i+1}),$$

where N is the total number of turns present on the path, $e_0 = \langle s, v_1 \rangle$ and $e_N = \langle v_N, d \rangle$. A path P is a ordered series of vertices given as $P = \langle v_0, v_1, \ldots, v_N, v_{N+1} \rangle$ with origin $s = v_0$ and destination $d = v_{N+1}$. The optimal values of λ_d and λ_u depend on the characteristics of the street network and have to be calibrated for each network.

5.1 Optimum Reliable Path Algorithm

The optimum reliable path algorithm, as presented in Algorithm 2, is again a modified form of Dijkstra's shortest path algorithm [6,3]. Note that the initial conditions at the onset of the algorithm are the same as those for the shortest most reliable path algorithm mentioned in Algorithm 1. The only addition is the graph optimum turn weighting function $op : \varepsilon \to \mathbb{R}^+$ and $op_s : E \to \mathbb{R}^+$, which stores the optimum weight of the optimal path from s.

The algorithm first initializes all edges connected to the source vertex with zero unreliability and distance equal to their respective distance from the source vertex. Based on these values, it calculates the optimum weight of these edges. It also sets the connected edge of these edges as NIL. Next, the algorithm iterates through each edge minimizing the cumulative optimum weight from the source and selecting the edge with minimum value of op_s. The cumulative optimum weight from the selected edge to all connected edges is recalculated. The algorithm assigns to each edge its predecessor edge at each iteration to facilitate reconstruction of the final path. The algorithm iterates until an edge is selected, which has the destination vertex d as its terminating vertex. Reconstructing the optimum reliable path is then a matter of iterating through the previous edges of d until the previous edge is undefined (NIL).

5.2 Computational Complexity

The computational complexity of the optimum reliable path algorithm is the same as that for the shortest most reliable path algorithm as discussed in Section 2.2. However, it is slightly faster than Algorithm 1, since it does not have to check all edges twice for extracting the edge with the minimum optimum weight from the source s.

6 Performance Evaluation of the Optimum Reliable Path Algorithm

6.1 Calibration of λ_d and λ_u

In order to calibrate the values of λ_d and λ_u for the street dataset of this specific part of Paris, we ran the optimum path algorithm for various values for λ_d and λ_u, each time for the entire dataset with over 122,000 paths. We compared the results for the different λ_d and λ_u values with those obtained from the shortest path and shortest most reliable path algorithm. This comparison was done on the basis of the ratio of the distances for the optimum path and the shortest path (given by shortest path algorithm) and the difference of unreliability given by the optimum path algorithm and the shortest most reliable path algorithm. The goal of this procedure was to find values that best approximate both, reliability and distance. While favoring distance by high λ_d values (i.e. up to 6) resulted in unreliability measures of up to 14 with the mean value around 4, favoring reliability by high λ_u values created outliers with respect to the distance to travel. For our set of simulations we settled on values of $\lambda_d = 1$ and $\lambda_u = 6$ (see Figure 5).

Algorithm 2. Optimum Reliable Path Algorithm

Input : Graph $G = (V, E)$ is a connected, simple and directed graph;
$s \in V$ is the origin vertex; $d \in V$ is the destination vertex;
ε is the set of pairs of (directed) edges that share their "middle" vertex;
$w : E \to \mathbb{R}^+$ is the graph edge weighting function;
$r : \varepsilon \to \mathbb{R}^+$ is the graph turn weighting function;
$w_s : E \to \mathbb{R}^+$ stores the edge weight of the reliable path from s;
$r_s : E \to \mathbb{R}^+$ stores the turn weight of the reliable path from s;
$op : \varepsilon \to \mathbb{R}^+$ is the graph optimum turn weighting function;
$op_s : E \to \mathbb{R}^+$ stores the optimum weight of the optimal path from s;
$S = \{\}$ is a set of visited edges; $P = \{\}$ is an ordered set of vertices
Output: The optimum reliable path $P = \langle s, v_1, \ldots, v_N, d \rangle$

1 $w_s(e) \leftarrow \infty; r_s(e) \leftarrow \infty; op_s(e) \leftarrow \infty;$ for all $e \in E$
2 **for** $(s, v_j) \in E$ **do**
3      $r_s(s, v_j) \leftarrow 0$
4      $w_s(s, v_j) \leftarrow w(s, v_j)$
5      $op_s(s, v_j) \leftarrow \lambda_d \times w(s, v_j)$
6   previousEdge$(s, v_j) \leftarrow$ NIL
7 **end**
8 **while** $|E \setminus S| > 0$ **do**
9   Find $e \in E \setminus S$ so that $op_s(e)$ is minimized
10   **if** $e = (v_j, d)$ **then**
11   terminate the loop
12   **end**
13   Add e to S
14   **for** $e' \in E$ **do**
15   **if** $(e, e') \in \varepsilon$ **then**
16   **if** $op_s(e') > op_s(e) + op(e, e')$ **then**
17              $r_s(e') \leftarrow r_s(e) + r(e, e')$
18              $w_s(e') \leftarrow w_s(e) + w(e')$
19              $op_s(e') \leftarrow op_s(e) + op(e, e')$
20   previousEdge$(e') \leftarrow e$
21   **end**
22   **end**
23   **end**
24 **end**
25 Add d to P
26 **while** $e \neq$ NIL **do**
27   Prepend starting vertex of e to P
28     $e \leftarrow$ previousEdge(e)
29 **end**

This combination offered advantages for both aspects: most paths have a length close to the shortest distance; additionally, most paths do not encounter unreliable intersections. The maximum difference of unreliability compared to the most reliable path was 4 for very few cases.

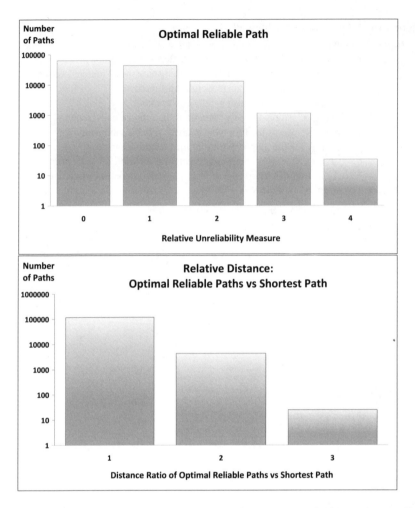

Fig. 5. (top) Number of paths vs difference between optimum path unreliability and most reliable path unreliability. (bottom) Number of paths vs rounded ratio of optimum path distance and shortest path distance. Displayed are the values for $\lambda_d = 1$ and $\lambda_u = 6$.

6.2 Experimental Results

For a comparative in depth analysis of the optimum reliable path algorithm with both, shortest path and shortest most reliable path algorithms, we used the same simulation environment, navigator characteristics, and performance metrics as discussed in Section 3. As parameters for the optimum path we chose the values determined in the study discussed in the previous section, $\lambda_d = 1$ and $\lambda_u = 6$. The aggregated results for 50 runs per route on the same set of 50 routes are presented in Tables 3 and 4.

The data shows that the distance traveled by the simulated navigator when traversing the optimum reliable path was 13.6% and 14.6% shorter than the distance it traversed following the shortest path, for straight angle 0° and straight angle 12°, respectively.

The analysis of the individual path runs revealed the following: for only 10 paths in the case that the straight angle was set to 0° and 12 paths in case of the straight angle was set to 12° was the distance traveled by the simulated navigator on the optimum reliable path greater compared to following the shortest path. Amongst these for only 4 paths in case the straight angle was set to 0° and 3 paths in the case the straight angle was set to 12° the distance was greater than 10% (the worst one being 20% greater in case of the straight angle being set to 12°). This, overall, can be regarded as a considerable saving of travel distance. However, the number of requeries done on optimum reliable path was greater than the number of requeries done while traversing the most reliable path (but still far less than those done on shortest path). The missed target travels, stopping distance from destination and actual unreliability were only slightly greater than those for most reliable path, indicating an overall better optimal performance.

Table 3. Aggregate results for various performance metrics (with no angle specified for straight)

Algorithm	Shortest Path	Shortest Most Reliable Path	Optimum Path $\lambda_d = 1, \lambda_u = 6$
Distance Travelled	125719.5	137118.5	113158.9
Stopping Distance	6134.1	374.6	696.1
Total Distance	131853.6	137493.1	113855
Requeries	6773	1291	2799
Missed target travel	566	36	63
Actual Unreliability	8.3	1.3	3.4

Table 4. Aggregate results for various performance metrics (with a straight angle of 12°)

Algorithm	Shortest Path	Shortest Most Reliable Path	Optimum Path $\lambda_d = 1, \lambda_u = 6$
Distance Travelled	106959.5	103114.5	93641.7
Stopping Distance	2888.5	104.9	123
Total Distance	109848	103219.4	93764.7
Requeries	5209	406	1076
Missed target travel	319	24	26
Actual Unreliability	5.8	0.5	1.1

7 Conclusion and Future Work

In this work, we developed wayfinding algorithms that favor reliable paths with lesser navigational complexity. Our expectation was that it would lead to fewer mistakes a wayfinder potentially could make during route following. The simulation results support our expectation. When traversing a reliable path, the simulated navigator was found to be getting at (or closer to) the actual destination more often than when traversing the shortest path. The simulation also showed an advantage in terms of the number of times the simulated navigator got lost and needed to re-contact the automated navigation system, as well as the overall distance traveled. The data set used for the simulations was a complex street network. More simulation based analyses need to be done, making use of a variety of street networks. Cognitive studies with human navigators would provide additional proof of our hypothesis.

Our method of unreliability measure calculation took into account the instruction equivalent choices encountered at complex intersections. This approach can be further refined by the following aspects:

Including Landmarks. Calculating the unreliability of a route is based on the ambiguity at intersections introduced by instruction equivalent branches. Salient features (landmarks) in the environment reduce this ambiguity [22]. The specific placement of a landmark at an intersection can be integrated into its measure of salience [15] and there are approaches which base—in more general terms—the calculation of routes on the availability of landmarks [1].

Incorporating User Preferences. Next generation route directions will be adapted to several characteristics of a user. Hotly debated is the familiarity of a user with her environment. Tracking user movements by GPS devices will allow for establishing patterns of frequently visited places which can be assumed to be known [17,20]. Knowledge about one's environment influences the amount of information provided in route directions. Further factors are the modality of travel and specific user characteristics.

Navigational Complexity. An additional refinement of the navigational complexity could take into account whether a turn is made to the left or to the right. Depending on the general side one is traveling on it would be taken into account if a street has to be crossed. Likewise, the error-free criterion (see Section 3) could be changed to allow errors right from the start.

Cognitive Load. A different perspective on route directions is offered by taking the spatial context of a route into account to reduce the number of instructions that has to be given for a specific route [21,13,5]. While these approaches aim to reduce the number of instructions by, for example, using complex intersection as landmarks a combination with the presented optimal algorithm would be feasible to calibrate the trade-off between the overall amount of information required to communicate a route and the unreliability associated with navigationally complex intersections.

A last point concerns the computational time complexity of the reliable path algorithms, which is worse than the one for the shortest path (quadratic in the worst case). This could be improved by using a binary heap as mentioned in Section 2.2. The optimum reliable path algorithm provides a configurable way of tuning the system towards either shortest or most reliable paths. However, calibrating the values of λ_d and λ_u for an entire street network as done in Section 6.1 might not always produce the best results for each and every path and sometimes these values might need to be calibrated for individual paths.

The very last point would take into consideration the modality that is chosen to communicate the action that has to be performed at an intersection depending on the navigational complexity of an intersection.

Acknowledgements

Alexander Klippel has been funded by the Cooperative Research Centre for Spatial Information, whose activities are funded by the Australian Commonwealth Cooperative Research Centres Programme and by the Faculty of Engineering, The University of Melbourne, Australia.

References

1. Caduff, D., Timpf, S.: The landmark spider: Representing landmark knowledge for wayfinding tasks. In: Reasoning with Mental and External Diagrams: Computational Modeling and Spatial Assistance, pp. 30–35 (2005)
2. Carlson-Radvansky, L.A., Logan, G.D.: The influence of reference frame selection on spatial template construction. Journal of Memory and Language 37, 411–437 (1997)
3. Cormen, T.H., Leiserson, C.E., Rivest, R.L., Stein, C.: Introduction to Algorithms, 2nd edn. MIT Press and McGraw-Hill, Cambridge (2001)
4. Coventry, K.R., Garrod, S.C.: Saying, Seeing, and Acting: The Psychological Semantics of Spatial Prepositions. Psychology Press, Hove (2004)
5. Dale, R., Geldof, S., Prost, J.-P.: Using natural language generation in automatic route description. Journal of Research and Practice in Information Technology 37, 89–105 (2005)
6. Dijkstra, E.W.: A note on two problems in connexion with graphs. Numerische Mathematik. 1, 269–271 (1959)
7. Duckham, M., Kulik, L.: "Simplest" paths: Automated route selection for navigation. In: Kuhn, W., Worboys, M.F., Timpf, S. (eds.) COSIT 2003. LNCS, vol. 2825, pp. 169–185. Springer, Heidelberg (2003)
8. Evans, G.W.: Environmental cognition. Psychological Bulletin 88, 259–287 (1980)
9. Franklin, N., Henkel, L.A., Zangas, T.: Parsing surrounding space into regions. Memory and Cognition 23, 397–407 (1995)
10. Golledge, R.G.: Path selection and route preference in humn navigation: A progress report. In: Kuhn, W., Frank, A.U. (eds.) COSIT 1995. LNCS, vol. 988, Springer, Heidelberg (1995)
11. Herskovits, A.: Language and Spatial Cognition: An Interdisciplinary Study of the Representation of the Prepositions in English. Cambridge University Press, Cambridge (1986)
12. Klippel, A.: Wayfinding choremes. In: Kuhn, W., Worboys, M.F., Timpf, S. (eds.) COSIT 2003. LNCS, vol. 2825, pp. 320–334. Springer, Heidelberg (2003)
13. Klippel, A., Tappe, H., Kulik, L., Lee, P.U.: Wayfinding choremes - a language for modeling conceptual route knowledge. Journal of Visual Languages and Computing 16, 311–329 (2005)
14. Klippel, A., Tenbrink, T., Montello, D.R.: The role of structure and function in the conceptualization of directions. In: revision
15. Klippel, A., Winter, S.: Structural salience of landmarks for route directions. In: Cohn, A.G., Mark, D.M. (eds.) COSIT 2005. LNCS, vol. 3693, pp. 347–362. Springer, Heidelberg (2005)
16. Landau, B.: Axes and direction in spatial language and spatial cognition. In: van der Zee, E., Slack, J. (eds.) Axes and direction in spatial language and spatial cognition, pp. 18–38 (2003)
17. Li, C.: User preferences, information transactions and location-based services: A study of urban pedestrian wayfinding. Computer, Environment and Urban Systems (forthcoming)
18. Mark, D.M.: Finding simple routes: 'Ease of description' as an objective function in automated route selection. In: Proceedings, Second Symposium on Artificial Intelligence Applications (IEEE), Miami Beach, pp. 577–581. IEEE, Los Alamitos (1985)

19. May, A.J., Ross, T., Bayer, S.H.: Drivers' informational requirements when navigating in an urban environment. Journal of Navigation 56, 89–100 (2003)
20. Patel, K., Chen, M.Y., Smith, I., Landay, J.A.: Personalizing routes. In: The 19th Annual ACM Symposium on User Interface Software and Technology, pp. 15–18. ACM Press, New York (October 2006)
21. Richter, K.-F., Klippel, A.: A model for context-specific route directions. In: Freksa, C., Knauff, M., Krieg-Brückner, B., Nebel, B., Barkowsky, T. (eds.) Spatial Cognition IV. LNCS (LNAI), vol. 3343, pp. 58–78. Springer, Heidelberg (2005)
22. Sorrows, M., Hirtle, S.C.: The nature of landmarks for real and electronic spaces. In: Freksa, C., Mark, D.M. (eds.) COSIT 1999. LNCS, vol. 1661, pp. 37–50. Springer, Heidelberg (1999)
23. Streeter, L.A., Vitello, D., Wonsiewicz, S.A.: How to tell people where to go: Comparing navigational aids. International Journal of Man/Machine Studies 22, 549–562 (1985)
24. Tenbrink, T.: Identifying objects on the basis of spatial contrast: An empirical study. In: Freksa, C., Knauff, M., Krieg-Brückner, B., Nebel, B., Barkowsky, T. (eds.) Spatial Cognition IV. LNCS (LNAI), vol. 3343, pp. 124–146. Springer, Heidelberg (2005)
25. Tversky, B., Lee, P.U.: How space structures language. In: Spatial Cognition, An Interdisciplinary Approach to Representing and Processing Spatial Knowledge, pp. 157–175 (1998)
26. Vorwerg, C., Rickheit, G.: Typicality effects in the categorization of spatial relations. In: Freksa, C., Habel, C., Wender, K.F. (eds.) Spatial cognition: An interdisciplinary approach to representing and processing spatial knowledge, pp. 203–222. Springer, Berlin (1998)
27. Zwaan, R.A., Radvansky, G.A.: Situation models in language comprehension and memory. Psychological Bulletin 123, 162–185 (1998)

Interpreting Route Instructions as Qualitative Spatial Actions

Hui Shi, Christian Mandel, and Robert J. Ross

Universität Bremen, Germany
{shi,cman,robertr}@informatik.uni-bremen.de

Abstract. In this paper we motivate the use of *qualitative spatial actions* as the fundamental unit in processing user route instructions. The spatial action model has been motivated by an analysis of empirical studies in human-robot interaction on the navigation task, and can be interpreted as a conceptual representation of the spatial action to be performed by the agent in their navigation space. Furthermore, we sketch out two distinct models of interpretation for these actions in cognitive robotics. In the first, the actions are related to a formalized *conceptual user modeling* of navigation space, while in the second the actions are interpreted as *fuzzy operations* on a voronoi graph. Moreover, we show how this action model allows us to better capture the points at which user route instructions become out of alignment with a robot's knowledge of the environment through a number of examples.

1 Introduction

Route navigation instructions allow one agent to instruct another to a particular location within their shared environment. While robotic agents in the near future may have access to extremely detailed environmental descriptions through technological application, e.g., GPS, a-priori mappings etc, the need for an artificial agent to be capable of processing route instructions remains an interesting research question for a number of reasons. Firstly, and arguably most importantly, cooperation with *naive* users in natural rather than technical interactions places onus on the agent to handle natural spatial instructions, e.g., route instructions, rather than forced goal selection through other means, e.g., hierarchical list selection. Furthermore, disparities between the robot's spatial representation and the ever-changing world are always a possibility, thus opening up the possibility of adding route instructions to goals which were not already known to the robotic agent.

The modeling and interpretation of route instructions has been addressed from a number of different research perspectives. In the spatial cognition and cognitive modeling communities, a great deal of effort has been applied to the analysis of route instructions as a reflection of a speaker's cognitive map (e.g., [26,8,28,10]). While such approaches are certainly interesting in terms of providing insight into the nature of the spatial models that motivate route instructions as verbalised by users, they are often somewhat abstracted from the detail required for online computational analysis. On the other hand, some in the robotics community have attempted to process route instructions as procedural information without reference to any explicit spatial representation [13]. In practice however, concrete robotic systems use very detailed spatial representations and reasoning processes that operate at a finer level of granularity than those models proposed by the cognitive modeling community [19,3,24], but yet little work has been done to date on unifying such approaches with the cognitive spatial representation and language processing efforts.

In this paper we propose an abstraction model for the interpretation of route instructions so as to unify a number of approaches to route instruction modeling and reasoning that have already been considered in the literature. We begin in Section 2 by reviewing a number of empirical studies involving the presentation of route instructions to artificial communicating partners. Such studies, showing that route instructions are most frequently, but certainly not always, presented by a human partner as a series of actions to be performed by an agent in a navigation space, lead us to propose a *Qualitative Spatial Actions* (QSA) as the fundamental modeling unit of verbal route instructions as intended by users. We introduce the elements of this QSA model in Section 3, detailing its coverage and relationship to linguistic and conceptual knowledge within an interpretation model. While the categories and coverage of the QSA model are driven by empirical analysis, computational systems must provide suitable interpretations of qualitative spatial actions in context. In Section 4 and 5 we present two distinct models for the interpretation of qualitative spatial actions. In the first, in Section 4, QSAs are interpreted as constructive operators within a formalized Conceptual User Model based on a *Route Graph* [27,9] augmented with spatial relations drawn from the *Double Cross Calculus* [7,8]. In the second, in Section 5, QSAs are interpreted using *fuzzy functions* applied against a voronoi graph representation as already applied in a concrete robotics application. We compare and contrast the two interpretation models in Section 6 by identifying a number of areas where we see *interpretation disparities* between the two interpretation models discussed. Before concluding in Section 8, we relate the approaches proposed in this paper to other work on the interpretation of route instructions in Section 7.

2 Empirical Results of Route Description Studies

As indicated above, the proposed QSA model is based on an analysis of concrete route instruction examples as given by users to mobile robots. We have built upon the results of two separate corpora of empirical results for this analysis. The first of these corpora was that due to the Instruction Based Learning (IBL) project [13], while the second was collected from our own project investigations of interaction between users and a semi-autonomous wheelchair [18].

2.1 Corpus 1: Instruction Based Learning

The Instruction Based Learning (IBL) project [13,2] attempted to construct robotic systems in which verbal commands including route instructions were interpreted into internal program code for execution by mobile robots. In designing the system, empirical studies were conducted to allow: (a) the construction of a program primitive set; and (b) the derivation of a domain model to allow construction of a language analysis grammar and the training of speech recognition software. Those studies, detailed by Bugmann in [2], are summarized here to elucidate the contents of the corpus.

```
1 okay take your first right
2 and continue down the street
3 past Derry's
4 past Safeway
5 and your parking lot will be on your right
```

Fig. 1. Short Route Instruction from the IBL Corpus

In the studies, 24 subjects were instructed to give route instructions which were to be processed by a small remote-controlled robot situated within a *toy town*. The subjects were divided into three equi-sized groups. The first two of these groups were informed that their route instructions would actually be post-processed by a remote human operator who observes the environment through a camera situated on the *head* of the robot. Subjects, who were also told to use previously defined routes where possible, had to describe six distinct routes consisting of three short and three long routes. Whereas the first two groups of subjects produced unconstrained monologic speech, the third group was forced into simplified dialogues with an operator to produce smaller chunked route instructions – albeit to describe the same underlying route instructions as the first group.

The resulting corpus contains a total of 144 routes instructions. Figure 1 shows one of the short monologue style route description segmented by hand into major or minor clauses.

2.2 Corpus 2: Shared-Control Wheelchair Studies

The second corpus which we base our analysis on was obtained in a series of experiments which were conducted to investigate the nature of language used by humans in communicating with robots about spatial concepts, such as route navigation, spatial relations between objects [22,6,21,18]. As with the IBL experiments, the reader is directed to the above sources for a detailed description of the experimental setup and results – here we simply identify the salient points of those studies for illustrative purposes. In particular it should be noted that while the experiments did involve a number of different robotic platforms,

```
1 dann äh , muss ich mich jetzt umdrehen
  I must turn around
2 aus der Tür fahren
  drive out the door
3 äh , dann nach rechts mich drehen
  then turn to the right
4 dann , - ziemlich lange geradeaus
  straight ahead for a relatively long way
5 äh , rechts vorbei an den Fahrstühlen
  pass by the lifts
6 vorbei an den Haupttreppen
  pass by the stairs
7 ähm, und dann, – sind wir eigentlich schon am Stugaraum
  and then we should be at the student-union room
```

Fig. 2. Example Route Instruction from the Rolland Corpus

e.g., Sony Aibo dogs, pioneer robots, wheelchair, and interaction scenarios, e.g., scene descriptions, route instructions, map annotations, we focus here on the results concerning the presentation of route instruction and directed movement commands to the semi-autonomous wheelchair.

The route navigation studies were performed using *Rolland*, a semi-autonomous wheelchair, which was situated in an office environment within a university. Over 40 subjects (a mix of native English and German speakers) participated in the two-phase experiments. In the first phase, users were situated in the wheelchair and asked to move around the partially known environment, *telling* the wheelchair about the locations of particular rooms within the environment. Having become more familiar with the environment and landmarks present in the environment, the second phase of the experiment then begins. In the second phase, users are asked to navigate Rolland to a destination through a verbal route description which makes use of some of the previously identified landmarks in the shared environment. Subsequent route instructions were then recorded by a recorder positioned on Rolland, and were later transcribed for analysis.

Figure 2 shows one sample route description given by a native German speaker to Rolland (with a descriptive rather than literal translation).

2.3 Discussion

From a quick scan of both route instruction examples we see that the route description is, to the most part, conveyed as a set of actions to be performed by the agent to achieve the goal of moving them from a start location to the goal. Thus, unsurprisingly, the route instructions are construed as something more akin to a plan rather than a static spatial description such as a scene description.

That said, the route instructions found in the corpus do of course move beyond well-behaved sequences of qualitatively or goal directed actions. In particular, we find the predictable occurrences of finer grained functional instructions, as well as a range of corrections, sanity checks, and other constraints, e.g.:

(1) a. turn about ninety degrees right – IBL: U20_GB_EH_19
     b. oh sorry erm from the hospital – IBL: U9_GC_HG_3
     c. if you go to the grand hotel you have gone too far – IBL: U17_GB_MD_5

Despite the presence of such utterances which may not at first seem applicable within an action driven modeling, it is our hypothesis that route instructions remain best modeled as actions with constraints which manipulate the location and movement of an agent within a navigation space. Furthermore, if this is the case we argue that a suitable and consistent approach to the modeling of route instructions as such actions should naturally lend itself to interpretations against distinct spatial representations. Working from these assumptions, we will sketch out the Qualitative Spatial Action model in the next section.

3 The Qualitative Spatial Actions Model

In the last section we noted that in empirical studies route descriptions manifest themselves in language as sequences of conditionals and declarative instructions where the primary processes present seem to map quite neatly to spatial movements and manipulations. In this section we propose the *Qualitative Spatial Action Model* as one means of capturing such route instructions.

3.1 Model Details

Informally we can define the *Qualitative Spatial Action Model* as a sequence of abstract spatial actions. To formally define the model, we must first introduce the sets of elementary types which model spatial orientations and places:

- $\mathcal{O}$ defines the set of orientation relations. There are different levels of qualitative direction granularity [17]. In this work we combine the front/back and the left/right dichotomy and distinguish 8 meaningful disjoint orientation relations: *front, rightFront, right, rightBack, back, leftBack, left, leftFront*. o, o', o_1, etc. are used for orientations.
- $\mathcal{M}$ consists of all landmarks in the spatial environment considered. Generally, a landmark is some view a human can perceive in the environment. m, m', m_1, etc. are used for landmarks.

A Qualitative Spatial Action Model instance is then a sequence of spatial actions drawn from the set, which can be extended for modelling new applications.

- *turn(o)*: reorient to o of $\mathcal{O}$ with respect to the agent's ego-centric frame of reference.
- *travel(c)*: move forward until condition c holds.

- *stop*: stop any movement.
- *annotate(m, c)*: relate landmark m according to condition c.

where condition c consists of a set of ego-centric relations defined by

ego_ori : $\mathcal{M} \times \mathcal{O}$		the orientation between a *landmark* and the current position
via :	$\mathcal{M}$	a *landmark* is on a travel path
passBy : $\mathcal{M}$		a *landmark* is on the left or on the right hand side of a travel path
passOn : $\mathcal{M} \times \mathcal{O}$		a specific case of the *passBy* relation, where the orientation of the *landmark* to the travel path is given.
until :	$\mathcal{M}$	a landmark is at the end of a travel path.

Applying the model, the route description from the Rolland Corpus depicted in Figure 2 can be modeled as the series of actions in Figure 3. Some important features to note here: (a) utterance 4 is not captured by our model since we do not yet have an approach to capturing extremely subjective features such as "a long way", "not far", and so forth; (b) it is possible to extend the set of spatial relations with new ones, such as "along", but in this paper we restrict the QSAM model to the above introduced relations in order to discuss its interpretations.

```
1  turn(back)
2  travel({via(door)})
3  turn(right)
4
5  travel({passBy(lifts)})
6  travel({passBy(mainStairway)})
7  annotate(stugaRaum, {ego_ori(stugaRaum, front)})
```

Fig. 3. The representations of the route description

4 Interpreting QSAM: Conceptual User Navigation Space

Knowledge that mental construals of space are schematized (cf. [25]) along with a systematic simplification of metric information, has led to the description of cognitively inspired mental space models, such as the mental space of navigation consisting of objects such as landmarks and paths (e.g., [20,5]). Such models in turn have given rise to cognitively inspired formalized models of space which often attempt to make use of qualitative descriptions of navigation space [1,4,28,11,27]. Such models are interesting since they are expressive enough to represent most knowledge humans used to carry out specific navigation tasks, while, on the other hand, supporting an efficient way to predicate and explain navigation and spatial behaviors.

In the following we review one such model of users conceptual navigation space and show how the QSAM actions can be directly constructed into such a model for reasoning purposes.

4.1 The Double-Cross Calculus and the Route Grpah

This conceptual model combines the Double-Cross Calculus [7,8,28], with the Route Graph [27,9] to provide an abstract topological representation of navigation with well defined spatial relations. The Double-Cross Calculus, put forward by Freksa, uses orientation information for qualitative spatial representation and reasoning. In this calculus, the concept *orientation grid* is introduced to represent qualitative orientation information. The grid is aligned to the orientation determined by two points in 2-dimensional space, the start point and the end point of a movement. Combining the front/back and the left/right dichotomy the Double-Cross Calculus can distinguish 15 meaningful disjoint orientation relations, see Fig. 4. There are six areas, seven on the lines, and two points. Moreover, the concepts with respect to route segments, i.e., *entry*, *exit* and *course*, are used to represent some positions, like *atEntry*, *onCourse*, *rightAtEntry*, etc. Note that we name all these positions, while in the original Double-Cross Calculus each position is associated with a pair of numbers between 0 and 7.

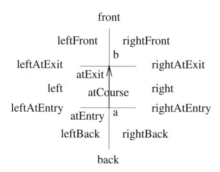

Fig. 4. Double-Cross orientation grids with 15 different positions

Route Graphs [27,9] have been introduced as a general concept for navigation by various agents in a variety of scenarios. They can be used as metrical maps with sensory input and additional metrical computation models to control the navigation of robotic agents in the environment, or be used at the cognitive level to model abstractly humans' topological knowledge while they act in the space. The most important character of Route Graphs is that different routes can be integrated into a graph-like structure, in which the information concerning these routes are composed. The most important concepts of Route Graphs are *route* and *route segment*. A route is a sequence of route segments, while a route segment consists of an *entry*, an *exit* and a *course*. For example, an entry or an exit in a route graph at the metrical level can be an angle in degree with respect to a global 2-D geometric system, the course is just characterized by metrical data for length and width; while an entry/exit at the cognitive level may contain qualitative orientation information (e.g., to the left/right), the course is then the path between two reorientations.

One major advantage to combining the Double-Cross Calculus and the Route Graph is the possibility to integrate different routes into a graph like structure, to apply spatial relations between routes, or to construct new routes from existing ones. Fig. 5 shows an example, after identifying the common place p and q where the entries and the exits of one route can be recomputed according to the other one, using the operations provided by the Conceptual Model [10].

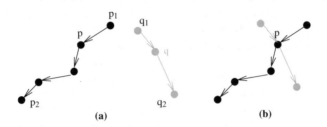

Fig. 5. Integration of two separate routes (a) into one route graph (b)

4.2 The Conceptual User Model

Topological maps like Route Graphs are suitable for human navigation using landmarks and passages, but the representation of orientations (e.g, turn left or right) and spatial relations (e.g., on the left or right) remain unspecified. On the other hand, most approaches to representing qualitative spatial knowledge consider the relationship between spatially extended objects, without the supplement of the topological information (e.g., the best route, the same location). Thus, our solution has been to combine the Double-Cross Calculus and the Route Graph into a conceptual model of navigation space. The foundation of the model can be found in [10].

Formally then, the Conceptual User Model is represented as a triple of types, functions and relations. The set of types consists of:

- $\mathcal{O}$ defines the set of orientation relations. In this model we use 15 disjoint orientations defined by the Double-Cross Calculus (see Fig. 4).
- $\mathcal{M}$ consists of all landmarks in the spatial environment considered, the same as defined in the Qualitative Spatial Action Model.
- $\mathcal{P}$ consists of all places in the spatial environment considered. A place can be marked with a landmark, such as a room. a door. p, p', p_1, etc. are used for places.
- $\mathcal{V}$ defines the set of all directed connections between different places, denoted as vectors. We will use **ab** for the vector between a and b. A vector can be a corridor or a street (or part there of). A subset of $\mathcal{V}$ is specified as $\mathcal{S}$ route segments, and will be denoted using s, s', s_1, etc.

The most important functions are *oEntry*, *oExit* and *oCourse* which takes a route segment and returns the entry, the exit (as an orientation) and the course (as a vector) of the route segment.

The two elementary relations are *at* and *ori*. The relation $at: \mathcal{M} \times \mathcal{P}$ marks a place with a landmark. The relation $ori: \mathcal{V} \times \mathcal{P} \times \mathcal{O}$ defines the orientation of a place with respect to a vector, such as $ori(\mathbf{bc}, p, rightFront)$ means p is on the right front of $\mathbf{bc}$. Based on these two relations a set of auxiliary relations can be defined, for example,

$on:$	$\mathcal{V} \times \mathcal{P}$	a *place* is on a *vector*
$left\_of:$	$\mathcal{V} \times \mathcal{P}$	a *place* is on the left of a vector
$right\_of:$	$\mathcal{V} \times \mathcal{P}$	a *place* is on the right of a vector
$begin\_of:$	$\mathcal{V} \times \mathcal{P}$	a *place* is the start place of a vector
$end\_of:$	$\mathcal{V} \times \mathcal{P}$	a *place* is the end place of a vector

The following axioms define these relations according to the elementary relation *ori*.

$$on(\mathbf{ab}, p) \Leftrightarrow ori(\mathbf{ab}, p, atCourse) \qquad (1)$$

$$left\_of(\mathbf{ab}, p) \Leftrightarrow ori(\mathbf{ab}, p, left) \qquad (2)$$

$$right\_of(\mathbf{ab}, p) \Leftrightarrow ori(\mathbf{ab}, p, right) \qquad (3)$$

$$begin\_of(\mathbf{ab}, p) \Leftrightarrow ori(\mathbf{ab}, p, atEntry) \qquad (4)$$

$$end\_of(\mathbf{ab}, p) \Leftrightarrow ori(\mathbf{ab}, p, atExit) \qquad (5)$$

4.3 Building Qualitative Route Representations

Given the above modeling of user navigation space, QSAM actions can be interpreted as constructive operators in building such Qualitative Route Represenations. Specifically, we interpret each *travel* action as a new route segment with appropriate relations as condition. Furthermore, each *turn* action then defines the entry of a route segment, while an *annotate* action is simply interpreted as defining a set of spatial relations on some route segment.

To illustrate, let us consider again the example route description which was presented as natural language in Figure 2, before being re-modeled in the QSAM in Figure 3. Figure 6 shows the resultant qualitative route representation following the interpretation of QSAM primitives as constructive operators.

As defined, a route is a sequence of route segments. Suppose $\langle s_0, s_1, \cdots, s_n \rangle$ is a route, then it should satisfy the following constrains, where $0 \leq i \leq n$ and $0 \leq j \leq n-1$:

- $oEntry(s_i), oExit(s_i) \in \mathcal{O}$
- $\exists \mathbf{pq} \in \mathcal{V}.oCourse(s_i) = \mathbf{pq}$
- $oCourse(s_j) = \mathbf{p_1 p_2} \wedge oCourse(s_{j+1}) = \mathbf{p_3 p_4} \rightarrow p_2 = p_3$

```
1  oEntry(s_0) = back
2  at(door, x), on(oCourse(s_0), x)
3  oEntry(s_1) = right
4
5  at(lifts, y), left_of(oCourse(s_1), y) ∨ right_of(oCourse(s_1), y)
6  oEntry(s_2) = front
   at(mainStairs, z), left_of(oCourse(s_2), z) ∨ right_of(oCourse(s_2), z)
7  at(stugaRaum, u), ori(oCourse(s_2), u, front)
```

Fig. 6. The representation of the sample route description

4.4 Reasoning on Spatial Relations

Qualitative route descriptions composed of QSAM actions may be reasoned on against an already existing construal of the users mental image of the environment where the navigation takes place. Suppose such an environment can be represented, in the Conceptual User Model, by a set of places possibly associated with landmarks, a set of routes, and a set of orientation relations, which are supposed to be known by the intelligent robot. Before a new route representation in the Conceptual User Model can be added to the environment, its consistence with respect to the environment is going to be proved, which means that the spatial relations in the route representation should be consistent with those hold in the environment. Furthermore, the new introduced conditions will be simplified according to the environment, if it is possible.

To illustrate this procedure, let us return again to the sample route description in Figure 6. Suppose that the stored environment contains the relation $right\_of(\mathbf{pp_1}, p_2)$, where p is the start place of the route description, at p_1 is the door, and at p_2 the lifts, i.e. the lifts on the right hand side of the vector between the start place and the door. Suppose $oCourse(s_0) = \mathbf{pb}$ and $oCourse(s_1) = \mathbf{bc}$, from $oEntry(s_1) = right$ we know $ori(\mathbf{pb}, \mathbf{bc}, right)$. Now, according to the composition table of the Double-Cross Calculus, we can conclude that

$$ori(\mathbf{bc}, p_2, rightFront) \vee ori(\mathbf{bc}, p_2, right) \vee ori(\mathbf{bc}, p_2, rightBack)$$

Thus, the lifts (at the place p_2) can only be passed on the right hand side. The alternation of "on the left" (i.e., $left\_of(oCourse(s_1), p_2)$) is inconsistent.

5 Interpreting QSAM: Fuzzy Operations

In this section we introduce a second interpretation approach which enables a mobile robot to directly interpret and execute a given route description modeled as a sequence of qualitative spatial actions. The interpretation model proposed is tailored for the spatial representations commonly used in cognitive robotics and can be used to effectively interpret spatial actions given by a naive user to a real-time robotic system without requiring the use of sometimes computationally expensive logical reasoning engines. In order to make such an interpretation against a global environment map, we develop and apply a number of fuzzy functions which provide interpretations of spatial relations against a voronoi graph based structure. These functions then allow us to evaluate interpretation hypotheses for individual sub-goals.

5.1 Spatial Relations as Fuzzy Functions

We begin by establishing a set of fuzzy functions in order to make the spatial relations established earlier algorithmicly useful. We do so by treating the task of evaluating spatial relations as a judgment of how well at least two given points out of the global workspace description (cf. Route Graphs in section 4.1) can be correlated via the spatial relation to be analyzed. In the following, we define a selection of the ego-centric relations used in the definition of Qualitative Spatial Actions (See Section 3), i.e., *ego_ori*, *via* and *until*. While this does not cover all possible relations, we believe that those definitions below can provide a basis for a number of other relations.

Ego-centric orientation. When humans use directional prepositions in natural language, they often use them from their current point of view. In this example we describe the binary relation *ego_ori*, in which the agent references a relatum (a landmark in our case) from their current pose ego. We now address the binary direction *rightFront* and start by computing the angle α between (a) the straight line that connects ego with relatum and (b) the vector that is based in the position of ego and aligned to the current heading of ego. Taking the most compatible angle β for the given direction *rightFront*, which is $-\pi/4$ w.r.t. the eight-valued level of granularity, and the normalizing constant c, we now compute the fuzzy rating of the spatial relation *rightFront* as can be seen in equation (6). Note that the given rating function is freely chosen out of a set of possible rating functions that map to the domain $[0..1]$ and converge against 1 for $|\alpha - \beta| \to 0$ and against 0 for $|\alpha - \beta| \to \pi$, respectively.

$$FR_{ego\_ori}(ego, relatum) = e^{-\frac{1}{2}(\frac{|\alpha - \beta|}{c})^2} \tag{6}$$

Passing Via Given Landmarks. The fuzzy function corresponding to the spatial relation *via* takes as its first argument the current pose of the agent, i.e. the ego. The second argument is given by the two landmarks m_1 and m_2, both labeled as relatum. The third argument *referent* is the global place we wish to reach by passing from ego via relatum. Starting from ego we choose an arbitrary relatum that is to be evaluated by the spatial relation, in the neighborhood of ego. In a first step we apply a standard graph search algorithm to compute the route ρ from the closest place nearby the ego to relatum. If ρ does not exist, we judge that relatum with the lowest possible value 0. Otherwise we continue with the computation of a circular region χ with diameter $d_\chi = \|\boldsymbol{m_2} - \boldsymbol{m_1}\|$ such that the centers of the landmarks lie on χ's circumference. The mentioned circle χ now segments ρ into the parts before, inside and after χ. Taking the length λ

of ρ after χ, and the normalizing constant c, we now compute the fuzzy rating of the spatial relation *via* as can be seen in equation (7).

$$FR_{via}(ego, referent, relatum) = e^{-\frac{\lambda}{c}} \qquad (7)$$

The fact that fuzzy ratings decrease for places that are farther away from χ, captures the intuitive nature of human route descriptions in that elements typically apply to the direct neighborhood of referenced landmarks.

Going Until Indexed Junction. In order to interpret the spatial relation *until* two parameters are necessary: (a) the current pose of the agent, and (b) a relatum given by the place which we wish to travel to from the ego, e.g., the fourth junction to the left. In order to look for an indexed place with a given attribute, we first have to define the fuzzy rating for the existence of that attribute in a given place ø. Therefore we model a junction to a given direction in analogy to the binary direction relation as described above, however with a different computation of the angle α. In this case the angle α is computed as the angle between two route segments sharing the common place ø. We now start with the computation of the route ρ from the closest place nearby ego to relatum. Interpreting ρ as an ordered set of places $\{p_0, p_1, ..., p_n\}$, we compute the ordered set of fuzzy ratings $FR(\rho) = \{fr_1, fr_2, ..., fr_n\}$, which represent the judgments that there exists a junction to the left in the corresponding places p_i. Considering the $j = \binom{n}{i}$ possible combinations to find in p_n the i_{th} place with a junction to the left, we can now compute $\{\mathcal{A}_1, \mathcal{A}_2, ..., \mathcal{A}_j\}$ as the $\binom{n}{i}$-subsets of $FR(\rho)$ (cf. [15]), yielding the overall fuzzy rating for the spatial relation "until" as can be seen in (8).

$$FR_{until}(ego, relatum) = \sum_{k=1}^{j} \prod_{l=1}^{n} \begin{cases} fr_l : fr_l \in \mathcal{A}_k \\ 1 - fr_l : fr_l \notin \mathcal{A}_k \end{cases} \qquad (8)$$

5.2 Interpretation of Qualitative Spatial Actions

After introducing fuzzy functions as the necessary primitives for interpreting spatial relations, we now present an algorithm that pushes us toward goal-oriented robot movement by calculating the most likely target pose given the current pose of the agent and the sequence of actions that represents the user's route description. We therefore utilize a search tree $\mathcal{S}$, the nodes of which represent fuzzy rated places that result from the evaluation of the involved spatial relations. Considering a single node of $\mathcal{S}$ as a compound of a pose and its fuzzy rating or score, the root $p1$ is given by the initial pose of the agent along with the highest possible score 1. Taking the first spatial action SA_1 that appears in the action sequence, one chooses the first candidate pose $p2$ that acts as the relatum for the fuzzy function associated with SA_1. After calculating $p2$'s own score, it is multiplied by the score of the parent node $p1$. This mechanism propagates the uncertainty of former fuzzy ratings down to lower levels of the search tree. The

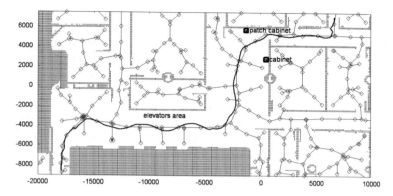

Fig. 7. Formalized and executed route according to the following route description: "...go to the elevators area and continue until the first junction to the left...turn left there and pass through the cabinet and the patch cabinet...then continue until the second door to the left...turn left, go through the door and stop...". Note that the spatial relations used in the above coarse route description are interpreted w.r.t. different layers of the global route graph. While *"until the first junction to the left"* and *"until the second door to the left"* are interpreted against the depicted nodes and edges, *"to the elevators area"* and *"through the cabinet and the patch cabinet"* are judged against region-nodes that abstract the area in front of the elevators and in-between the two cabinets.

algorithm now recursively continues in a depth-first branch-and-bound manner with the evaluation of the following spatial relations.

In order to keep the size of the search tree small, we apply two distinct bound-criteria. First, we do not construct nodes from the evaluation of a single spatial relation if the resulting score is lower than a given threshold c_1. In our implementation we chose $c_1 = 0.25$. Second, we stop the evaluation at a given node if its cumulative score is already lower than the highest score of a leaf resulting from the evaluation of the final spatial relation. In addition to these rules, we only address candidates for the relatum of the spatial relation to be analyzed that are closer to the current ego than a given distance c_2. In our implementation we chose $c_2 = 15m$. Experiments showed that the selection of relata within a given radius c_2 around the referent is the computational most expensive operation during the construction of the search tree. We therefore implemented a quad-tree representation of the global route graph search-space in order to make fast spatial queries.

The example in Figure 7 shows the executed path of our automated wheelchair according to the proceeding interpretation of an action sequence that matches the given coarse route description.

5.3 Keynotes of Approach

By applying fuzzy functions that map places to the co-domain of [0..1] we take account for the coarse qualitative nature of spatial relations that reside in natural language route descriptions. In contrast to the qualitative conceptual approach, the search for a solution of a given navigation task explicitly allows for ambiguous situations by the assignment of equal probability values for similar configurations. Taking the *passBy* example from above, we interpret the spatial relation *passBy* occurring in the action *travel*({*passBy*(lifts)}) by using the appropriate fuzzy function, where places at both the left and right hand side should be judged with respect to the relatum "lifts". According to the resultant probabilities, we might conclude that the "lifts" can only be passed on the right hand side. Thus, the two interpretations produce the same result in this example, although the processes applied to achieve the result are totally different.

One of the main benefits of interpreting QSAM via the evaluation of fuzzy functions leads directly to its main disadvantage. Because situations with small probabilities may be taken into account as part of the solution (as long as they are judged to be more probable than a given threshold, cf. section 5.2), the search through the whole search space is a computationally expensive operation. Another shortcoming is the need to provide an implemented fuzzy function for each spatial relation that may appear in the corpus of the selected domain. While the logic-oriented reasoning on QSAM gets by with the definition of a normal form of spatial relations e.g. *on*, *left_of*, *right_of*, *begin_of* and *end_of*, fuzzy functions come along as very specialized implementations of the corresponding spatial relation, e.g. *through*. Up to now it remains unclear how to translate specific spatial relations to the set of relations from the normal form in an automatic way.

6 Interpretation Disparities

In this section, we discuss situations in which both interpretations deliver different results, i.e., *interpretation disparities*.

6.1 Different Levels of Abstractions

The Voronoi Graph describes a network of places and connections between those places with maximal clearance to surrounding obstacles. At the conceptual level on the other hand, the main interest is in qualitative spatial information and abstract places. For example, a room in the Voronoi Graph is represented by a set of places, whereas connections between places at the conceptual level are vectors containing no metric information. Thus, an environment in the Conceptual User Model is in fact an abstract view of the Voronoi Graph.

Since the action model can be interpreted w.r.t. two different models within different abstractions, disparities may occur in interpreting the same action. Consider for example an area including a group of lifts and the region before them (a situation in our office building scenario). Within the Voronoi Graph the region and each lift are represented by a set of places, that are connected by edges. While at the conceptual level it may be abstracted into two places, one for the region and one for the group of lifts, which are connected with two vectors with opposite directions. in order to interpret a "turn" action occurring in this area that intends to take a lift, there is at most one possible interpretation at the conceptual level. The fuzzy interpretation may result several possible solutions, since more than one lift can be taken.

6.2 Errors in Route Descriptions

Navigation tasks belong to high-level cognitive processes that involve the judgment of environment information, the localization of spatial objects, and the analysis of spatial relations between these objects. Describing a route to a robot is a special case of navigation, and may be subject to *knowledge-based mistakes* ([16]) by users.

Unsolvable spatial information. Among the corpus examples we collected, a considerable number of utterances contained references to landmarks, such as doors, rooms, etc. An example is: "Drive past the copier room and the mailbox room.". Here, the robot's knowledge of the intended entities, e.g., the "copier room" and the "mailbox room", is presupposed. Suppose that the robot has no knowledge about them. In this case the fuzzy function based approach will search for all possible places before eventually detecting missing information. The conceptual approach will first add new places with related landmarks into the environment, and then use the place identification function to find out whether the place with the given landmark already exists in the environment or not.

Spatial relation mismatches. In case where a user relates spatial objects incorrectly, taking the example "Drive past the copier room and the mailbox room." again, and now we suppose that the mailbox room should be passed first, with the copier room after that. The Conceptual User Model might detect this mismatch, if there exist relations in the conceptual environment, such that the correct spatial relation between the copier room and the mailbox room can be inferred. During integration of the new route description, an inconsistent situation will be detected using logical reasoning. While, to interpret this action the search algorithm in the fuzzy function based approach might first reach the place for mailbox room without memorising the landmark, and then find the place for the copier room. After finding the place marked with the "copier room", it continues to search for a place marked with the "mailbox room". A zigzag route would be found in such a case.

Orientation mismatches. Moreover, mistakes can also be caused by using incorrect orientation, a special case of deciding upon spatial relations. Now consider again the scenario described in Section 4 where the lifts may only be passed on the right given the intended route from the user. On the other hand, suppose the utterance "pass by the lifts on your left" is given by the user. At the conceptual level this inconsistency can easily be detected using logical reasoning on spatial relations. While, the search algorithm of the fuzzy interpretation will search all possible places on the left without any success.

Localization mismatches. Now we consider a situation where a user believes themselves to be at a location, which is different from the one localized by the robot's sensor system, thus the route description may contain landmarks or directional orientations which do not match the robot's internal spatial representation. If this situation occurs, both the conceptual and the fuzzy interpretations will detect the irresolvable landmark or the impossible orientation. While, sometimes a route description can be "correctly" interpreted, though the locations do not match with each other. In this case, the mismatch remains undetected.

7 Relation to Other Work

Over the past ten years there has been considerable interest in verbal HRI within the spatial domain. For reasons of space, we limit ourselves to those efforts specifically concerning the processing and modeling of route instructions.

In particular, there have been a number of proposed models which attempt to capture route instructions as spatial models [28,11,14]. The Route Graph, referenced in Section 4 and detailed in [27,9], is one prominent example. The route graph is essentially a graph-based abstract representation of space which can be instantiated to a number of different *kinds* including voronoi-like low-level representations to more abstract representations such as the conceptual user model descried in Section 4. While Route Graphs are very interesting as a method for representing complete navigation spaces, little work has been performed to date on showing how individual route graphs can be composed from verbal descriptions, or on how individual route descriptions can be used against complete route graphs in computational systems. We would hope that this paper meets, at least in part, such questions.

Another approach to the interpretation of route instructions has come more directly from the robotics community [12,23]. In the IBL project [13,2] a number of *functional primitives* were extracted from the IBL corpus to be implemented as behaviors directly upon a robot which had a local-only rather than a global conceptualization of space. The IBL functional primitives are in some ways quite similar to the actions defined within the QSAM model. However, we have attempted to give them a grounding as actions which may be applied in a range of navigation spaces, thus allowing us to concretely relate them to the forms of spatial representation actually used in hybrid rather then only behavior based robots. Godot on the other hand [23] made use of an internal map and used landmark locations within that map to aid the route interpretation process.

8 Summary, Conclusion, and Future Work

In this paper we have attempted to re-affirm actions as the fundamental unit of verbal route descriptions given by a user to an artificial agent. We have based our approach on existing route-instruction corpora, and proposed *Qualitative Spatial Action Model* for representing humans' route instructions. Moreover, we have concretely linked qualitative spatial actions to two existing approaches to the representation of navigation space, thus facilitating the *interpretation* of route instructions by artificial agents to carry out them efficiently. Since the *Conceptual User Model* and the *Fuzzy Operation Model* represent the navigation space at different abstraction levels, they can serve to analyse and identify various disparity problems, which enables efficient human-robot interaction.

We feel that linking these different approaches to the modelling and interpretation of routes is valuable since it allows a more widespread understanding of the practical issues involved in interpreting and reasoning on route descriptions in robotic systems. Furthermore, we also believe that the introduction of an intermediate model, the Qualitative Spatial Action Model in this paper, is useful beyond the understanding of full route instructions, since they can also serve as a basis of the conceptual or metric spatial interpretation into a selected spatial representation model, which should be implemented by a robot supporting natural verbal interaction with humans. The primitive set chosen for the model was based on a subset of the situations encountered in the empirical studies described above, and is extensible and modifiable to capture the requirements of different applications.

In future work we would like to extend the QSAM model to cover more of both the IBL and Rolland corpa. In particular this would involve the accommodation of quantitative or at least pseudo-quantitative movement instructions, and region related spatial relations. Having established the QSAM model, we also wish to apply the abstractions as an interface to on-line route instructions, e.g., users issuing instructions to the wheelchair as they move down the corridor.

Acknowledgments. We gratefully acknowledge the support of the Deutsche Forschungsgemeinschaft (DFG) through the Collaborative Research Center SFB/TR 8 Spatial Cognition - Subprojects I3-SharC and A1-RoboMap. We would also like to thank Prof. John Bateman and Prof. Bernd Krieg-Brückner for useful discussions on some of the issues presented in this paper.

References

1. Allen, J.F.: Maintaining knowledge about temporal intervals. CACM 26(11), 832–843 (1983)
2. Bugmann, G., Klein, E., Lauria, S., Kyriacou, T.: Corpus-Based Robotics: A Route Instruction Example. In: Proceedings of IAS-8 (2004)
3. Chronis, G., Skubic, M.: Robot navigation using qualitative landmark states from sketched route maps. In: Proceeding of the IEEE 2004 International Conference on Robotics and Automation, pp. 1530–1535. IEEE Computer Society Press, Los Alamitos (2004)
4. Cohn, A.G., Bennett, B., Gooday, J., Gotts, N.M.: Qualitative spatial representation and reasoning with the region connection calculus. Ceoinformatics 1, 1–44 (1997)
5. Denis, M.: The description of routes: A cognitive approach to the production of spatial discourse. Cahiers de Psychologie Cognitive 16, 409–458 (1997)
6. Fischer, K.: What Computer Talk Is and Is not: Human-Computer Conversation as Intercultural Communication. Computational Linguistics 17 (2006)

7. Freksa, C.: Qualitative spatial reasoning. In: Mark, D.M., Frank, A.U. (eds.) Cognitive and Linguistic Aspects of Geographic Space, Kluwer Academic Publishers, Dordrecht (1991)
8. Freksa, C.: Using orientation information for qualitative spatial reasoning. In: Frank, A.U., Formentini, U., Campari, I. (eds.) Theories and Methods of Spatio-Temporal Reasoning in Geographic Space. LNCS, vol. 639, pp. 162–178. Springer, Heidelberg (1992)
9. Krieg-Brückner, B., Frese, U., Lüttich, K., Mandel, C., Mossakowski, T., Ross, R.J.: Specification of route graphs via an ontology. In: Freksa, C., Knauff, M., Krieg-Brückner, B., Nebel, B., Barkowsky, T. (eds.) Spatial Cognition IV. LNCS (LNAI), vol. 3343, pp. 989–995. Springer, Heidelberg (2005)
10. Krieg-Brückner, B., Shi, H.: Orientation calculi and route graphs: Towards semantic representations for route descriptions. In: Raubal, M., Miller, H.J., Frank, A.U., Goodchild, M.F. (eds.) GIScience 2006. LNCS, vol. 4197, Springer, Heidelberg (2006)
11. Kuipers, B.: The spatial semantic hierarchy. Artificial Intelligence 119, 191–233 (2000)
12. Lauria, S., Bugmann, G., Kyriacou, T.: Training personal robots using natural language instruction. IEEE Intelligent Systems 16(3), 38–45 (2001)
13. Lauria, S., Kyriacou, T., Bugmann, G., Bos, J., Klein, E.: Converting natural language route instructions into robot executable procedures. In: Proceedings of the 2002 IEEE International Workshop on Human and Robot Interactive Communication, pp. 223–228. IEEE Computer Society Press, Los Alamitos (2002)
14. MacMahon, M.: A framework for unterstanding verbal route instructions. In: Proceedings of AAAI Fall Symposium on the Intersection of Cognitive Science and Robotics: From Interfaces to Intelligence (2004)
15. Nijenhuis, A., Wilf, H.: Combinatorial Algorithms. Academic Press, London (1978)
16. Reason, J.: Human Error. Cambridge University Press, Cambridge (1990)
17. Renz, J., Mitra, D.: Qualitative direction calculi with arbitrary granularity. In: Zhang, C., W. Guesgen, H., Yeap, W.-K. (eds.) PRICAI 2004. LNCS (LNAI), vol. 3157, pp. 65–74. Springer, Heidelberg (2004)
18. Shi, H., Tenbrink, T.: Telling rolland where to go: Hri dialogues on route navigation. In: WoSLaD Workshop on Spatial Language and Dialogue (October 23-25, 2005)
19. Skubic, M., Matasakis, P., Forrester, B., Chronis, G.: Extracting navigation states from a hand-drawn map. In: Proceeding of the IEEE 2001 International Conference on Robotics and Automation, IEEE Computer Society Press, Los Alamitos (2001)
20. Talmy, L.: How language structures space. In: Pick, H.L., Acredolo, L.P. (eds.) Spatial Orientation: Theory, Research and Application (1983)
21. Tenbrink, T.: Identifying objects in english and german: Empirical investigations of spatial contrastive reference. In: WoSLaD Workshop on Spatial Language and Dialogue (October 23-25, 2005)
22. Tenbrink, T.: Identifying objects on the basis of spatial contrast: an empirical study. In: Freksa, C., Knauff, M., Krieg-Brückner, B., Nebel, B., Barkowsky, T. (eds.) Spatial Cognition IV: Reasoning, Action, Interaction. International Conference Spatial Cognition 2004, Frauenchiemsee, Germany, October 2004, pp. 124–146. Springer, Heidelberg (2005)
23. Theobalt, C., Bos, J., Chapman, T., Espinosa-Romero, A.: Talking to godot: Dialogue with a mobile robot. In: Proceedings of the 2002 IEEE International Conference on Intelligent Robots & Systems, pp. 1338–1343. IEEE Computer Society Press, Los Alamitos (2002)

24. Thrun, S.: Robotics mapping a survey. In: Lakemeyer, G., Nebel, B. (eds.) Exploring Artificial Intelligence in the New Millenium, Morgan Kaufmann, San Francisco (2002)
25. Tversky, B.: Structures of mental spaces – how people think about space. Environment and Behavior 35(1), 66–80 (2003)
26. Tversky, B., Lee, P.: How space structures language. In: Freksa, C., Habel, C., Wender, K.F. (eds.) Spatial Cognition. LNCS (LNAI), vol. 1404, pp. 157–175. Springer, Heidelberg (1998)
27. Werner, S., Krieg-Brückner, B., Hermann, T.: Modelling navigational knowledge by route graphs. In: Habel, C., Brauer, W., Freksa, C., Wender, K.F. (eds.) Spatial Cognition II. LNCS (LNAI), vol. 1849, pp. 259–316. Springer, Heidelberg (2000)
28. Zimmermann, K., Freksa, C.: Qualitative spatial reasoning using orientation, distance, and path knowledge. Applied Intelligence 6, 49–58 (1996)

Knowledge Based Schematization of Route Directions

Samvith Srinivas and Stephen C. Hirtle

School of Information Sciences, 135 N. Bellefield Ave, Pittsburgh, PA, 15213
{sas29,hirtle}@pitt.edu

Abstract. It is common for a wayfinding task to involve travel across a familiar and an unfamiliar region that encompass different parts of the same route. Routes of this kind would entail schematized descriptions and the schematization would directly depend on the familiarity of the region being described. This paper presents a new formalization that identifies key conceptual elements of such routes and introduces a principle of "knowledge chunking" that enables their schematization. This is followed by empirical evidence that supports this schematization of route directions for wayfinder's who may perform such a task. The evidence suggests the need for future wayfinding systems to produce schematized route descriptions based on the user's prior knowledge of a route. The formal approach presented is useful in implementing such a system and possible methods for its implementation are discussed.

1 Introduction

A common problem in spatial domains is that of giving and following route direction to get from one place to another. Over the past several decades, there has been a considerable amount of research conducted on the nature and quality of route directions. While some of the initial research was done by psycholinguists (e.g. [1], [2], [3], [4]), more recent research on various facets of route directions have since been studied by geographers, psychologists and computer scientists (e.g. [5], [6], [7], [8], [9], [10], [11], [12], [13], [14], [15]). Some of these studies required participants to navigate an unfamiliar environment using route directions prepared and presented by the researchers ([6], [9], [10], [11]). Other studies required participants to produce route directions for familiar environments, which were collected and later analyzed by the researchers using various measures ([7], [8], [15]). Some work involved the combination of two kinds of environments. For instance, Fontaine and Denis required participants to describe routes in a part urban and part underground (Paris subway) environment [15], and Lovelace, Hegarty and Montello, were some of the earlier researchers who considered the relationship between quality of route directions and the overall familiarity of an environment [12]. In particular, they investigated the correlation between the levels of spatial knowledge and the quality of route descriptions provided by participants. More recently, Tomko and Winter addressed the issue of representing elements of a city from a hierarchical city structure by using what they term "granular route directions" [16].

While there are many instances where wayfinding tasks take place in a region that is completely familiar or unfamiliar, it is also interesting to consider situations where wayfinding tasks take place in an environment that is composed of a familiar part and an unfamiliar part along the same route. It is a common occurrence that wayfinder's sometimes travel from a region of familiarity to a region of unfamiliarity, or vice versa, from an unfamiliar region to a familiar region. An example would be a wayfinder's first time visit to a neighboring town (the unfamiliar region) from her residential neighborhood (the familiar region), or returning home (the familiar region) having been driven by colleagues to a restaurant in an unfamiliar neighborhood (the unfamiliar region).

In this paper, we consider the case of a partially familiar route. We hypothesize that routes of this kind would entail schematized descriptions and the schematization of these descriptions would directly depend on the familiarity of the region being described. We believe that in order to be cognitively adequate, route descriptions would have to be schematized on the basis of the individual wayfinder's prior knowledge. While schematization of maps has been the focus of earlier studies ([17], [18]), we introduce the concept of schematization of routes descriptions based on prior knowledge of a wayfinder. Our work begins by describing a formalization that models routes of this nature. We then present a small empirical study that aims to determine whether humans indeed prefer schematized route directions for partially familiar routes. In the empirical section, we consider two cases of route directions, walking directions for a medium scale space (college campus) and driving directions for a large scale space (city). The paper concludes with several observations about the usefulness of the formal approach for future implementation in a wayfinding system.

2 The Model—Knowledge Based Schematization of Route Directions

2.1 Basic Concepts

In this section we present a conceptualization of the problem introduced above. While there may be various approaches proposed for modeling routes, we use a graph theoretic approach. The terminology we use to represent routes is derived from work on Wayfinding Choremes by Klippel et al., [19], [20], [21]. This approach is based on the *RouteGraph theory,* initially presented by Werner et al., as a common conceptual framework to represent spatial navigation in general [22]. Klippel, refines this model in order to represent the concepts related to a humans movement through an environment. He presents the following terminology that is used to represent wayfinding and route directions : route, route segment, decision point, origin and destination [19]. For the representation of a route in our model, we use some of the terminology introduced by Klippel in [19] and further introduce a few key elements.

Klippel [19] defined a basic route (<route>) as one which is composed of an origin (<O>), destination (<D>) and route segment (<seg>) (Figure 1).[1]

---

[1] While the terminology remains conceptually consistent with Klippel [19], we introduce our own naming conventions.

<route> ::= <O> <seg> <D>

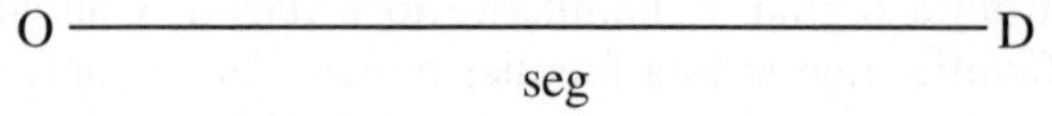

O ————————————————— D
           seg

Fig. 1. A basic route with origin (<O>), destination (<D>) and route segment (<seg>)

From the generic structure, Klippel [19] adds the concept of decision points. A decision point (<DP>) is a point along a route where the traveler has to make a choice between at least two possible directions (Figure 2). Addition of a decision point along a route would necessitate the addition of a second route segment. A combination of the two concepts constitutes a route part (<RP>). The square brackets indicate the optional nature of this element.

<route> ::= <O> <seg> [<RP>] <D>
<RP> ::= <DP> <seg>

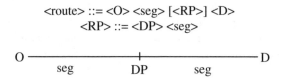

Fig. 2. A route with a decision point, two route segments, origin and destination. The five basic concepts necessary for the characterization of routes as listed by Klippel in [19].

Klippel defined two kinds of decision points, decision points with direction change (DP+), and decision points without direction change (DP-) [19], [20], [21]. This distinction, while interesting, will not be part of our characterization but the difference will become relevant again when generating natural language descriptions. This concept is based on previous studies in the field where similar approaches were presented. For example, Lovelace et al., presented a similar concept in terms of 'potential choice point' landmarks and 'choice point' landmarks [12].

2.2 Concepts for Coding Route Familiarity

It must be noted that the elements presented so far are the basic elements necessary for the characterization of routes. While Klippel and colleagues extend these basic elements as part of their work, we provide our own extensions that will help us in the characterization of partially familiar routes. In the following section, we present our extensions to the basic concepts we have just reviewed, then we use these concepts in presenting the various models for knowledge based schematization of routes.

To Klippel's original framework, we introduce the concept of a partially familiar route, or knowledge route (<kroute>). A partially familiar route is one which incorporates a familiar route segment (<K>) within a known region and an unfamiliar route segment (<N>) within an unknown region along the same route (Figure 3). This gives the most basic form of a partially familiar route. The braces indicate that the order of <K> and <N> can be interchanged.

$$\text{<kroute> ::= <O> \{<K> <N>\} <D>}$$

```
        K              N
O ─────────────────+──────────────── D
```

Fig. 3. The most basic form of a partially familiar route

Points along the familiar portion of the route will consist of one or more points called known locations (KLs). A KL can be one of three types of points: (1) a well-established landmark within a neighborhood, (2) a familiar building that is often frequented, even if it does not rise to 'landmark' status, and (3) the intersection of two segments along on a route that the user is able to locate during navigation. Thus, a KL is a point along a route that a person is confident of being able to navigate to while in the K section of the route. We will use the concept of KLs in producing schematized route directions. We list three broad categories of KLs, one is a local landmark (e.g. "The Capitol"), the second is a building or address that an individual may frequent (e.g. "Hillman Library"), and the third is a decision point (e.g. "Bates Street entrance ramp to I-376 "). While decision points and landmarks have been studied extensively (e.g., [10], [12], [13], [14], [19], [23], [24]), concepts relating to the second category of KLs have been the focus of relatively fewer studies in the past (e.g. [25], [26], [27]).

The third concept we introduce is a special case of a KL which is the KL that is closest to or at the intersection of a K and N segment of a route, called a known decision point and is denoted as (DPk) (Figure 4). DPk's are the transition points between a known region and an unknown region. Upon inclusion of this concept, the basic form of a <kroute> can be further represented as.

$$\text{<kroute}_{f:u}\text{> ::= <O> <K> <N> <D>}$$
$$\text{<K> ::= <seg> <DPk>}$$
$$\text{<N> ::= <seg>}$$

In this representation, the braces have been removed since interchanging the order of <K> and <N> will change the order of <seg> <DPk> within <K>. Here, we represent the alternative case.[2]

$$\text{<kroute}_{u:f}\text{> ::= <O> <N> <K> <D>}$$
$$\text{<N> ::= <seg>}$$
$$\text{<K> ::= <DPk> <seg>}$$

```
              K                    N
         ⌢⌢⌢⌢⌢⌢⌢⌢      ⌢⌢⌢⌢⌢⌢⌢⌢⌢⌢⌢
O────+────+────+──────────+────── D
    seg₁ KL₁ seg₂ KL₂ ... KL_{n-1} seg_n DPk
```

Fig. 4. The basic partially familiar route with the 'n' KLs including the DPk

---

[2] We use a subscript here to distinguish knowledge route ordered fam:unf from a knowledge route ordered unf:fam. However, in the future the subscripts will be left off as the ordering will be clear from the context.

The final concept we introduce is the concept of a 'knowledge chunk'. The concept of knowledge chunking is similar to the concept of landmark chunking, presented by Klippel in [19]. Klippel's conceptual view of a chunk is that it is "made up from individual entities that are grouped together under a given perspective or according to grouping principles" [19]. In our model, knowledge chunking is carried out after the DPk has been determined. Knowledge chunking involves grouping all the segments in the section of K into one 'knowledge chunk' that we term as the 'proceed to (<argument>)' statement or <PS(arg)>. In the case of the basic <kroute> presented in Figure 4, directions describing the route along the 'n' segments in the section of K can be chunked into one 'proceed to DPk' statement or <PSd> (figure 5).

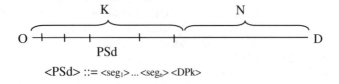

<PSd> ::= <seg$_1$> ... <seg$_n$> <DPk>

Fig. 5. The <PSd> statement used to represent the DPk and the n segments preceding it

For example, consider a DPk-Hillman Library. By definition of a DPk, the wayfinder would know the directions to Hillman Library, hence the <PSd> statement in this case would be "Proceed to Hillman Library". All directions up to Hillman Library can therefore be avoided, effectively reducing the cognitive load on the wayfinder. This would include references to intermediate known locations. Thus, in Figure 4, one only needs to explicitly mention KL_n, the DPk, and not KL_1 to KL_{n-1}. We will present specialized cases of the concepts introduced above in the following sections. This analysis will result in suggested guidelines for determining the appropriate DPk for a given situation.

3 Models of Partially Familiar Routes

Having introduced the key elements in our approach, we now proceed with the characterization of partially familiar routes (<kroute>). The methodology presented aims at addressing the issue of producing cognitively adequate route descriptions for routes of type <kroute>. We present three distinct cases and in each case we present a set of primitives which serve in their characterization.

3.1 Case 1: The KN-NK Models

Case 1 includes routes with exactly one K and N section. There are two possible models, the KN model, wherein the K section immediately follows the origin O (Figure 6a) and the NK model, wherein the N section immediately follows the origin O (Figure 6b).

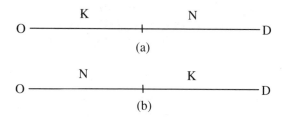

Fig. 6. (a) The KN model and (b) The NK model

KN Model. A key factor when producing route directions is the inclusion of sufficient information at a decision point. It may be recalled that the general definition of DPk given in Section 2.2 is the KL which is nearest to or at the intersection of a K and an N section of a route. In the case of the KN model, the DPk is the outermost KL (moving toward the destination) (Figure 7).

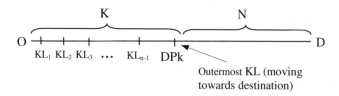

Fig. 7. A KN route with the DPk (*outermost KL*)

If the DPk is a point, the elements necessary for the characterization of a route are the point, the succeeding segment along with the orientation. In Figure 8, we present the example of the DPk-Hillman Library located on Schenley Drive, where upon arrival at Hillman Library (point), the wayfinder must proceed East (orientation) on Schenley Drive (succeeding segment).

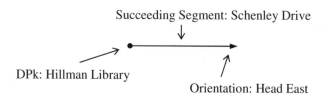

Fig. 8. The elements included in the description of a route in the KN point model

With regards to knowledge chunking in the KN model with DPk as a point, the <PS(arg)> directive will be a "proceed to DPk" or <PSd>. Hence the elements to be included in the description of a <kroute> in the KN point model are, the proceed to DPk statement (<PSd>), the succeeding segment with orientation (<Sseg + o>), the unknown segment(s) of the route (<N>) and the destination (<D>). It follows that a <kroute> of type

$$\text{<kroute>} ::= \text{<O> <K> <N> <D>}$$
$$\text{<K>} ::= \text{<seg}_1\text{> <seg}_2\text{> ... <seg}_n\text{> <DPk>}$$
$$\text{<N>} ::= \text{<seg}_{n+1}\text{> <seg}_{n+2}\text{> ... <seg}_{n+m}\text{>}$$

Can be represented as

$$\text{<kroute>} ::= \text{<PSd> <Sseg + o> <N> <D>}$$
$$\text{<N>} ::= \text{<seg}_{n+1}\text{> <seg}_{n+2}\text{> ... <seg}_{n+m}\text{>}$$

If DPk is an intersection of two segments, elements necessary for the characterization of a route are both the segments with the orientation of direction of travel for the succeeding segment. In Figure 9, we present the example of Bates Street (preceding segment1) meeting I-376 E (succeeding segment2). Upon arrival at Bates Street, the wayfinder heads *South* on Bates (*orientation1*), takes a Left onto I-376 and heads East (orientation2) (Figure 9).

It is important to note that knowledge of *orientation1* is implicit to the wayfinder, since the segment itself (preceding segment1) and the segment following it (succeeding segment2), are both part of a familiar route segment (<K>). By the definition of a familiar route segment (<K>), it follows that *orientation1* is not a necessary element in the characterization. However, we include it's representation in instances where we believe it plays a key role in the explication of the navigation process. We make the distinction by representing such elements in italics. Orientation2 on the other hand is a necessary element, since it involves the travel to a segment in an N section.

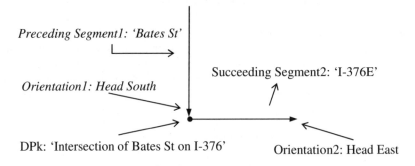

Fig. 9. The elements included in the description of a route in the two segment KN model

With regards to knowledge chunking in the KN model with DPk as an intersection of two segments, the <PS(arg)> directive will be a "proceed to DPk" or <PSd> Hence the elements to be included in the description of a <kroute> in the KN two segment model are the proceed to DPk statement (<PSd>), the succeeding segment2 with orientation (<Sseg2 + o>), the unknown segment(s) of the route <N> and the destination <D>. It follows that a <kroute> of type

$$\text{<kroute>} ::= \text{<O> <K> <N> <D>}$$
$$\text{<K>} ::= \text{<seg}_1\text{> <seg}_2\text{> ... <seg}_n\text{> <DPk>}$$
$$\text{<N>} ::= \text{<seg}_{n+1}\text{> <seg}_{n+2}\text{> ... <seg}_{n+m}\text{>}$$

Can be represented as

$$\text{<kroute> ::= <PSd> <Sseg2 + o> <N> <D>}$$
$$\text{<N> ::= <seg}_{n+1}\text{> <seg}_{n+2}\text{> ... <seg}_{n+m}\text{>}$$

NK Model. One can imagine reverse directions from an unfamiliar location to one's home as a typical real life example of the NK model. In general, the NK model characterizes a wayfinder moving from an unfamiliar section of the route to a familiar section of a route. Thus the wayfinder is returning to a point of familiarity. In the case of the NK model, the DPk is the outermost KL (moving toward the origin) (Figure 10).

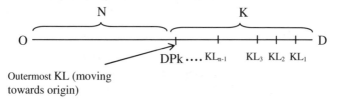

Fig. 10. A NK route with the DPk (*outermost KL*)

With regards to knowledge chunking in the NK model with DPk as a point, the <PS(arg)> directive will be a "proceed to destination" or <PSdest> and will appear at the end of the route direction. The elements necessary for the characterization of a route are the preceding segment along with a 'proceed to destination' directive (<PSdest>). We present the example of the Hillman Library located on Schenley Drive. Upon arrival at Schenley Drive (preceding segment) the wayfinder must proceed *West* (*orientation*) to arrive at *Hillman Library (point)* and then proceed to the destination (<PSdest>) (Figure 11).

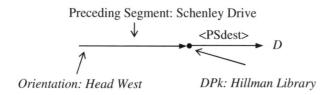

Fig. 11. The elements included in the description of a route in the NK point model

Hence the elements to be included in the description of a <kroute> in the NK point model are, the origin (<O>), the unknown segment(s) of the route (<N>), the preceding segment with *orientation* (<Pseg + *o*>), followed by the proceed to destination statement (<PSdest>), It follows that a <kroute> of type

$$\text{<kroute> ::= <O> <N> <K> <D>}$$
$$\text{<N> ::= <seg}_1\text{> <seg}_2\text{> ... <seg}_n\text{>}$$
$$\text{<K> ::= <DPk> <seg}_{n+1}\text{> <seg}_{n+2}\text{> ... <seg}_{n+m}\text{>}$$

Can be represented as

<kroute> ::= <O> <N> <Pseg + o> <PSdest>
<N> ::= <seg_{n+1}> <seg_{n+2}> ... <seg_{n+m}>

If the DPk is an intersection of two segments in the NK model, the problem is geometrically similar to the two segment KN model depicted in Figure 9. However the description of a <kroute> in the NK two segment model differs from the KN model, the fundamental difference being the <PS(arg)> statement. In the case of the two segment NK model, the <PS(arg)> statement will be a "proceed to destination" or <PSdest>. The other conceptual difference is that the <Sseg2 + o> element (succeeding segment along with its orientation) can be eliminated (since the segment following it also lies in a K section). And so it follows that a <kroute> in the NK two segment model, of type

<kroute> ::= <O> <N> <K> <D>
<N> ::= <seg_1> <seg_2> ... <seg_n>
<K> ::= <DPk> <seg_{n+1}> <seg_{n+2}> ... <seg_{n+m}>

Can be represented as

<kroute> ::= <O> <N> <Pseg1 + o> <PSdest>
<N> ::= <seg_1> <seg_2> ... <seg_n>

3.2 Case 2: The KNK-NKN Models

Case 2 includes routes with exactly one K and two N sections or vice versa. There are two possible models, the KNK model, wherein the K section immediately follows the origin O followed consequently by another K section (Figure 12a) and the NKN model, wherein the N section immediately follows the origin O followed consequently by another N section (Figure 12b).

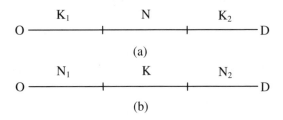

Fig. 12. (a) The KNK model and (b) The NKN model

KNK Model. The model builds on the KN and NK models presented earlier and we represent the KNK problem conceptually in Figure 13, we refer to two DPks a (KN)DPk and an (NK)DPk. The (KN)DPk is present at a transition from a K to an N section and an (NK)DPk is present at a transition from an N section to a K section.

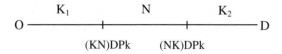

Fig. 13. The KNK problem with the two DPks

If the DPk is a point, the kroute of type

$$\text{<kroute>} ::= \text{<O>} \text{<K}_1\text{>} \text{<N>} \text{<K}_2\text{>} \text{<D>}$$
$$\text{<N>} ::= \text{<seg}_1\text{>} \text{<seg}_2\text{>} \ldots \text{<seg}_n\text{>}$$
$$\text{<K}_1\text{>} ::= \text{<seg}_{11}\text{>} \text{<seg}_{12}\text{>} \ldots \text{<seg}_{1n}\text{>} \text{<(KN)DPk>}$$
$$\text{<K}_2\text{>} ::= \text{<(NK)DPk>} \text{<seg}_{21}\text{>} \text{<seg}_{22}\text{>} \ldots \text{<seg}_{2n}\text{>}$$

Can be represented as

$$\text{<kroute>} ::= \text{<PSknd>} < K_1Sseg + o> \text{<N>} < K_2Pseg + o> \text{<PSdest>}$$
$$\text{<N>} ::= \text{<seg}_1\text{>} \text{<seg}_2\text{>} \ldots \text{<seg}_n\text{>}$$

And similarly, in the KNK two segment model, a <kroute> of type

$$\text{<kroute>} ::= \text{<O>} < K_1\text{>} \text{<N>} < K_2\text{>} \text{<D>}$$
$$\text{<K}_1\text{>} ::= \text{<seg}_{11}\text{>} \text{<seg}_{12}\text{>} \ldots \text{<seg}_{1n}\text{>} \text{<(KN)DPk>}$$
$$\text{<K}_2\text{>} ::= \text{<(NK)DPk>} \text{<seg}_{21}\text{>} \text{<seg}_{22}\text{>} \ldots \text{<seg}_{2n}\text{>}$$
$$\text{<N>} ::= \text{<seg}_1\text{>} \text{<seg}_2\text{>} \ldots \text{<seg}_n\text{>}$$

Can be represented as

$$\text{<kroute>} ::= \text{<PSknd>} < K_1Sseg2 + o> \text{<N>} < K_2Pseg1 + o> \text{<PSdest>}$$
$$\text{<N>} ::= \text{<seg}_1\text{>} \text{<seg}_2\text{>} \ldots \text{<seg}_n\text{>}$$

It must be noted that a combination of a point model and a two segment model is also possible. In this case the appropriate DPk (point or two segment) can be combined within the same route.

NKN Model. The model again builds on the previous KN and NK models. The transition from N to K in this case cannot be followed by a "proceed to destination" (<PSdest>) statement as in the KNK model. This is because in the KNK model, the NK transition took place to a K section that composed of the destination. This is not the case here, in this model, the K is sandwiched between two N sections. Hence the <PS(arg)> statement in this model will be a "Proceed to (KN)DPk)" or (<PSknd>). Since there is only one K section in this model, we number the segments of each DPk in order, giving <KPseg1> as the preceding segment of the (NK)DPk and the <KSseg4> as the succeeding segment of the (KN)DPk. Hence it follows that in the NKN point model, the <kroute> of type

$$\text{<kroute> ::= <O> <N}_1\text{> <K> <N}_2\text{> <D>}$$
$$\text{<K> ::= <(NK)DPk> <seg}_1\text{> <seg}_2\text{> ... <seg}_n\text{> <(KN)DPk>}$$
$$\text{<N}_1\text{> ::= <seg}_{11}\text{> <seg}_{12}\text{> ... <seg}_{1n}\text{>}$$
$$\text{<N}_2\text{> ::= <seg}_{21}\text{> <seg}_{22}\text{> ... <seg}_{2n}\text{>}$$

Can be represented as

$$\text{<kroute> ::= <O> <N}_1\text{> <PSknd> <KSseg4+ o> <N2> <D>}$$
$$\text{<N}_1\text{> ::= <seg}_{11}\text{> <seg}_{12}\text{> ... <seg}_{1n}\text{>}$$
$$\text{<N}_2\text{> ::= <seg}_{21}\text{> <seg}_{22}\text{> ... <seg}_{2n}\text{>}$$

And the NKN two segment model can be represented as

$$\text{<kroute> ::= <O> <N}_1\text{> <PSknd> <KSseg4 + o> <N}_2\text{> <D>}$$
$$\text{<N}_1\text{> ::= <seg}_{11}\text{> <seg}_{12}\text{> ... <seg}_{1n}\text{>}$$
$$\text{<N}_2\text{> ::= <seg}_{21}\text{> <seg}_{22}\text{> ... <seg}_{2n}\text{>}$$

3.3 Case 3: KNKN$^+$-NKNK$^+$ Models

We present this case in the interest of completeness. The components of this model can be formed by combining individual concepts from the KN, NK and the KNK models. No new concepts or concept extensions are required as a sequence of NKNKNK… or KNKNKN… can be solved by using the NK and KN transitions from the previous models discussed and the appropriate <PS(arg)> statements from the NK, KN and KNK models. Also, there is no theoretical possibility of a <kroute> containing the subsequence NN or KK, since the two consecutive N's can be considered a single N and the two consecutive K's can be considered a single K.

4 Experimental Support

Our aim is to establish whether schematized directions are preferred over complete directions given that the subject has prior knowledge of the area. This study examines two different scales. The first experiment looks at walking directions in a college campus and surrounding area and is denoted a medium scale space [28]. The second experiment looks at driving directions for locations within a 30 minute driving time of campus and is denoted a large scale space.

4.1 Experiment 1: Schematized Walking Directions for Medium Scale Spaces

The first experiment deals with walking directions. We use the University of Pittsburgh campus and the neighboring area for our experiment, as the campus reflects a suitable space for which walking directions will be useful. The directions we created directed a subject from the origin (location where the experiment was conducted) to a destination. Six destinations, presumed to be unknown to the participants, were chosen. All destinations were chosen such that there was at least one 'Known Decision Point' (DPk) between the origin and the destination. Each destination had one of four specified DPks. The DPks were used to produce schematized routes and the subject's preference for schematized or complete routes was recorded. The schematization carried out was consistent with the model presented in Section 2. We investigated the following hypothesis: There are significantly more schematized route directions chosen by subjects.

Method
Participants. Twenty paid participants (11 Females, 9 Males) were recruited through flyers posted around the University of Pittsburgh's Oakland campus. The only requirement was that subjects were above the age of 18. The mean age of participants was 24 years. The average number of years that participants were students at the University of Pittsburgh was 2.08 years. Subjects were paid for their participation at the fixed rate of US$15 for a one-hour study, which consisted of both Experiment 1 and Experiment 2.

Materials and Procedure. Subjects were first shown a map of the university campus and asked if they knew how to locate four familiar campus buildings that were to serve as DPks in the study. In case the subjects were not sure of the locations, they were briefed; very few subjects needed briefing as popular buildings on the campus were chosen as DPks. Subjects average rating of confidence across all the four DPks was 6.7 on a seven point scale where seven denoted that the subject was 'extremely confident' in the location of the DPk. Once knowledge of the DPks was established, subjects were allowed to proceed with the experiment.

They were presented with a system which was accessible online. The subject's task was to use the system to view and possibly print out directions to each of six destinations near campus which they would take with them. For each destination, a screen was presented to the subjects with the option of schematized directions or complete directions. An example is shown below:

> The directions to 125 Pier Street are listed below in two parts, Since the Hillman Library is a popular building that falls directly along the route from the IS building to 125 Pier Street, we have listed the directions from Hillman Library to the 125 Pier Street below,
>
> We have also listed the directions from the IS building to Hillman Library. You can make your choice as to what directions you would like to print out in order to help you navigate to your destination.
>
> *IS building to Hillman Library*
> *Hillman Library to 125 Pier Street*

In the example above, Information Science (IS) building (location of the experiment) is the origin of the route. The text in italics are hyperlinks that link to the respective directions. Clicking on either link produces directions for that section of the route. The user can choose to make a printout of that section of the route. The user also has the option of selecting the alternative section of the route for viewing, and can make a printout of that section as well or return to the welcome page. Once subjects were familiar with the system, they were allowed to continue with the experiment. The subjects were asked to print whichever section(s) of the route they would like to have if they were to navigate to the destination. Hence the subjects had a choice of printing directions from the DPk to the destination alone, from the origin to the DPk alone, or both. Subjects were told that they must be prepared to navigate to any one of the six destinations. Their choice of printouts was recorded.

Results and Discussion. Overall, for all six destinations, there were significantly more schematized route directions (t = 3.76, p < 0.001) chosen. The choices were also significant for each of the six destinations, where there were significantly higher proportions of schematized directions chosen (p's < 0.05). Subjects were asked to rate (on a seven point scale) their preference for schematized directions, where four denoted that the subject found the approach 'somewhat preferable' and seven denoted 'extremely preferable;' The average rating of the subjects was 5.25. Comments about the approach indicated that subjects preferred this approach but there was emphasis that their prior knowledge was an important aspect that determined their preference for this approach.

The results strongly support the hypothesis formulated at the beginning: There are significantly more schematized route directions chosen by subjects. These results suggest the need to consider the familiarity of a person's environment, while producing walking route directions that may include areas of familiarity and unfamiliarity within the same route traversed.

4.2 Experiment 2: Schematized Driving Directions for Large Scale Spaces

The second experiment dealt with driving directions. The aim was to establish whether subjects preferred schematized driving directions if they are familiar with a section of the route. The boundary of the large scale space represented in this experiment was restricted to the city of Pittsburgh. The DPk in this experiment was the intersection of two segments, namely, 'Bates Street' and 'I-376'. Subjects were presented with a learning task and their confidence of locating 'I-376' was verified. The subjects were then presented with two sets of driving directions to six unknown destinations; the two sets included the schematized route directions and complete route directions. The origin was the IS building (location of the experiment) and was the same for all six directions. The subject's choice of route direction type was recorded.

Hypothesis We investigate the following hypothesis: The proportion of schematized directions for the routes that include 'I-376' is significantly greater than the proportion of schematized directions for the other routes.

Method
Participants. The same group of participants from Experiment 1 was used for this experiment.

Materials and Procedure. We wanted to provide subjects with a learning task in order to reinforce the location of the DPk—'Bates Street to I-376'. We provided the subjects with eight route directions, which were divided into two groups. Four of the directions involved traveling through the DPk and the other four directions involved travel through other parts of the city. The origin for all eight routes was the IS building (the location of the experiment). The subjects were asked to draw the routes on a map, alternating between one set of route directions from each of the two groups. Each direction was to be marked on a new map—both for the sake of clarity and so as to not make the object of the learning task too obvious. We also wanted to measure the rate at which the subjects learnt the directions to the DPk. In order to do so, we asked the subjects to use two different colors to mark sections of the route that they were familiar with and unfamiliar with. The subjects were informed that a segment or point along the route was considered familiar to them if they could effectively navigate to the location.

Once the learning task was completed, subjects were tested on their knowledge of the DPk. The subjects average rating of confidence of locating the DPk was 5.9 on a seven point scale where seven indicated that the subject was 'extremely confident' in locating the DPk. All subjects answered 'yes' when queried about whether they would be able to locate the DPk given the starting location was the IS building. Once confidence in locating the DPk was established, subjects were allowed to proceed with the experiment.

Subjects were provided with two sets of route directions to six destinations around the city of Pittsburgh. The origin was fixed as the IS building. All six route directions had one set of directions that were complete and another set of route directions that were schematized. Two of these directions included the DPk. The two routes that included the DPk were schematized with the DPk as the starting point of the route direction. For example, under the schematized instruction, the first step might read:

1. Head East on I-376 taking the Bates St entrance ramp (Ext 7) to Monroeville - go 11 mi.

even though there are approximately four segments required to get from the IS building to the entrance ramp. The full directions had these four segments explicitly listed.

The rest of the routes were schematized over points that were potentially unfamiliar to the subjects. The average rating for familiarity of all four of these unfamiliar points was 3.58, on a seven-point scale where seven indicated that the subjects were 'extremely confident' in traveling to the destination. The directions were presented online and the subjects were asked to make a choice of route directions that they would prefer if they were to travel to each of the six destinations. The route direction choice of the subjects was recorded.

Results and Discussion. The proportion of schematized directions for routes schematized over the DPk was found to be significantly greater than the proportion of schematized directions for the other routes which were schematized over unfamiliar points ($t = 3.84$, $p = 0.0001$). Subjects were asked to rate their preference for this approach (on a seven-point scale), the average rating of the subjects was five, where four denoted that they found this approach 'somewhat preferable', and seven denoted that they found this approach 'extremely preferable'.

The results strongly support the hypothesis formulated at the beginning: The proportion of schematized directions for the routes that include 'I-376' is significantly greater than the proportion of schematized directions for the other routes. These results suggest the need to consider the familiarity of a person's environment, while producing driving route directions that may include sections of familiarity and unfamiliarity within the same route traversed.

In general, the results suggest the need to consider the subjects prior knowledge in producing route directions to a user. The section to follow explores various practical implementation possibilities, followed by the conclusion in Section 6.

5 Open Issues

5.1 Generating Known Locations

How should one go about picking the known decision points? One reasonable assumption would be to pick well-known landmarks. In fact, it has been well established that landmarks play a key role in the production of good route descriptions ([10], [12]). As a result, landmarks have subsequently been the subject of much research and various studies have focused on incorporating landmarks in route descriptions ([13], [19]). However, comprehensive surveys conducted as part of work in spatial knowledge, reveals that often times, locations that are considered "best known" or "landmarks", are locations that are tied to an individuals activity pattern, that is best known locations could be buildings that the individual may frequent ([25], [26]).

To address this question in part, a survey was conducted as part of our experiments. Before beginning the data collection, we asked participants to list four or five locations that they considered as "landmarks" in and around the University of Pittsburgh's Oakland campus and at the same time asked participants to list four or five buildings that they frequent. The scatter plot, shown in Figure 14, displays the responses of subjects to these two questions. Each point in the plot indicates the number of times that a building was judged to be a landmark and was also regularly frequented. There was one location, the Cathedral of Learning, which is a tall and dominant academic building on campus, which was judged to be landmark and regularly frequented building by over half the subjects. The remaining buildings show no strong relationships. The presence of a strong conditional relationship would have all the points on one side of the diagonal. In fact, there are buildings visited on a regular basis that are not considered landmarks and buildings considered landmarks that are not regularly visited.

Hence based on the results of this study and the earlier studies conducted on spatial knowledge, we make a distinction between the concepts of "landmarks" and "known buildings." In choosing known decision points, an ideal system would be personalized to include regularly visited locations of the traveler, rather than just established landmarks in the overall space.

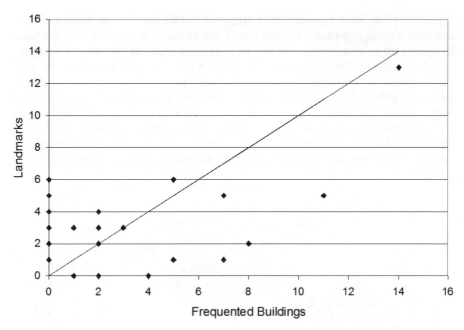

Fig. 14. The responses of subjects with the y-axis displaying the subjects who responded that a building was a landmark and the x-axis represents the buildings that were regularly frequented

5.2 Implementation in Wayfinding Systems—The Learning Factor

A practical implementation of the model presented will have to access an individual's knowledge of individual segments or points of a route. In order to do so, we introduce the concept of a "learning factor." A learning factor is the number of times a point along a route has to be traversed in order to be considered learnt or familiar to a user. In the case of a KL which is a building or a landmark, it will be a measure of the number of times a person visits a particular location in order for the location to be considered learnt. In case of segments along a route, it will be a measure of the number of times a route is traversed in order for it to be considered learnt. While there are various studies of route and survey knowledge present in the literature (e.g. [29], [30]), we look at this issue from a system perspective and present a methodology that might enable a system to infer this knowledge.

In Figure 15, three routes are depicted; the first three segments of all these three routes are common. If the learning factor has been determined as three, then the next time a route is requested, the system can verify if the requested route includes any of the three segments and model them as a K section in the models presented earlier. The system can then treat the route as a <kroute> and apply the knowledge chunking rules to the route directions, before presenting the schematized route to the user. As the reader may recall as explained in Experiment 2, we made an attempt to determine a rough estimate of what the learning factor might be by asking participants to change the color of the marker they used while marking the routes (use a different color for sections they were familiar or unfamiliar with). We determined that on average (excluding those who knew the DPk from the beginning) it took 2.4 instances of sketching a route that passes through a DPk before subjects felt they were confident of locating it. In other words, in less than three attempts of sketching a route that passed through the DPk, subjects felt they knew how to locate it. It must also be noted that they were made to alternate between the two types of routes (i.e. the ones that included the DPk and the ones that did not). This may be considered as a preliminary study, because the effectiveness of route and survey knowledge is a complex issue (see [30]). However, it is possible to imagine an automated wayfinding system that implements the principle discussed, tracks a user's movement, and subsequently, over time, be able to determine with certain confidence that a section of a route is familiar to a user.

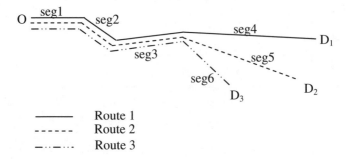

Fig. 15. The Figure depicts three routes, with the same origin and three different destinations. The first three segments of each route being common across the three routes.

Another possible implementation is the combination of these principles with the use of Map Gestures [31]. A map gesture can be used to direct a user from a DPk to the destination in the NK model, or from the origin to the DPk in the KN model.

6 Concluding Remarks

The formalization presented in our work concerns route descriptions but some of the concepts identified can be naturally extended and applied to the knowledge based schematization of maps. Empirical evidence confirmed a wayfinder's preference for schematized route directions based on her knowledge. Earlier studies suggest that schematization of maps is possible, and important [17], [18], knowledge based schematization of maps seems to be a natural extension of this work, albeit a challenging issue. For instance, in route descriptions, the conceptual element that promotes schematization of the route is the knowledge chunk; one can imagine the concept of knowledge chunking being applied to maps.

This relates to other recent work by Tomko and Winter where they use granular route directions to represent the elements of a city from a hierarchical city structure [16]. However, an important issue that does arise is the manner in which this may be carried out. We also believe that map gestures [31] might play an important role in presenting maps that represent such environments however this issue will have to be confirmed with empirical studies. While we focused on the relationship between the level of detail of route directions that a wayfinder prefers and her familiarity of a section, a key extension to this work is an investigation of the appropriate modalities for various levels of knowledge. Hirtle and Sorrows address the issue of different modalities by designing a multimodal tool for locating buildings on a college campus [32]. The tool that was built is an online system that incorporates three modes of information, maps, verbal directions and images. The main purpose of the multimodal tool developed as part of their work was to build on existing knowledge of the wayfinder. An interesting issue that calls for investigation is to determine the best modalities for various levels of user's knowledge.

References

1. Klein, W.: Local deixis in route directions. Wiley, New York (1982)
2. Klein, W.: Deixis and spatial orientation in route directions. Plenum Press, New York (1983)
3. Talmy, L.: Semantics and syntax of motion. Academic Press, New York (1975)
4. Wunderlich, D., Reinelt, R.: How to get there from here. Wiley, New York
5. Allen, G.L.: From Knowledge to Words to Wayfinding: Issues in the Production and Comprehension of Route Directions. In: Frank, A.U. (ed.) COSIT 1997. LNCS, vol. 1329, pp. 363–372. Springer, Heidelberg (1997)
6. Allen, G.L.: Principles and Practices for Communicating Route Knowledge. Applied Cognitive Psychology 14, 333–359 (2000)
7. Mark, D.M., Gould, M.D.: Wayfinding directions as discourse: A comparison of verbal directions in English and Spanish. Multilingua 11, 267–291 (1992)
8. Tversky, B., Lee, P.U.: Pictorial and verbal tools for conveying routes. Spatial Information Theory 1661, 51–64 (1999)
9. Streeter, L.A., Vitello, D., Wonsiewicz, S.A.: How to tell people where to go: comparing navigational aids. International Journal of Man-Machine Studies 22, 549–562 (1985)
10. Daniel, M.-P., Denis, M.: Spatial Descriptions as Navigational Aids: A Cognitive Analysis of Route Directions. Kognitionswissenschaft 7, 45–52 (1998)
11. Denis, M., Pazzaglia, F., Cornoldi, C., Bertolo, L.: Spatial Discourse and Navigation: An Analysis of Route Directions in the City of Venice. Applied Cognitive Psychology 13, 145–174 (1999)
12. Lovelace, K.L., Hegarty, M., Montello, D.R.: Elements of good route directions in familiar and unfamiliar environments. Spatial Information Theory 1661, 65–82 (1999)
13. Raubal, M., Winter, S.: Enriching Wayfinding Instructions with Local Landmarks. In: Egenhofer, M.J., Mark, D.M. (eds.) GIScience 2002. LNCS, vol. 2478, pp. 243–259. Springer, Heidelberg (2002)
14. Tom, A., Denis, M.: Referring to landmark or street information in route directions: What difference does it make? In: Kuhn, W., Worboys, M.F., Timpf, S. (eds.) COSIT 2003. LNCS, vol. 2825, pp. 362–374. Springer, Heidelberg (2003)
15. Fontaine, S., Denis, M.: The Production of Route Instructions in Underground and Urban Environments. In: International Conference on Spatial Information Theory: Cognitive and Computational Foundations of Geographic Information Science, pp. 83–94. Springer, Heidelberg (1999)
16. Tomko, M., Winter, S.: Recursive Construction of Granular Route Directions. Journal of Spatial Science 51, 101–115 (2006)

17. Klippel, A., Richter, K.-F., Barkowsky, T., Freksa, C.: The Cognitive Reality of Schematic Maps. In: Meng, L., Zipf, A., Reichenbacher, T. (eds.) Map-based Mobile Services - Theories, Methods and Implementations, pp. 57–74. Springer, Heidelberg (2005)
18. Agrawala, M., Stolte, C.: Rendering Effective Route Maps: Improving Usability Through Generalization. In: Conference on Computer Graphics and Interactive Techniques, pp. 241–250 (2001)
19. Klippel, A.: Wayfinding Choremes - Conceptualizing Wayfinding and Route Direction Elements. In: Department of Mathematics and Informatics, Vol. Doctorate of Philosophy, p. 195. University of Bremen, Bremen, Germany (2003)
20. Klippel, A.: Wayfinding Choremes. In: Kuhn, W., Worboys, M.F., Timpf, S. (eds.) COSIT 2003. LNCS, vol. 2825, pp. 320–334. Springer, Heidelberg (2003)
21. Klippel, A., Tappe, H., Kulik, L., Lee, P.U.: Wayfinding Choremes - A Language for Modeling Conceptual Route Knowledge. Journal of Visual Languages and Computing 16, 311–329 (2005)
22. Werner, S., Krieg-Brückner, B., Herrmann, T.: Modelling Navigational Knowledge by Route Graphs. In: Habel, C., Brauer, W., Freksa, C., Wender, K.F. (eds.) Spatial Cognition II. LNCS (LNAI), vol. 1849, pp. 295–316. Springer, Heidelberg (2000)
23. Sorrows, M.E., Hirtle, S.C.: The nature of landmarks for real and electronic spaces. Spatial Information Theory 1661, 37–50 (1999)
24. Presson, C.C., Montello, D.R.: Points of reference in spatial cognition: Stalking the elusive landmark. British Journal of Developmental Psychology 6, 378–381 (1988)
25. Golledge, R.G., Spector, A.: Comprehending the Urban Environment. Geographical Analysis 10, 403–426 (1978)
26. Gale, N., Golledge, R.G., Halperin, W.C., Couclelis, H.: Exploring Spatial Familiarity. The Professional Geographer 42, 299–313 (1990)
27. Tom, A., Denis, M.: Language and spatial cognition: Comparing the roles of landmarks and street names in route instructions. Applied Cognitive Psychology 18, 1213–1230 (2004)
28. Garling, T., Golledge, R.G.: Environmental perception and cognition. In: Zube, E., Moore, G. (eds.) Advances in Environment, Behavior and Design, Plenum, New York, vol. 2, pp. 203–236 (1987)
29. Hirtle, S.C., Hudson, J.: Acquisition of Spatial Knowledge for Routes. Journal of Environmental Psychology 11, 335–345 (1991)
30. Golledge, R.G., Dougherty, V., Bell, S.: Acquiring Spatial Knowledge: Survey versus Route-Based Knowledge in Unfamiliar Environments. Annals of the Association of American Geographers 85, 134–158 (1995)
31. Hirtle, S.C.: The use of maps, images and "gestures" for navigation. Spatial Cognition II 1849, 31–40 (2000)
32. Hirtle, S.C., Sorrows, M.E.: Designing a multi-modal tool for locating buildings on a college campus. Journal of Environmental Psychology 18, 265–276 (1998)

Map Use and Wayfinding Strategies in a Multi-building Ensemble

Christoph Hölscher[1], Simon J. Büchner[1], Tobias Meilinger[1,2], and Gerhard Strube[1]

[1] University of Freiburg, Centre for Cognitive Science
Friedrichstr. 50, 79098 Freiburg, Germany
{hoelsch,buechner,strube}@cognition.uni-freiburg.de
[2] Max-Planck-Institute for Biological Cybernetics
Spemannstr. 34, 72076 Tübingen, Germany
tobias.meilinger@tuebingen.mpg.de

Abstract. This experiment investigated the role of familiarity, map usage and instruction on wayfinding strategies and performance. 32 participants had to find eight goals in a multilevel building ensemble consisting of two distinctive vertical segments. Generally users who were familiar with the building ensemble outperformed first-time visitors of the setting. We tested if the standard wall-mounted floor maps found in the majority of public buildings can help navigation in a complex unknown environment. Unfamiliar users tried to make use of these plans more frequently, but were not able to compensate for spatial knowledge deficits through them. Two strategies of across-level wayfinding are compared with respect to a region-based hierarchical planning approach. Strategy selection relied largely on task and instruction characteristics. Overall, the strategy of moving horizontally into the target section of the building prior to vertical travel was shown to be more effective in this multi-building setting.

1 Introduction

Finding one's way around public buildings such as airports, hospitals, offices, or university buildings often proves to be a tedious and frustrating task. Wayfinding in a complex setting with less than perfect knowledge requires decision making under uncertainty and we aim to identify the behaviours and strategies that people employ to navigate in such environments.

Several researchers [e.g. 5, 12] stressed the role of familiarity with a building, and illustrated how training of sequential routes or survey knowledge can boost orientation performance in complex buildings like hospitals. Most studies on wayfinding performance and building complexity have limited themselves to investigations in the horizontal plane of isolated floor levels. Soeda et al. [16] were able to show that wayfinding performance in tasks involving floor level changes is largely hindered by disorientation during vertical travel on stairs or elevators, extending earlier observations [13].

Hölscher et al. [7] report findings about wayfinding difficulties observed in a complex multi-level conference centre: identifying incongruent floor layouts, disorienting staircases and lack of visual access to important level-related building features as main causes of difficulty [cf. 18].

The current study extends this research to a setting which introduces an additional dimension of complexity: It comprises of both multiple levels, and multiple building parts that differ in their vertical and horizontal configuration. Hölscher et al. [7] have pointed both to the importance of familiarity with a building and the use of specific strategies to support route choice decisions in multi-level wayfinding. The building setting in the current paper provides the opportunity to shed further light on the relationship between these aspects, and how the configuration of the building and task characteristics contribute to it.

1.1 Wayfinding Strategies in Complex Buildings

When people have only an incomplete or uncertain representation about a spatial setting, they need to rely on navigation strategies to fill in the gaps. Hölscher et al. [7] identified three wayfinding strategies that are used to support route choice decisions in three-dimensional multi-level buildings: The *central point strategy* of sticking as much as possible to main corridors and main places in the building – predominantly used by first-time visitors – and the *direction strategy* of choosing routes that head towards the horizontal position of the goal as directly as possible, irrespective of level-changes (cf. least-angle strategies [2, 6]). Both proved to be inferior to a third strategy based on forming a hierarchically organized navigation plan. Such a strategy seems to be based on cognitively segmenting the setting into regions which guide navigation decisions (the approach was introduced by Wiener et al. [19], albeit for two-dimensional outdoor settings). This later strategy reduces the complexity of planning and navigation by first entering the target region before starting fine-tuned search. In the Hölscher et al. study [7], this strategy of first moving to the correct floor in the building was characterized as a *floor strategy,* since in that building floor levels were the predominant hierarchical aspect of the building.

Research questions & predictions. For the building complex in the present study, it is not immediately obvious how a route choice preference would look if a hierarchy-based, region-oriented strategy is employed: The building consists of at least two identifiable building parts which could be interpreted as regions by a navigator. At the same time, however, the building has a multi-level structure, which may also foster a level-based mental regionalization of the building complex. Thus it is not clear how the participants will segment the building when applying a hierarchy-based strategy. Also, it is not clear if individuals idiosyncratically prefer a certain strategy or if they adapt it according to external constraints such as the task, the instruction or familiarity.

Based on the results in Hölscher et al. [7] we will test the hypothesis that a floor-based strategy will be a) more successful than a direction-based strategy and that b) users familiar with the building will rely on it to a greater extent.

Besides the characteristics of the building, wayfinding tasks can differ especially regarding the amount and type of goal directed information they provide and thus influence wayfinding strategy choices. Rooms in public buildings are generally identified by a room number, a propositional descriptive scheme that implies information about the general region of a goal location, i.e., in a task like "go to room no. 1215" the participant knows on which floor he will find the target room. The room numbering scheme in this setting also provides information about the building part in which the room is located. In contrast to this, a task instruction that only provides directional information (and no propositional) may not foster a hierarchical planning process to a similar extent. In this case people may try to minimize the angular deviation from the linear distance and follow a direction-based least angle strategy [2, 6]. Thus, the kind of information that is provided in the task instruction may elicit route choice and performance differences. We expect the number-based task instruction to foster a floor-based strategy rather than a direction-based strategy.

1.2 The Role of Maps

Thorndyke & Hayes-Roth [17] found evidence that the information learned from a map is quite distinct from information learned while navigating in an environment, since the map provides survey knowledge rather than accurate, direct route knowledge. Gärling et al. [5] presented evidence that showing a floor map to participants immediately prior to testing reduced the effects of familiarity with the building and improved wayfinding performance. Learning from a map can be equally effective as having actually visited a building [14] or even lead to better performance compared to long-term users of a building [12].

Yet it is not unequivocally clear that access to floor maps does indeed have a positive impact on real-life wayfinding performance. Butler et al. [1] present evidence that you-are-here maps in a similarly complex setting had no positive effect at all, in fact, wayfinders attending to such wall-mounted maps lost time without gaining any navigational advantage. It is also well-documented that using a map that is misaligned with one's current orientation can be detrimental [11], a feature of many standard floor maps in office buildings.

Independent from map alignment all maps have to be transferred from survey perspective into route perspective. This transfer is associated with switching costs [10, 15]. Learning an environment from direct experience or an oblique perspective drawing, therefore, can lead to better performance than learning from a floor plan, or from a floor plan and a cross-section [4].

We have chosen a specific type of map for our study: In all public buildings in Germany – as well as most of the Western world – fire regulations specify exactly how the floor plans on display must be designed. These plans are defined by law and are omnipresent. One would naturally hope that these maps adequately support wayfinding tasks in the building especially in case of an emergency.

Research questions & predictions. We have several expectations towards the outcome of the experiment. In general familiar participants should perform better than unfamiliar ones due to their enhanced knowledge about the environment. Also, participants using a map should perform better than those not using a map. Map users receive additional information about the environment which participants not using a map don't have available. In order to built up a spatial representation of the environment they have to rely on the knowledge they have acquired by navigating through it. However, the study by Butler et al. [1] would suggest no effect of map use.

Concerning the interaction of familiarity and map use the studies of Richardson et al. [14] and Moeser [12] led us to the expectation that the availability of maps will allow unfamiliar users to - at least partially - compensate for their spatial knowledge deficit compared to familiar users. Thus, unfamiliar participants in the map condition should use maps more frequently than familiar participants but with respect to their performance they should be as successful as familiar participants.

2 Method

In the experiment, navigational performance and verbalisation of visitors with prior experience with the adjacent university buildings was compared to the behaviour of novices visiting for the first time. Map usage was controlled by either, allowing or disallowing the participants to use the frequently displayed fire plans of the building.

Generally, the task consisted of finding a goal specified by room number. In two of the tasks this instruction was systematically compared to giving exact location information by showing the room from the outside, without mentioning the room number.

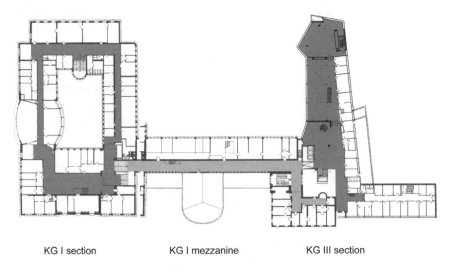

KG I section   KG I mezzanine   KG III section

Fig. 1. Plan view of the building ensemble. The left and the central part belong to the original building from 1911, the right part was added later.

2.1 Participants

16 women and 16 men, aged between 18 and 34 ($M = 23$, $SD = 3.0$), participated in the experiment. The 16 participants familiar with the building had visited it on a weekly basis for at least two semesters. The 16 unfamiliar participants, by contrast, had never been to the building at all. Participants were randomly assigned to the map and instruction condition, yielding experimental groups of equal size.

2.2 Material

The design of Freiburg University's main building (see Fig. 1) is composed of two distinct building sections, built in 1911 (KG I) and 1961 (KG III). The KG I section consists of a squarish main section with a linear outbuilding (mezzanine, half-height level change), that at its end meets the open-space design of the KG III section.

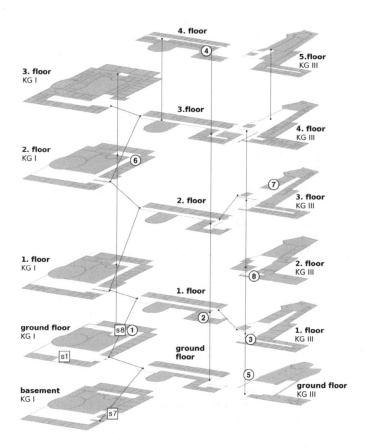

Fig. 2. Configuration of floors with circulation areas[1]

The floors of the building parts are connected by various passages and stairways (except 2nd floor, KG III). Due to different heights between the floors, and different numbers of floors (four at KG I compared to five at KG III) the user has to change the floors in an irregular and often unexpected way (see Fig. 2). The numbering system in the building consists of 4-digit room numbers. The first digit codes whether a room is

---

[1] Starting points and goals of the navigation tasks are marked by numbers (example: "s1" is the starting point of the first task, "1" its goal location. As tasks are linked, "1" marks the starting point for task 2 and "2" marks its goal.). The figure also documents the mismatching floor numbering scheme across buildings parts.

located in the KG I or KG III part (e.g. 1019 vs. 3047), treating the main part of KG I and its mezzanine floors as one. The second digit corresponds to the floor level, counted separately for KG I and KG III, leading to a mismatch on upper levels (see Fig. 2).

The wall-mounted maps in our experiment are standard fire plans that, by law, are required to be posted at specified points in all public buildings. Each plan depicts the layout of the current floor, but is limited to either KG I or KG III. Plans are located next to main staircases and elevators on each level. 11 out of these 20 wall-mounted fire maps were misaligned by 90° or 180°, the rest were properly aligned with a viewer's facing direction.

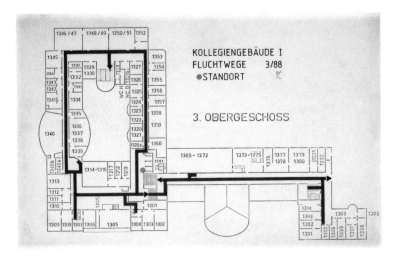

Fig. 3. Example fire plan, located in the KG I main section, 3rd floor

2.3 Procedure

The experiment lasted about 90 minutes, including the instruction and debriefing as well as post-experimental written tests (to be reported in a later publication, as they are beyond the scope of this paper). Participants were instructed to think aloud and thus verbalize their thoughts and considerations [3, 13]; performance was video-taped. Half of the participants were allowed to use the floor maps of the building, whereas the other half were not. Due to the size of the fire plans it was not possible to read the maps without standing close to it, so in the no map condition it was impossible for the participants to look at the maps without the experimenter taking notice. The participants' main task was to subsequently find eight locations in the building. All participants received the tasks in the same order, and the tasks were linked such that one task's goal location was the next task's starting point (see Fig. 2).

Tasks were selected in a way that their difficulty increased (with respect to the number of floor changes and the requirement to change the buildings). This was done in order to accommodate for the participant's increasing knowledge about the building while she performs the tasks. In order to provide a fairly realistic situation, the wayfinding tasks were instructed by verbally providing the room number such as "Find room number 1019".[2] This type of instruction provides the same information as a student might have when he gains the room number from his timetable.

While tasks 1-3 were designed to foster only one wayfinding strategy each, the remaining tasks were planned to equally allow for both a floor-based and a direction-based strategy, i.e., changing first to the floor of the assumed goal location (floor strategy) or walking first horizontally into the direction where the goal is assumed to be located (direction strategy). In order to investigate the effect of instruction we varied the type of instruction (propositional vs. directional) in two tasks: For one half of the participants it was changed in task 6, for the other in task 7. This was done in order to balance between the directions of movement: The two tasks were matched with respect to complexity of the path, level changes and length. Instead of a number-based instruction, the target room was shown visually to the participant in such a way that the experimenter pointed to the target room, while the participant was standing in the door leading to the courtyard. In addition the target room was marked by a black "X" in the window. This type of instruction provided both vertical and horizontal direction information *non-verbally*. It explicitly avoided any reference to room-numbering or level information (as an instruction like "it is the third window from the bottom" might have done). The order of instruction, i.e., whether participants received the X instruction on task 6 *or* on task 7, was randomly assigned.

2.4 Measures

Performance. For each task, the shortest route as well as a list of reasonable route alternatives was determined beforehand. Navigation performance was measured with a number of variables:

- Duration: time to complete the task
- Stops: number of stops other than stops at maps
- Complete stop time: sum of the time of all stops
- Detour distance: the length of detours, where detour is defined as "leaving the path and returning to it at the same point".
- Distance ratio: the ratio of the walked distance to the shortest distance possible

Specific behaviors and verbalization measures. The second group of measures classified the participants' verbal comments. The traversed route of each task was first drawn into the plans of the building. This was also used to determine route distances and superfluous way after getting lost (see above). The verbal codes and stops were written beside this drawn trajectory at the location where they were mentioned. The coding scheme for classifying the verbal comments was fine-tuned until all categories could be reliably recognized with a sufficient inter-rater reliability (kappa value of .70; for a detailed description of verbal data handling, see [7]).

---

[2] The target locations were: Room 1019 (task 1), r. 1192 (task 2), r. 3120 (task 3), r. 1486 (task 4), courtyard (task 5), r. 1234 (task 6), r. 3301 (task 7), r. 3220 (task 8).

The following measures were collected based on verbal comments:
- Looked at room no.: frequency of looking at room numbers in order to orientate oneself.
- Orient on outside: frequency of looking to the outside in order to orientate oneself.
- Forgot room number: frequency of forgetting the target room number and asking the experimenter for it.

3 Results

The wayfinding tasks were analyzed with respect to both performance measures and corresponding verbalization. First, we present results regarding the influence of familiarity with the building and map usage on wayfinding performance. Second, we report strategy-related results. A multivariate ANOVA was performed with the within-factor task and the between factors familiarity (familiar, unfamiliar) and map usage (map used, no map used). The MANOVA yielded a significant main effect of familiarity, $F(12, 17) = 2.37$, $p = .05$ and a trend for map usage $F(12, 17) = 2.02$, $p = .09$. The interaction was not significant, $F(12, 17) = 0.85$, $p = .60$.

Table 1. Selected means of performance and behavior measures in the analysis of familiarity and map usage effects

Measures	Familiarity			Map usage		
	Unfamiliar	Familiar		No map used	Map used	
Duration [s]	214	160		182	194	
Stops [n]	3.29	1.66	*	2.51	2.42	
Complete stop time [s]	41	20	*	23	39	*
Detour distance[m]	39	21	*	33	26	
Distance ratio	1.90	1.52	*	1.75	1.66	
Looked at room no.	1.07	0.81		1.19	0.63	*
Orient on outside	0.20	0.06	*	0.11	0.15	
Forgot room number	0.60	0.36	*	0.51	0.45	
Failure of a plan	0.09	0.09		0.10	0.07	
Stops at maps [n]	1.09	0.59	*			
Time/stop at maps [s]	20	9	*		(* p<.05)	

3.1 Familiarity

Participants familiar with the building were expected to perform better than those who were not. They completed the task faster, the paths they chose were shorter, they stopped less often, and detours were shorter and less frequent. The total distance they walked was shorter and their ratio of walked distance and shortest distance possible was smaller. In addition they oriented on the outside less often. Familiar participants forgot the number of the target room less often than unfamiliar participants (all

$F(1, 28) > 5.83$, $p < .025$, $\eta^2_{partial} > .17$; for means see Table 1). The result, that familiar participants performed better than unfamiliar participants, was expected and probably due to their deeper spatial knowledge about the test environment.

According to our hypothesis, participants who were familiar with the building should be able to rely on their knowledge about the building and not be dependent on information from maps. We did a separate analysis of the number of stops and the time spent at maps including only participants from the map condition since only they were allowed to use maps[3]. Here, the overall effect of familiarity, $F(2, 11) = 6.13$, $p = .016$, was significant. The main effect was significant for both the number of stops at maps, $F(1, 12) = 10.36$, $p = .007$, $\eta^2_{partial} = .46$ and the time spent at maps $F(1, 12) = 11.33$, $p = .006$, $\eta^2_{partial} = .49$. Thus, unfamiliar participants used maps more often and for a longer time, probably trying to compensate for their lacking knowledge about the building. For familiar participants it was not necessary to use maps very often due to their elaborate knowledge about the setting. However, even familiar participants who visit the setting regularly performed far from perfect.

3.2 Maps

Unfamiliar participants used maps more often, but map use did not lead to any performance advantages: No significant difference could be found in any of the performance measures except for total stop time, $F(1, 28) = 7.10$, $p = .013$, $\eta^2_{partial} = .20$ (for means see Table 1). This particular result is caused by the map users spending extra time reading the map while non-map users did not. Participants with map access used room numbers less often than participants who didn't use a map, $F(1, 28) = 9.57$, $p = .004$, $\eta^2_{partial} = .26$.

No interaction between familiarity and map usage was observed in any of the measures. So contrary to our hypothesis, map usage did not enable unfamiliar participants to compensate for the lack of knowledge that they were facing.

Yet, not all maps were aligned with their surrounding. Could the lack of performance advantages in maps be due to improper map alignment, generally known to hamper performance? To answer this, we separately analyzed task 3 & 7 in which only correctly aligned maps were encountered by the participants: again neither a significant main effect of map use, nor an interaction with familiarity was established. Thus, the lack of advantage for map users cannot be ascribed to misaligned maps.

3.3 Strategies

The analysis of strategies in this study was twofold: On the one hand we investigated strategy *choice* as a function of different factors like task instruction and familiarity. On the other hand we investigated the success (in terms of wayfinding performance) of a chosen strategy.

---

[3] For this and the following analyses, two participants (one familiar, one unfamiliar) were moved from the map condition to the no-map condition, as they never used a map in any of the tasks. They did not gain any information from external representations and thus may be treated in the same way as participants who were not allowed to use a map.

Our hypothesis stated that strategic choice would vary as a function of task, instruction and familiarity, not by intrinsic preferences of individuals. This could be clearly established (cf. Table 2).

Table 2. Task properties (left panel) and frequencies of strategy choice (right panel)

Task	Instruction	Task properties		Frequencies		
		Floor change required (No. of floors)	Building change required	Floor strategy	Direction strategy	Sum
1		no (0)	no	-	-	16
2		yes (1)	no	25	5	30
3		no (0)	yes	2	24	26
4		yes (3)	yes	18	10	28
5		yes (4)	optional	20	8	28
6	X	yes (3)	yes	5	10	15
	Room No.	yes (3)	yes	10	7	17
7	X	yes (3)	yes	5	11	16
	Room No.	yes (3)	yes	7	6	13
8		yes (2)	yes	7	25	32

In the tasks where both strategies were applicable (4-8) only two participants chose the same strategy throughout all tasks. In general, in different tasks different strategies were applied ($\chi^2(7, N = 221) = 68.8, p < .001$)[4].

Also, the variation of instruction type in task 6 and 7 led to the choice of different strategies (Fig. 4): If participants had to reach the room marked with an 'X' they more often changed the building first, whereas if they had to reach the same room identified by its number they more often changed the floor first (task 6 & 7 added together, $\chi^2(1, N = 61) = 3.68, p < .028$, one-tailed).

Across all eight tasks, familiar and unfamiliar participants acted very similarly: No general difference in their choice of strategy across all tasks was revealed ($\chi^2(1, N = 221) = 1.15, p = .283$). Similar to familiar and unfamiliar participants, map users and participants not using a map also did not differ in their general strategy preference ($\chi^2(1, N = 221) = 0.03, p = .956$), nor did women and men ($\chi^2(1, N = 221) = 0.17, p = .392$). So the two main factors determining the choice of a strategy are task and instruction. Familiarity, gender and individual preferences play a minor role.

---

[4] Not all subjects could be clearly assigned to one strategy in every task. For comparing the influence of task, both instructions in task 6 & 7 were added together.

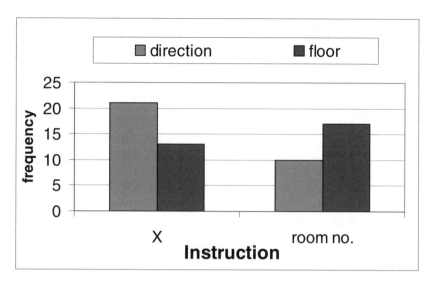

Fig. 4. Choice of strategy (frequency) as a function of instruction. The figure shows the data from tasks 6 & 7 combined.

But which strategy was more useful? In order to answer this question, we performed a MANOVA on the above mentioned measures with factors strategy, instruction and task. The latter two factors were included in order to statistically control the influence of task and instruction. As it was not possible to reasonably apply both strategies in task 1, 2 or 3, these tasks were excluded from this analysis.

Table 3. Selected means of performance and behavior measures in the analysis of the post-hoc identified strategy choices

Measures	Strategy		
	Floor	Direction	
Duration [s]	212	153	*
Stops [n]	2.76	1.82	
Complete stop time [s]	31	24	
Detour distance [m]	29	16	*
Distance ratio	1.86	1.33	*
Looked at room no.	1.11	0.65	
Orient on outside	0.17	0.08	*
Forgot room number	0.56	0.39	
Failure of a plan	0.15	0.02	*
		(* $p<.05$)	

For the remaining five tasks, participants who changed the floor before changing the building performed worse: There was a main effect of strategy $F(12, 191) = 4.05$, $p < .001$ in the way that people who changed the floor first performed worse than those who changed the building first. They had to detour more, their detours were longer, they walked longer distances, and took longer to reach their goal. Their ratio of the walked distance and shortest possible distance was higher, they oriented on the outside more often and their plans failed more often (all $F(1, 202) > 5.42$, all $p < .025$, all $\eta^2_{partial} > .03$, for means refer to Table 3).

4 Discussion

We have identified major wayfinding performance differences between participants who were familiar with the building and those who were not. This finding was expected as it confirms the finding of familiarity effects by other researchers. It demonstrates how difficult indoor wayfinding in a realistically complex setting is for untrained users, the target audience of most public buildings. Even the familiar participants performed far from perfect in this setting, pointing to usability challenges in the multi-level multi-building ensemble. The observation that participants frequently forgot target room numbers suggests a severe memory load imposed by the complex building structure. This stresses a practical need to better understand wayfinding challenges in architecture.

4.1 Maps

When given the opportunity, unfamiliar participants tried to compensate for their lack of prior knowledge longer and more often than familiar users by making use of the wall-mounted maps. However, surprisingly it turned out that the floor maps did not have any positive impact on wayfinding *performance*, neither for first-time nor for regular visitors.

In fact, reading the maps required additional time and did not receive any payback in performance enhancement [1]. One might suspect that the map deficits relate to a lack of alignment, but we were able to show that this is not a decisive factor here. Why did the floor maps not help in wayfinding? Despite violating several design principles [8], the fire plans provide detailed survey information about single floors. However, the wayfinding tasks require across-level navigation. Participants may have had trouble to integrate information from single floor maps into a coherent multi-level representation of the whole building or even parts of it. In this particular building ensemble, the floor numbering is inconsistent between the main parts of the building (KG I vs. KG III) and the floor numbering is not transparent with respect to the mezzanine levels. Nonetheless, the main deficit of the fire plans clearly generalizes to other multi-level multi-building environments: In most real-world cases the fire plans of complex buildings do not cover the area beyond a single building section or floor and thus fail to supply the necessary information for route planning across floors or sections.

The fire plans were not only designed to help fire fighters entering the building to quickly assess its topography in case of a fire but also to help visitors to quickly find a way out in case of an emergency. In our study participants were instructed to find the target room as quickly as possible. At the same time they were instructed to solve the task in a calm manner without any hurry, so they actually had as much time as they wanted to study the map and orient themselves in the building. This would clearly not be the case in an emergency where people are under stress and tend to panic. If the maps didn't help the participants in a common wayfinding task, how can they ever be useful in an emergency situation? This as well as other research [8] shows that the deployment and design of emergency maps have to be re-considered and be subject to empirical testing.

4.2 Strategies

When people have incomplete knowledge about a path network, they rely on strategies to guide their path choices. We found that strategy choice is largely influenced by characteristics of the building, task features and the instruction rather than being an idiosyncratic preference of the individual participant.

The strategies tended to follow a pattern of regionalization: The building can be mentally divided into a dual hierarchy of building parts and floor levels. When only one hierarchical aspect is involved (the floor change in task 2 or the building change in task 3), people relied very strictly on it. Thus, they changed either the floor or the building before searching for the room's horizontal position.

When participants had to change building and floors within a single task, the picture was more ambiguous. If the participants knew that the goal was further away horizontally than vertically, they first walked horizontally – changing the building – and then went for the correct floor. This was the case with the visual instruction (X in the window) in task 6 and 7 as well as in task 8 where the participants first had to cross the whole building KG I. If the participants otherwise had to assume that the horizontal distance was – at least *potentially* – shorter, compared to the vertical one, they first changed the floor and only then the building section. That was the case in task 4 and 5 and in the room number instruction in task 6 and 7. We must assume that participants did not know where exactly in a floor the room they were looking for was located and that they therefore first tried to reach the closest part of the floor in the correct part of the building (KG I vs. KG III), to start their search. This sub-goal was considered the relevant reference for strategy selection, especially since this heuristic avoids unnecessary backtracking in many cases. In other words, if two options for a hierarchical strategy were equally available in a task, the vertical and horizontal direction information was used to disambiguate the path choice.

Strategy choice also seems to be strongly influenced by the type of information provided with the task (tasks 6 & 7): At least in our setting, providing a room number provoked a floor-first strategy choice, while a non-propositional instruction in the identical task, providing direction information initiated a change-building-first choice.

In the latter case, the behavior we observed is in line with an attempt of the participant to minimize the angular deviation from the linear distance, since the target location is further away horizontally than vertically. This pattern can be interpreted as a variant of what for the strictly horizontal case has been termed a *least-angle-strategy* [6].

It is an open issue for further studies to identify whether the information provided in the task description or the dual hierarchy account discussed above has greater impact, as we have not systematically juxtaposed these aspects in the present study. It will be especially relevant to see how regionalisation and a least-angle heuristic might interact, e.g., whether people initially regionalise the building (initial strategy) and then apply the least-angle heuristic at decision points further on the way.

Finally, which strategy was more successful? In the case of this multi-level *multi-building* setting, the strategy of first moving into the correct part of the building before changing to the proper level was much more effective across a number of performance measures. On first sight, this observation is at odds with the performance boost found in Hölscher et al. [7] for exerting a floor-first strategy in a multi-level building. But in fact, in both studies the effective strategy relies on hierarchical planning to overcome difficulties in the building. The present study extends the complexity of the multi-level building in Hölscher et al. (2005) to a multi-building setting. This multi-building setting has a mismatching number of floors, so that, e.g., for the second and third floor, changing between the buildings was not easily possible and the 4th floor of KG III had the same elevation as the 3rd floor of KG I. Across all tasks the change-building-first strategy is best suited to overcome the additional burden of the present setting, namely the inconsistencies in floor numbering and relative floor elevations between parts of the building.

Also in contrast to our previous findings on strategy choice, no general differences between strategy choices of participants who were familiar with the building and novices was found, a fact that is likely attributable to the strong influence of task- and instruction-related effects. Yet, for the one task where familiarity differences were clearly identified, the familiar participants definitely preferred the more successful strategy. This is a direct replication of the earlier finding, the experts' preference for a strategy corresponding with the strategy's success rate.

4.3 Future Research

We will continue our efforts to extend the indoor wayfinding research to further types of public buildings with different spatial layouts and properties. Both the settings in Hölscher et al. [7] and in the present study owe their complexity mostly to vertical structure, while within a single level the path choices were relatively limited. It remains to be investigated how the hierarchically-based wayfinding strategies that we have identified extend to layouts that provide greater within-floor horizontal extend and complexity. Hospital or airport terminal buildings would be prime examples of such settings to validate the scope of applicability as well as the adaptive value of human wayfinding strategies.

A follow-up study has been conducted in the same building in which we have re-designed maps and installed a signage system considering cognitive factors in navigation. The study involved the same tasks in order to compare results. The data is currently analyzed, but first results indicate that even large, salient maps which were designed particularly for the purpose of easy navigation may not allow first-time visitors to reliably compensate for their lack of knowledge in such a complex real-life building setting. Instead, signs appear to be the main attractors of attention and people tended to follow them instead of intensively studying the maps. The reasons for these results and its theoretical implications will need to be carefully investigated.

Acknowledgements. The research is supported by the German Research Foundation (DFG) in the Transregional Collaborative Research Center "Spatial Cognition" (SFB/TR-8). We are grateful to Martin Brösamle and Leonard Reinecke for support with data collection and analysis, and to Lara Webber for proof-reading and comments.

References

1. Butler, D.L., Acquino, A.L., Hissong, A.A., Scott, P.A.: Wayfinding by newcomers in a complex building. Human Factors 35(1), 159–173 (1993)
2. Conroy Dalton, R.: The secret is to follow your nose: Route path selection an angularity. Environment and Behavior 35(1), 107–131 (2003)
3. Ericsson, K.A., Simon, H.A.: Protocol analysis: Verbal reports as data. MIT Press, Cambridge (1993)
4. Fontaine, S.: Spatial cognition and the processing of verticality in underground environments. In: Montello, D.R. (ed.) COSIT 2001. LNCS, vol. 2205, Springer, Heidelberg (2001)
5. Gärling, T., Lindberg, E., Mäntylä, T.: Orientation in Buildings: Effects of familiarity, Visual Access, and Orientation Aids. Journal of Applied Psychology 68(1), 177–186 (1983)
6. Hochmair, H., Frank, A.U.: Influence of estimation errors on wayfinding decisions in unknown street networks - analyzing the least-angle strategy. Spatial Cognition and Computation 2(4), 283–313 (2002)
7. Hölscher, C., Meilinger, T., Vrachliotis, G., Brösamle, M., Knauff, M.: Finding the Way Inside: Linking Architectural Design Analysis and Cognitive Processes. In: Freksa, C., Knauff, M., Krieg-Brückner, B., Nebel, B., Barkowsky, T. (eds.) Spatial Cognition IV. LNCS (LNAI), vol. 3343, pp. 1–23. Springer, Heidelberg (2005)
8. Klippel, A., Freksa, C., Winter, S. (i.p.): You-are-here maps in emergencies – The danger of getting lost. Spatial Science
9. Lawton, C.A.: Strategies for indoor wayfinding: The role of orientation. Journal of Environmental Psychology 16(2), 137–145 (1996)
10. Lee, P.U., Tversky, B.: Costs of Switching Perspectives in Routed and Survey Descriptions. In: Proceedings of the twenty-third Annual Conference of the Cognitive Science Society, Edinburgh, Scotland (2001)
11. Levine, M., Jankovic, I., Palij, M.: Principles of spatial problem solving. Journal of Experimental Psychology: General 11, 157–175 (1982)
12. Moeser, S.D.: Cognitive Mapping in a Complex Building. Environment and Behavior 20(1), 21–49 (1988)
13. Passini, R.: Wayfinding in architecture, 2nd edn. Van Nostrand Reinhold Company, New York (1992)
14. Richardson, A.R., Montello, D.R., Hegarty, M.: Spatial knowledge acquisition from maps, and from navigation in real and virtual environments. Memory and Cognition 27, 741–750 (1999)
15. Shelton, A.L., McNamara, T.P.: Orientation and Perspective Dependance in Route and Survey Learning. Journal of Experimental Psychology: Learning, Memory and Cognition 30, 158–170 (2004)

16. Soeda, M., Kushiyama, N., Ohno, R.: Wayfinding in Cases with Vertical Motion. In: Proceedings of MERA 1997. Intern. Conference on Environment-Behavior Studies, pp. 559–564 (1997)
17. Thorndyke, P.W., Hayes-Roth, B.: Differences in spatial knowledge acquired from maps and navigation. Cognitive Psychology 14, 560–589 (1982)
18. Weisman, J.: Evaluating architectural legibility: Way-finding in the built environment. Environment and Behavior 13(2), 189–204 (1981)
19. Wiener, J.M., Schnee, A., Mallot, H.A.: Use and Interaction of Navigation Strategies in Regionalized Environments. Journal of Environmental Psychology 24(4), 475–493 (2004)

How Much Information Do You Need? Schematic Maps in Wayfinding and Self Localisation

Tobias Meilinger[1,2], Christoph Hölscher[1], Simon J. Büchner[1], and Martin Brösamle[1]

[1] University of Freiburg, Centre for Cognitive Science
Friedrichstr. 50, 79098 Freiburg, Germany
{hoelsch,buechner,martinb}@cognition.uni-freiburg.de
[2] Max-Planck-Institute for Biological Cybernetics
Spemannstr. 40, 72076 Tübingen, Germany
tobias.meilinger@tuebingen.mpg.de

Abstract. The paper is concerned with the empirical investigation of different types of schematised maps. In two experiments a standard floor plan was compared to three strongly schematised maps providing only route knowledge. With the help of one of the maps, the participants had to localise themselves in two tasks and performed two wayfinding tasks in a multi-level building they didn't know before. We recorded map usage time and a range of task performance measures. Although the map provided much less information, participants performed better in wayfinding with an unambiguous schematic map than with a floor plan. In the self localisation tasks, participants performed equally well with the detailed floor plan and with the schematised map versions. Like the users of a schematic map, users of a floor map presumably oriented on the network structure rather than on local geometric features. This allows them to limit the otherwise potentially very large search space in map-based self localisation. In both types of tasks participants looked at the schematised maps for a shorter time. Providing less than standard information like in a highly schematised map can lead to better performance. We conclude that providing unambiguous turning information (route knowledge) rather than survey knowledge is most crucial for wayfinding in unknown environments.

Keywords: Schematisation, map, wayfinding, self localization, route knowledge, survey knowledge, multilevel building.

1 Introduction

Maps are a common tool for orienting ourselves in our environment, may they come in paper form or be displayed on our mobile device. Comparing a paper hiking map with one displayed on a mobile device or a subway map the amount of information provided in those maps can vary tremendously. In the paper map you might see individual houses whereas in the mobile or subway map only the coarse direction of routes is displayed. The question of this study is how much information in a map is necessary, how much is superfluous? Is a highly schematised map sufficient for orientation or do we need further details? Does this depend on the goal we want to achieve with our map: Is a schematic map sufficient for finding our goal, but not for locating our current location after getting lost?

To address these questions we, first, review several theoretic approaches to schematisation. We, second, try to classify these approaches by the distinction of route and survey knowledge and identify the relevance of this knowledge from empirical studies. Third, we propose cognitive processes underlying wayfinding and self orientation with maps. From these assumptions we derive hypotheses predicting performance in wayfinding and self orientation for normal and highly schematised maps. Last, we test these predictions in two experiments and discuss the results with respect to the literature.

1.1 Theories of Schematisation

The question of what information is necessary for locating ourselves and finding our goals has found different answers. In cognitive science, this is often referred to as schematisation; the abstraction from unnecessary detail to concentrate on the essential information [e.g. 8]. For maps this involves omitting details e.g. the corner of a house or omitting dimensions e.g. colour information. We will introduce several approaches of schematisation. The reference point for all these approaches is the *topographic map*. In our terms a topographic map is a map which displays correct distances and angles between locations. Common hiking maps and also most city maps are topographic. It is important to notice that all maps, also topographic maps, do not display all spatial information available in our environment and therefore are schematic [27]. However, as metric relations are kept constant, a topographic map can be seen as a reference point to (more) schematised maps.

In a *topological map* only information about the network structure can be obtained. As a consequence a user located at B (see Table 1) can only determine to go into direction C, but not whether this implies turning right or left, as this information might not be displayed correctly in a map. Not knowing whether your path turns right or left is all right when taking the subway as your destination is written on a sign on the train, but it is beyond the pale for walking to your goal: Standing at an intersection the information of having to navigate to the city hall does not help at all, if you don't know in which direction to walk in order to do so. Consequently, mere topology can be sufficient for using a subway, but not for walking to our goal.

One approach to schematise maps comes from discrete curve evolution [2]. The shape of routes is simplified. Curvature between two locations is *straightened* (see Table 1). Local arrangements are to be kept constant. E.g. there is a house adjacent on the right side of the street. When the street is straightened the new position of the house should not be far away from the street, not on the street nor to the left of the street, but again on the right side adjacent to the street.

Another approach to schematisation is to *categorise* the environment [12]. Categorisation doesn't include the whole continuum of a route, but is focused on only e.g. the intersections. These intersections can be categorised again by reducing the possible angles of two intersecting streets to, say, only 90° or 90° and 45° (see Table 1). Especially for route maps which provide information about how to get from the start to the goal this is a feasible approach. In a second step it is also possible to

Table 1. Schematisation principles and pictorial examples for these principles. From a topographic, i.e., a metrically correct map (left of the arrow) a schematised map (right of the arrow) is derived. The amount of survey and route information preserved in the schematised map is roughly described in the columns on the right side.

Schematisation principle	Survey information	Route information
Topologic map	incorrect	incorrect
Straighten	rather correct	correct
Categorise junctions	rather incorrect	correct
Enhance relevant information	start & goal correct	correct
Route knowledge map	incorrect	correct

cluster intersections [20]. E.g. if you have to turn right at a T-intersection you could omit all intersections before the T-intersection where you have to walk straight, like we often do in verbal directions like "at the T-intersection turn right".

Another route map approach for cars is based on the principle to *enhance relevant information* and reduce or remove irrelevant information [1]. Most often when driving a car we have to cruise around several streets to reach a highway. Then we cover most of the distance on this highway before cruising again small streets in order to reach our goal. In a topographic map most parts will be occupied by the long highway. This approach enlarges the important parts of the map at the start and at the end of the route and shrinks the long distances on a highway (see Table 1). In doing so, the relative location of the goal with respect to the start is kept constant.

There is no general theory of schematisation yet [cf. 13]. All approaches omit certain information from the environment. This involves the curvature of a segment (straighten & enhance information), the length of a segment (enhance information), exact angles at branching points or even streets not necessary on a specific route (categorise junctions). As a consequence the exact metric locations displayed on a map, which provide so called survey knowledge, are distorted to a smaller (straighten) or stronger extent (categorise junctions) or do not provide meaningful information at all, as in the case of a topological map (see Table 1). We will review the importance of survey knowledge and its counterpart route knowledge in the next section.

1.2 Route and Survey Knowledge in Schematised Maps

The distinction between route and survey knowledge is fundamental in spatial orientation research [e.g. 24]. Route knowledge includes knowledge about a series of actions that have to be taken in order to reach the goal independent from knowing the exact position of the goal, e.g. turn right at the church, then the second street to the left. Survey knowledge on the other hand includes knowledge about the direction and distance between locations independent from knowing a path that leads there, e.g. the train station is about 300 Meters east from here.

Previous research strongly suggests that route knowledge is the crucial factor in finding your goal: In our daily life we recall route knowledge rather than survey knowledge. About 80% of all mentioned descriptions in verbal directions are concerned with actions and landmarks [6]. People very familiar with an environment have been shown to express only little survey knowledge of this environment [e.g. 16]. For reaching a goal in cities and buildings, survey knowledge was shown to play only a minor role [11, 15]. Orienting on survey relations could even be detrimental for performance [15]. Topographic maps displaying route *and* survey information have been found to provide no additional help in wayfinding compared to signs which only display route information [3, 10] or to verbal directions not providing survey information [15, 22]. So at least for wayfinding route knowledge seems fundamental, whereas survey knowledge can be omitted.

Transferring these results to schematic maps, one could think of a map only concentrating on providing correct route knowledge while omitting all survey knowledge. For our experiment we constructed such a schematic route knowledge map and compared it to a topographic map additionally providing survey knowledge. In constructing such a map (see bottom row of Table 1), we applied two principles: (1) Each junction in the map was connected to the closest junction by a straight line of normalised length, no matter whether the real distance was 5 or 50 meters. Mere turns between intersections were not considered. When walking between two junctions multiple changes in direction could occur. (2) All angles at junctions were changed to 90° or 180° angles. A turn to the right remained a turn to the right, but the turning angle in the map was always 90°. Local orientation of junctions in the map was correct. A T-intersection in the map always corresponded to a T-intersection in reality, a turnoff in the map always to a real turnoff, although the exact angles between streets might have differed. Despite the topologic network structure, only the local orientation of intersections was represented in such a map. The map was metrically incorrect. Route information was preserved: at any point the participant was able infer from the map whether to turn left, right or walk straight on in order to reach the next junction. Contrary to that survey information was omitted: no correct inference regarding distances and overall orientation could be drawn.

1.3 Wayfinding and Self Localisation with Schematised Maps

Starting from empirical findings we described two principles for constructing a strongly schematised map, which we compared to a standard topographic map. The basic idea is that the schematised map is sufficient for orientation, despite providing much less information compared to the topographic map. Does this hold true for all spatial orientation tasks? In the following we differentiate between finding a goal and localising oneself e.g. after getting lost. We propose that for wayfinding a schematised map is sufficient, whereas for self localisation participants lack important information and therefore should perform worse.

Wayfinding. When we want to reach a goal using a map, we usually know our current location and the location of our goal. Following Passini [19] we assume three steps in solving the wayfinding problem. The steps can be iterated several times:

Planning: The first step is planning a route from the start to the goal (or the general area of the goal) in the map. We could encode or learn the whole map, throw it away and plan the route based on our representation of the map. It has been shown, however, that planning is much easier using external representations like a map than using our own internal representations – possibly one reason why maps exist [21].

Transformation and encoding: When we have settled for one route, we have to encode, i.e., memorise the route. Only very few people walk around looking constantly at the map. Even if they do so, they have to transform the information from the map in order to use it for moving around. This transformation involves aligning the map mentally or physically with the environment so that "up" in the map corresponds to "forward" in the environment [e.g. 14]. For a transportable map, this could be accomplished by rotating the map. The transformation, however, also involves a perspective switch from the top down perspective of the map to the ground-level (egocentric) perspective in which we encounter the environment [e.g. 23]. As our memory capacities are limited we probably won't encode the whole route, but only a part of it and start with this.[1]

Walking and monitoring progress: After transforming and encoding the map, we use our internal representation to guide our locomotion. E.g. we walk straight on to the next intersection and turn left there. In doing so we have to monitor our progress, i.e., to mach our internal representation e.g. of an intersection with our environment, before executing a behaviour e.g. turning left and then access the information of what to do next and where. Matching locations of the environment with corresponding internal representations helps us with monitoring our progress, identifying our goal and keeping us oriented. When we reach the end of the memorised (sub-)path and/or feel unsure, we look into the map again and go back to the planning stage or to encoding and transforming the upcoming part of our already planned route. When making a mistake (or using an erroneous map) we can get lost, i.e., our actual location does not correspond to our assumed location in the map or in the representation formed from it. After that, we have to localise ourselves again, before being able to plan, encode and execute a new route. Self localisation will be described in the next section.

---

[1] The transformation process can also happen online during walking the route. For this, the map would need to be encoded beforehand. Again as argued for planning we assume that the transformation is much easier, when having access to the external representation of the map, than when having to rely on an internal representation of the map.

We described our assumptions regarding the process of wayfinding using a map. Within this model alternative strategies can be imagined. Our examples described a route strategy which includes a one dimensional string of actions at decision points. However, also a survey or least angle strategy is possible [9, 11]. This strategy includes identifying the direction and distance to a (sub-)goal (planning), encoding and transforming this into a horizontal perspective and trying to walk directly to this spot (walking and monitoring progress). The survey strategy is only applicable using a topographic map as the route knowledge map does not provide correct survey information.

Taking this model of wayfinding we assume that the schematised map provides sufficient information for all stages of the wayfinding process applying a route strategy. As lots of detail information is missing we predict that the planning, encoding and transformation process could be performed faster and less error prone than with a topographic map. Participants therefore should be quicker in consulting the map. For wayfinding itself we assume that participants with a schematised map perform at least as good as participants with a topographic map although the topographic map provides much more information.

Self localisation. When we are disoriented, i.e., when we do not know where in relation to our memory or a map we are, we try to localise ourselves. To regain our orientation in an unknown environment we have to *compare features of our surrounding with features of a map* [e.g. 26]. For example, when standing at a T-intersection we can search for all T-intersections in the map. Based on the individual geometry of our T-intersection we might distinguish this T-intersection from other T-intersections. In doing so, we localise ourselves using *local cues* which are visible from our current location. These cues could be the geometry, or landmarks displayed in a map e.g. churches, street sizes or doors in the map of a building. The literature on self localisation is very much focused on such local cues and emphasises the importance of geometric features [e.g. 7]. In contrast we can also orient on the *network structure* of our surrounding, i.e., only taking decision points into account, e.g. "if I am here in the map, then there should be a T-intersection straight ahead and a crossroads to the left". Localising on local cues or on the network structure is probably best described as a hypothesis testing procedure, i.e., we generate a hypothesis about our current location and try to confirm or reject this hypothesis by collecting more information.

Our experiments took place in a multi-level building. Compared to single layer spaces like cities, the relation and representation of multiple layers poses difficulties. Humans have trouble correctly aligning vertical spaces in pointing tasks [18]. Soeda et al. [25] observed wayfinding performance in tasks involving vertical level changes. They found people losing their orientation due to vertical travel, supporting more informal results of Passini [19].

Our schematised map only preserves the network structure of the environment and the raw layout of intersections e.g. T-intersections, but lacks exact local geometry. As geometry is considered an important cue for self localisation [7], we assume participants to localise better if using a topographic map which preserves geometry.

1.4 Hypotheses

We proposed a map schematisation approach providing route knowledge and omitting survey knowledge. Such a highly schematised map was compared to a standard topographic map additionally providing information about survey relations as well as local geometry. Due to local geometry which was shown to be important in self localisation, we predicted that participants with a topographic map would perform better in localising themselves. Due to the central importance of route knowledge for wayfinding, we predicted that participants with a schematised map would perform at least as good as participants with a topographic map. This would be despite the fact that the topographic map provides much more information. Due to the sparser information in the schematic map, we predicted that participants would be faster in encoding information from the schematic map than from the topographic map. This was expected in both types of tasks, wayfinding and self localisation.

In Experiment 1 we compared a topographic map, i.e. a floor plan of a multilevel building with our highly schematised map. In Experiment 2 we investigated the relevance of ambiguity, an issue which occurred in Experiment 1, with a set of two new schematized maps. Conducting both experiments with the same tasks and in the same setting allowed us to compare results between the experiments.

2 Experiment I

2.1 Methods

Participants were asked to participate in two self localisation tasks. They had to locate the position in a map corresponding to their actual position in a building unknown to them. They also performed two wayfinding tasks in the same building. For this they were shown their actual position in the map and had to find a goal also shown to them on the map. All tasks were either conducted with a topographic floor plan or with a highly schematised map.

Participants. Participants were attendees of an annual summer school for human and machine intelligence which takes place at a conference centre in Günne, near Düsseldorf, Germany. They were recruited from the list of participants of the summer school via e-mail, before the event started. 5 women and 13 men agreed to participate in the experiment. The participants were at the end of their twenties ($M = 28.6$; $SD = 5.7$), all were native German speakers.

Material. The conference centre was built in 1970 (see Figure 1). It consists of four floors connected with five staircases. Its complexity causes many visitors to get lost. For further discussion of the building see [11].

The participants either got a floor plan or a schematised map for the task. In the *floor plan* each level of the building was seen from birds eye view (see Figure 2 left side). Symbols for staircases were added and connected with dashed green lines. The metric distances in the floor plan were correct. Doors were not displayed. Participants were not allowed to enter rooms. The display of rooms enabled participants to judge the outlines of the building.

The schematic or *simple map* (Figure 2 right side) was derived from the floor plan following the principles described in 1.2. Each junction and staircase (node) in the map was connected to the closest staircase or junction (node) by a straight line of normalised length. Turns between nodes were ignored, except for one turn in the square in the middle of the basement, where this was not possible. All angles at junctions were changed to 90° or 180° angles. In comparison to the floor plan, the simple map provided route knowledge and omitted survey knowledge. Turning information was correct, but distance and global orientation information were not to be relied on. Despite the topologic network structure of nodes, only the local orientation of intersections was represented in the simple schematised map.

Both floor plan and simple map were presented on an A4 paper (29.7 cm x 21 cm) in an opaque folder which had to be opened in order to see the plan or map.

Procedure. The participants performed two *self localisation* tasks. They were taken to the starting points blindfolded (number 1 and 4 in Figure 1). In order to reach the start of the first task they entered the building from outside. They were able to guess that they were on the ground floor or in the basement. For the second task they were

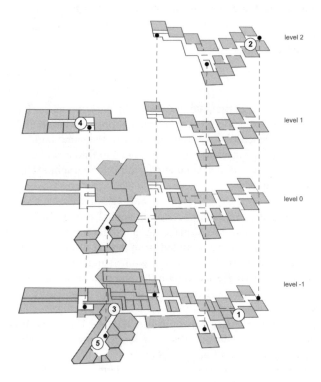

Fig. 1. The conference centre where the experiment took place. Starting points for the self localisation tasks (number 1 and 4) are shown. In the wayfinding tasks the participants had to walk from number 2 to 3 and from number 4 to 5. The numbers correspond to the order in which all tasks were performed.

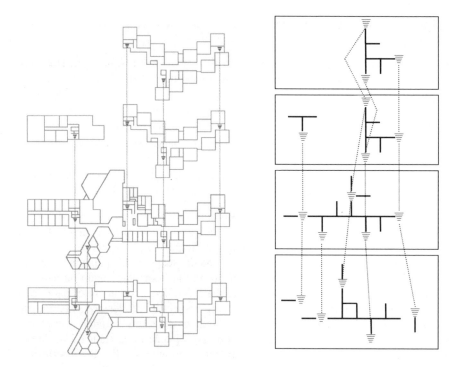

Fig. 2. The floor plan on the left and the schematised "simple" map on the right. Corresponding staircases on different floors are connected with green dashed lines.

also disoriented and brought to the correct floor via corridors only accessible to staff and by an elevator starting from the basement. They could, therefore, infer not being in the basement any more. The participants' task was to locate themselves in the map, i.e. to show their actual position on the map. For this they were allowed use the map and walk around, but not both at the same time. The experimenter instructed them to only answer when they were certain and not simply to guess where they were.

For the two *wayfinding* tasks the experimenter brought the participants to the starting point without blindfold (number 2 and 4 in Figure 1). He showed them their current position in the map and their goal. In the building the goals were marked with a red square on the floor (number 3 and 5). The participants had to find the goal as quickly as possible while moving with normal walking speed.

The participants started with the first self localisation task (number 1) followed by the first wayfinding task (from number 2 to 3). These tasks took place in the *large area*: For the self localisation task it was necessary to more or less consider the whole building as their possible actual location. In the wayfinding task they covered about ⅔ of the building. After the tasks in the large area they performed a self localisation and a wayfinding task in the *small area*. In the self localisation task (number 4) the participants could exclude the parts of the building already known to them reducing the number of available alternatives. The adjacent wayfinding task (from number 4 to 5) covered only about ⅓ of the building and was therefore much shorter than the first wayfinding task.

The participants were videotaped. From the video we derived the following dependant measures:

- *Time* to complete the task, taken from the video. Extra time, e.g., stops with explanations because of experimental issues was subtracted
- *Distance* covered
- *Detours* to locations visited before (only in wayfinding tasks)
- *Average detour distance* per detour (only in wayfinding tasks)
- *Map usage:* Number of stops to use map (participants were asked only to use the map while standing)
- *Average map time* per map usage, i.e., average time between opening the folder containing the map and closing it again.

Two experimenters conducted the tasks in parallel. During the experiment the participants were asked to verbalise their thoughts. They also accomplished two pointing tasks before and after the second wayfinding task (at number 4 and 5 in Figure 1). Pointing and verbalisations are beyond the scope of this paper and therefore reported in a later publication. One participant in the floor plan condition had to be excluded due to not being able to complete the tasks. The assignment of participants to experimental conditions was controlled with respect to gender and experimenter. Parameter values deviating more then three standard deviations from the overall mean were replaced by the most extreme value observed inside three standard deviations. We computed independent t-test to compare between maps, gender and experimenter. Since nonparametric U-tests revealed very similar results only the more common parametric t-tests are reported.

2.2 Results

The results did not differ due to experimenter (all 20 $t(16) < 2.1$, $p > .096$, $d < 0.87$). The data was therefore collapsed for further analysis.

Table 2. Average performance in self localisation for Experiment 1 and 2. Means and (standard deviations) are shown. Means displayed in italics differ in direct comparison at $p < .05$.

	Experiment 1		Experiment 2	
	Floor plan	Simple map	Comb map	Square map
Large area				
Time [s]	352 (240)	330 (218)	246 (82)	236 (99)
Distance [m]	87 (112)	116 (59)	80 (31)	66 (23)
Map usage [n]	4.7 (4.7)	6.0 (4.0)	4.6 (0.7)	4.1 (1.5)
Av. map time [s]	*65 (36)*	*32 (16)*	36 (17)	40 (15)
Small area				
Time [s]	106 (63)	100 (49)	165 (90)	141 (55)
Distance [m]	10 (10)	18 (15)	31 (13)	22 (17)
Map usage [n]	2.3 (1.0)	2.1 (0.8)	3.1 (1.4)	2.8 (1.3)
Av. map time [s]	31 (16)	39 (41)	37 (22)	35 (9)

Table 3. Wayfinding performance in Experiment 1 and 2. Means and (standard deviations) are shown. Means displayed in italics differ in direct comparison at p < .05.

	Experiment 1		Experiment 2	
	Floor plan	Simple map	Comb map	Square map
Large area				
Time [s]	305 (144)	264 (128)	286 (65)	266 (85)
Distance [m]	183 (62)	159 (28)	153 (17)	139 (23)
Detours [n]	1.8 (1.5)	1.2 (0.8)	1.0 (0.7)	0.3 (0.7)
Av. detour dist. [m]	25 (18)	20 (21)	21 (10)	22 (2)
Map usage [n]	5.3 (3.9)	5.4 (2.2)	6.6 (2.1)	4.9 (1.5)
Av. map time [s]	28 (7)	24 (21)	21 (10)	25 (11)
Small area				
Time [s]	143 (76)	165 (52)	191 (164)	199 (125)
Distance [m]	66 (18)	94 (35)	81 (25)	86 (31)
Detours [n]	0.9 (1.4)	1.6 (1.1)	1.0 (0.5)	1.4 (1.5)
Av. detour dist. [m]	*10 (4)*	*24 (9)*	18 (13)	14 (7)
Map usage [n]	3.1 (1.8)	4.4 (1.3)	4.9 (2.0)	4.9 (1.7)
Av. map time [s]	*26 (12)*	*14 (8)*	16 (10)	19 (10)

Self localisation. In the large area participants with a floor plan looked per stop twice as long at their plan than participants with the simple map (see Table 2 left side, $t(11.0)^2 = 2.57$, $p = .026$, $d = 1.21$). We did not find any further significant differences regarding self localisation neither in the large nor the small area (all seven $t(16) < 1.22$, $p > .243$, $d < 0.61$). No *gender* differences in self localisation performance were found (all eight $t(16) < 1.15$, $p > .270$, $d < 0.62$).

Wayfinding. In the *small area* participants with a floor plan performed better than participants with a simple map (see Table 3 left side). Their average distance of detours was smaller ($t(10) = 2.67$, $p = .024$, $d = 1.86$). There was also a trend to stop less often ($t(16) = 1.77$, $p = .097$, $d = 0.83$) and cover less distance ($t(16) = 2.07$, $p = .055$, $d = 0.98$). When using the floor plan they, however, stopped for longer times than participants with a simple map ($t(16) = 2.64$, $p = .018$, $d = 1.24$). The groups did not differ significantly with respect to time or number of detours ($t(16) < 1.13$, $p > .276$, $d < 0.54$). In the *large area* participants with a simple map performed numerically better. These differences, however, never reached the level statistical significance (six $t(16) < 1.08$, $p > .300$, $d < 0.51$). Wayfinding performance did not differ due to *gender* (all twelve $t(16) < 1.84$, $p > .078$, $d < 1.33$).[3]

---

[2] Both experimental groups differed in their variance. The degrees of freedom were therefore adjusted from 16 to 11.0).
[3] Due to unequal group sizes and adjustment of the degrees of freedom to account for unequal variances, some rather large effect sizes in favour of men did not become significant.

2.3 Discussion

When using their maps, participants with the floor plan looked longer in the map than participants with the simple map – both in wayfinding and in self localisation. Consistent with our predictions they encoded more information from the floor plan or they needed more time to extract the relevant information from the floor plan.

For *self localisation* we predicted a better performance in participants with a floor plan. Only the floor plan not the schematic map provided local geometry which was considered an important cue in self localisation. Contrary to our prediction both groups performed equally well in localising themselves. The performance measures did not even show a consistent numerical advantage for the floor plan, excluding a lack of statistical power as an explanation. We conclude that both groups mainly used the network structure available in both maps for localising themselves. Why was that? Using local geometric features might offer too many opportunities to look for in the floor plan. E.g. there were a lot of bends in a corridor to check for in the map. Focusing on nodes in the network structure instead reduced the possible search space to a reasonable size. Fewer hypotheses had to be tested and kept in memory.

For *wayfinding* we expected participants with the simple schematic map to perform at least as good as participants with the floor plan. Despite containing much less information, the schematic map should provide the relevant information for wayfinding. While participants using the simple map performed numerically better in the large area, participants using a floor plan performed better in the small area. Why did they perform better in one task? We assume that the simple schematic map provided ambiguous turning information after floor changes. After walking down the stairs, participants with the simple map could not know whether they should turn left or right next. Participants with a floor plan could disentangle this ambiguity by local geometric features e.g. the form of a corridor. Indeed in the small area task almost all detours using the simple map had their origin after exiting stairs. In the large area task no ambiguity occurred, as all stairs were located at the end or very close to the end of a corridor. Here, no advantage of the floor plan was observed. In order to address this problem we conducted a second experiment in which we varied the ambiguity of two schematic maps.

3 Experiments II

3.1 Methods

The goal of this experiment was to determine the influence of ambiguity in schematic maps. In the simple schematic map of Experiment 1 we identified an ambiguity for participants after floor changes. Especially in the small area participants could not know from the map which direction they had to go when exiting stairs. To disentangle this ambiguity we placed the symbols for staircases to the side of a corridor and oriented them facing the direction towards the corridor (see Figure 3 right side). Also the lines connecting floors via the staircases were changed and entered the stair from the back additionally indicating ones orientation when exiting a staircase. For the

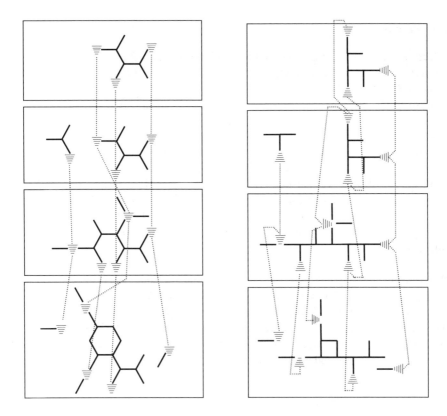

Fig. 3. The maps used in Experiment 2. On the left side the ambiguous "comb map", on the right side the unambiguous "square map".

ambiguous map the staircases and connections between the floors were the same as in the simple map of Experiment 1 (see Figure 3 left side). Additionally the structure of intersections was changed from 90° angles to 120° angles in order to provide two path alternatives branching off at the same angles. At an intersection the map always provided a left and a right alternative, no matter whether in the building this was a T-intersection or a corridor branching off to the left. In the later case the main straight corridor was indicated in the map not with a straight corridor, but with a turn to the right. Due to its honeycomb structure we called this map "comb map". In contrast to that we called the unambiguous map "square map".

The experiment took place at the same annual summer school as Experiment 1 one year later. Nine women and nine men agreed to participate in the experiment. Again, the participants were around the end of their twenties ($M = 27.4$; $SD = 12.6$) and spoke German fluently. One participant in the square map condition not reported here had to be excluded due to not being able to complete the tasks. The selection and assignment of participants, the tasks, procedure, instruction and data analysis were identical to Experiment 1.

3.2 Results

Except for detours in wayfinding in the small area ($t(16) = 2.39, p = .029, d = 1.13$), the results did not differ with respect to experimenter (all 19 $t(16) < 1.51, p > .150, d < 0.60$). No different results were obtained when including the experimenter in the analysis of this parameter. Therefore, only the collapsed data are reported.

Self localisation. The participants performed equally well in localising themselves, no matter whether they used the comb map or the square map (see Table 2 right side, all eight $t(16) < 1.24, p > .236, d < 0.59$).

In the small area women outperformed men. They were faster (109s vs. 197s, $t(16) = 3.12, p = .007, d = 1.47$), covered less distance (15m vs.. 37m, $t(16) = 4.17, p = .001, d = 1.96$) and used the map less often (2.2 vs. 3.7, $t(16) = 2.78, p = .014, d = 1.31$) before correctly localising their position. Women and men did not differ in time per stop or in the in the large area (all five $t(16) < 1.27, p > .224, d < 0.60$).

Wayfinding. Participants with the comb map and the square map did not differ in their general wayfinding performance (see Table 3 right side). In the large area there was a trend for participants with the square map to make less detours ($t(16) = 2.0, p = 0.63, d = 0.94$) and use the map less often ($t(16) = 1.94, p = .070, d = 0.91$; all ten other measures $t(16) < 1.42, p > .176, d < 0.68$). Wayfinding performance did not differ with respect to gender (all twelve $t(16) < 1.84, p > .302, d < 0.51$).

3.3 Discussion

Participants with the ambiguous comb map and participants with the unambiguous square map did not differ in localising themselves. As in Experiment 1 this findings indicate that the network structure was the main source of information used for self localisation. Unambiguous local intersections and staircases only provided in the square map did not lead to a significantly better performance. Local unambiguousness did not play a crucial role in these tasks.

We do not have an explanation for the better performance of women in self localisation. This did not occur in Experiment 1 where there were no significant differences, men even performed numerically better. Generally, men are known to perform slightly better than women in spatial orientation tasks (for a recent review see [5]). For the wayfinding tasks used in Experiment 1 and 2 which were the same as tasks used in another experiment [11] no advantage for women was observed.

We did not observe any significant differences in wayfinding performance with respect to the maps. There was a trend for participants in the large area to perform better with the unambiguous square map. With nine participants per group this difference, however, did not reach the level of significance. Ambiguity could therefore *not* be a crucial factor for wayfinding with schematic maps. A minor importance of ambiguity could, however, not be ruled out. To see how participants with ambiguous *and* unambiguous schematic maps perform in relation to a floor plan we compared the results of Experiment 1 and 2.

4 Comparison of Experiment I and II

4.1 Methods

In order to compare a floor plan with ambiguous and unambiguous schematic maps over both experiments we used three groups: a) The floor plan, b) the unambiguous square map and c) the two ambiguous schematic maps, consisting of the simple map from Experiment 1 and the comb map from Experiment 2. We compared these three groups using a one-way ANOVA with pair-wise planned contrasts between the groups when an overall difference was observed. Especially the contrast between the floor plan and the unambiguous square map was of interest as the other two contrasts were partially contained in the data already presented in section 2 and 3.

4.2 Results

Self localisation. In the *large area* the time participants looked at the map per stop differed as a function of the kind of map (see Figure 4 left side $F(2, 33) = 6.07$, $p = .006$; $\eta^2 = .27$). Participants with a floor plan looked longer at the map compared to participants with schematic maps (floor plan vs. ambiguous maps: $t(33) = 3.45$, $p = .002$, $d = 1.14$; floor plan vs. unambiguous square map see also Table 2 outer columns: $t(33) = 2.41$, $p = .022$, $d = 0.93$). In the *small area* there was a trend for the participants to differ in the distance covered before locating oneself ($F(2, 32) = 2.62$, $p = .088$; $\eta^2 = .14$). Here participants with a floor plan covered less distance compared to participants with ambiguous maps ($t(32) = 2.26$, $p = .031$, $d = 1.11$; two other contrasts: $t(32) < 1.67$, $p > .106$, $d < 0.86$). We did not reveal any further reliable differences in other parameters (six $F(2, 33) < 1.0, p > .380, \eta^2 < .06$).

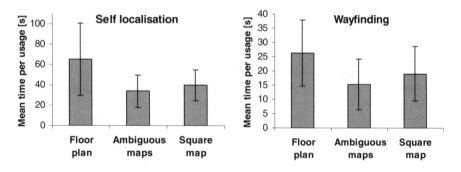

Fig. 4. Mean time of map usage in participants with the floor plan, ambiguous maps and the unambiguous square map in both experiments. Means and standard deviations are displayed for self localisation in the large area (left side) and for wayfinding in the small area (right side).

Wayfinding. In the *large area* task participants differed in the number of detours ($F(2, 33) = 5.0, p = .013, \eta^2 = .23$) and the distance covered ($F(2, 33) = 3.31, p = .049, \eta^2 = .17$) as a function of the kind of map they used (see Figure 5). Participants with the unambiguous square map performed better than participants with the floor plan (detours: $t(33) = 3.14$, $p = .003$, $d = 1.24$; distance $t(33) = 2.54$, $p = .016$, $d = 0.93$, four other contrasts: $t(33) < 1.96$, $p > .059$, $d < 1.06$). We observed no further differences in the four other parameters (all $F(2,33) < 0.61$, $p > .553$, $\eta^2 < .04$).

In the *small area* task the participants differed in the time of their average map usage (see Figure 4 right side, $F(2, 33) = 3.87$, $p = .031$, $\eta^2 = .19$). Participants with ambiguous schematic maps looked shorter at their maps than participants with the floor plan ($t(33) = 2.78$, $p = .009$, $d = 1.08$; two other contrasts $t(33) < 1.59$, $p > .122$; $d < 0.70$). There was a tendency for floor plan users to stop less often compared to participants with schematic maps ($F(2, 33) = 3.14$, $p = .056$, $\eta^2 = .16$). We observed no differences in the other four parameters (all $F(2, 23/33) < 2.44$, $p > .109$, $\eta^2 < .18$).

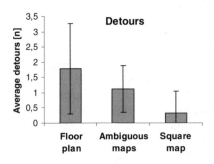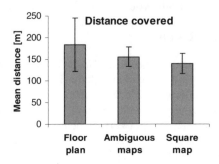

Fig. 5. Wayfinding performance in participants with the floor plan, ambiguous maps and the unambiguous square map in the large area compared over both experiments

4.3 Discussion

Participants with an unambiguous schematic map performed better in *wayfinding* than participants with a floor plan. In the large area task they made less detours and covered shorter distances. Despite its lower information content, using an unambiguous schematic map can lead to better performance than using a floor plan. This result is consistent with our prediction.

Why can it be better to use a schematic map? Time for planning and encoding alone could not be the reason. Participants looked at the schematic maps for shorter times, but this could only influence the overall wayfinding time and could not explain the fewer errors, i.e., fewer detours and the less distance covered. We assume that participants either encoded better information in concentrating on route knowledge or that they applied a better strategy, i.e., a route strategy. The schematic maps were constructed to provide only correct route knowledge which was shown to be central to wayfinding (see section 1.2). From such a map a user could learn where to turn at an intersection, but a user could not learn survey knowledge as distances and directions between locations on the schematic map did not correspond to the real distances and directions in the building. This fact was known to the participants. They were thus forced to encode and use only route knowledge. In contrast, participants with a floor plan could also encode survey knowledge or even local geometry. This information was less useful for the task (see section 1.2). Considering that memory capacity is limited, concentrating on relevant information should lead to fewer errors and therefore less detours and less distance covered.

The survey information from the floor plan might, however, not just distract from the more relevant route knowledge, it also enables the application of a survey strategy [9, 11]. Within a floor a participant could encode only the direction and distance of a (sub-)goal and try to walk there directly. In doing so, a route not leading to the (sub-)goal could be chosen resulting in more detours and distance covered. Although the survey strategy needs very little information to be transformed and encoded from the map, the encoded vector pointing to the (sub)goal has to be updated constantly. Consequently, the survey strategy implies a higher memory load during walking through the building which also could lead to more errors. With the route strategy the turning information at decision points had to be transformed and encoded from the map and maintained until it was used, but nothing had to be updated. So the memory load for the route strategy was high only during transforming and encoding when participants could use the map as an external representation to ease their tasks [21]. Further research has to clarify whether the advantage of a schematic route-knowledge map stems from applying a different strategy or rather from a memory effect, e.g., not encoding irrelevant survey information or additionally relying on verbal memory to encode route knowledge. In the comparison of route and survey knowledge it might also be interesting to produce a map only providing survey knowledge and no route knowledge. Such a map would include the correct topographical location of decision points, but connecting paths would not be visualised. But very likely such a map would be practically useful only for navigation tasks in open terrain, not for indoor environments.

One factor in route knowledge is ambiguity. Route knowledge should be unambiguous. We did not find any significant differences when varying ambiguity in the second experiment. Maybe our variation was not strong enough. If we had used a topological and therefore a completely ambiguous map for comparison, we probably would have found stronger effects. However, when comparing the floor plan to the schematic maps only the contrasts to the unambiguous square map became significant. Ambiguity, therefore, has to be relevant in some way, although other factors might be more central.

Consistent with our prediction, participants needed less *time for encoding* information from a schematic map than from a floor plan. This holds true for both self localisation and for wayfinding. With less information provided in the schematic maps, participants are limited in the amount of information to encode. Additionally, they do not have to search for the relevant information within irrelevant information.

In *self localisation* no general advantage of the floor plan compared to schematic maps could be revealed. There was a trend for participants with the floor plan to cover less distance. This, however, only holds true for comparing floor plans with ambiguous schematic maps. No reliable difference or even trends between floor plan and the unambiguous square map could be revealed. We conclude that most participants relied on the network structure rather than the geometric layout. Searching the floor plan for locations with a specific geometric layout probably offered too many possible alternatives. Limiting the search space to easily identifiable configurations of nodes like intersections and staircases reduced the number of possible alternatives to a reasonable amount that can be handled by humans. The higher importance of network structure over geometry stands in contrast to results from self localisation studies which emphasise the importance of geometric features [e.g. 7]. There are, however, substantial differences between these studies and our experiment: Our participants had to localise themselves in an unknown environment using a map providing an *external representation*. In research on self localisation participants most often had seen the environment before and judged their current location based on their memories of this environment which are *internal representations*. A further difference besides internal vs. external representations is the kind of space the experiments took place in. In most self localisation experiments only room sized environments were used. According to Montello [17] these spaces can be called *vista spaces* as all the space is visible from one point of view which also holds true for open places or even small valleys. Contrary to that, to understand *environmental spaces* we have to move around and take several views of the space into account which is the case for towns or buildings like in our experiment. So at least the kind of space (vista vs. environmental space) and the representation system on which the self localisation task was based (internal vs. external) differ between our experiment and most self localisation studies. Identifying, which factor or combination of factors do in fact cause participants to localise on the *network structure* rather than on *geometry* is subject to future research.

When comparing *gender* over both experiments (not shown) we did not observe significant differences like in Experiment 2. Reliable differences of any kind should be even stronger when comparing more participants. We, therefore, do not think that the gender differences observed in Experiment 2 should be emphasised too much.

5 General Discussion

Despite containing much less information, using a highly schematic map can lead to better wayfinding performance than using a topographic floor plan. Providing unambiguous route knowledge is central for this performance benefit. Self localisation with such a map is generally at least not worse than with a floor plan. Like the users of a schematic map, the users of floor map orient on the network structure rather than on local geometric features which would imply a very large search space. Both in wayfinding and self orientation participants are faster to encode information from the schematised map.

How do these results generalise to other situations and maps? All significant results in this study are based on large effect sizes with respect to Cohen [4]. In this field experiments we are not dealing with a highly artificial laboratory effect only observed under very specific conditions. The practical application for the results in self localisation will, however, be limited. In many maps today our current location is already marked, e.g. when using a wall-mounted you-are-here-map or a GPS-based system. For an old fashioned city map we often localise using street names or landmarks rather than comparing the network structure of our surrounding with the one in our map. Also the wayfinding results are not transferable to all situations: Our highly schematic maps can only be constructed for non-circular street layouts, as can be seen in the schematic maps of the basement. For the common route maps which only display one route this is not a problem. For other maps, the rather strict construction constraints have to be relaxed and e.g. turns have to be allowed, too.

From a navigational point of view, the floor plan might also be improved e.g. by marking corridors in a different colour, probably leading to faster encoding times and maybe also less errors. Our results might therefore depend on the kind of maps we used. The point we wanted to make, however, was how little information is sufficient for good performance. In any case, this information about where to turn at a decision point should never be omitted! More information could be helpful, but less information will probably be detrimental. This point also applies to generalising to other settings than multi-level buildings, e.g. cities. The building used was rather complex. For more simple environments, other results might be expected, but at the same time any map might be obsolete in a simple setting. The complexity of our building, however, shows the importance of unambiguous turning information for non-trivial wayfinding tasks. Unlike in a building, in a city there is only one "floor", but this floor extends much further horizontally. Both are rather complex and they both are environmental spaces according to Montello's [17] definition, therefore the results probably can be generalised.

Our results as well as wayfinding literature regarding maps in comparison to signs [3, 10] and verbal directions [15, 22] suggest: When trying to reach a goal in an unknown environment unambiguous turning information at decision points is more important than survey knowledge. May your route knowledge be communicated by signs, verbal directions or maps – this is the type knowledge you need!

Acknowledgements

The research was supported by grants to Christoph Hölscher and Markus Knauff from the DFG within the Transregional Collaborative Research Center, SFB/TR 8. The authors are very grateful to the participants of the experiment, to Gregor Wilbertz, Henrike Sprenger, Lara Webber, Wibke Hachmann, Kerstin Garg-Hölscher and Kai-Florian Richter for discussion and support with data acquisition and data analysis.

References

1. Agrawala, M., Stolte, C.: Rendering effective route maps: Improving usability through generalization. In: Proceedings of the AAAI Spring Symposium on Smart Graphics (2001)
2. Barkowsky, T., Latecki, L.J., Richter, K.-F.: Schematizing maps: simplification of geographic shape by discrete curve evolution. In: Freksa, C., Brauer, W., Habel, C., Wender, K.F. (eds.) Spatial Cognition II, pp. 41–53. Springer, Berlin (2000)
3. Butler, D.L., Acquino, A.L., Hissong, A.A., Scott, P.A.: Wayfinding by newcomers in a complex building. Human Factors 35(1), 159–173 (1993)
4. Cohen, J.: Statistical power analysis for the behavioral sciences. Erlbaum, Hillsdale, NY (1988)
5. Coluccia, E., Louse, G.: Gender differences in spatial orientation: A review. Journal of Environmental Psychology 24, 329–340 (2004)
6. Denis, M.: The description of routes: a cognitive approach to the production of spatial discourse. Current Psychology of Cognition 16, 409–458 (1997)
7. Hermer-Vasquez, L., Spelke, E.S., Katnelson, A.S.: Sources of Flexibility in Human Cognition: Dual-Task Studies of Space and Language. Cognitive Psychology 39, 3–36 (1999)

8. Herskovits, A.: Schematization. In: Olivier, P., Gapp, K.P. (eds.) Representation and processing of spatial expressions, pp. 149–162. Erlbaum, Mahwah, NJ (1998)
9. Hochmair, H., Frank, A.U.: Influence of estimation errors on wayfinding decisions in unknown street networks - analyzing the least-angle strategy. Spatial Cognition and Computation 2, 283–313 (2002)
10. Hölscher, C., Büchner, S., Meilinger, T., Strube, G.: Wayfinding Strategies and Map Use in a Multi-Building Ensemble (in press)
11. Hölscher, C., Meilinger, T., Vrachliotis, G., Brösamle, M., Knauff, M.: Finding the Way Inside: Linking Architectural Design Analysis and Cognitive Processes. In: Freksa, C., Knauff, M., Krieg-Brückner, B., Nebel, B., Barkowsky, T. (eds.) Spatial Cognition IV - Reasoning, Action, Interaction, pp. 1–23. Springer, Berlin (2005)
12. Klippel, A.: Wayfinding choremes - Conceptualizing wayfinding and route direction elements. Universität Bremen, Bremen (2003)
13. Klippel, A., Richter, K.-F., Barkowsky, T., Freksa, C.: The Cognitive Reality of Schematic Maps. In: Meng, L., Zipf, A., Reichenbacher, T. (eds.) Map-based Mobile Services - Theories, Methods and Implementations, pp. 57–74. Springer, Berlin (2005)
14. Levine, M., Jankovic, I., Palij, M.: Principles of spatial problem solving. Journal of Experimental Psychology: General 11, 157–175 (1982)
15. Meilinger, T., Knauff, M.: Ask for your way or use a map: A field experiment on spatial orientation and wayfinding in an urban environment (submitted)
16. Moeser, S.D.: Cognitive Mapping in a Complex Building. Environment and Behavior 20, 21–49 (1988)
17. Montello, D.R.: Scale and multiple psychologies of space. In: Frank, A.U., Campari, I. (eds.) Spatial information theory: A theoretical basis for GIS, pp. 312–321. Springer, Berlin (1993)
18. Montello, D.R., Pick, H.L.: Integrating knowledge of vertically aligned large-scale spaces. Environment and Behavior 25, 457–484 (1993)
19. Passini, R.: Wayfinding in architecture, 2nd edn. Van Nostrand Reinhold Company, New York (1992)
20. Richter, K.-F., Klippel, A.: A Model for Context-Specific Route Directions. In: Freksa, C., Knauff, M., Brückner, B.K., Nebel, B., Barkowsky, T. (eds.) Spatial Cognition IV - Reasoning, Action, Interaction, pp. 58–78. Springer, Berlin (2005)
21. Scaife, M., Rogers, Y.: External cognition: how do graphical representations work? International Journal of Human-Computer Studies 45, 185–213 (1996)
22. Schlender, D., Peters, O.H., Wienhöfer, M.: The effect of maps and textual information on navigation in a desktop virtual environment. Spatial Cognition and Computation 2, 421–433 (2000)
23. Shelton, A.L., McNamara, T.P.: Orientation and Perspective Dependance in Route and Survey Learning. Journal of Experimental Psychology: Learning, Memory and Cognition 30, 158–170 (2004)
24. Siegel, A.W., White, S.: The development of spatial representations of large-scale environments. In: Reese, H.W. (ed.) Advances in Child Development and Behaviour, Academic Press, New York (1975)
25. Soeda, M., Kushiyama, N., Ohno, R.: Wayfinding in Cases with Vertical Motion. In: MERA 1997, pp. 559–564 (1997)
26. Warren, D.H.: Self-Localization of Plan and Oblique Maps. Environment and Behavior 26, 71–98 (1994)
27. Tversky, B.: Some ways that maps and diagrams communicate. In: Freksa, C., Habel, C., Wender, K.F. (eds.) Spatial Cognition II, pp. 72–79. Springer, Berlin (2000)

Wayfinding Strategies in Behavior and Language: A Symmetric and Interdisciplinary Approach to Cognitive Processes

Thora Tenbrink[1] and Jan M. Wiener[2]

[1] FB10 Linguistics and Literary Sciences, University of Bremen,
Postfach 330440, 28334 Bremen, Germany
tenbrink@sfbtr8.uni-bremen.de
[2] Collège de France
Laboratoire de Physiologie de la perception et de l'action
11, place Marcelin Berthelot, 75231 Paris, France
jan.wiener@college-de-france.fr

Abstract. We present an interdisciplinary approach to the investigation of strategies and heuristics reflecting the cognitive processes underlying human wayfinding. To achieve this, we symmetrically investigate navigation behavior and associated language. This novel approach combines two completely different and independent directions of research that complement each other naturally and necessarily, but which have seldom been directly combined so far. The current focus on wayfinding strategies and heuristics is a fairly new scientific goal both in behavioral and linguistic research areas; also, the methods of discourse analysis have rarely been directly adopted to systematically investigate parallels between natural discourse and navigation behavior. In this paper, we outline and motivate our approach and present first results gained in combined empirical investigation.

1 Introduction

Research on wayfinding deals with the cognitive processes involved in decision making and planning that allow to reach destinations beyond the current sensory horizon. Wayfinding includes tasks such as cognitive mapping, orienting towards nonvisible locations, route planning, or deciding on the sequence of destinations to visit. Within this field, both linguistic and behavioral aspects of wayfinding have been investigated. However, this is typically done separately; research methods and experimental paradigms are seldom aligned in order to investigate just how language relates to behavior, pointing to shared underlying cognitive processes. Although psychological approaches sometimes involve thinking-aloud protocols along with behavioral data, these are not analyzed using discourse-analytic methods, and further language data are typically not elicited. Within the field of wayfinding, the only area in which both aspects have been combined systematically, yielding highly interesting results, concerns route knowledge and its verbal description. However, the investigation of route directions constitutes only a subpart of human wayfinding knowledge. Our joint research expands these findings in that we aim at systematically studying a wider variety of wayfinding tasks, such as search and exploration, route learning and optimization, and route following and route planning, both on behavioral and linguistic levels. In this paper, we motivate and sketch our general approach and present first qualitative results of symmetric experimental work, pointing at the identification of systematic parallels between wayfinding behavior and language.

1.1 Strategies, Mechanisms, and Heuristics Underlying Wayfinding

A crucial aspect involved in the process of wayfinding is that humans employ strategies and heuristics to approach their tasks, either because the required spatial knowledge is imprecise or incomplete, or the effort of actually calculating correct solutions is too high or requires too much time. For example, when navigating unfamiliar environments, it is impossible to decide about future actions based on spatial knowledge alone. Nevertheless, humans proceed in systematic ways in solving their current task. For instance, if the target direction is known but the environment is unknown, navigators have been shown to apply the *least angle heuristic*: at intersections, they simply select the street which has the least deviation from the target direction (e.g., Hochmair & Frank, 2002). Christenfeld (1995) showed that human subjects, choosing from a set of alternative routes that were identical with respect to metric length, target point, and number of turns, reliably preferred the route that allowed to delay the turning decision as long as possible, conceivably due to a tendency to minimize mental effort. This strategy also offers a possible explanation for the fact that people's route choices are often asymmetric: that is, subjects often choose different routes from A to B than from B to A (e.g., Stern & Leiser, 1988). Gärling & Gärling (1988) investigated wayfinding strategies by monitoring pedestrian shopping behavior with respect to distance minimization. Most shoppers first chose the location farthest away, probably to diminish the effort to carry bought goods, and then minimized distances locally between shopping locations (see also Gärling et al., 1986). This minimization heuristics is often referred to as the nearest neighbor (NN) heuristic.

Wiener & Mallot (2003) investigated the role of environmental regions for human route planning and wayfinding behavior in a series of navigation experiments using virtual reality technology (VR), showing that regions influenced route planning and navigation behavior: subjects preferred paths that minimized the number of region boundaries crossed during a navigation. They also preferred paths that allowed for fastest access to the region containing the target place. These results demonstrate that environmental regions are explicitly represented in spatial memory and are used during route planning. Results of this study were used to develop a cognitive model of region-based route planning, referred to as the '*fine-to-coarse* planning strategy'. Planning a route using a focal representation results in a detailed plan for the close surrounding, i.e., a plan that allows for immediate movement decisions. For distant locations only coarse information at the region level is given.

In addition to the *fine-to-coarse* planning strategy, Wiener et al. (2004) identified two further wayfinding strategies, the *cluster* strategy and the *least-decision-load* strategy. Essentially, the cluster strategy states that route planning takes into account the distribution of target locations within an environment, predicting that subjects preferred to visit as many targets as fast as possible. The least-decision-load strategy states that subjects given the choice between alternative paths choose the path that minimizes the number of possible movement decisions. The latter strategy could be employed because the risk of getting lost is smaller on less complex routes. In an additional experiment, Wiener et al. (2004) studied the formation and use of regional knowledge during the learning of an environment. They could show that regional information is encoded in spatial memory very early, most probably (i.) because by creating regions space becomes structured, thus facilitating the learning of an environment, and (ii.) because regional knowledge (coarse spatial information) allows to partly compensate for missing or imprecise fine-grained spatial knowledge, for example, by limiting the search for a specific object to the corresponding region.

To develop a deeper understanding of the strategies and heuristics underlying wayfinding it appears inevitable to also study which environmental features influence navigation behavior. According to O'Neill (1992), for example, wayfinding performance decreases with increasing floor plan complexity. Werner & Long (2003) demonstrate that the misalignment of local reference systems impairs the participants' ability to integrate spatial information across multiple places. Janzen et al. (2000) investigated the influence of oblique angled path intersections within an environment on wayfinding performance. Subjects' error rate depended on which branch they used to enter such an intersection. Furthermore, interrelations between visuo-spatial properties of architectural environments and experience of space as well as navigation behavior were investigated by Wiener & Franz (2005) and Wiener & Franz (submitted). They demonstrated that certain geometrical features of indoor spaces allow do predict both, experience of the spaces and navigation behavior. Recent experiments in complex buildings (Hölscher et al., 2005) involving the 'thinking aloud' method led to the identification of specific strategies for navigation such as the (efficient) floor strategy (first walking to the assumed floor of the goal) and the (less efficient) central point strategy (walking back to a central point and trying to find the goal from there). These experiments already point to the usefulness of language to investigate participants' strategies. We believe, however, that deeper insights can be gained by a more systematic linguistic investigation of the language associated with wayfinding tasks, going beyond thinking-aloud data and employing discourse-analytic methodology. In the following, we briefly discuss linguistic approaches to spatial cognition.

1.2 Spatial Language and Cognition

The idea that language may be viewed as a 'window' to cognition, in the sense that linguistic phenomena reflect underlying concepts, has been around ever since the beginnings of interdisciplinary research in cognitive science. Most research investigating this relationship has focused on knowledge and performance, on the inventory of language and the semantics of interesting (e.g., spatiotemporal) linguistic expressions (Herskovits, 1986), and on cross-cultural parallels and differences between linguistic and non-linguistic spatial representations (Levinson, 2003), often aiming at identifying the direction of causality between language and cognition (are differences in language responsible for differences in non-linguistic task solving or vice versa; e.g., Li & Gleitman, 2002; Levinson et al., 2002). Central contributions highlighting the relationship between linguistic and non-linguistic spatial encoding were provided, for example, by Landau & Jackendoff (1993), Hayward & Tarr (1995), Munnich et al. (2001), and in collections by Bloom et al. (1996), van der Zee & Slack (2003), and Carlson & van der Zee (2005). Some studies furthermore address the impact of linguistic abilities on spatial performance (Hermer-Vazquez et al., 1999), and the processes of building mental models from linguistic representations (Johnson-Laird, 1983; Taylor & Tversky, 1992; Tversky et al., 1994; Avraamides et al., 2004).

With respect to wayfinding in large-scale space, the relationship between spatial language and cognition has been investigated intensively only in tasks addressing route knowledge. Here, speakers' linguistic representations are analyzed as to the extent to which they reflect central cognitive wayfinding elements such as landmarks, spatial chunking and segmentation, decision points, perspective choice, basic and deviating spatial axes, ordering phenomena, and the like, all of which have been identified for cognition and language alike (e.g., Habel, 1988; Tversky, 1996; Couclelis, 1996; Denis, 1997; Allen, 2000; Klippel, 2003). Taylor & Tversky (1996) (for English) and Tappe (2000) (for German) investigated speakers' descriptions of sketch maps in terms of different kinds of perspectives (route vs. survey vs. gaze perspective) taken on the scene. These are distinguished, for instance, by lexical and conceptual choices such as using an observer as relatum.

A number of studies furthermore deal with the efficiency and adequacy of verbal route descriptions (e.g., Daniel & Denis, 1998; Denis et al., 1999; Lovelace et al., 1999). Here, besides varying degrees of correctness and consistency, central questions involve strategies of information conveyance, e.g., the amount of knowledge and ability that is presupposed on the part of the listener, the level of granularity chosen for description, memory load, and repair strategies in case of misunderstanding. Important variability arises with respect to discourse tasks and settings, i.e., monological or dialogical, in-advance or online, written or spoken, degrees of familiarity, etc. The impact of each of these factors has not been differentiated in much detail so far (but see Schober, 1993; Filipi & Wales, 2004), though a number of authors point out their importance (Daniel & Denis, 1998; Habel, 2003; Visser & Wolff, 2003); further evidence for the role of this variability comes from more general research in the wider field of discourse analysis (Clark, 1996; Schober & Brennan, 2003; Pickering & Garrod, 2004).

1.3 Comparing Spatial Behavior and Language

Although investigations directly comparing wayfinding behavior and language in tasks other than route navigation are rare, a number of studies address speakers' strategies in describing spatial environments (see Tenbrink et al., 2002; Tenbrink, 2007, for overviews). An early study (Linde & Labov, 1975) highlighted ways of describing the interior of buildings; the authors differentiate between static and dynamic spatial reference and show that speakers tend to adopt a strategy of mentally following a continuous route through the building (see also Ehrich, 1985; Herrmann & Grabowski, 1994). Later studies differentiate between local and global strategies of description (e.g., Carroll & von Stutterheim, 1993; Rieser, 1997), investigate the ways in which speakers choose strategies and switch between them throughout a discourse, for example, with respect to perspective (Tversky, 1996; Mainwaring et al., 2003), and identify various levels of description (Fischer & Moratz, 2001).

For a systematic investigation of parallels between spatial behavior and language, the methods of discourse analysis can be adopted usefully by drawing on knowledge about cognitive processes as a motivation and background. For example, a finding which is currently receiving increasing attention mainly aroused by Pickering & Garrod (2004) is the idea that speakers in spatial dialogue not only interactively align their language to each other but also jointly develop 'situation models' which are constructed for the specific purposes of the present discourse, similar to the notion reviewed by Zwaan & Radvansky (1998). The basic idea is that the conceptualization of a scene plays a role in referring to it. Thus, in experiments reported by Garrod & Anderson (1987), speakers developed consistent description schemes for reference to objects in a maze.

A crucial feature of the approach of 'cognitively motivated discourse analysis' as distinct from psycholinguistic research paradigms is to enable speakers to spontaneously produce language in precisely defined tasks but without external influence as far as linguistic choices are concerned. The investigation of systematic differences in relation to the given task and discourse setting leads to a better understanding of the relationship between language and underlying conceptions. Among the differences that have proved to be crucial in spatial language (cf. Tenbrink, 2007) is the explicit mention of specific elements such as perspective and relatum, the choice of spatial axis, and the level of granularity (Hobbs, 1990), detail, or precision. In addition, a number of syntactic and semantic distinctions in linguistic expressions point to underlying conceptual differences which can be explained by features of the given setting as well as by the inventory of the language involved. The systematic patterns identified in language reflect the underlying conception in ways that can be related directly to previous findings from other (non-linguistic) sources. Our previous investigations carried out in various scenarios (Tenbrink, 2007) point to the flexibility of speakers' spatial strategies (Tenbrink et al., 2002) and the nature of the underlying conceptualizations which may be expressed in a broad range of ways, highlighting certain aspects of the spatial configuration while leaving others underspecified (Tenbrink, 2003). A general conclusion that can be drawn from these previous findings is that wherever there is a difference in conceptualization, there will be a difference in linguistic expression if the speaker is allowed freedom to express the spatial relationships intuitively, i.e., in an open experimental design that does not require the speaker to use a specific linguistic (e.g., syntactic) choice. Such conceptual differences become apparent with any conceivable change in the discourse setting, be it the spatial configuration, the nature of the interaction partner (human or robot), or details of the task at hand.

2 General Approach

As summarized above, a number of cognitive components and processes that contribute to spatial cognition in behavior and language have already been identified. However, often such investigations remain sporadic and one-sided, focusing on particular aspects of either spatial behavior or language. Crucially, the designs of behavioral and linguistic experiments are seldom aligned to allow for a direct comparison of results. Our aim, in contrast, is to systematically vary the wayfinding task, subjects' knowledge, and properties of the environments, using virtual reality simulations or small-scale real world scenarios; and to apply the method of cognitively motivated discourse analysis in investigating the patterns underlying speakers' linguistic representations of the spatial tasks. As an outcome, the relationship between cognitive processes and their externalizations in spatial behavior and spatial language can be precisified systematically, leading to the development of hypotheses about the use, function, and interplay of different cognitive components and processes involved in specific wayfinding tasks and reflected in language usage.

Our approach is to confront participants with different wayfinding tasks. While solving the tasks, their navigation behavior is recorded at the level of trajectories, coding position and heading over time. Any interrelation between wayfinding task, spatial properties of the environment, and subjects' knowledge will allow the inference of strategies applied when solving a specific wayfinding task. The behavioral and cognitive patterns identified by the behavioral experiments should be reflected in language, although it is not in all cases obvious beforehand in which ways the reflection may appear, and which discourse situation is best suited to uncover the underlying concepts. Therefore, a number of different discourse situations will be investigated, varying between spoken and written language, self-talk, and monologic versus dialogic scenarios, depending on the suitability to the task in each case.

While the analysis of the behavioral data is deeply rooted in established methods of psychology and biology, the linguistic analysis takes a slightly different stance. Importantly, the behavioral results will, together with previous linguistic findings, lead to specific hypotheses that can be tested directly on the linguistic data collected. In addition to this, a number of systematic patterns will emerge from the natural language data which could not be predicted on the basis of previous results, but which then lead to the formulation of post-hoc interpretations that can again be tested in subsequent work. This approach differs from most previous investigations of spatial language and cognition in that speakers are given a high degree of freedom for language production, and in that neither linguistic performance nor the effects of a linguistic representation on the receiver are addressed. Furthermore, not the semantics and structure of linguistic expressions are in focus but rather their applicability in certain precisely devised discourse tasks.

Specifically, the analysis of linguistic data may concern explicit mention of certain aspects of the spatial task, e.g., landmarks and detailed spatial relationships, the choice of spatial axes, granularity levels, and underlying perspectives, the presentation of linguistic items as given or as new (Halliday & Matthiessen, 1999), the usage of presuppositions, order of mention, etc. Previously known as well as marginal kinds of landmarks are described in a different way (if at all) than newly perceived, or specifically important, landmarks (Talmy, 2000). Crucial factors of the wayfinding process receive greater linguistic prominence. The level of granularity chosen for a description reflects the underlying conceptualization dependent on the task (see Timpf & Kuhn, 2003) and its significance for the speaker (and hearer). The general processes governing linguistic choices in various kinds of discourse situations, and the ways in which they reflect underlying cognitive processes and mental conceptualizations, have been investigated within the wider field of discourse analysis for several decades; the methods and findings developed in this field are decisive for the investigation of the language produced along with and paralleling spatial behavior. By comparing spontaneous non-linguistic spatial behavior with linguistic strategies, the specific goals and methods of linguistic discourse analysis are directly cognitively motivated. Also, participants' utterances often contain explicit information about underlying strategies (Hölscher et al., 2005); such information is specifically interesting since speakers provide direct access to aspects that are important to them (Sacks et al., 1974). This kind of evidence is accessible to some degree by the elicitation of thinking-aloud protocols, which is an established method (Fonteyn et al., 1993). However, typically such data are not analyzed using linguistic, discourse-analytic methods; therefore, only those aspects that the participants themselves were aware of can be accessed. In our approach, we aim at detecting systematic linguistic features in the data (collected in more than one way); these point to underlying cognitive processes that may not be immediately obvious.

In the following, we exemplarily present a first experiment which was originally carried out with a focus on spatial behavior (Wiener et al., submitted). Informal questioning of the participants then opened up a number of aspects and pointed to underlying strategies that could not be directly derived from the data, highlighting the significance of taking linguistic representations into account. To enhance this idea, the experiment was replicated with a small number of participants, using three variations of linguistic representations: self-talk, a written account of the spatial experience, and a written instruction formulated for an unexperienced friend. The language data were analyzed qualitatively using the method of cognitively motivated discourse analysis.

3 Experiment 1

Depending on the number of targets, route planning can be a very complex and computationally expensive task. This is best demonstrated by the well-known traveling salesman problem (TSP). Here, the task is to find the least costly path (usually the shortest path) in order to visit multiple target locations and to return to the start place. The number of possible solutions is calculated as the factorial of the number of target locations, i.e. for visiting 4 target locations there are 24 alternative solutions, and for visiting 9 target locations there are already 362 880 alternative solutions.

Obviously, if human navigators are faced with path planning problems comparable to the TSP, for example on typical shopping routes on which multiple locations have to be visited, they rely on planning strategies and heuristics rather than calculating and comparing all possible path alternatives. One possible strategy to solve TSPs that has received considerable attention is the so-called nearest neighbor algorithm (NN). From its current location, which at first corresponds to the starting place, the NN algorithm visits the closest target location that has not been visited before. By simply repeating this procedure until all target locations have been visited and by returning to the starting place, the NN algorithm usually finds good or near-optimal solutions for TSPs of small sizes.

In a recent study, Wiener et al. (submitted) investigated human performance as well as the cognitive strategies involved in solving traveling salesman tasks by active navigation. For this, 25 places were arranged on a regular grid in a large experimental room and each place was marked by a unique symbol (see Figure 1). Subjects were asked to

Fig. 1. Left: schematic drawing of the experimental layout of Experiment 1; middle: subject solving one navigation tasks; right: schematic drawing of the experimental layout of Experiment 2

solve a total of 36 different TSPs with 4 to 9 target places. For each trial, they were given a so-called 'shopping list' (see Figure 2) depicting the symbols of the start place and the target places. Their task was to navigate the shortest closed loop connecting the start place with all target places. During navigation, subjects were allowed to mark the visited target places with little black markers. For each navigation task, the experimenter recorded subjects' trajectories, i.e. the sequence in which the target places were visited. From this, the length of the chosen path was calculated. By comparing the length of the chosen path with the length of the optimal path, the performance measure *overshoot* was calculated. 100% overshoot corresponds to a path with twice the length of the optimal path.

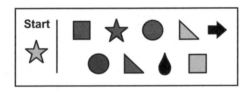

Fig. 2. Exemplary shopping list for a navigation task with 9 target places

The general outcome of the experiment was that subjects showed decreasing performance as the number of target locations increased. While subjects showed almost optimal performance, i.e. almost no overshoot, with 4 target locations, overshoot gradually increased to just under 10% above the optimal path for TSP with 9 target places. Accordingly, subjects' start time, i.e. the time after first looking at the shopping list until initiating the movement, increased with increasing number of target locations to visit (see Figure 3).

Regarding the route planning strategies, Wiener et al. (submitted) analyzed whether humans deployed the nearest neighbor (NN) strategy and the cluster strategy (see Section 1.1) when solving the TSPs. For this, the navigation tasks that subjects were asked to solve in Experiment 1 could be assigned to different route types, the NN-adequate tasks and the NN-inadequate tasks. For NN-adequate tasks, the actual NN-algorithm predicts the optimal, i.e. the shortest possible, route. For NN-inadequate tasks the NN-algorithm predicts sub-optimal routes. In addition to the NN-adequate and the NN-inadequate tasks, a further group of tasks were labelled cluster tasks. On these navigation tasks, the target places were distributed in two distinct target clusters of unequal size (e.g, on a navigation tasks with 6 target places, there was one target cluster of 2 places and one target cluster of 4 places). Cluster tasks were NN-ambiguous as the NN algorithm did not make clear predictions: the closest target places were equidistant from the starting place, and similar situations also re-occurred during navigation.

Results showed that subjects outperformed the NN algorithm on navigation tasks in which the NN algorithm did not predict the optimal solution (NN-inadequate tasks), i.e. on average subjects found significantly shorter routes than the NN algorithm. Furthermore, when navigating cluster tasks, subjects showed a preference to first visit the larger target cluster. The former result demonstrates that the NN algorithm alone is not sufficient to explain human path planning behavior, at least in this particular situation. The latter result demonstrates that during route planning not only the closest target place but the distribution of target places in the environment is taken into account. The observed bias to visit as many target places as fast as possible has been termed cluster strategy (see Wiener et al., 2004).

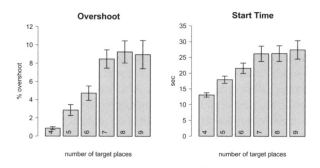

Fig. 3. Results of Experiment 1: left: subjects' performance in finding the optimal solution decreased, i.e. overshoot increased with increasing number of targets; right: subjects' start time increases with increasing number of targets

As the NN algorithm was not sufficient to explain subjects' navigation behavior, and as the planning strategies applied were not apparent from the trajectories recorded, subjects were informally interviewed after the experiment. Most of the 24 subjects reported to have applied one of two strategies when faced with large TSPs. The first strategy was based on a regionalization of the environment. Subjects reported that they individually subdivided the 25 target places into different regions. During route planning they then assigned the target places to be visited to these regions, allowing them to first plan a coarse route on the basis of the regions. Such a coarse route plan is simple, easily remembered, and the fine detailed route plan is then created by inserting close target places during navigation. Applying such a region-based route planning strategy will result in routes that minimize the number of region transitions. The second strategy reported was to first select a subset of the target places depicted on the shopping list according to some criteria, for example, the color of the corresponding symbols. Then a coarse route plan was generated taking into account only the selected subset of target places. This route plan was then refined either before or during navigation by inserting the missing target places into the route.

Both of the reported navigation strategies follow essentially the same logic: they simplify the route planning task by applying a hierarchical planning scheme. First, a coarse route that is simple and easy to remember is generated on basis of an abstraction of the environment. This route plan is then refined during navigation by inserting target places. Such a hierarchical scheme not only reduces computational effort during route planning but also minimizes working memory load for the route plan.

4 Experiment 2

In order to empirically validate whether subjects applied the reported region-based strategy, Wiener et al. (submitted) carried out a second experiment. In this experiment, the environment was subdivided into 5 different, objectively identifiable, regions by placing symbols of equal color next to each other (see Figure 1). As in Experiment 1, subjects had to solve 36 TSPs with 4 to 9 target places. These navigation tasks could again be assigned to one of two types that differed with respect to the trajectories of the optimal solution on the region level. In Region-Strategy-adequate tasks (RS-adequate) the optimal solution also minimized the number of required region transitions. In Region-Strategy-inadequate tasks (RS-inadequate) navigation tasks the optimal solution required more region transitions than other, sub-optimal (i.e., longer) routes. It was predicted that if subjects applied a region-based strategy during navigation, they first visited all target locations in one region before entering the next region containing targets. By these means subjects were expected to minimize the number of region-transition on both types of navigation tasks.

Results of this experiment showed that when navigating RS-inadequate tasks, subjects clearly preferred sub-optimal routes that minimized the number of region transitions. This result provides support for the region-based navigation strategy that subjects reported in Experiment 1. While the experiments presented here took place in small scale space, Wiener & Mallot (2003) and Wiener et al. (2004) reported a very similar hierarchical planning scheme to account for subjects' route planning and navigation behavior in (virtual) large scale environments that were divided into different regions (see Section 1.1). This suggests that the applied hierarchical planning strategies are scale-independent and constitute general means to solve complex and computationally expensive route planning tasks.

5 Experiment 3

Since the informal questioning of participants in Experiment 1 already turned out to be valuable as shown by the results of Experiment 2, the question arises whether more information could be gained by investigating linguistic representations systematically and using linguistic expertise. For this purpose, Experiment 1 was repeated with a small number of participants who were asked to perform a number of linguistic tasks in addition. This small experiment can be regarded as a pilot study serving to test and exemplify our general approach.

The experimental setup and navigation tasks mirrored those of Experiment 1, except that each participant received only 6 navigation tasks, namely, one simple TSP (i.e., with 5 target places) and one complex TSP (with 9 target places) of each type of navigation tasks (cluster, NN-adequate, and NN-inadequate). This decision was taken in order to keep the total experiment time on a similar level as the behavioral experiments. Since sufficient behavioral data had already been collected, only a smaller amount of behavioral data was needed this time to ensure comparability of the data. The participants were asked not only to perform the same task as the subjects in experiment 1, i.e. to actively navigate the shortest possible route to visit all target places depicted on the shopping lists, but also, to try to 'think aloud' during the process. The subjects' trajectories were recorded as in Experiment 1. Furthermore, after solving each of the 6 tasks, they were seated at a computer and asked to type in the answers to two questions designed to trigger linguistic representations of the spatial task. The questions were the following (translated here from German):

"1. Now that you have solved this task several times you are an expert concerning the question how it can best be solved. I now ask you for two more things. First: Please describe in as much detail as possible how you have solved the task, what you were considering while you solved it, and why you performed precisely in this way rather than another. Also, describe what, or which of the particular tasks, was easy and what was difficult for you during the task.

2. Second: Please write an instruction for a good friend of yours that allows him/her to solve the task as well as possible."

These questions were designed to trigger linguistic representations of the mental processes associated with the spatial task from memory, i.e., after performing it, and in the manner of an instruction, which implies that the results should reflect the participants' conscious solutions as to how best to solve this task. Thus, we used three different ways of accessing subjects' cognitive processes via language, corresponding to three temporal perspectives: during the task, after the task, and before (having an imaginary friend perform) the task.

Seven subjects participated in the experiment. They were university students and were paid 8 Euro an hour.

5.1 Results

An analysis of the navigation behavior of Experiment 3 demonstrates that subjects reached similar performance levels as compared to Experiment 1 (see Figure 4), suggesting that the data are indeed comparable. Such a symmetry between pure navigation behavior and navigation coupled with linguistic tasks is essential for the reliability of the analysis, since it is conceivable that subjects develop distinct strategies even with seemingly negligible changes to the task.

The three different linguistic tasks yielded three substantially different kinds of linguistic representations. They will first be reported in some detail; after that, the shared features and most relevant aspects will be extracted. Due to the small number of participants, the linguistic analysis will be entirely qualitative and exemplary.

Fig. 4. Subjects' overshoot performance in Experiment 3

Thinking aloud. The verbal task of thinking aloud during the behavioral task itself turned out to be problematic for some of the subjects. They had to be prompted several times, and some mentioned that they did not know what to say. As an outcome, some of the recorded files do not contain much more than a simple listing of the items to be collected, such as 'the blue star, and now the yellow triangle' etc. Although such data are not particularly revealing with respect to underlying strategies, they do inform about thought processes on the grounds of the way in which items are mentioned prior to and while performing the task. For instance, sometimes items are mentioned together and in a coherent way, as in 'dann haben wir hier das schwarze Dreieck, den roten Pfeil, das gelbe Dreieck, den blauen Stern, den roten Tropfen' (then, we have here the black triangle, the red arrow, the yellow triangle, the blue star, the red drop), reflecting a cognitive grouping process that is also represented by the subjects' collecting the relevant items relatively rapidly after one another. The utterance is followed by 'so, der blaue Bogen ist da vorne' (okay, the blue bow is ahead), representing another item separately, as indicated by 'so'. Other participants directly express the underlying process by recurring utterances such as 'ich mach jetzt erstmal alle in der linken Ecke oben' (I first do all those in the left corner above), and explicitly differentiating, for example, between the left and right sides of the grid, pointing to a (vague) regionalization of the spatial scene. Another aspect reflected in the utterances is to establish a path leading from the present position via another item to a goal: 'dann weiter über den blauen Bogen zum gelben Kreis' (then further via the blue bow to the yellow circle). Furthermore, spatial vicinity played a role: 'vom grünen Kreuz aus der nächste ist der gelbe Bogen' (from the green cross the next one is the yellow bow).

A few participants were quite at ease with this verbal task, and they verbalized their reasoning processes clearly. One person explicitly started off by stating that he thought it would be easiest to walk roughly around the scene and simply depart into the center if there are items to be collected there. Later, he remarked that many of the items were situated together in one corner so that for these items the path was virtually predetermined; after collecting these items, he stated that those grouped together were over but there were still some missing. These thoughts explicitly reflect the processes of grouping and conceptualizing as separate, and consequently dealing with the items separately. Interestingly, this person seemed to infer that this was the most obvious way of solving this task, by using the term 'vorgegeben' (predetermined). Considering this idea, it is not surprising that other participants had greater difficulties in verbalizing these 'natural' cognitive processes, as they were not aware of possible alternatives.

Analysis of the written texts. The written tasks did not pose a problem for any of the participants. They even managed to spontaneously develop a consistent text structure suitable for the discourse task at hand, which mirrors the temporal structure of the spatial task. In both tasks, temporal markers were frequently used, starting out with expressions such as 'first of all' and 'initially' and continuing with connectives, e.g., 'dann' (then) and 'als nächstes' (next). In the 'instruction' task, three of the seven instructors further structured their text by using paragraphs and numbers. This is an interesting result in light of the fact that the spatial task did not directly require a particular temporal order, and the written tasks were also not restricted to one particular TSP task but addressed the general procedure. Thus, participants clearly subdivide the cognitively demanding main task into smaller steps and subtasks in order to deal with the involved difficulties. Sometimes the temporal structure within the text is directly correlated with the spatial structure, as reflected by 'von dort aus' (from there). Causal connectives serve to mark reasons for specific decisions, such as, 'in the complex tasks it was difficult to plan the whole route, therefore I searched for groups of items and searched for the shortest route between the groups.' In this way, the participants marked their underlying strategies explicitly.

The instructions are generally shorter and less detailed than the descriptions. They contain a number of encouraging statements, e.g., 'just try it', and tips such as 'maybe it could be helpful to sort the symbols according to shape or colour', 'you could try using reference points as support'. Thus, the 'instruction' task induced some of the participants to reflect even more about how the task could be solved, and to suggest strategies that they had not tried out themselves.

Subtasks. In addition to explicit marking by temporal and causal expressions, the subdivision of the main task into smaller subtasks is reflected by the mention of a number of systematically recurring verbs:

1. Verbs of searching, looking, and finding ('ansehen, suchen, finden, vergleichen', etc.) reflect the subprocess of identifying objects. Here, the focus is generally on the visual process; the finding itself is often expressed together with an evaluation such as 'difficult to find'. Thus, the identification process is in itself a non-trivial task involving some amount of time and effort.
2. Verbs of thinking and orienting ('orientieren, überlegen') as well as verbs of memorizing ('einprägen, merken', etc.) and verbs of planning and deciding ('Route festlegen, planen, koordinieren, ordnen', etc.) point to the fact that the cognitive processes involved are non-automatic and conscious to a high degree, and that the participants included specific planning processes in their spatial behavior.
3. Verbs of action ('gehen, laufen, markieren', etc.) describe the actual performance of the spatial task.
4. Finally, a number of verbs or verbal groups clearly reflect underlying strategies: 'von Symbol zu Symbol hangeln' (proceed from symbol to symbol), 'dazustückeln' (add remaining pieces), 'zusammenfassen' (summarize), 'in Grüppchen sortieren / gruppieren' (sort into groups), 'die Gruppen miteinander verbinden' (connect the groups). Here, the

main strategy that emerges from the analysis of the verbs is the cluster strategy already identified in the behavioral data: participants conceptualize the items as groups and profit from this by mentally, linguistically, and behaviorally treating them collectively, thereby in effect reducing the number of items to be memorized.

Relevant spatial aspects. Apart from the verbs, other linguistic items also point to the relevance of a number of spatial aspects that play a role in solving the tasks:

1. Spatial structure: There are many ways in which the confusing assembly of shapes and colours in the scene is structured. Probably the most frequent lexical item reflecting a conceptual structuring process is 'Gruppe' (group) and 'Gruppierung' (grouping); altogether these occur 32 times in the 14 files (containing little more than 2.000 words). Another label for the same phenomenon is 'Häufung' (accumulation). Thus, the clustering of items obviously plays a major role for the conceptualization. Another way of imposing structure on the scene is to search for particular shapes or colours (typically colours) and to memorize them as 'reference points', as explicitly stated by one participant. Furthermore, participants sometimes refer to sides ('Seite'). Finally, some participants conceived of the route to be taken as something that should best have the shape of a circle ('Kreis'). In all of these cases, the basic procedure is to start from this rough structuring of the scene and to include the residuals (single remaining items that do not belong to the imposed structure) by departing from this original idea. The residuals themselves are referred to linguistically as 'restlich/übrig' (remaining), 'Rest' (remainder), 'letzte' (last), 'fehlende' (missing), 'einzeln' (individual), and (as mentioned above) by the verb 'dazustückeln' (add remaining pieces). It is interesting to note that the reference to the residuals is by far not as consistent (linguistically) as the reference to the groups; this may reflect a clearcut notion of the grouping process but a less straightforward process of dealing with the remaining items.

2. Spatial vicinity: Many lexical items point to the significance of the notion of distance. Since the task was to find the shortest route, clearly participants wished to avoid greater distances and departures from their envisioned route. Thus, it appears natural that they referred to the advantage of items being positioned in close vicinity to each other, as reflected by frequent mention of 'am nächsten / nächstgelegen/ nächste Häufung / Seite' (closest (group/side)), 'nahe beieinander' (closely together), 'nebeneinander' (beside each other), and a few others. Note, however, that this notion does not always relate to single items, but very often to whole groups or sides of the scene or to the next frequently occurring category. This reflects the fact that the strategy of searching for the nearest item is not sufficient by itself.

Strategies. The participants' strategies involved in solving this spatial task are reflected, on the one hand, by directly mentioning them, and on the other hand, by the usage of linguistic forms in particular ways. On the basis of the analysis so far, the following aspects could be identified by the linguistic data to be relevant for participants' strategies:

1. Clustering: Participants conceptually group the items and often mention (and visit) larger clusters first.

2. Regionalization: Participants mentally divide the scene into regions (with fuzzy borders) and order their route according to the vague directions to be taken.

3. Categorization: Participants pick out categories of items, preferably on the basis of frequent occurrence, and focus on the places of those items first that share a given feature (such as colour or shape).

4. Neighborhood: Spatial vicinity doubtless plays a role for participants, though not as a global or exhaustive strategy (as suggested, for instance, by the Nearest Neighbor Algorithm), but rather, as the local procedure at a given place, e.g., within a previously identified cluster of objects.

5. Shape of trajectory: Participants conceptualize a route, preferredly by imagining a circle-like trajectory that aids in mentally ordering the items to be visited.

6. Insertion of detail: After having devised a coarse route plan on the basis, for example, of any of the previously mentioned aspects which serve to abstract away from the complex details of the scenario, participants decide on the specific steps to take by inserting detours from the general trajectory.

5.2 Discussion

The results of our experiments highlight the value of an iterative process in investigating spatial strategies by analyzing behavioral data as well as verbal data, and testing hypotheses directly in subsequent targeted experimentation. The informal questioning of participants in Experiment 1 led to the identification of a number of interesting aspects such as underlying strategies that might motivate the participants to solve the given task in a particular way. However, this method does not allow for a systematic analysis of linguistic representations, because the language was not collected and recorded systematically; additionally, only those aspects that the subjects themselves are aware of can be approached in this way. A linguistic approach, in contrast, also allows for systematically investigating features of language that reveal factors that are relevant to the participant without requiring conscious awareness of these processes. Our analyses of various linguistic tasks enabled us, on the one hand, to specify just how speakers verbalize their cognitive processes in solving a complex wayfinding task. Their strategies are reflected both by explicit reasoning and by systematic linguistic patterns, such as consistent lexical choices as well as the absence of clearcut means of linguistically capturing strategic steps. Such results both confirm the existence of already identified strategies (such as the cluster strategy and region-based strategies, Wiener & Mallot 2003 and Wiener et al. 2004), and they establish the relationship to natural language which had hitherto been unexplored. On the other hand, the linguistic data revealed further underlying complexities that could not have been revealed by the analysis of the behavioral data alone, including categorization processes, a relationship between the Nearest Neighbor strategy and cluster strategies, and an interplay between the overall shape of a coarse route plan with the insertion of details. These results reflect underlying cognitive processes such as human conceptualization (Rosch & Lloyds, 1978) and hierarchical structuring processes, as reflected, for example, in linguistic localization sequences (Herrmann & Grabowski, 1994, p 152). Furthermore, some of the concepts that became apparent by the linguistic analysis, such as the relevance of boundaries, spatial vicinity, and trajectory shapes, are well-known in the field of spatial cognition (e.g., Lynch, 1960; Kuipers et al., 2003) and have been explored from various perspectives based on the notion of *image schemata* (Johnson, 1987; Rüetschi & Timpf, 2005). However, the details of how these concepts and cognitive processes become relevant when confronted with tasks such as the Traveling Salesman Problem still need to be explored further. Based on our experience from the present pilot study, it seems promising to pursue the linguistic approach in systematic ways that allow for valid generalizations.

Linguistic representations reflect underlying cognitive processes on several levels by highlighting those aspects that are crucial for the spatial problem at hand and neglecting others. Some aspects of the linguistic representation may be due to the specific discourse task given to the participant. This necessitates, on the one hand, the variation of discourse tasks in order to differentiate between spatial task-related and discourse-related factors; and on the one hand, linguistic knowledge about particularities of discourse types that reveal themselves in the textual representation in various ways. With respect to the present exemplary task, one result of the contrastive analysis of three kinds of texts is that the 'description' variant was the most detailed and therefore informative, while the 'thinking aloud' task mostly only allowed for the inference of underlying processes which most of the participants were not able to express directly during the performance of the task. These results are revealing in light of the fact that 'thinking aloud' tasks are often employed as an established method of gathering information concerning cognitive processes. Due to these limitations, it can be concluded that the 'thinking aloud' method should generally be supplemented by other kinds of verbal task. By eliciting further linguistic data in other ways, participants get a better chance of expressing their thought processes linguistically if the effort of producing language during the task itself turns out to be problematic. In all cases, a systematic linguistic analysis is likely to reveal more substantial results than those provided by explicit mention of strategies by participants. This latter conclusion naturally applies to a higher degree to the more informative kinds of discourse tasks; however, it should be noted that some aspects (in this case, the mention of spatial regions by referring to left or right sides or corners) may occur only while (not after) the spatial task is performed. Therefore, further on-line linguistic tasks such as instructing a fellow participant while performing the task may be a good addition to thinking-aloud tasks and off-line accounts such as those analyzed here.

6 Conclusion

In this paper, we have presented our interdisciplinary approach for the investigation of strategies and heuristics underlying human wayfinding. The approach is based on the symmetric investigation of navigation behavior and associated language. We have reported a linguistic replication of a behavioral wayfinding study that is to be regarded as a first pilot study, in which the suitability of the general approach was tested by a close and detailed view on the linguistic features of a few participants' representations. The preliminary analysis of the linguistic representations collected in Experiment 3 revealed not only strategies and mechanisms that were already identified by the behavioral experiments, but furthermore pointed to complexities in combining spatial strategies and additional aspects that were relevant for solving the navigation tasks. From here, the next step will be to design further behavioral experiments allowing for the empirical validation of these new insights gained from the linguistic version of the navigation experiment. Such an iterative procedure, symmetrically combining findings from both behavioral and linguistic experiments, is essential for the presented approach. By these means, we will be able to obtain more profound insights into the strategies and mechanisms underlying human wayfinding than behavioral or linguistic experiments alone could produce. Furthermore, the relationship between cognitive processes and their reflections in behavior and language can be established systematically by this approach, showing in some detail how underlying concepts are represented externally.

Acknowledgements

The authors gratefully acknowledge funding by the Volkswagen Foundation for a Tandem research project granted to Jan Wiener and Thora Tenbrink. Furthermore, thanks is expressed to the DFG (WI 2729/1-1 and SFB/TR 8 Spatial Cognition). We also thank our participants for their efforts.

References

Allen, G.: Principles and practices for communicating route knowledge. Applied Cognitive Psychology 14, 333–359 (2000)

Avraamides, M.N., Loomis, J.M., Klatzky, R.L., Golledge, R.G.: Functional Equivalence of Spatial Representations Derived From Vision and Language: Evidence From Allocentric Judgments. Journal of Experimental Psychology: Learning, Memory, and Cognition 30(4), 801–814 (2004)

Bloom, P., Peterson, M., Nadel, L., Garrett, M.: Language and Space. MIT Press, Cambridge, MA (1996)

Carlson, L.A., van der Zee, E. (eds.): Functional features in language and space: Insights from perception, categorization and development. Oxford University Press, Oxford (2005)

Carroll, M., von Stutterheim, C.: The representation of spatial configurations in English and German and the grammatical structure of locative and anaphoric expressions. Linguistics 31, 1011–1041 (1993)

Christenfeld, N.: Choices from identical Options. Psychological Science 6, 50–55 (1995)

Clark, H.H.: Using Language. Cambridge University Press, Cambridge (1996)

Couclelis, H.: Verbal directions for way-finding: space, cognition, and language. In: Portugali, J. (ed.) The construction of cognitive maps, pp. 133–153. Kluwer Academic Publishers, Dordrecht (1996)

Daniel, M.-P., Denis, M.: Spatial Descriptions as Navigational Aids: A Cognitive Analysis of Route Directions. Kognitionswissenschaft 7(1), 45–52 (1998)

Denis, M.: The description of routes: A cognitive approach to the production of spatial discourse. Cahiers de Psychologie Cognitive 16(4), 409–458 (1997)

Denis, M., Pazzaglia, F., Cornoldi, C., Bertolo, L.: Spatial discourse and navigation: an analysis of route directions in the city of Venice. Applied Cognitive Psychology 13(2), 145–174 (1999)

Ehrich, V.: Zur Linguistik und Psycholinguistik der sekundären Raumdeixis. In: Schweizer, H. (ed.) Sprache und Raum: Ein Arbeitsbuch für das Lehren von Forschung, pp. 130–161. Metzler, Stuttgart (1985)

Filipi, A., Wales, R.: Perspective-taking and perspective-shifting as socially situated and collaborative actions. Journal of Pragmatics 36(10), 1851–1884 (2004)

Fischer, K., Moratz, R.: From Communicative Strategies to Cognitive Modelling. In: Workshop Epigenetic Robotics Lund (2001)

Fonteyn, M.E., Kuipers, B., Grobe, S.J.: A Description of Think Aloud Method and Protocol Analysis. Qualitative Health Research 3(4), 430–441 (1993)

Gärling, T., Gärling, E.: Distance minimization in downtown pedestrian shopping. Environment and Planning A 20, 547–554 (1988)

Gärling, T., Säisä, J., Böök, J., Lindberg, E.: The spatiotemporal sequencing of everyday activities in the large-scale environment. Journal of Environmental Psychology 6, 261–280 (1986)

Garrod, S.C., Anderson, A.: Saying what you mean in dialogue: a study in conceptual and semantic co-ordination. Cognition 27, 181–218 (1987)

Habel, C.: Prozedurale Aspekte der Wegplanung und Wegbeschreibung. In: Schnelle, H., Rickheit, G. (eds.) Sprache in Mensch und Computer, pp. 107–133. Westdeutscher Verlag, Wiesbaden (1988)

Habel, C.: Incremental Generation of Multimodal Route Instructions. In: Freedman, R., Callaway, C. (eds.) Working Papers of the 2003 AAAI Spring Symposium on Natural Language Generation in Spoken and Written Dialogue, pp. 44–51. AAAI Press, Menlo Park, California (2003)

Halliday, M.A.K., Matthiessen, C.M.I.M.: Construing experience through meaning: a language-based approach to cognition. Cassell, London (1999)

Hayward, W., Tarr, M.: Spatial language and spatial representation. Cognition 55, 39–84 (1995)

Hermer-Vazquez, L., Spelke, E.S., Katsnelson, A.S.: Sources of Flexibility in Human Cognition: Dual-Task Studies of Space and Language. Cognitive Psychology 39(1), 3–36 (1999)

Herrmann, T., Grabowski, J.: Sprechen: Psychologie der Sprachproduktion. Spektrum Verlag, Heidelberg (1994)

Herskovits, A.: Language and Spatial Cognition: an interdisciplinary study of the prepositions in English. Studies in Natural Language Processing. Cambridge University Press, London (1986)

Hobbs, J.R.: Granularity. In: Weld, D.S., de Kleer, J. (eds.) Qualitative Reasoning about Physical Systems, pp. 542–545. Morgan Kaufmann, San Mateo, California (1990)

Hochmair, H., Frank, U.: Influence of estimation errors on wayfinding-decisions in unknown street networks - analyzing the least-angle strategy. Spatial Cognition and Computation 2(4), 283–313 (2002)

Hölscher, C., Meilinger, T., Vrachliotis, G., Broesamle, M., Knauff, M.: Finding the Way Inside: Linking Architectural Design Analysis and Cognitive Processes. In: Freksa, C., Knauff, M., Krieg-Brückner, B., Nebel, B., Barkowsky, T. (eds.) Spatial Cognition IV: Reasoning, Action, Interaction, pp. 1–23. Springer, Heidelberg (2005)

Janzen, G., Herrmann, T., Katz, S., Schweizer, K.: Oblique Angled Intersections and Barriers: Navigating through a Virtual Maze. In: Habel, C., Brauer, W., Freksa, C., Wender, K.F. (eds.) Spatial Cognition II. LNCS (LNAI), vol. 1849, pp. 277–295. Springer, Heidelberg (2000)

Johnson, M.: The body in the mind. University of Chicago Press, Chicago, Il (1987)

Johnson-Laird, P.: Mental Models: Towards a Cognitive Science of Language, Inference, and Consciousness. Cambrige University Press, Cambridge (1983)

Klippel, A.: Wayfinding choremes: Conceptualizing wayfinding and route direction elements. Ph.D. thesis, Universität Bremen (2003), http://www.uni-bremen.de

Kuipers, B., Tecuci, D., Stankiewicz, B.: The skeleton in the cognitive map: a computational and empirical exploration. Environment and Behavior 35(1), 80–106 (2003)

Landau, B., Jackendoff, R.: 'What' and 'where' in spatial language and cognition. Behavioral and Brain Sciences 16, 217–238 (1993)

Levinson, S.C.: Space in language and cognition: explorations in cognitive diversity. Cambridge University Press, Cambridge (2003)

Levinson, S.C., Kita, S., Haun, D.B.M., Rasch, B.H.: Returning the tables: language affects spatial reasoning. Cognition 84, 155–188 (2002)
Li, P., Gleitman, L.: Turning the tables: language and spatial reasoning. Cognition 83(3), 265–294 (2002)
Linde, C., Labov, W.: Spatial networks as a site for the study of language and thought. Language 50(4), 924–939 (1975)
Lovelace, K.L., Hegarty, M., Montello, D.R.: Elements of Good Route Directions in Familiar and Unfamiliar Environments. In: Freksa, C., Mark, D.M. (eds.) COSIT 1999. LNCS, vol. 1661, pp. 65–82. Springer, Heidelberg (1999)
Lynch, K.: The Image of the City. MIT Press, Cambridge (1960)
Mainwaring, S.D., Tversky, B., Ohgishi, M., Schiano, D.J.: Descriptions of simple spatial scenes in English and Japanese. Spatial Cognition and Computation 3(1), 3–42 (2003)
Munnich, E., Landau, B., Dosher, B.A.: Spatial language and spatial representation: a crosslinguistic comparison. Cognition 81(3), 171–208 (2001)
O'Neill, M.J.: Effects of familiarity and plan complexity on wayfinding in simulated buildings. Journal of Environmental Psychology 12, 319–327 (1992)
Pickering, M.J., Garrod, S.: Towards a mechanistic psychology of dialogue. Behavioural and Brain Sciences 27(2), 169–190 (2004)
Rieser, H.: Repräsentations-Metonymie, Perspektive und Koordination in aufgabenorientierten Dialogen. In: Umbach, C., Grabski, M., Hoernig, R. (eds.) Perspektive in Sprache und Raum, Studien zur Kognitionswissenschaft, pp. 1–26. Deutscher Universitätsverlag, Wiesbaden (1997)
Rosch, E., Lloyds, B.: Cognition and categorization. Lawrence Erlbaum, Hillsdale, NJ (1978)
Rüetschi, U.J., Timpf, S.: Using Image Schemata to Represent Meaningful Spatial Configurations. In: Meersman, R., Tari, Z., Herrero, P. (eds.) On the Move to Meaningful Internet Systems 2005: OTM 2005 Workshops. LNCS, vol. 3762, pp. 1047–1055. Springer, Heidelberg (2005)
Sacks, H., Schegloff, E., Jefferson, G.: A Simplest Systematics for the Organization of Turntaking for Conversation. Language 50, 696–735 (1974)
Schober, M., Brennan, S.: Processes of interactive spoken discourse: The role of the partner. In: Graesser, A., Gernsbacher, M., Goldman, S. (eds.) Handbook of Discourse Processes, pp. 123–164. Lawrence Erlbaum Associates, Mahwah, NJ (2003)
Schober, M.F.: Spatial perspective taking in conversation. Cognition 47, 1–24 (1993)
Stern, E., Leiser, D.: Levels of spatial knowledge and urban travel modeling. Geographical Analysis 20, 140–155 (1988)
Talmy, L.: Towards a cognitive semantics. MIT Press, Cambridge, MA (2000)
Tappe, H.: Perspektivenwahl in Beschreibungen dynamischer und statischer Wegeskizzen. In: Habel, C., von Stutterheim, C. (eds.) Räumliche Konzepte und sprachliche Strukturen, Niemeyer, pp. 69–95 (2000)
Taylor, H., Tversky, B.: Spatial mental models derived from survey and route descriptions. Journal of Memory and Language 31, 261–292 (1992)
Taylor, H., Tversky, B.: Perspective in spatial descriptions. Journal of Memory and Language 35, 371–391 (1996)
Tenbrink, T.: Conveying spatial information in linguistic human-robot interaction. In: DiaBruck, 7th Workshop on the Semantics and Pragmatics of Dialogue, Proceedings, September 4th-6th, pp. 207–208 (2003)
Tenbrink, T.: Space, time, and the use of language: An investigation of relationships. Mouton de Gruyter, Berlin (2007)
Tenbrink, T., Fischer, K., Moratz, R.: Spatial Strategies in Linguistic Human-Robot Communication. In: Freksa, C. (ed.) KI-Themenheft 4/02 Spatial Cognition, pp. 19–23. arenDTaP Verlag (2002)

Timpf, S., Kuhn, W.: Granularity Transformations in Wayfinding. In: Freksa, C., Brauer, W., Habel, C., Wender, K.F. (eds.) Spatial Cognition III. LNCS (LNAI), vol. 2685, pp. 77–88. Springer, Heidelberg (2003)

Tversky, B.: Spatial Perspective in Descriptions. In: Bloom, P., Peterson, M., Nadel, L., Garrett, M. (eds.) Language and Space, pp. 109–169. MIT Press, Cambridge, MA (1996)

Tversky, B., Franklin, N., Taylor, H.A., Bryant, D.J.: Spatial Mental Models from Descriptions. JASIS 45(9), 656–668 (1994)

van der Zee, E., Slack, J.: Representing direction in language and space. In: Explorations in language and space, Oxford University Press, Oxford (2003)

Visser, W., Wolff, M.: A cognitive approach to spatial discourse production: Combining manual and automatic analyses of route descriptions. In: Proceedings of EuroCogSci 2003: The European Cognitive Science Conference, Osnabrueck, Germany, September 10-13, pp. 355–360 (2003)

Werner, S., Long, P.: Cognition meets Le Corbusier - Cognitive principles of architectural design. In: Freksa, C., Brauer, W., Habel, C., Wender, K.F. (eds.) Spatial Cognition III. LNCS (LNAI), vol. 2685, pp. 112–126. Springer, Heidelberg (2003)

Wiener, J.M., Ehbauer, N., Mallot, H.A.: Path planning and optimization in the traveling salesman problem (submitted)

Wiener, J.M., Franz, G.: Isovist analysis captures properties of space relevant for locomotion and experience (under revision)

Wiener, J.M., Mallot, H.A.: Fine-to-Coarse Route Planning and Navigation in Regionalized Environments. Spatial Cognition and Computation 3(4), 331–358 (2003)

Wiener, J.M., Schnee, A., Mallot, H.A.: Use and Interaction of Navigation Strategies in Regionalized Environments. Journal of Environmental Psychology 24(4), 475–493 (2004)

Wiener, J.M., & Franz, G.: Isovists as a means to predict spatial experience and behavior. In: Freksa, C., Knauff, M., Krieg-Brückner, B., Nebel, B., Barkowsky, T. (eds.) Spatial Cognition IV. LNCS (LNAI), vol. 3343, pp. 42–57. Springer, Berlin, Germany (2005)

Zwaan, R.A., Radvansky, G.A.: Situation Models in Language Comprehension and Memory. Psychological Bulletin 123(2), 162–185 (1998)

A Spatial Cognitive Map and a Human-Like Memory Model Dedicated to Pedestrian Navigation in Virtual Urban Environments

Romain Thomas[1] and Stéphane Donikian[2]

[1] Formerly PhD student at IRISA/INRIA,
Campus de Beaulieu, 35042 Rennes Cedex, France
abberom@gmail.com
[2] IRISA/CNRS, Campus de Beaulieu, 35042 Rennes Cedex, France
donikian@irisa.fr

Abstract. Many articles dealing with agent navigation in an urban environment involve the use of various heuristics. Among them, one is prevalent: the search of the shortest path between two points. This strategy impairs the realism of the resulting behaviour. Indeed, psychological studies state that such a navigation behaviour is conditioned by the knowledge the subject has of its environment. Furthermore, the path a city dweller can follow may be influenced by many factors like his daily habits, or the path simplicity in term of minimum of direction changes. It appeared interesting to us to investigate how to mimic human navigation behavior with an autonomous agent. The solution we propose relies on an architecture based on a generic model of informed environment, a spatial cognitive map model merged with a human-like memory model, representing the agent's temporal knowledge of the environment, it gained along its experiences of navigation.

1 Introduction

One of the most important skills for a virtual human is its ability to navigate inside an environment, as it is part of a large number of behaviours. Agent navigation in a virtual environment is an important problem of interest, and a lot of researches has been dedicated to it in various research fields, such as cognitive science [2,30], robotic [13,16,15], and behavioural animation [17,23]. Reproducing the navigating activity requires more information than the geometric representation of the environment. It is necessary to provide additional data such as mereotopological and semantic information. Gibson states in his theory of affordance [10] that *animals perceive the environment in terms of what they can do with and in it*. The *what ... with* aspect has been addressed by M. Kallmann [14] with the notion of smart objects. The *in it* aspect has been addressed by N. Farenc [9] and G. Thomas [25] with their respective models of informed urban environments. This kind of approach is based on an omniscient point of view, where the agent can obtain any information from the informed environment to plan its path, which limits the realism of the simulation. Another approach based on artificial vision is using information retrieval from an image captured by a camera located on the head of the virtual human. This process, using the well known Z-buffer, has been introduced by O. Renault et al. [24] to compute the perception of a virtual human. More recently, N. Courty [6] has extended this kind of approach by blurring the peripheral vision area in the perceived image and by introducing a saliency map calculation. Salient points are extracted from the perceived image and used in the visual attention guidance of a virtual human navigating without any goal in an urban environment. C. Peters et al. [22] use also an artificial vision model, but in their case it is coupled to a human memory model (based on the Memory Model of Atkinson and Shiffrin [3]) that allow to manage scanning and object retrieval techniques inside an apartment.

Furthermore, in simulation, heuristics employed to guide the navigation focus most of the time on the shortest path computation. However, Duckham et al. [8] propose a path planning algorithm based on the most simple journey instead of the shorter one, while Hochmair et al. [12] investigate least-angle and initial segment strategies for route selection in an unknown environment. The navigation behaviour of an agent in classical simulations differs a lot from what is observed in studies involving human subjects [30], in the sense that agents simulation exhibits standardized behaviours which lack the apparent "fuzziness" and plurality of human behaviours. A human being navigating in an environment is confronted with the problem of his own spatial localization. Indeed, most of the cognitive works on that subject state that the knowledge a pedestrian can have of his environment differs a lot from what is really the environment: his perception and memory are most of the time incomplete and distorted [29]. Then it seemed interesting to us to study what should be added to the classical navigation architecture in behavioural animation, to simulate a more realistic navigation behaviour. After a careful study of formerly performed research in the field of cognitive psychology, the introduction of a spatial cognitive map and a human-like memory appeared necessary. The map is necessary to restrict the omniscience of the agent, due to the fact that planning will be computed with incomplete knowledge of the environment. As the memory is dedicated to take into account a temporal factor in the simulation, the major consequence is that, paths taken to go from one location to another will not be the same at different moment during the simulation as the state of the agent's memory is continuously evolving.

In the rest of this paper, we will present the different components of our architecture, as "add-on" to the classical perception-decision-action architecture. Section 2 will describe our architecture. Section 3 to 5 will describe respectively the model of environment we use, the model of spatial cognitive map we have created and will sketch our memory model. Then, finally, Section 6 will explain the agent's navigation process using these various elements.

2 The Overall Architecture

Let us introduce the overall architecture (cf. figure 1) of our system. An environment is modeled by an Informed Hierarchical Topological Graph, containing

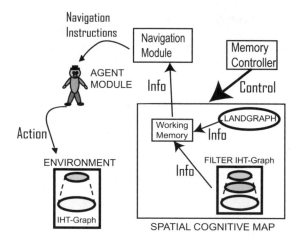

Fig. 1. The navigation system architecture

two layers: the Simple Space Layer and the Global Area Layer. Simple spaces are small-surfaced spaces where the agent navigates. They come from the convex partitioning of the environment. Practically a simple space can be either a building or a navigable space. The Spatial Cognitive Map contains subjective information the agent has acquired on its environment. It is composed of a Filter Informed Hierarchical Topological Graph, a Memory Controller, a Working Memory and a Landgraph. The Filter Informed Hierarchical Topological Graph contains three layers which are topological graphs: the Filter Simple Space Layer, the Filter Global Area Layer and the Local Space Layer. A navigation module, managing the navigation algorithms, uses the data collected from the working memory to elaborate a navigation plan. All these components are presented in the next sections.

3 The Environment

The virtual environment M where the simulation takes places is modeled with an Informed Hierarchical Topological Graph named IHT-Graph. It has two layers: the Simple Space Layer (named as SSL) and the Global Area Layer (named as GAL), corresponding to the two types of objects contained in the environment.

$$M = \{E, F, >, \xrightarrow{\tau}\}$$

where

- $E = E_s \cup Z$ with
  E_s the set of simple spaces of the world M
  Z the set of global areas of the world M
- $F = \phi_F \cup \phi_{NF}$ the set of boundaries associated to E. ϕ_F and ϕ_{NF} are respectively the sets of passable and impassable boundaries of E

- $\triangleright$ the *part of* relation in M, linking simple spaces to global areas they belong to.
- $\leadsto$ the connexity relation, that can be applied among both simple spaces and global areas.

Let us define $Ref(E)$, the set of identifiers, such as $Ref(E) = Ref(E_s) \cup Ref(Z)$, with:

- $Ref(E_s) \subset \mathbb{N}$ is the subset of $\mathbb{N}$ that contains the set of identifiers of elements of E_s.
- $Ref(Z) \subseteq \Big(Ref(E_s) \times \mathcal{P}(Ref(E_s))\Big)$ is a set of couples, whose first component is the identifier of the kernel and the second component is the set of identifiers of the ring of the global area (with $\mathcal{P}$ the set of parts).

It contains a unique identifier for each space of E.

Let us define also the association function p such as:

$$p : E \longrightarrow Ref(E)$$
$$e \longrightarrow i$$

For a space $e \in E$, it returns its identifier number $i \in Ref(E)$.

Likewise, we define the inverse function p^{-1}:

$$p^{-1} : Ref(E) \longrightarrow E$$
$$i \longrightarrow e$$

For an identifier $i \in Ref(E)$, it returns the associated space $e \in \mathbb{E}$.

A simple space is an object which is pictured by a convex polygon. Those polygons can be either navigable portions of the environment or building of the urban layout. The set of simple spaces E_s is obtained by partitioning the layout with a constrained Delaunay triangulation.

E_s is defined as follows:

$$E_s = \{e = (i, a, o, s) / i \in Ref(E_s),\ a \in B \cup N\ ,\ o \subseteq O_p\ ,\ s \in \Re\} \quad (1)$$

with

B the set of convex polygons corresponding to buildings of M
N the set of navigable convex polygons of M
O_p the set of punctual objects part of M
s is the saliency associated to the space e

An example of the layout partitioning is shown in fig. 2. Furthermore a simple space contains three types of information:

- *Geometrical*: within each simple space, the associated polygon is stored as the set of its borders.
- *Topological*: the SSL is a topological graph and each simple space has at least one neighbour. Therefore, the connections between a simple space and its neighbours are stored in the simple space.
- *Semantic*: the type value of simple space is stored. It can be a building or a navigable zone.

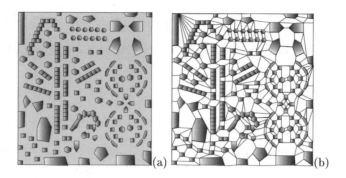

Fig. 2. (a) An urban layout, representing only buildings (b) the simple spaces of the layout

Moreover, each simple space is gifted a *saliency parameter s*. This parameter represents the visual and thematic importance a space can have for the majority of the city dwellers. A train station will be more salient than a dead end located in a suburb. At this stage of our work, this parameter is set empirically for each space.

Let us define also four functions for $e = (i, a, o, s) \in E_s$:

- $Ct(e) = a$. It gives the boundary of e, and is represented as an ordered list of points.
- $Ob(e) = o$. It gives the set of objects associated to e
- $\phi(e) = f$. It gives the set $f \in (\phi_F \cup \phi_{NF})$ of boundary segments of e.
- $S(e) = s$. It gives the saliency parameter of the space e.

A global area(**GA**) is an object which is a conceptual blend of a Lynch quarter [19], and a Penn local area [21]. The implementation of quarters for real and virtual cities in our model was crucial. Indeed, the route planning and navigation in an environment is a multi-level planning in term of abstraction of spaces. For instance, Lynch [19] states that a pedestrian who is not familiar with a town will be prone to position himself and navigate, using the major subdivisions of space like quarters and general orientation. Similarly the more the pedestrian's knowledge of the environment grows, the more he will navigate using very specific landmarks related to his past experiences. In our model, global areas simulate the notion of big area exhibiting a single identity.

The set Z of global areas of M is defined as follows:

$$Z = \{z_g = (i, n, c)/i \in Ref(Z), n \in E_s, c \in Z_C\}$$

Practically a GA gathers simple spaces of the environment which can be related to another simple space of interest. This particular simple space is called the **kernel** of the global area (n). It symbolizes the identity of the GA. Then, the set of spaces linked to this kernel is called the **ring** of the global area (c). So each global area has a ring and a kernel.

Let us define four functions operating on a global area.

- $N(z_g) = n$. It returns the kernel of z_g
- $C(z_g) = c$. It returns the ring of z_g

- $\mathcal{R}(z_g) = r$. It returns the gathering potential of the kernel of the global area z_g
- $Ct(z_g) = a$: It returns the contour of the ring of z_g, as a list of boundaries belonging to $\phi_F \cup \phi_{NF}$.

Definition 1. The Part of Relation:

$$\forall z_g \in Z, \forall b \in E_s, \ z_g \triangleright b \Leftrightarrow b \in N(z_g) \cup C(z_g)$$

A simple space b belongs to the global area z_g if and only if it is part of the ring or the kernel of z_g.

Definition 2. The Connexity Relation:

For simple spaces: $\xrightarrow{\mathcal{T}}$: $\forall a, b \in E_s : a \xrightarrow{\mathcal{T}} b \Leftrightarrow \exists f \in F, (f \in \phi(a)) \wedge (f \in \phi(b))$
For global areas: $\xrightarrow{\mathcal{T}}$: $\forall a, b \in Z : a \xrightarrow{\mathcal{T}} b \Leftrightarrow C(a) \cap C(b) \neq \emptyset$

Let us define the topological distance between two simple spaces $a, b \in E_s$. a and b are at a topological distance n, as defined by the following relation $a \xrightarrow[n]{\mathcal{T}} b$:

$\xrightarrow[n]{\mathcal{T}}:$
- $a \xrightarrow[1]{\mathcal{T}} b \Leftrightarrow a \xrightarrow{\mathcal{T}} b$
- $a \xrightarrow[n]{\mathcal{T}} b \Leftrightarrow \exists c \in E, (a \xrightarrow{\mathcal{T}} c) \wedge (c \xrightarrow[n-1]{\mathcal{T}} b)$

The first step in computing a GA is to determine its kernel. We have been inspired by the work of Cutini [7] and his K index, whose role is to evaluate the gathering potential of city dwellers an open space can have. We adapted the index K_{ga}, and for each simple space of the urban layout, we compute the value of this index defined by:

$$K_{ga} = \frac{\mathcal{T}(s)}{\mathcal{I}(s)}$$

where $\mathcal{I}(s)$ is the spatial integration, a very common measure in the Space Syntax community.

$$\mathcal{I}(s) = \frac{\sum_{k=1}^{card(N)} i_k}{card(N)}$$

with $i_k = min(\{i \in \mathbb{N}/e \xrightarrow[i]{\mathcal{T}} e_k\})$ and $N \subset E_s$ the set of navigable simple spaces.

It represents the mean value of the topological minimum distance between the space s and all the other simple spaces of the environment. Practically, if the topological distance between spaces $s1$ and $s2$ equals 3, it means that, there is at least 2 simples spaces lying on the path from $s1$ to $s2$.

For $e_1, e_2 \in E_s$. The visibility relation between e_1 and e_2 is represented by the following relation $\mathcal{V}(e_1, e_2)$. This relation is true if and only if:

$$\exists p \in e_1, \exists f \in \phi(e_2) \forall f_B \in \phi_{NF}, \left(\overline{B(e_2)p} \cap f_B = \emptyset\right) \wedge \left(\overline{M(f)p} \cap f_B = \emptyset\right)$$

and

$\mathcal{V}(e_1, e_2) = 1$ if e_1 is visible from e_2
$\mathcal{V}(e_1, e_2) = 0$ otherwise

$\mathcal{T}(s)$ is the neighbourhood size. It is defined as the ratio of the number of spaces from which the simple space s is visible to the total number of simple spaces of the environment.

$$\mathcal{T}(s) = \frac{\sum_{k=1}^{card(N)} \mathcal{V}(s, e_k)}{card(N)}$$

Once the index K_{ga} is computed for each simple space, the algorithm selects spaces having the highest K_{ga} (the number of spaces selected depends on the user will). Those spaces will be kernels of all the global areas of the environment. Once the kernels are selected, their corresponding rings are computed. We derive a **radiation measure** for each kernel, depending on the kernel K_{ga} and its neighbours ones and on the kernel surface. Then, we compute the radius of the ring for each kernel. The more the kernel radiation grows the more the ring radius will be important.

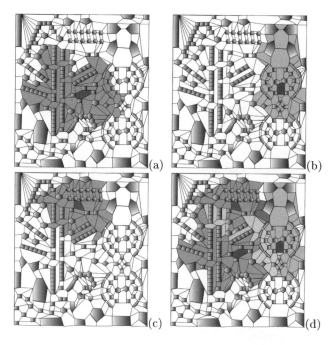

Fig. 3. (a)(b)(c) Three different Global areas computed in our sample environment (d) overlapping of Global areas (pink and green regions)

Fig. 3 shows three global areas computed in our sample environment. In the figure, the kernel of each global area is darker than the spaces taken from the ring. The rings of global areas can overlap, as shown in Fig. 3.d: a space can belong to the rings of many global areas. Let's precise that a global area is connected to another in the global area layer if and only if their ring overlap.

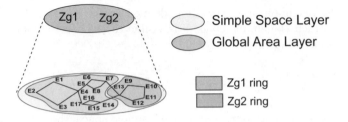

IHT-Graph of the environment

Fig. 4. An example of Hierarchical Topological Graph

The SSL and GAL form a hierarchical graph. Fig. 4 schematically illustrates an IHT-Graph modelling two non-connected GAs and their respective rings in the SSL.

4 The Spatial Cognitive Map

In the previous section, we described the model of environment where the simulation takes place. The Spatial Cognitive Map (SCM) will have almost the same shape like the environment, given that it will model the knowledge on the environment the agent has already acquired during the simulation. The main idea is that the SCM acts as a filter between the environment and the agent's decisional module. In our architecture, the perception is simulated by a query to the database representing the environment: if the agent wants to obtain information on a space, the decisional module will query the IHT-Graph for this information. A limitation of this type of model is that it gives the agent an omniscience which impairs the realism of simulations. Limiting this omniscience to the set of spaces the agents previously discovered, allows them to exhibit navigation behaviours more related to the ones human subjects can show: the agent navigating from an origin point to a destination one, will be able to build its route using spaces it

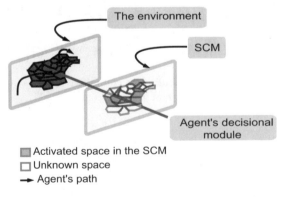

Fig. 5. The Spatial Cognitive Map acting as a filter on the environment

has already discovered or is currently discovering. More details on the navigation process will be given in section 6.

Fig. 5 shows a sample of simple spaces taken from an environment and the way the SCM can act as a filter. An important concept of our model is the **Activation**. Once the agent has visited a space of the environment, the corresponding filter simple space is activated is the SCM. A **filter simple space** is a "light version" of a simple space. It does not contain the geometrical, topological nor semantic information stored in the simple space, but only a reference to its corresponding space in the environment (we will see in the next section that filter spaces are endowed with memory parameters). The activation of a filter simple space implies that the agent's decisional module can obtain all information from the corresponding simple space in the IHT-graph. We mentioned that the SCM has a structure very similar to the one of the environment (IHT-Graph) that we will call **Filter IHT-Graph**. It is a hierarchical topological graph containing:

- **The Filter Simple Space Layer (FSSL):** this layer is a topological graph containing all the filter simple spaces corresponding to the simple spaces of the environment already visited by the agent.
- **The Filter Global Area Layer (FGAL):** this layer is the one corresponding to the global area layer of the IHT-graph of the environment. The activation of a global area is a bit more tricky than the simple space one: a filter global area **(FGA)** can only be activated if the kernel of the corresponding global area has been visited. Then, the ring of the FGA, will contain all the filter simple spaces already visited by the agent.
- **The Local Space Layer (LSL):** this layer does not have any corresponding one in the IHT-Graph. We added the concept of local space in our architecture to model a local vision of the space surrounding the agent. As we will see below, local spaces are employed to determine visual landmarks which take part in the planning and navigation processes of the agent. A local space is a set of simple spaces, that lye in the agent's visual field.

Fig. 6 illustrates the query process through the SCM. When the agent wants to access information about a distant simple space or global area, it has to query the SCM first. If the filter object is activated in the SCM, access to the IHT-graph is allowed and the information is transferred to the agent's decisional module.

Local Spaces are set of simple spaces that model a specific vision of the environment an agent can have (cf. figure 7). Simple spaces contained in a local space are spaces which can be seen by the agent from a particular space called the **Generator** of the local space. The set of perceived spaces is called the **Surface** of the local space. All spaces that can be observed from the generator within the bounds of the agent's visual field (which the user sets at the start of the simulation), represent all the possible points of view one can have from a generator. A local space can be seen as a model of isovist [28]. The only problem in building a local space is the choice of the generator space: a generator is a simple space whose saliency value is above a saliency threshold set at the start of the simulation and is corresponding to the degree to which the agent can be

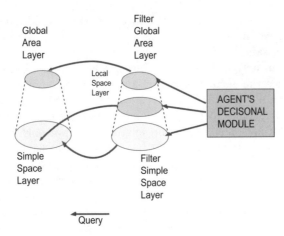

Fig. 6. Querying the environment through the Spatial Cognitive Map

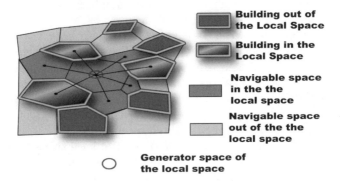

Fig. 7. An example of local space

impressed by its environment. Another constraint of the generator selection is given by the **agent's perceptive modes**. We adapted the Chopra and Badler [4] attentional mode, to calibrate the perception of our agents. It consists of three modes: exogenous, endogenous and passive. In the exogenous mode, the attention is spread over a high number of things, exceptional and peripheral events are noticed (The attention is high). In the endogenous mode, the agent is supposed to be thoughtful and focused on a plan to execute (It is not prone to pay attention to its environment). In the passive mode, the perception is attracted by highly contrasted and salient zones but the attention is quite low. We can order the perception modes relatively to the quality of the agent's attention they imply, from the strongest to the weakest: $Exogenous \Rightarrow Passive \Rightarrow Endogenous$. Then we can state that, a generator can only be selected if the agent perception is in a *passive* or an *exogenous* mode: we suppose that an agent in *endogenous* mode is too thoughtful to perceive the importance of a potential generator. Once a generator is perceived, its surface is computed, and the local space is added to local space layer. As described in the next section, a local space allows to identify visual landmarks. In the contrary of a global area that gathers spaces linked by a common identity given by the kernel of the GA, but which are not visible, a local space regroups spaces that are partially mutually visible. The underlying idea of global areas is identity, whereas local spaces relies on visibility.

5 The Memory Model

The memory model of our system is described in detail in [27]. In this section, we will sketch shortly the way it works. The memory model uses the Filter IHT-Graph as the static part of the agent's long-term memory. It can be seen as a blend of the works of Yeap [13], Atkinson and Schiffrin [3] and Lieury[18], that we have adapted to the constraint of our architecture. Each element of the IHT-Graph is endowed with a **recall** and a **recognition** parameter. Fig. 8 shows the way object retrieving works. We have been inspired by the Anderson activation and retrieval model [1]. The navigation module will ask for a specific space to be retrieved in the SCM. Then rehearsal of recognition or recall tests will be triggered. The agent is endowed with a recall threshold and a recognition one. The memory values of the object to be retrieved must be above the corresponding threshold. If they are not, the threshold is weakened and a new test is run. The number of tests depends on the agent's profile, defined at the start of the simulation.

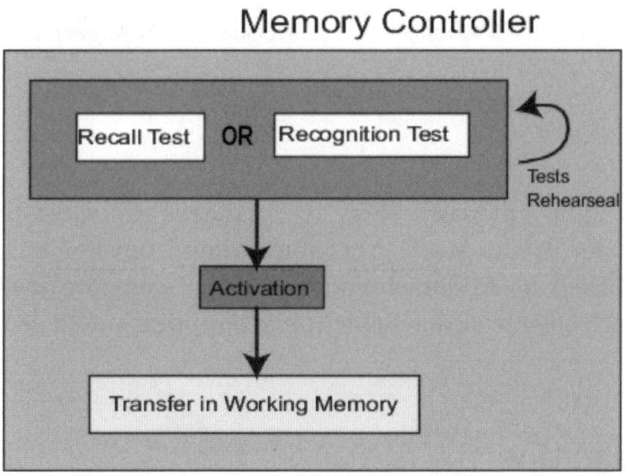

Fig. 8. The retrieving process

The Landgraph, which is a graph linking all the landmarks the agent knows, is our associative memory model. A landmark is the generator of a local space, as a local space represents a *memory image* in the Gillund and Schiffrin theory [11]. Spaces around the landmark are linked to it (Fig. 9), and the edges linking them are endowed with memory parameters, so that it is possible to have an associative chain of landmarks symbolizing an abstract path in the environment.

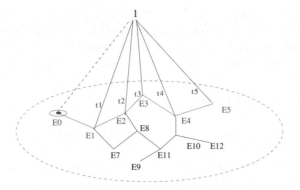

Fig. 9. A landmark l connected to its neighbour simple spaces

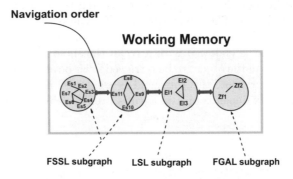

Fig. 10. A working memory example

At last, the working memory (Fig. 10) manages the short-term processes of our memory system. It can store subgraphs taken from FSSL, LSL, and FGAL, which can be viewed as Miller chunks [20]. The elements inside the WM are ordered by the navigation order, which is a temporal order as presented in the next section.

6 The Navigation

The environment, the spatial cognitive map and the memory system, described in the previous sections, allow to design the navigation system. The central idea of this system is that a human being navigates through an urban environment, using a plan it previously made. This plan contains a very small number of elements, due to the limiting properties of the working memory. Therefore many planning phases will occur in the process of navigation, we name this type of navigation the **planned navigation**. Our model of planned navigation shows many similarities with the one developed by Wiener and Mallot [30], in the sense that it generates a plan based on several levels of space abstraction, and that the plan evolves during the navigation. However, it is sometimes impossible for the agent to plan a path correctly, then it will have to navigate following an immediate heuristic [16,30,15], until it recognizes a space allowing to replan its path to the next place. That is what we call the **reactive navigation**. Because of the very high number of subcases in the algorithm, depending on the number of heuristics, we will not explain in detail how reactive navigation works in our system, but we will simply show how it is triggered during the planning and the navigation process.

First, we explain the planned navigation algorithm. The first step is to run the planning algorithm. This algorithm generates a **navigation plan**. A navigation is composed of a succession of **Beacons**. The beacons are global areas kernels and Landmarks taken from the SCM of the agent. It can be either **complete** or **partial**. Given a starting point and a destination one, a complete plan contains all the information for the agent to navigate between these two points. A partial plan is produced when the agent cannot establish a continuous chain of beacons between the two points. Complete navigation plan triggers a planned navigation whereas partial navigation plan triggers a reactive navigation.

The first step in elaborating a plan is to compute a path of global areas. A continuous succession of recalled filter global areas is searched in the FGAL of the IHT-graph as shown in Fig. 11. This simulates the fact that a human subject will first identify the different major zones involved in the path computation. The FGA selection criterion is the minimum number of GA employed to make a path. GAs with high recall parameter values are privileged.

Fig. 12 shows how a complete plan is computed. Three global areas are selected from the FGAL. Then the LSL is searched to find a chain of landmarks joining the point D to the second zone. Landmarks with highest recall values will be

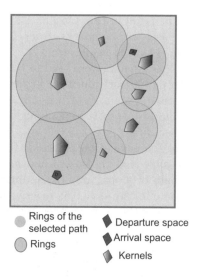

Fig. 11. Global areas selections

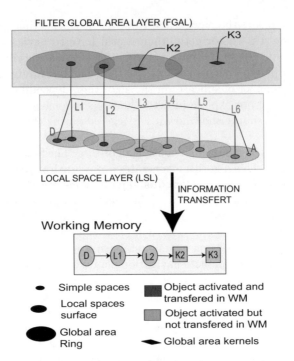

Fig. 12. A plan elaboration from the departure point D to the arrival point A

selected. Then the information selected is transferred by the memory controller in the working memory. The planned navigation is executed until the agent reaches L2. The way landmarks are linked together by the means of the surface of local spaces is described in detail in [26]. Once in L2, the agent searches local spaces of the second global area for landmarks to reach the third and last GA, and the planned navigation is started again.

From now the only recall parameters were used to retrieve information in the cognitive map. Recognition parameter values are higher than recall ones. So, if during the navigation process the agent recognizes a space that is susceptible to be the starting point of a shortest path, a re-planning process will be executed, exactly like the one described above.

During the computing of the navigation plan, due to the uncompleteness of the cognitive map, it is possible that the system reaches a configuration where it is unable to obtain a chain of global area or landmarks between the origin point and the destination one. In that case, the reactive navigation algorithm is executed. Its principle is to make the agent wander in the environment, following various heuristics like "minimum deviation angles" [5], "the maximum number of borders" [16], etc... until a space is recognized, which will eventually trigger the recall of neighbour spaces. A plan is elaborated using the newly recalled and recognized spaces, until the agent loses itself again or reaches its destination. Fig. 13 illustrates three different paths taken by an agent to reach a destination

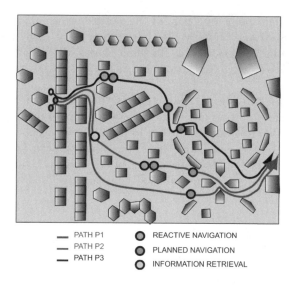

Fig. 13. Three different paths taken by the agent, depending on the state of its memory

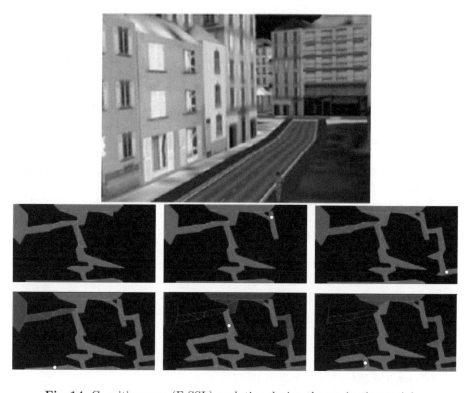

Fig. 14. Cognitive map (F-SSL) evolution during the navigation activity

point from an origin point in a part of the sample environment, depending on the state of its SCM and Memory. The figure shows that many planning phases can occur in the navigation process. In some phases, the agent could not recall enough elements to create a navigation plan and is forced to switch into a reactive navigation mode.

Fig. 14 shows an autonomous character travelling inside a virtual environment and illustrates the evolution of the spatial cognitive map during its navigation process. The white circle illustrates the location of the agent in the map. The different colors given to spaces of the Filter Simple Space Layer are used to illustrate the value of the recall parameter of the corresponding area. A linear interpolation between red (higher value of the parameter) and blue (lower value) colors is used.

7 Conclusion

In conclusion, we elaborated our navigation architecture which complete the classical *perception-decision-action* architecture. We presented how it was necessary to endow the agent with such structures, to allow it to exhibit navigation behaviors which are not standardized, due to the fact that each agent of the simulation plans its route using the personal and incomplete information taken from its SCM, which gained along the simulation. The planing process is a multi-level planning. An agent with a weak knowledge of the environment will still be able to compute a path, and will eventually reach its destination, in a different way than an agent endowed with a better knowledge of the environment. Furthermore, the fact that this information can not be available at any time due to the constraints implied by the agent's memory, allows us to use different navigation modes (planned and reactive ones), and to multiply the type of paths exhibited by the agent during the simulation. We built our model using theoretical results from the cognitive science field dedicated to human navigation, so most of our construct are cognitively justified but not validated. The model is highly parameterizable (for sake of simplicity, we have only given few of the control parameters of the model) so it would be an interesting future work to calibrate results of our model comparing to studies involving human subjects in a large scale environment.

References

1. Anderson, J.R., Schooler, L.J.: Reflections of the environment in memory. Psychological science 2, 396–408 (1991)
2. Arthur, P., Passini, R.: Wayfinding: People, Signs and Architecture. McGraw-Hill Ryerson and Yohimbine, Toronto (1992)
3. Atkinson, R.C., Shiffrin, R.M.: Human memory: A proposed system and its control processes. In: Spence, K.W., Spence, J.T. (eds.) The Psychology of Learning and Motivation, vol. 2, pp. 89–95. Academic Press, London (1968)

4. Chopra-Khullar, S., Badler, N.: Where to look? automating attending behaviors of virtual human characters. Autonomous Agents and Multi-Agent Systems 4(1/2), 9–23 (2001)
5. Conroy, R.: Spatial Navigation In Immersive Virtual Environments. In: Philosophy and architecture, The Faculty of the Built Environment, London (January 2001)
6. Courty, N., Marchand, É., Arnaldi, B.: A new application for saliency maps: synthetic vision of autonomous actors. In: ICIP (3). International Conference on Image Processing, pp. 1065–1068 (2003)
7. Cutini, V.: Lines and squares: towards a configurational approach to the morphology of open spaces. In: 4th International Space Syntax Symposium, London, vol. 2, pp. 49.1–49.14 (2003)
8. Duckham, M., Kulik, L.: Simplest paths: Automated route selection for navigation. In: Kuhn, W., Worboys, M.F., Timpf, S. (eds.) COSIT 2003. LNCS, vol. 2825, pp. 169–185. Springer, Heidelberg (2003)
9. Farenc, N., Boulic, R., Thalmann, D.: An informed environment dedicated to the simulation of virtual humans in urban context. In: Brunet, P., Scopigno, R. (eds.) EUROGRAPHICS 1999, pp. 309–318. Blackwell, Oxford (September 1999)
10. Gibson, J.J.: The ecological approach to visual perception. Lawrence Erlblaum Associates, Hillsdale, USA (1979)
11. Gillund, G., Schiffrin, R.M.: A retrievial model for both recognition and recall. Psychological Review 91(1), 1–66 (1984)
12. Hochmair, H.H., Karlsson, V.: Investigation of preference between the least-angle strategy and the initial segment strategy for route selection in unknown environments. In: Freksa, C., Knauff, M., Krieg-Brückner, B., Nebel, B., Barkowsky, T. (eds.) Spatial Cognition IV. LNCS (LNAI), vol. 3343, pp. 79–97. Springer, Heidelberg (2005)
13. Jefferies, M.E., Yeap, W.K.: The utility of global representation in a cognitive map. In: Montello, D.R. (ed.) COSIT 2001. LNCS, vol. 2205, pp. 233–246. Springer, Heidelberg (2001)
14. Kallmann, M.E., Thalmann, D.: A behavioral interface to simulate agent-object interactions in real-time. In: Computer Animation 1999, pp. 138–146. IEEE Computer Society Press, Los Alamitos (1999)
15. Kettani, D., Moulin, B.: A spatial model based on the notions of spatial conceptual map and object's influence areas. In: Freksa, C., Mark, D.M. (eds.) COSIT 1999. LNCS, vol. 1661, pp. 401–416. Springer, Heidelberg (1999)
16. Kuipers, B.J., Byun, Y.T.: A qualitative approach to robot exploration and map-learning. In: Proc. of Workshop on Spatial Reasoning and multi-sensor fusio, pp. 390–404. IEEE Computer Society Press, Los Alamitos (1987)
17. Lamarche, F., Donikian, S.: Crowd of virtual humans: a new approach for real time navigation in complex and structured environments. Computer Graphics Forum 23(3), 509–518 (2004)
18. Lieury, A.: Lamémoire, 4th edn. Mardaga (1992)
19. Lynch, K.: The image of the city. MIT Press, Cambridge, Massachusetts (1960)
20. Miller, G.: The magical number seven, plus and minus two: some limits of our capacity for processing information. Psychological Review 63, 81–97 (1956)
21. Penn, A.: Space syntax and spatial cognition: Or, why the axial line? Environment and Behaviour 35, 30–65 (2003)
22. Peters, C.: Synthetic vision and memory for autonomous virtual humans. Computer Graphics forum 21(4), 743–752 (2002)
23. Soraia Raupp-Musse. Human crowd modeling with various levels of behaviour control. PhD thesis, EPFL, Lausanne, Switzerland (2000)

24. Thalmann, N.M., Thalmann, D., Renault, O.: A vision-based approach to behavioral animation. Journal of Visualization and Computer Animation 1(1), 18–21 (1990)
25. Thomas, G., Donikian, S.: Modelling virtual cities dedicated to behavioural animation. Computer Graphics Forum 19(3), 71–80 (2000)
26. Thomas, R.: Modèle de mémoire et de carte cognitive spatiales: applicationà la navigation du piéton en environnement urbain. PhD thesis, University of Rennes 1 (February 2005)
27. Thomas, R., Donikian, S.: A model of hierarchical cognitive map and human memory designed for reactive and planned navigation. In: 4th International Space Syntax Symposium, London, vol. 2, pp. 1–72. University Colledge (2003)
28. Turner, A., Doxa, M., OSullivan, D., Penn, A.: From isovists to visibility graphs: a methodology for the analysis architectural space. Environment and Planning B: Planning and Design 28, 103–121 (2001)
29. Tversky, B.: Structures of mental spaces. In: Proceedings of the 3th international space syntax symposium, Atlanta, pp. 12.1–12.5 (2001)
30. Wiener, J.M., Mallot, H.A.: fine-to-coarse. route planning and navigation in regionalized environments 3, 331–358 (2003)

The Influence of Scale, Context and Spatial Preposition in Linguistic Topology

Anna-Katharina Lautenschütz[1], Clare Davies[2,*], Martin Raubal[1], Angela Schwering[1,2], and Eric Pederson[3]

[1] Institute for Geoinformatics, University of Münster, Robert-Koch-Str. 26-28, 48149 Münster, Germany
annakl18@hotmail.com, raubal@uni-muenster.de
wiansc@wi.uni-muenster.de
[2] Ordnance Survey, Research & Innovation C530, Romsey Road, Southampton SO16 4GU, UK
clare.davies@ordnancesurvey.co.uk
[3] Linguistics Department, University of Oregon, Eugene, OR 97403-1290, USA
epederso@uoregon.edu

Abstract. Following a similar method to that of Mark and Egenhofer (1994), a questionnaire-based experiment tested for possible effects of scale, context and spatial relation type on the acceptability of spatial prepositions. The results suggest that the previous assumption of scale invariance in spatial language is incorrect. The physical world as experienced by humans, and described by human language, is not a fractal: scale appears to change its very physical nature, and hence the meaning of its spatial relations. The experiment demonstrated how scale influences preposition use, and how different prepositions appeared to evoke different levels of acceptability in themselves. Context, in terms of object type (solid or liquid), interacted with these factors to demonstrate specific constraints upon spatial language use. The results are discussed in terms of figure-ground relations, as well as the role of human experience and the classification of the world into 'objects' in different ways at different scales. Since this was a preliminary and artificially-constrained experiment, the need for further research is emphasized.

1 Introduction

Spatial relations are considered to be one of the most distinctive aspects of spatial or geographical information. Despite occasional use of maps, diagrams and models, humans inevitably use language to communicate where objects are, and this is most commonly done by reference to their relation to other objects. Similarly, almost any GIS (geographic information system) query uses spatial relations to analyze or describe the constraints of spatial objects.

In order for GIS to be based on models of relevant geographic concepts, so as to improve usability and task relevance, in the early 1990s it was deemed important to develop formal models of the spatial relations that tend to exist within geographic

---

* Corresponding author.

space. Egenhofer and Franzosa [1] argued that spatial relations can be grouped into three different categories:

- Topological relations, which are invariant under topological transformations of the reference objects [2];
- Metric relations in terms of distances and directions [3];
- Relations concerning the partial and total order of spatial objects as described by prepositions such as 'in front of', 'behind', 'above' and 'below' [4].

Topological relations describe the spatial configurations of two objects, without reference to metric distance. For human spatial cognition, topology has long been considered the most important type of spatial relation, since Lynch [5] pointed out that humans remember urban topology and use it when wayfinding or navigating through space. According to Mark and Egenhofer [6], people capture and use topology more frequently and accurately than metric properties such as distance and shape. A recent analysis of the relations mentioned or implied in a national topographic dataset (Ordnance Survey of Great Britain's OS MasterMap® [7]) also showed that topological relations such as connection, intersection and adjacency were among the most commonly defined relations among geographic features. These relations are therefore the focus of this paper.

1.1 Spatial Language

How then are topological relations expressed in spatial language, such as through the use of prepositions? How predictable is the match between a given topological relation and people's choice of spatial preposition for describing it? This question is more important than it may at first appear. If we do not properly understand how spatial language is selected and interpreted by human speakers and listeners, then our GIS, robots and other technologies will not be able to reliably match the expectations and intentions of human listeners and speakers to linguistic spatial descriptions.

For example, in order to simplify the usage of GIS for non-experts it would be enormously helpful to be able to use natural language expressions; e.g. Riedemann [8] emphasized the importance of GIS terminology reflecting the user's language. Query languages can be improved when the predicates are chosen according to user needs, and the underlying cognitive understanding of spatial relations needs to be taken into account when defining user-appropriate semantics. In turn, this can improve the development of ontologies [9,10].

To progress towards this, we have to understand the factors that influence the choice of spatial terms (verbs and prepositions, and similar forms) when a speaker of a given language attempts to describe, or evaluate a description of, a specific spatial relation. Mark and Egenhofer [11] investigated this via an experiment with human participants. This examined the influence of geometric factors on people's acceptance of (agreement with) a sentence such as "The road crosses the park", when accompanied with drawings showing various configurations of the road, represented as a line feature, and a region (the park, represented as a featureless 2D region) that was partly or wholly 'crossed' by it. This study demonstrated that the notion of 'crossing' is not an all-or-nothing concept: rather there are degrees of acceptability of the tested linguistic term. For example, people would be less convinced of the sentence's validity where the line did not continue right across the region (e.g. if it doubled back).

More recently, a program of research by Kenny Coventry and colleagues (e.g. [12]) has demonstrated that the choice of preposition in describing spatial relations depends on other factors besides geometric configuration, at least at the immediate scale often referred to as 'tabletop' or 'figural' space. Two types of influential factor have been identified: 'functional' relations between two objects (such as a coffee pot and a cup), and 'dynamic-kinematic' relations (e.g. the apparently most likely movements that the objects will make relative to each other). This body of work suggests that geometry is not enough: the choice of spatial linguistic terms may depend on the nature of the objects under consideration.

One factor that has previously been ruled out of such considerations is that of spatial scale. Talmy [13,14] argued that when spatial language references topological relations, shape and magnitude are irrelevant to the appropriateness of the expression. Essentially, this manifests a claim that spatial language is scale invariant or *scale-neutral*: the same linguistic terms would describe the same spatial relations at any scale. This claim has been widely discussed by Talmy and others, e.g. at the NCGIA Specialist Meeting for Research Initiative 2 "Languages of Spatial Relations" where other attendees including Mark and Zubin questioned the results of scale neutrality [15].

It is important to know whether Talmy's claim of scale neutrality for spatial language is in fact a safe assumption, for the following reasons:

1. If the language used to describe topological spatial relations varies at different scales, this may suggest that people's underlying cognitive models of those relations may also differ in content, and hence in their availability for analogical reasoning and other aspects of problem solving [16].
2. Geographic information scientists need to know whether, and when, we can generalise from cognitive and linguistic studies of spatial language using figural spaces (e.g. items on a table) to environmental and geographical spaces [17]. If not, then research studies need to take scale explicitly into account before generalising about spatial language use.
3. This is also true for key findings such as those described by Coventry and Garrod - do non-geometric factors also come into play at larger scales? In particular, does the *context* of use, including the functional nature of the objects themselves, make a difference to the description of their spatial relations at some scales and not others?

We test Talmy's claim in the present study, and simultaneously examine the potential role of object type as just one, quite easily isolated, aspect of the context of use. Since we have no reason to assume that all spatial terms are equally affected by any given influential factor, we included three different topological relations of intersection, adjacency and connection. Following the same experimental paradigm as Mark and Egenhofer's seminal study, we focused on the relation between a line and an area (region) feature, as described in a sentence and illustrated by a simple diagrammatic drawing (which could be interpreted as a simple map at various scales). The rating of sentence 'acceptability' was the dependent variable. Talmy's claim forms a null hypothesis - that scale does not make a difference. Similar null hypotheses may be advanced for the effects of context and type of relation. In the next three sections we will explain how we distinguished among scales and object types, and how we chose which examples of spatial relations to test.

1.2 Scales in Spatial Perception

What do we mean by the concept of 'scales', and how can we differentiate among them? In general, a scale defines "the ratio between the dimensions of a representation and those of the thing it represents" ([17], p.313). Whereas this definition is meant for maps, scale in human perception is defined as "the size of a space relative to a person" (ibid).

Montello distinguishes between figural, vista, environmental and geographical space. *Figural* space is smaller than the human body and is apprehended without any locomotion, e.g. pictures, small objects and distant landmarks. *Vista* space, as the term implies, can be apprehended without locomotion, just by sight, but is larger than the human body. These spaces are usually single rooms, town squares and small valleys, but also include the surface of the earth as viewed from a plane. An *environmental* space (e.g. a city) is too big to be apprehended without locomotion and is learned over time; it does not need to be learned through models or maps, but these are often used as aids. *Geographical* space is much larger than the human body and is perceived over time mainly through symbolic representations, e.g. maps. Maps thus represent geographical and environmental space but are themselves a part of figural space, because they are much smaller than the human body.

The present experiment used Montello's scale distinctions, but whereas he distinguished four spaces, only three spaces are used here. The reason for this was the difficulty in representing scenes that were clearly and uniquely 'vista' space, as opposed to figural or environmental.

Montello's analysis suggested that human beings' perceptual experiences differentiate between these types of spatial scale. These different scales may therefore be reflected in people's choice of topological spatial language, if Talmy's claim of scale neutrality is false.

1.3 The Role of Context: Object Kinds or Geographic Feature Types

One of the most obvious aspects of the context surrounding a spatial relation is the nature of the two objects whose relationship is being described. Experience with geographic features and terms suggests that one of the most fundamental of these may be the distinction between solid and liquid features - in other words, between dry land and hydrology. The ways in which liquid objects behave, and hence their relations with each other and with solid objects, obviously differ from those among solid objects. In the geographic context in particular, but also at small scales where this distinction has generally been overlooked, this may be expected to make a difference to the way that spatial language terms are employed.

This distinction is also of interest to the domain of geospatial ontologies. Research into the potential of geospatial ontologies at Ordnance Survey and elsewhere is aimed at increasing the interoperability of geographical datasets, by adding semantics to enable comparison of concepts. Hydrology has been used as a major source of domain-specific concepts within this work to date [18], and it is important to know whether relations in this domain are likely to be distinctive in their manner of description.

2 Method

2.1 Experimental Design

As stated earlier, three factors were examined within this experiment – scale, context and type of topological relation – in an effort to begin to identify whether and when they influence spatial term use in natural language. The factor *scale* had three treatment levels: figural, environmental and geographic space. Figural space is the smallest scale used in the experiment: the examples chosen for this scale were string and trickle (line objects), for which the sentences described spatial relations to a leaf or a puddle (region objects). The next largest scale was environmental space, for which we used old road and stream (line objects), relating them to the region objects park and lake. For the largest scale, geographic space, we related the line objects gas pipeline and canal to the region objects country and sea.

The factor *context* (object type) has two levels, representing the two types of object – liquid and solid. The solid line objects at the three scales were string, old road and gas pipeline; the solid region objects were leaf, park and country. The corresponding liquid objects were trickle, stream and canal, and puddle, lake and sea. Each line feature at each scale was paired with the two different region features at the same scale.

In this initial experiment we wished to focus on spatial relations which were primarily topological, and of strong relevance to geographic information. Accordingly, spatial relations and prepositions were chosen that occurred frequently among the geographic features listed in the OS MasterMap™ Real-World Object Catalogue[1]. Prepositions that implied the third dimension or that were used in non-spatial contexts (e.g. 'on') were avoided, in order to focus clearly and simply on unambiguous spatial relations that could easily be represented both linguistically and diagrammatically. The *intersection* and *connection* relations were represented most often among such spatial prepositions in the catalogue, and were therefore chosen for analysis, along with the similarly common *adjacency*. It appeared from the catalogue evidence that these represented three of the most common exemplars of the various line-region relations described by the 'nine-intersection' model previously referenced by Mark and Egenhofer [6]. The three chosen relations also allowed us to keep a consistent sentence structure: all the relations were amenable to the same simple "The [line] runs [preposition] the [region]." This consistency of structure avoided introducing linguistic complications into the experiment, allowing a stronger and clearer test of the experimental factors of interest.

Each of these three relations was represented by two prepositions, to check whether different prepositions within a relation might yield different results. Again drawing on the spatial relations that we found to be most commonly used in the GIS context, the spatial relation *connection* was represented by the prepositions *from* and *to*; *intersection* was represented by the prepositions *across* and *through*; *adjacency* was represented by *next to* and *alongside*. The semantics of English prepositions are controversial and challenging to describe. Prepositions are either assumed to be highly polysemous [19] or semantically more general with complex rules of application [20]. However, for the purposes of the current design, every use of each

---

[1] See http://www.ordnancesurvey.co.uk/oswebsite/products/osmastermap/

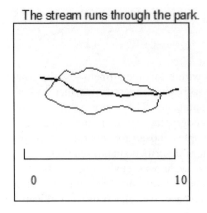
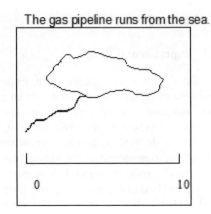

Fig. 1. Two examples from the questionnaire

preposition was designed to be as constant as possible, beyond the independent variable factors of scale and context. Accordingly, the results should be interpretable regardless of one's fuller semantic analysis for these English prepositions.

The participants were presented with 68 sentences describing a spatial relation. Each sentence was accompanied by a drawing to illustrate the spatial relation, in a similar style to those of Mark and Egenhofer [6]: see Figure 1. However, in the present experiment each of the three spatial relations was represented through only one drawing, used throughout the experiment. The drawings were intended to help to visualize the relation, and to demonstrate that despite the changes of scale and object, we were considering the objects more in terms of their spatial relations than their visual identities: this should have encouraged any tendency towards scale neutrality, and hence made a stronger test of our alternative hypothesis of scale dependence. It was explained that the drawings were not the key stimuli but merely illustrations, and participants were urged to focus on rating the sentence rather than the drawing itself[2].

For every combination of scale and spatial relation eight sentences were generated, to use all possible combinations of the line and region features and prepositions, leading to 3 x 3 x 8 = 72. However, this led to a problem with the spatial relation *intersection* because two liquid features cannot intersect, e.g. "the stream runs across the lake" would obviously be deemed unacceptable due to the physical properties of liquids. These 6 examples were therefore removed, while leaving the non-intersection cases of liquid to liquid relations, i.e.. adjacency and connection. This left 66 sentences in total. Two dummy (nonsense) sentences, deliberately mixing objects of different scales, were added in to test people's concentration (e.g. "The old road runs next to the leaf"): if participants rated these highly then their results were excluded from analysis. The stimuli were presented on paper; the experiment took about 20 minutes to complete. To check for potential order effects, participants were randomly

---

[2] Piloting the study showed that by making the drawings deliberately sketch-like, and choosing objects that could be fairly unconstrained in their shape and outline, the same drawing could serve as a reasonably convincing supporting diagram across all scales and objects, given that the instructions were to evaluate the sentence only.

assigned to complete the questionnaire either in its original form (within which the questions were randomly ordered), or with the question order reversed.

Table 1 illustrates how the various factors and examples were combined.

Table 1. Combinations of features in the sentences presented in the experiment: all line features were combined with all region features at their respective scales, using all six spatial prepositions apart from liquid-liquid intersections

	Line features	Region features
Figural	String/Trickle	Leaf/Puddle
Environmental	Old road/Stream	Park/Lake
Geographic	Gas pipeline/Canal	Country/Sea

The participants marked their agreement on a continuous line from 0 to 10: zero meant no agreement with the sentence; 10 meant that the sentence was perfectly accepted. Figure 1 shows two example sentences and drawings.

2.2 Participants

26 adults completed the final version of the questionnaire. These were all volunteers living or working in the Southampton area of England. Three participants made unexpectedly high ratings of one or other of the two 'dummy' questions, a fourth participant was apparently only 10 years old, and two others did not have British English as their first language. These six were therefore excluded from analysis. This left 20 (an adequate sample in terms of statistical power for this entirely within-subjects design, given the expected effect sizes which were confirmed through piloting). The final sample included 9 females. Age ranged from 22 to 50 (mean=37 years, standard deviation=8). One female did not give her age. Otherwise the males were slightly but significantly older than the females (mean=40 as opposed to 32; $t_{17}=2.89$, $p=0.01$).

3 Results

Participants completed the experiment with no apparent problems or omissions, giving a mean rating of less than 1 to the two 'dummy' sentences but a mean rating of 7.0 (standard deviation=2.6) to the 66 test ones.

The results were analysed using a repeated-measures analysis of covariance (ANCOVA). Age was included as a continuous covariate, gender and question order as between-subject factors, and scale, preposition and object types as within-subject factors. None of the between-subject factors or the age covariate had any significant effects, so these will not be discussed further.

The main effects[3] of scale ($F_{2,1169}=32.47$, $p<0.0001$) and preposition ($F_{5,1169}=7.41$, $p=0.0005$) were both very strongly significant, as was the interaction between them ($F_{10,1169}=3.52$, $p=0.0033$). The main effect of object type was insignificant

---

[3] Conservative Greenhouse-Geisser significance values are quoted here, since sphericity was apparently violated for some main effects and interactions. However, this made no difference to the outcomes.

($F_{3,1169}$=1.87, p=0.15). However, all of its interactions with the other variables were significant: scale x object type ($F_{6,1169}$=3.14, p=0.018), preposition x object type ($F_{13,1169}$=4.70, p=0.0021), and the second-order interaction scale x preposition x object type ($F_{26,1169}$=2.46, p=0.025).

The effect of scale showed that overall, across all object and relation types, sentences were rated higher at the environmental scale (mean=7.5, s.d.=2.4) than at the geographical (mean=7.1, s.d.=2.4) or figural (mean=5.9, s.d.=3.2) scales. The most highly rated preposition, across all scales and object types, was *through* (mean=7.5, s.d.=2.7), while the least was *from* (mean=6.6, s.d.=2.6) closely followed by *to* (mean=6.7, s.d.=2.7). As shown in Table 2, the interaction between the two factors was also particularly stark for *through*, which scored the lowest of all at figural scale, but the highest at geographical scale. However, this effect was not as strong for *across*, the other 'intersection' term. To a much lesser extent, *next to*, *to* and *from* also showed lower ratings at figural scale than at larger (especially environmental, as opposed to geographical) scales.

Table 2. Mean *(sd)* sentence ratings by scale and preposition

	figural	environmental	geographic
alongside	6.5 *(2.7)*	7.7 *(2.1)*	7.2 *(2.3)*
next to	6.0 *(3.0)*	7.2 *(2.5)*	6.8 *(2.4)*
to	6.2 *(2.8)*	7.2 *(2.5)*	6.4 *(2.6)*
from	6.1 *(2.6)*	7.2 *(2.3)*	6.3 *(2.6)*
across	7.0 *(2.5)*	7.2 *(2.7)*	7.3 *(2.3)*
through	6.0 *(3.2)*	8.0 *(2.4)*	8.5 *(1.4)*

The interaction of scale with object type suggested that at figural scale the sentences referencing a solid area (leaf) were rated much lower than at environmental scale (e.g. with solid-solid (string-leaf), mean=6.3, s.d.=2.8, but at environmental scale (old road-park), mean=8.1, s.d.=1.9). This was also true to a lesser extent for the other two object pairings. At environmental scale the solid-solid (old road-park: mean=8.1, s.d.=1.9) sentences were rated considerably higher than the liquid-liquid (stream-lake: mean=6.9, s.d.=2.8) and solid-liquid (old road-lake: mean=7.0, s.d.=2.8) pairings, with the liquid-solid (stream-park: mean=7.6, s.d.=2.2) sentences falling between the two. At geographical scale, however, all the object type pairings had similar average ratings of around 6.8-7.3.

The interaction of object type with preposition, across scales, suggested that the strongest-rated sentences were those using *across* with solid-solid object pairings (mean=8.2, s.d.=1.5), followed by those using *through* with liquid-solid (mean=7.7, s.d.=2.8). By contrast, the weakest sentences were those with a liquid-solid pairing and either *from* (mean=5.8, s.d.=2.7) or *to* (mean=6.0, s.d.=2.7), or with a liquid-liquid pairing and *next to* (mean=6.0, s.d.=3.0), or with solid-liquid and *across* (mean=6.3, s.d.=2.8). Thus *across* was shown to differ substantially between object types regardless of scale, whereas *through* differed substantially across scales regardless of object types.

The significant second-order interaction (scale x object type x preposition) reflects the fact that certain individual sentences were particularly highly or poorly rated. Very low-rated sentences (i.e. averaging more than half a standard deviation below the overall mean, i.e. below 5.7) were as follows:

- *The trickle runs from the leaf* (figural, liq-sol, mean=4.9, s.d.=2.7)
- *The string runs through the leaf* (figural, sol-sol, mean=5.1, s.d.=3.3)
- *The trickle runs through the leaf* (figural, liq-sol, mean=5.2, s.d.=3.3)
- *The canal runs to the country* (geographical, liq-sol, mean=5.2, s.d.=2.5)
- *The trickle runs next to the puddle* (figural, liq-liq, mean=5.3, s.d.=3.6)
- *The old road runs across the lake* (environmental, sol-liq, mean=5.3, s.d.=3.4)

Very high-rated sentences (i.e. averaging more than half a standard deviation above the overall mean, i.e. above 8.3) were as follows:

- *The stream runs through the park* (environmental, liq-sol, mean=9.1, s.d.=0.7)
- *The old road runs through the park* (environmental, sol-sol, mean=8.9, s.d.=0.9)
- *The canal runs through the country* (geographical, liq-sol, mean=8.8, s.d.=0.9)
- *The old road runs across the park* (environmental, sol-sol, mean=8.5, s.d.=1.5)
- *The canal runs across the country* (geographical, liq-sol, mean=8.4, s.d.=1.4)

4 Discussion

The results of the ANOVA suggested that the participants did differentiate between scales in their response to specific prepositions. The overall findings show that more sentences were rated inappropriate in *figural* space than in *environmental* or *geographical* space. Therefore, the null hypotheses of scale invariance in use of spatial relational terms can be rejected. Prepositions which are accepted at the scale of local geographic space do not scale down easily to tabletop-scale items. Similarly, we can reject the assumption that connection, adjacency and intersection terms have similar acceptability across scales and contexts: *from* seems to be more problematic overall than *through*. Yet the ratings of *through* varied with scale more than any other preposition we tested, while *across* did not vary across scale but did vary across object type.

A moment's reflection by the reader will confirm that in fact *across* and *through* indicate very different relations when used for manipulable objects, and are only frequently interchanged in English at the environmental or geographical scale. Since *across* implies that the ground is a surface, while *through* tends to more strongly imply the third dimension, acceptability ratings can be expected to vary more for the latter in relation to changes in the perceived shape/substance of the 'ground' object (i.e. that through which the figure object 'runs'). Later we will further discuss the relevance of such figure-ground distinctions.

Environmental space was the best-rated scale overall: most sentences were rated positively, and two out of the three best rated sentences belong to this scale. By looking at the three best and three worst rated sentences in this scale, it becomes obvious that *context* has a major effect on the participant's rating. All of the three best rated sentences have a solid region feature, namely park, and all of the three worst rated sentences have a liquid region feature, namely lake.

What factors might cause these effects? First, regarding the differences among prepositions, it appears that the 'connection' terms *to* and *from* were least transferable across object types. This is probably because people expect edge-to-end connections to be specified only where the two connecting objects are of the same type. For instance, a stream may run from/to the edge of a lake (or a canal from/to a sea), but for it to run 'from' a park one would assume that it began somewhere within the park, not at its border; similarly with canal and country, and with trickle and leaf.

To and *from* may also be problematic with solid-solid connections where the objects have no obvious causal relationship, e.g. the string running to/from the leaf, or the gas pipeline appearing to start/end abruptly at the border of the country. Yet the old road running from or to the park - perhaps because one might easily imagine a road that ends at the park's edge - is less obviously problematic. The sentences are rated negatively if a canal or gas pipeline runs *to* or *from* a country. Sentences are not rated that negatively when the canals or gas pipelines run *to* or from the sea. Here people's reasoning may be shaped through their experience with maps and models. Images of world maps or country maps show very often how a line feature ends or starts at the sea (or appears to do so, e.g. if it subsequently runs underwater). Conversely, they do not tend to end or start at a country. This suggests that *to* and *from* would only have been deemed acceptable if the region object was a physically different kind of thing from its surroundings - i.e. sea - which is apparently not a constraint that people expected at the small scale. So again, scale made a difference to the expected physical circumstances implied by a given preposition.

This links to the most difficult issue we encountered when designing and analysing this study. It is hard to assess the effect of scale on preposition use, independently of object type, because *objects at different scales have different characteristics.* At figural scale, it proved extremely hard to find an irregularly shaped 'area' object that might be deemed topologically equivalent to country or park - i.e. something relatively two-dimensional, that could be traversed over and beyond within the same plane. The smaller the space, the more items tend to be understood as individual, non-continuous objects with a salient third dimension. By contrast, in larger spaces solid areas tend to be easier to consider as a flat surface, not least because they tend to be considerably wider than they are tall. Thus *through* was considered unacceptable with leaf (implying penetration or depth perpendicular to the plane of the leaf), but unproblematic both with park and country (implying only traversal). In addition, larger-scale objects tend to be more continuous with others, and their boundaries increasingly abstractly rather than physically defined; thus the apparent unlikelihood of a gas pipeline stopping abruptly at a country's border.

The results concerning object type pairs have important implications for the use of spatial prepositions in describing object relations between hydrology and other domains. The characterisation of these relations is a key factor in ongoing work at Ordnance Survey and at Münster, to formally model hydrological and topographic data ontologies (e.g. [18]) The apparent lack of overall differences between object type pairs seems to have been due to different effects (at different scales and relation types) cancelling each other out, rather than to an indifference of spatial semantics to the physical state of objects. The same spatial relationship is not implied by the same term, when the objects are water rather than solid features. A liquid connecting to a solid is not deemed realistic if it is depicted as connecting with the edge, rather than showing a termination point inside the solid area. The issue here is people's expectation of *physical realism*.

For *across*, a solid line object crossing a solid area was seen as plausible at all scales, but not when it crossed a body of water (puddle, lake or sea). Again, this relates to expectations of physical realism: the old road or gas pipeline (but perhaps not the string, which was less harshly rated) would require some form of suspension over the water. Since the form of this suspension (e.g. a bridge) was not explicitly drawn or stated, participants found the relationship less plausible. Similarly, when a liquid line feature (trickle, stream or canal) was declared to be *next to* or *alongside* a water body (puddle, lake or sea), participants may have expected to either see a clear spatial separation between the two, or to have one mentioned in the sentence.

This has implications for cartographic representation, as well as for choice of linguistic terms when describing adjacent hydrological features. Physical structures that must be in place to avoid two water features merging together, or a solid line feature collapsing into a water body, need to be specified explicitly if the relevant spatial relations are to be deemed plausible.

Another issue concerning scale is the way in which different spaces tend to be apprehended. Geographic space is not only apprehended through physical actions such as manipulation and locomotion, but also with the help of models and maps. In fact, at both environmental and geographic scales, the images we used would have been interpreted as small-scale maps rather than near-lifesize pictures. This may have made it more likely that participants would think of these features differently, and perhaps as more two-dimensional than three-dimensional. However, to some extent this must be people's perception of geographic-scale objects anyway, even in cultures where maps are less prevalent, since they are inevitably much wider and longer than they are tall.

Where does this leave the hypothesis of scale-invariant spatial relations? Naturally, we should express caution about the generalisability of this study to more realistic contexts of use: like all experiments, there was some potential for artefactual results. However, it is hard to believe that an artificial context, and one in which the same illustration was used across scales (which should have encouraged participants to respond similarly as well), would show *more* scale variance than in a richer and more complex context. Further research is clearly needed to confirm and expand these findings: while it seems likely that some prepositions may maintain scale invariance, others do not. Prepositions such as *near* seem genuinely indifferent to scale precisely because they have no semantic specification of object properties (e.g., *line near a circle / planet near the sun*). However, we cannot conclude from such isolated examples that all such spatial language will be scale invariant.

Conversely, given the different physical entities and structures implied by the above analysis, it would be possible to maintain that scale invariance has not been disproved for situations where objects at different scales are more closely matched in their physical properties. However, our frustrating search for such equivalent examples across different scales and domains demonstrates in itself that the statement of scale invariance is ultimately meaningless. The physical world as experienced by humans, and described by human language, is not a fractal: scale changes its very physical nature, and hence the meaning of its spatial relations.

If this is the case, then it is arguably not scale *per se* that determines spatial preposition choices, but the nature and relative importance of the 'figure' and 'ground' objects which vary unavoidably with it. 'Ground' usually refers to the larger and more immobile object in a spatial relation, and to the object of a sentence or prepositional phrase, whereas the smaller object (and the sentence's subject) is more often the 'figure' [21]. The figure and ground objects of different scales will have different granularity, or degree of topological and metric precision. Some spatial prepositions also carry information that implies the objects' shape (e.g. line or area), such as *along* or *across*. Golledge [22] states that in describing topological relations, "the choice of nouns and prepositions conveys conventional information that is often fuzzy or inaccurate" (p.411). Yet convention has taught us which prepositions to use for certain concepts, which makes it possible for humans to understand spatial relations and spatial concepts. The present experiment demonstrates significant variance among spatial prepositions and their implied relations, suggesting that spatial prepositions strongly reflect people's reasoning about figure-ground relations in space.

Landau [21] has demonstrated that the characteristics of the 'ground' object tend to have a greater impact on spatial language use than those of the 'figure'. This appears to be borne out to some extent by the current study - it was apparently the nature of the leaf, as opposed to the nature of a park or country, that caused a problem for *through*. Although a string is as much a single three-dimensional solid object as a leaf is, whereas an old road is arguably more two-dimensional and 'pathlike', even a closer equivalent to a road (e.g. a slug trail) would not be deemed to run 'through' a leaf in the way that a road can run through a park. A vein within the leaf could do so, but this forms part of the leaf itself. Even if a road could be argued to be part of a park in a similar way, the fact that the traversal of a non-component object such as a car or highway can still be described with either *across* or *through* indicates that the difference is real. Similarly with the gas pipeline and the country (as opposed to the string and the leaf or the old road and the park), it was the nature of countries that seems to have caused a problem: no line object is expected to stop dead at a country's abstractly defined borders, whereas two *physically* distinct objects may well lie separately in space.

It should be noted that the finding that the same spatial terms are considered differently within English at different scales does not imply that this would generalise to all other languages. We know, for instance, that speakers of some languages apply to smaller scales some linguistic terms that in English we mostly[4] reserve for larger ones (e.g. [23]), although it is currently unclear whether this also applies to preposition use. However, the number of spatial prepositions in other languages varies

---

[4] Since US and British English can show differences in usage of prepositions (e.g. Clark 1968), we also checked these results with a second sample of 11 US English speakers. We found the same pattern of main effects, although without any strong interactions; however, a combined ANOVA still suggested significance for the interactions as before, with no significant influence of English dialect. Overall this strengthens our view that scale and type of relation do affect people's expectations of spatial preposition use, independently of potential dialect variations.

greatly compared to English (e.g. [21]). Therefore we certainly cannot assume that scale is equally variant in every language: we might expect it to vary more where there is an atypically large and finely differentiated range of prepositions, as is the case for English and for other Germanic languages. Nevertheless, it seems likely (because of the physical difference in existent object types at different scales, as discussed) that some scale effects will still be found even in languages with reduced prepositional inventories. Alternatively, other grammatical and lexical markings which express distinctions similar to Germanic prepositions might be expected to exhibit scale variance.

Overall, although obviously further research is needed to extend these findings, this experiment has given us some insight into the variable status of some key spatial prepositions at different scales, and into some of the factors that may influence their use. The results lead to the conclusion that scale is not neutral as Talmy [13,14] suggested, but rather plays a significant role in linguistic topology. Reasons for these differences may include humans' spatial experience, the abstractions that they use when reasoning about space, and the variable nature of our classification of the world into 'objects' at different scales.

References

1. Egenhofer, M.J., Franzosa, R.: Point-Set Topological Spatial Relations. International Journal of Geographical Information Systems 5, 161–174 (1991)
2. Egenhofer, M.J., Herring, J.R.: Categorizing Binary Topological Relations Between Regions, Lines, and Points in Geographic Databases. Technical Report, Department of Surveying Engineering. University of Maine, Orono, Maine (1992)
3. Peuquet, D.J.: Representations of Geographic Space: toward a conceptual synthesis. Annals of the Association of American Geographers 78, 375–394 (1988)
4. Hernandez, D.: Relative Representation of Spatial Knowledge – The 2D Case. In: Mark, D.M., Frank, A.U. (eds.) Cognitive and Linguistic Aspects of Geographic Space, pp. 373–385. Kluwer, Dordrecht (1991)
5. Lynch, K.: The Image of the City. MIT Press, Cambridge, MA (1960)
6. Egenhofer, M.J., Mark, D.M.: Naïve Geography. In: Kuhn, W., Frank, A.U. (eds.) COSIT 1995. LNCS, vol. 988, pp. 1–15. Springer, Heidelberg (1995)
7. Lautenschütz, A.-K.: Spatial Relations: the influence of scale, context and spatial relation in linguistic topology. Unpublished Master's thesis, Institut für Geographie, Westfälische Wilhelms-Universität Münster, Germany (2006)
8. Riedemann, C.: Naming Topological Operators at GIS User Interfaces. In: Toppen, F., Painho, M. (eds.) 8th Conference on Geographic Information Science, Estoril, Portugal, pp. 307–315 (2005)
9. Guarino, N.: Formal Ontology and Information Systems. In: Guarino, N. (ed.) Formal Ontology in Information Systems, pp. 3–15. IOS Press, Trento, Italy (1998)
10. Schwering, A., Raubal, M.: Spatial Relations for Semantic Similarity Measurement. In: Akoka, J., Liddle, S.W., Song, I.-Y., Bertolotto, M., Comyn-Wattiau, I., van den Heuvel, W.-J., Kolp, M., Trujillo, J., Kop, C., Mayr, H.C. (eds.) Perspectives in Conceptual Modeling. LNCS, vol. 3770, pp. 259–269. Springer, Heidelberg (2005)

11. Mark, D.M., Egenhofer, M.J.: Calibrating the Meanings of Spatial Predicates from Natural Language: Line-Region Relations. In: Sixth International Symposium on Spatial Data Handling, Edinburgh, Scotland, pp. 538–553 (1994)
12. Coventry, K.R., Garrod, S.C.: Saying, Seeing and Acting: The Psychological Semantics of Spatial Prepositions. Essays in Cognitive Psychology Series. Psychology Press, Hove, New York (2004)
13. Talmy, L.: How language structures space. In: Pick, H.L., Acredolo, L.P. (eds.) Spatial Orientation - Theory, Research and Application, pp. 225–282. Plenum Press, New York (1983)
14. Talmy, L.: Toward a Cognitive Semantics. Concept Structuring Systems, vol. I. MIT Press, Cambridge, MA (2000)
15. Mark, D.M., Frank, A.U., Egenhofer, M.J., Freundschuh, S.M., McGranaghan, M., White, R.M.: Languages of Spatial Relations. Initiative Two specialist meeting report (NCGIA Technical Paper 89-2). National Center for Geographic Information and Analysis (1989)
16. Pederson, E.: How Many Reference Frames? In: Pierre, S., Barbeau, M., Kranakis, E. (eds.) ADHOC-NOW 2003. LNCS, vol. 2865, pp. 287–304. Springer, Heidelberg (2003)
17. Montello, D.: Scale and Multiple Psychologies of Space. In: Campari, I., Frank, A.U. (eds.) COSIT 1993. LNCS, vol. 716, pp. 312–321. Springer, Heidelberg (1993)
18. Hart, G., Temple, S., Mizen, H.: Tales of the River Bank: first thoughts in the development of a topographic ontology. In: Toppen, F., Prastacos, P. (eds.) Proceedings of 7th AGILE Conference, pp. 165–168. Crete University Press, Heraklion, Greece (2004)
19. Brugman, C.: The Story of "Over". Unpublished Master's thesis, Linguistics, University of California (1981)
20. Herskovits, A.: Language and Spatial Cognition: an interdisciplinary study of the prepositions in English. Cambridge University Press, Cambridge (1986)
21. Landau, B.: Multiple Geometric Representations of Objects in Languages and Language Learners. In: Bloom, P., Peterson, M., Nadel, L., Garrett, M. (eds.) Language and Space, pp. 317–363. MIT Press, Cambridge, MA (1996)
22. Golledge, R.G., Stimson, R.J.: Spatial Behavior – A Geographic Perspective. The Guilford Press, New York (1997)
23. Pederson, E.: Geographic and Manipulable Space in Two Tamil Linguistic Systems. In: Frank, A.U., Campari, I. (eds.) Spatial Information Theory, pp. 294–311. Springer, Berlin (1993)

Before or After: Prepositions in Spatially Constrained Systems

Kai-Florian Richter[1] and Alexander Klippel[2]

[1] Transregional Collaborative Research Center SFB/TR 8 Spatial Cognition
Universität Bremen, Germany
richter@sfbtr8.uni-bremen.de

[2] Cooperative Research Centre for Spatial Information
Department of Geomatics, The University of Melbourne, Australia
aklippel@unimelb.edu.au

Abstract. Cognitive agents use different strategies to identify relevant spatial information in communication. The chosen strategy depends on the agents' conceptualization of the spatial situation at hand. This situation is determined by structural and functional aspects that are induced by the environment and the actions performed or intended therein. In this paper, we are interested in conceptualizations in the context of route directions. We focus on the meaning of prepositions used to characterize movements (actions) in spatially constrained systems such as street networks. We report on different strategies employed by people to disambiguate turning actions at intersections and demonstrate how these can be reflected in automatically generated route directions, again concentrating on the assignment of prepositions for anchoring movement. Including methods that focus on the most successful strategies people use in computational systems is a prerequisite for route directions that respect for human conceptualizations of spatial situations and that become, thus, cognitively ergonomic route directions.

1 Introduction

Cognitive agents communicate about space through various modalities [1]. Each modality, such as natural language expressions or sketch maps, has their own representational characteristics. While the problem of graphic, map-like representations is their commitment to one out of potentially many instantiations of a spatial situations [2], the problem of linguistic expressions is to determine their meaning as they are inherently underspecified. This problem manifests itself in the large body of research literature on spatial prepositions (e.g., [3]). Prepositions are still not well handled in formal systems as a limited number of prepositions can have manifold meanings; the semantics of a preposition changes with the context in which the preposition is used [4,5]. This makes prepositions a highly challenging research topic.

In this paper, we focus on the semantics of prepositions in the context of route directions. More specifically, we are interested in assigning a preposition to anchor movements in a spatially constrained system, i.e. a city street network.

Most studies on the semantics of spatial prepositions take place in unconstrained, or better naturally constrained, environments, in that, for example, they rely on the influence of gravity and the characteristics of the objects in question. In contrast, we focus on the influence of structural aspects of an environment on the semantics of spatial prepositions. This approach is part of a larger research effort investigating the role of structure and function in the conceptualization of events.

We proceed with a short overview on route directions and landmarks introducing the key concepts used in this work and clarifying the distinction between *structure* and *function* in the next section. We continue discussing some empirical findings on how people's choices depend on a decision point's structure and on the action that has to be performed (Section 3). We then present a computational approach that takes these findings into account and selects spatial prepositions based on configurational knowledge and neighborhood relations of the involved functionally relevant entities (Section 5). Modeling these entities is explained in Section 4.

2 Route Directions and Landmarks

Route directions provide a wayfinder with information on how to reach her current destination; they are task-oriented specifications of the actions to be carried out [6,7]. Route directions are not merely a description of what a wayfinder will encounter along the route; they include references to actions at decision points, to landmarks, and confirmative information whether the correct route is still followed. Since they focus on the actions to be performed, verbal route directions are procedural discourse [8]. Routes are inherently linear; therefore, the linearization problem, i.e. the problem of choosing among many possible sequences when organizing spatial information that extends over two or three dimensions, does not apply here. With routes, the sequence of spatial information is determined by the spatial sequence of locations encountered during route following (see also Section 2.1). On the other hand, the linearization problem exists in a modified form [9]: structuring route knowledge comprises the spatial chunking of individual decision points into larger units. This is a process on the conceptual level to organize spatial knowledge. In that, it differs from and precedes the linguistic process of aggregation as explained, for example, in [10], which is used to form sentences that communicate several pieces of information at once.

The core of route directions is a sequential description of actions, like "walk" or "turn [left]" [11]. These actions are termed *route-following actions* in the following. The core is enhanced with descriptions of the spatial situations that will be encountered during route following, especially references to landmarks [12]. Landmarks foster the identification of decision points, the origin and destination of a route, provide verification of route progress, provide orientation cues for homing vectors, and suggest regional differentiating features. Landmarks are pertinent for route directions [8,13]. They are especially useful when anchoring actions in space, i.e. if the action that has to be performed at a decision point is anchored by the reference to a landmark, such as "turn right after the post office".

For the inclusion of landmarks in automatically generated route directions the spatial relation between the landmark and the action performed at a decision point has to be determined to provide anchoring in space. The parameters taken into account are the decision point itself, the route-following action performed at the decision point (i.e., the angle of the change in direction), and the position of a landmark (Fig. 1).

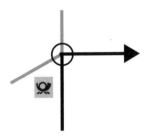

Fig. 1. Parameters determining assignment of a spatial preposition: the encircled decision point, a route-following action (right turn) indicated by the bold arrow-lines, and a landmark (a post office) positioned in proximity to the decision point

2.1 Structure and Function

Klippel's distinction of *structure* and *function* [14] is based on Montello's differentiation between the environment and the actions or movement that is performed in that environment [15][1]. This distinction is reflected terminologically by *path* and *route*.

Structure denotes the layout of elements physically present in the spatial environment. This comprises, for example, the number of branches at an intersection, the angles between those branches, and a landmark's position (relative to the action performed at an intersection). *Function* is related to the actions that take place in spatial environments. The functional characterization demarcates parts of the environment, i.e. those parts of the structure that are necessary for the specification of the action to be performed (see Fig. 2). Structure has an impact on the conceptualization of an action, in our case the change of direction at a decision point, and in return on the assignment of a *spatial projective term*, its modification, and the choice of a *spatial preposition* (see next section).

A trajectory, such as the movement of an agent, can be discretized by focusing on the direction changes the agent makes. Considering route following, the critical points of direction change are the decision points: here, a wayfinder needs to decide on the further way to take. Consequently, they are pertinent for route following [13]—even if no change of direction occurs, i.e., the agent

---

[1] Montello's book chapter has been available as 'in press' for several years on his homepage before publication; this explains the seemingly contradicting publication dates.

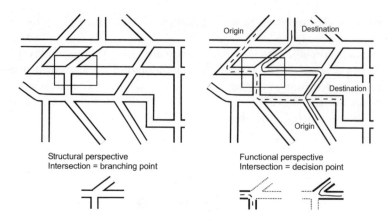

Fig. 2. Distinction of structural and functional aspects of route information (from [14])

is going straight at an intersection. A route may be represented as a sequence of decision points. Depending on the structure in which these direction changes take place, the conceptualization of the same change of direction as a discrete step in a route can differ. Accordingly, describing the turn at a decision point, for example, linguistically, may differ depending on the structure and action to be performed. In the next section, we outline some strategies people employ in giving directions at decision points. These strategies depend both on structural and functional aspects.

3 Strategies for Describing Route-Following Actions

Klippel, Tenbrink, and Montello [16] propose a characterization of aspects that influence the specification of spatial relations in the context of route directions. These aspects include the structure of an intersection, the route-following action to be performed at an intersection that demarcates functionally relevant parts, the availability of additional features that can be used to anchor the action to be performed, like landmarks, and the conceptualization of this action as a result of structure and function and the features available. For example, Figure 3 illustrates how an intersection's structure may change the linguistic characterization of a change of direction.

An analysis of a route direction corpus, sampled as reported in [9], reveals several options to communicate which branch to take at an intersection. The following categories reflect the conceptualization of turns at decision points and, thereby, correspond to different kinds of spatial knowledge: qualitative direction concepts expressed by projective terms, references to absolute directions, and direction-indicating verbs, for example, "turn right", "go west", "veer left"; qualitative modifications (hedges) specifying the direction [18,19], as in "slightly right", quantitative measures of directions in degrees, for instance, "turn exactly 90 degrees"; clock directions ("turn three o'clock"); references to the structure,

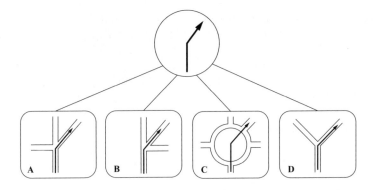

Fig. 3. According to the intersection in which it takes place, a change of direction is associated with different meaning. While the 'pure' change may be characterized as "veer right" at the intersection (A), at intersection (B) it may change to "the second right"; at the roundabout (C) it becomes "the third (or second) exit"; and at (D) "fork right" (from [17]).

for example, "dead end", "fork"; ordering concepts, like "the first exit"; reliance on landmarks to indicate a direction, as in "where the statue is".

In this paper, we concentrate on the aspect of relying on landmarks to indicate directions. Depending on their geometry and size in relation to the environment, landmarks can be considered point-like, linear, or area-like (cf. [20,21]). Here, we discuss landmarks conceptualized as being point-like, i.e. those located in a small, restricted area of an environment, that are functionally relevant only for a small part of the route, usually a single decision point. These landmarks may be located at a decision point or in-between two decision points.

Landmarks in-between two decision points may either be related to the decision point passed after them or to the one just passed. Those related to the decision point after them may be used to identify this decision point ("you'll pass a gas station; after that, turn right at the next intersection"). A landmark related to a decision point just passed can be used to confirm that the correct direction has been chosen ("if you pass the church, you picked the correct turn."). In the following, we focus on landmarks being conceptualized at a position *at* a decision point. These landmarks are frequently used to identify the intersection, especially if other intersections are nearby and, more importantly in the context of this paper, they are frequently used to anchor the direction change to be performed at an intersection [8]. Linguistically, the anchoring is encoded as *turn [right, left] [before, after, at]{landmark X}*. The different prepositions correspond to different relations of landmark X relative to the action to be performed at the decision point: the landmark is either passed before or after the route-following action, or not directly passed at all (cf. Section 5.2). These possible spatial relations between landmark and decision point are represented as $lm^<$ (landmark passed *before* decision point), $lm^>$ (landmark passed *after* decision point), lm^- (landmark *at* decision point, but not directly passed) (cf. also [22]).

Clearly, the preposition *at* is less constrained than *before* or *after*, i.e. its semantic scope is larger. In the context of this paper it potentially incorporates the other two. Still, we choose *before* or *after* if applicable, since they relate the position of a landmark directly to the action performed at the respective decision point and are, thus, more specific. The rational for this is rooted in considerations of cognitive ergonomics that strive to provide as much detail as possible with as little information as necessary. From the perspective of cognitive science we could say that the situation model [23] that is instantiated, for example, by a linguistic instruction, has to be rendered as precise as possible by using a preposition with a smaller semantic scope. We take a Griceian perspective [24] in generating references.

Function determines which structural aspects are relevant for the situation at hand. That is, the actions performed in route following induce spatial configurations and the sequence in which entities are encountered along the route. Thus, even if we consider static configurations of intersections in the following to determine the spatial relation between landmark X and the decision point, these result from performing an (imagined) action at this intersection, i.e. the entities can be considered events in this case.

4 Modeling Routes

As stated in Section 2.1, we consider decision points most pertinent for route directions. Following a route comprises two basic processes: getting to a decision point and, there, determining the further direction to take [25]. Accordingly, routes are represented as sequence of decision point/action pairs; each pair representing the route-following action required at a given decision point. We represent the environment, which in our work is a spatially constrained system, like a street network, as a graph reflecting its layout. The graph is annotated with additional information, for example, the geometry and position of landmarks.

Even though we model the structural aspects of wayfinding in an environment as a graph decision points are not merely a node in the graph, i.e. not just a point-like entity. Instead, a decision point denotes a certain area around an intersection, which contains not only the point where the branches meet but also parts of the branches itself (see Section 5.1). This way, it represents the configuration of path-segments at a branching point. The two functionally relevant path-segments, the route-segments, are part of the decision point, which corresponds to human mental conceptualization. This modeling also reflects another action-oriented aspect: people usually do not turn on the spot like some robots do, but rather in an extended process with deciding on the direction to turn and then changing direction gradually while keeping their forward movement [14,26]; especially when traveling by bike or car.

For the purpose of assigning a preposition to describe the spatial configuration, we abstract from this gradual direction change and model decision points such that all path-segments of the decision point meet in a single point, which we denote with the technical term *turning point* in the following (Fig. 4). This way,

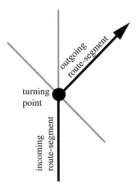

Fig. 4. A decision point with the two functionally relevant branches: the *incoming* and *outgoing* route-segment. The branches' meeting point is the *turning point*.

at the turning point the area around an intersection gets divided into a *region before action* and a *region after action* as a wayfinder's direction of motion induces a directedness of the functionally relevant branches. Landmarks located at a decision point are represented as a point inside the decision point's area whose position relative to the turning point can be determined based on a shared local or global coordinate system.

5 Determining References to Landmarks

In this section, we present a computational method to determine the spatial relation of a landmark relative to an action performed at a decision point. Thereby, we provide a formal characterization of a possible resulting mental conceptualization. Based on this relation, we can anchor the action performed at that decision point with the landmark. This method is based on configurational knowledge and neighborhood relations. It is employed in our model for generating *context-specific route directions* [20,27]. This model explicitly adapts the generated directions to the context, i.e. to the current action to take in the current surrounding environment (see Section 6).

Linear or circular orderings represent information about the sequence and neighborhood of entities without specifying any further metric information, such as distances between these entities (e.g., [28]). This kind of information is a powerful structuring means; it is also easy to determine as it only requires knowledge about a neighborhood relation between the relevant entities. This holds especially for routes: routes are linear and directed entities; the movement direction of the wayfinder induces an orientation on the point-like entities along the route, like decision points or landmarks. This orientation, in turn, induces an ordering on these entities [28].

We exploit ordering information to compute a landmark's position at a decision point. In the following, we explain how to determine the neighborhood relation of functionally relevant entities at decision points and how to generate references to landmarks based on this information.

5.1 Landmarks at a Decision Point

To determine references to landmarks at decision points, first of all those landmarks located at that decision point need to be identified. Whether a landmark can be considered to be at a decision point depends on its distance to the decision point. Distance can be based on different kinds of information: most typical, metric information is used, i.e. calculating distance between two entities based on coordinates. However, in the context of our work, the crucial information for referencing a landmark is what will be encountered between passing a landmark and a decision point (or vice versa). Therefore, next to metric distance between decision point and landmark, we also rely on ordering information.

In a first step, landmark candidates are identified. To this end, we determine which landmarks are within a decision point's region. The size of this region depends on different parameters, like a wayfinder's travel mode. Most importantly, we choose a decision point's region such that no other decision point falls within this region. This way, we prevent conflicting association of landmarks with decision points. This may, however, lead to different sizes of this region for different decision points (e.g., smaller regions for decision points that are close together). Additional information, like a landmark's saliency can be taken into account to further restrict the set of landmark candidates (see Section 6; cf. also, e.g., [29]).

From these landmark candidates, we check which landmarks actually are applicable as reference to a landmark at a decision point, i.e. we need to decide which of the landmarks from the candidate set is to be used in the instructions. For a landmark to be *at* a decision point, no other landmark candidate may be between the entity used as landmark and the decision point. For example, a post office may be referenced as a landmark at a decision point if no other building referable as landmark is passed between post office and decision point (see Fig. 5). That is, in the linear ordering of landmarks and decision point, decision point and landmark *at* the decision point need to be neighbors. This way, we achieve the (spatially) closest possible coupling of landmark and decision point, which fosters identification of the place where to perform the required action.

5.2 References to Landmarks at a Decision Point

Landmarks located at a decision point can either be passed before the route-following action takes place, after the action, or may not be located at a functionally relevant branch. In order to determine which relation between landmark and action holds and, consequently, which preposition to use preferably, we need to know a landmark's position relative to the functionally relevant branches. We distinguish three relations—$lm^<$, $lm^>$, lm^-—to denote a landmark's position. They are termed *reference relations*. If relation $lm^<$ holds, i.e. if a landmark is passed before performing an action at a turning point, the landmark must be *next to* the incoming route-segment. If the relation $lm^>$ holds the landmark is

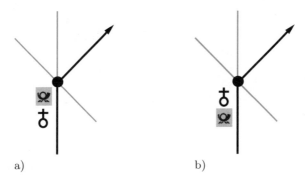

Fig. 5. Distance of a landmark—the post office—based on ordering information: a) landmark at a decision point: the post office is directly passed before the turning point; b) landmark not at a decision point: the church is passed after the post office, but before the turning point

passed after the route-following action took place and is next to the outgoing route-segment. If a landmark is neither next to the incoming nor the outgoing route-segment, it is not at a functionally relevant branch. We use the relation lm^- to describe this situation where a landmark is at a decision point which position with respect to the route-segments cannot be further specified.

But what does *next to* mean in this context? It refers to the neighborhood relation that underlies the ordering information of the entities at a decision point. For locating a landmark, the *next to* relation describes a region of a landmark's possible locations. These locations are chosen such that the landmark is located before or after the turning point and no other branch is between the functionally relevant branch and the landmark (see Fig. 6). For the incoming route-segment the region that satisfies these conditions is termed *before-region*, for the outgoing route-segment it is termed *after-region*. The landmark needs to be located in the before-region for the relation $lm^<$ to hold and in the after-region for the relation $lm^>$. To determine whether a landmark is in the before- or after-region of a decision point, we can exploit ordering information. The branches of a decision point form a circular ordering as they are neighbored according to their bearing—with the last branch in the ordering being neighbored to the first one (see Fig. 7a).

The ordering can be calculated by determining the angle of a branch relative to some reference direction, for example north representing a bearing of 0, and then sorting all branches according to their direction angle. By creating a new virtual branch ranging from the turning point to the landmark in consideration, we can introduce the landmark's location to the branches' ordering (see Fig. 7b). We recompute the ordering and can, this way, determine the landmark's position with respect to the newly established ordering. Since for the reference relations $lm^<$ or $lm^>$ to hold the neighborhood relation *next to* needs to hold, and for *next to* to hold no other branch may be between a route-segment and a landmark, we check whether the newly introduced virtual branch is a direct neighbor of either of the functionally relevant branches in the ordering.

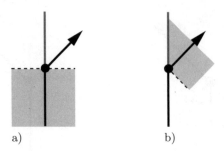

Fig. 6. Regions of a landmark's possible location relative to the functionally relevant branches: a) *before*, b) *after*

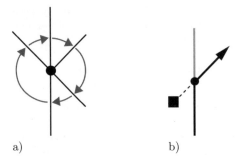

Fig. 7. a) An intersection with five branches. The arrows indicate the neighborhood relation between the branches; the branch pointed at is a neighbor of the branch the arrow starts from. This relation is symmetric. b) Virtual branch (dotted line) introducing a landmark's location to the branches' ordering.

Fig. 8. Example for a situation in which a landmark's virtual branch (the dotted line) is next to the incoming route-segment, but the landmark is located after the turning point. Here, $lm^<$ is not a valid result.

This allows determining the reference relation to use for a landmark at a decision point. If the virtual branch is a neighbor of the incoming route-segment, $lm^<$ holds, if it is a neighbor of the outgoing route-segment, $lm^>$ holds. This procedure is very simple, but does not cover all cases: consider, for example, the configuration shown in Figure 8. Here, the landmark's virtual branch is next to the incoming route-segment in the branches' ordering, but located after the

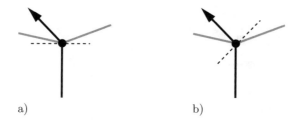

Fig. 9. Virtual branches (dotted lines) delimiting (a) the before-region and (b) the after-region, respectively

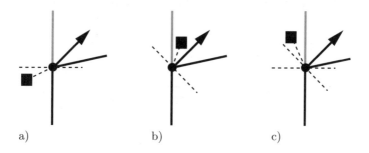

Fig. 10. Examples of a decision point's configuration and the resulting landmark's reference relation: a) $lm^<$, b) $lm^>$, c) lm^-

turning point. This violates the second condition of our definition of the before-region, namely that a landmark needs to be before the turning point. Therefore, calculating the neighborhood relation yields $lm^<$, but it is not a valid result in this case.

To account for this case, we introduce two additional virtual branches. These virtual branches originate at the turning point and head perpendicularly to the functionally relevant branch to the left and right (Fig. 9). That is, to create the before-region, we introduce two virtual branches perpendicular to the incoming route-segment; their calculation is straightforward. For the after-region we apply the same procedure with the outgoing route-segment.

This way, we ensure that the functionally relevant branch and a landmark's virtual branch cannot be direct neighbors if the landmark is not located in the correct region. Put differently, if they are direct neighbors, route-segment and landmark are on the same side of the turning point, as required in the definition of before- and after-region. Figure 10 shows examples for determining a landmark's position at a decision point using this approach. Note that we cannot introduce both sets of virtual branches—for the before- and after-region—to the ordering at the same time. This would result in overlaps in certain cases and, hence, unwanted restrictions of the two regions (cf. Fig. 10c).

For relation $lm^>$ we need to consider a further restriction. The corresponding preposition *before* is the most restricted in its semantic scope, i.e. it underlies the

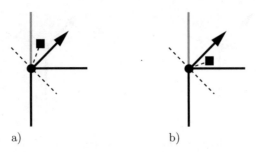

Fig. 11. "Turn right *before* landmark": a) landmark correctly indicating outgoing branch; b) ambiguous situation where landmark may mislead wayfinder

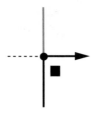

Fig. 12. Neighbored incoming and outgoing route-segment. Here, before- and after-region may overlap.

most constraints for being applicable to identify the branch to take. To indicate the correct branch, the landmark needs to be left of the outgoing branch when turning into the right half-plane and right of the outgoing branch for turning left (see Fig. 11). Otherwise, references to a landmark will lead to wrong references: in Figure 11b the default assumption would be to take the branch heading off 90 degrees to the right as this branch is directly in front of the landmark as seen from the direction of movement. However, the intended branch is the one heading off 45 degrees to the right. The order of branches and the landmark is calculated relative to some reference bearing. Accordingly, determining whether a landmark is left (or right) of the outgoing branch is straightforward: in this case it is directly before (or after) this branch in the order, i.e. the landmark is predecessor (or successor) of the outgoing branch in the ordered sequence of entities under consideration.

Finally, if incoming and outgoing route-segment are neighbors in the branches' circular ordering, the before- and after-region possibly overlap. In this case, both $lm^<$ and $lm^>$ are in principle valid results of determining the landmark's position at the decision point. However, in configurations as shown in Figure 12 the preposition *before* is not applicable, i.e. conceptually does not cover the situation model and is, therefore, never used. Hence, $lm^<$ represents this configuration; $lm^>$ is excluded (lm^- would be applicable, as well, as it incorporates $lm^<$ and $lm^>$, but is not chosen as it is less specific; see Section 3).

6 References to Landmarks in Instructions

With the presented method we are able to determine a landmark's relation to the route-following action to be performed at a decision point, and, hence, to suggest a corresponding preposition. These prepositions are used in generating descriptions of spatial situations to be communicated to a human user. We employ them in our computational model generating route directions that explicitly considers a route's properties and environmental characteristics. Accordingly, the resulting directions are termed *context-specific* as they adapt to the current action to take in the current surrounding environment [20].

The model generates abstract, modality-free specifications of the actions to be performed at decision points. It is based on a systematics of elements that can be used in route directions. According to the systematics, the model generates for each decision point all possible instructions that unambiguously describe the action to be performed there. Each such instruction represents a decision point and the action as a relational term, for example (DP_1, left). The direction relations correspond to the route-following actions a wayfinder needs to perform, for example, *left* corresponds to "turn left", *north* to "head north", etc. Generating directions for the complete route, then, is realized as an optimization process. This process decides for each decision point which instruction of a set of possible instructions to use; in that, it relies on spatial chunking, i.e. on mechanisms to combine a sequence of instructions for single decision points into a single instruction for several consecutive decision points [9,10]. With that, it accounts for another important principle of cognitively ergonomic route directions.

To generate instructions that link a landmark to the action to be performed, we annotate the direction relation with the landmark's identifier and corresponding reference relation determined as explained above. For example, this may result in a specification like (DP_1, left/church $lm^<$). This notation focuses on the most relevant information, i.e. which action to perform at which decision point (here, turning left at decision point 1). The action can then be further specified by, for example, linking it with a landmark present at the decision point (here, a church located before the decision point). This way, the general structure of our abstract route directions is always the same with the crucial information coming first; this is independent of whether the instruction is annotated or which type of annotation is to be used.

The specification does not need (and is not intended) to be externalized as it is. For example, literally, the specification given above corresponds to an instruction like "pass the church *before* you turn left". However, usually instructions focus on the route-following action to be performed at a decision point, not the landmark's location; i.e. the focus is on "turn left" in this example. Accordingly, an appropriate verbalization would be "turn left *after* the church". Thus, the relation between spatial relation and preposition to use is as follows: $lm^<$ corresponds to the preposition *after*, $lm^>$ to the preposition *before*, and lm^- to the preposition *at*.

There might be several landmarks present at a decision point. This results in several options for anchoring movement in space by reference to a landmark. In such cases, the computational system needs to decide on which reference to use. One possibility is to integrate preference rules in the model on which preposition, and accordingly landmark, to choose: a simple rule may, for example, state to prefer determinate references (*before, after*) over indeterminate (*at*) to adhere to the Gricean principles called for above. A more elaborate approach is to take a landmark's saliency into account. Existing approaches determine saliency, for example, with data mining methods [30], or based on visual, semantic, and structural features of an environment's enitities [29]; the latter approach has been extended to also account for the structure of the street network wayfinding takes place in [22]. These approaches allow for calculating a landmark's saliency value, which can then be used to determine the most salient landmark at a decision point in order to refer to this one in the instructions.

Furthermore, the model implements additional strategies for disambiguating references in direction giving (see Section 3). It employs ordering concepts in situations where several branches head in the same direction, like in "take the second left". Also, it analyses an intersection's structure based on the geometric configuration of its branches and the movement direction of the wayfinder. This reveals salient intersections, like T-intersections, that may be referred to in an instruction and, consequently, may function as a landmark [31]. Such references to the structure, which can be found in human direction giving, restrict which direction relations are applicable. For example, an angular direction change that maps to *veer right* in the employed direction model is coarsened to *right* when used in combination with a reference to a fork-intersection, as in "fork right".

7 Conclusion and Outlook

To disambiguate and to identify relevant information in a route direction task, people employ different strategies in describing spatial situations. The strategy used depends on the conceptualization of the situation, which in turn is determined by structural and functional aspects. In the context of route directions, the actions to be performed at decision points are the focus of interest. We report results of a linguistic analysis that reveals several strategies used by speakers to communicate which branch to take at an intersection. Choice of the strategy depends on several aspects including the action to be performed, the conceptualization of this action as a result of structure and function, and the availability of landmarks. In this paper, we especially focused on landmarks used to identify intersections.

Such landmarks are a powerful means to anchor the direction change to be performed at an intersection. The position of the landmark relative to the decision point determines how this anchoring can be done. We presented a computational approach to determine the proper spatial relation between landmark and decision point; this approach exploits qualitative knowledge at an intersection. The structure of an intersection, i.e. the configuration of its branches, constraints possible landmark's positions and, hence, the spatial relation that is applicable. Applicability of a relation can be determined by ordering information and neighborhood relations of an intersection's branches. Depending on a landmark's position in the branches' order, the landmark is either passed before ($lm^<$) or after ($lm^>$) the route-following action or not directly passed at all (lm^-). Based on the spatial relation that holds, a preposition can be determined that reflects the relation between landmark and the action to be performed at the decision point: in case of $lm^<$ the action is performed *after* the landmark; with $lm^>$ it is performed *before* the landmark. lm^- denotes a configuration with an action performed *at* a landmark. Our computational method allows for generating appropriate references to landmarks and, this way, anchoring actions in route following with a landmark. Thus, it implements an important principle of cognitive ergonomics for route directions.

The presented approach is part of a larger research effort investigating the role of structure and function in the conceptualization of events, and to develop a model that generates cognitively ergonomic route directions. Including such methods, which are similar to people's strategies to describe spatial situations, in a computational system is a prerequisite for automatically generated route directions that respect for human conceptualizations of spatial situations. This way, users' confidence is increased, and the route directions become easier to understand. This holds especially for the actions to be performed at complex intersections.

Acknowledgments

This work has been supported by the Transregional Collaborative Research Center SFB/TR 8 Spatial Cognition, which is funded by the Deutsche Forschungsgemeinschaft (DFG), and by the Cooperative Research Centre for Spatial Information, whose activities are funded by the Australian Commonwealth Cooperative Research Centres Programme. We like to thank Thora Tenbrink for valuable comments and suggestions on an earlier draft.

References

1. Wahlster, W., Reithinger, N., Blocher, A.: SmartKom: Towards multimodal dialogues with anthropomorphic interface agents. In: Wolf, G., Klein, G. (eds.) Proceedings of the International Status Conference Human-Computer Interaction, DLR, Berlin, pp. 22–34 (2001)
2. Habel, C.: Representational commitment in maps. In: Duckham, M., Goodchild, M., Worboys, M. (eds.) Foundations of Geographic Information Science, Taylor and Francis, London (2003)
3. Retz-Schmidt, G.: Various views on spatial prepositions. AI Magazine 9(2), 95–105 (1988)
4. Regier, T.: The Human Semantic Potential: Spatial Language and Constrained Connectionism. The MIT Press, Cambridge, MA (1996)
5. Coventry, K.R., Garrod, S.C.: Saying, seeing and acting. The psychological semantics of spatial prepositions. In: Coventry, K.R., Garrod, S.C. (eds.) Essays in Cognitive Psychology Series, p. 201. Psychology Press, Hove, New York (2004)

6. Schweizer, K., Katz, S., Janzen, G.: Orientierung im Raum - kognitive Grundlagen und sprachliche Realisierung. Tourismus Journal 4(1), 79–104 (2000)
7. Tversky, B., Lee, P.U.: How space structures language. In: Freksa, C., Habel, C., Wender, K.F. (eds.) Spatial Cognition, pp. 157–175. Springer, Berlin (1998)
8. Denis, M.: The description of routes: A cognitive approach to the production of spatial discourse. Cahiers Psychologie Cognitive 16(4), 409–458 (1997)
9. Klippel, A., Tappe, H., Habel, C.: Pictorial representations of routes: Chunking route segments during comprehension. In: Freksa, C., Brauer, W., Habel, C., Wender, K.F. (eds.) Spatial Cognition III, Springer, Berlin (2003)
10. Dale, R., Geldof, S., Prost, J.P.: Using natural language generation in automatic route description. Journal of Research and Practice in Information Technology 37(1), 89–105 (2005)
11. Lovelace, K.L., Hegarty, M., Montello, D.R.: Elements of good route directions in familiar and unfamiliar environments. In: Freksa, C., Mark, D.M. (eds.) COSIT 1999. LNCS, vol. 1661, pp. 65–82. Springer, Heidelberg (1999)
12. Golledge, R.G.: Human wayfinding and cognitive maps. In: Golledge, R.G. (ed.) Wayfinding Behavior — Cognitive Mapping and Other Spatial Processes, pp. 5–46. John Hopkins University Press, Baltimore (1999)
13. Michon, P.E., Denis, M.: When and why are visual landmarks used in giving directions? In: Montello, D.R. (ed.) Spatial Information Theory, pp. 400–414. Springer, Berlin (2001)
14. Klippel, A.: Wayfinding Choremes — Conceptualizing Wayfinding and Route Direction Elements. PhD thesis, Universität Bremen (2003)
15. Montello, D.R.: Navigation. In: Shah, P., Miyake, A. (eds.) Handbook of Visuospatial Thinking, pp. 257–294. Cambridge University Press, Cambridge (2005)
16. Klippel, A., Tenbrink, T., Montello, D.R.: The role of structure and function in conceptualization of directions (submitted)
17. Klippel, A., Hansen, S., Davies, J., Winter, S.: A high-level cognitive framework for route directions. In: Proceedings of the SSC 2005 Spatial Intelligence, Innovation and Praxis: The National Bienneal Conference of the Spatial Science Institute (2005)
18. Lakoff, G.: Hedges: A study in meaning criteria and the logic of fuzzy concepts. Journal of Philosophical Logic 2(4), 458–508 (1973)
19. Vorwerg, C.: Raumrelationen in Wahrnehmung und Sprache. Deutscher Universitäts-Verlag, Wiesbaden (2001)
20. Richter, K.F., Klippel, A.: A model for context-specific route directions. In: Freksa, C., Knauff, M., Krieg-Brückner, B., Nebel, B., Barkowsky, T. (eds.) Spatial Cognition IV. Reasoning, Action, Interaction: International Conference Spatial Cognition 2004, pp. 58–78. Springer, Berlin (2005)
21. Hansen, S., Richter, K.F., Klippel, A.: Landmarks in OpenLS - a data structure for cognitive ergonomic route directions. In: Raubal, M., Miller, H.J., Frank, A.U., Goodchild, M.F. (eds.) GIScience 2006. LNCS, vol. 4197, Springer, Heidelberg (2006)
22. Klippel, A., Winter, S.: Structural salience of landmarks for route directions. In: Cohn, A.G., Mark, D.M. (eds.) COSIT 2005. LNCS, vol. 3693, pp. 347–362. Springer, Heidelberg (2005)
23. Zwaan, R.A., Radvansky, G.A.: Situation models in language comprehension and memory. Psychological Bulletin 123(2), 162–185 (1998)
24. Grice, H.P.: Logic and conversation. In: Cole, P., Morgan, J.L. (eds.) Speech Acts. Syntax and Semantics, vol. 3, pp. 41–58. Academic Press, New York (1975)

25. Daniel, M.P., Denis, M.: Spatial descriptions as navigational aids: A cognitive analysis of route directions. Kognitionswissenschaft 7, 45–52 (1998)
26. Tversky, B., Lee, P.U.: Pictorial and verbal tools for conveying routes. In: Freksa, C., Mark, D.M. (eds.) COSIT 1999. LNCS, vol. 1661, pp. 51–64. Springer, Heidelberg (1999)
27. Richter, K.F., Klippel, A., Freksa, C.: Shortest, fastest, - but what next? A different approach to route directions. In: Raubal, M., Sliwinski, A., Kuhn, W. (eds.) Beiträge zu den Münsteraner GI-Tagen 2004, Münster, IfGIprints, pp. 205–217 (2004)
28. Schlieder, C.: Reasoning about ordering. In: Frank, A.U., Kuhn, W. (eds.) Spatial Information Theory, Springer, Berlin (1995)
29. Raubal, M., Winter, S.: Enriching wayfinding instructions with local landmarks. In: Egenhofer, M.J., Mark, D.M. (eds.) GIScience 2002. LNCS, vol. 2478, pp. 243–259. Springer, Heidelberg (2002)
30. Elias, B.: Extracting landmarks with data mining methods. In: Kuhn, W., Worboys, M.F., Timpf, S. (eds.) COSIT 2003. LNCS, vol. 2825, pp. 375–389. Springer, Heidelberg (2003)
31. Klippel, A., Richter, K.F., Hansen, S.: Structural salience as a landmark. In: Workshop Mobile Maps 2005, Salzburg, Austria (2005)

Discourse Factors Influencing Spatial Descriptions in English and German

Constanze Vorwerg and Thora Tenbrink

Universität Bielefeld & Universität Bremen, Germany
constanze.vorwerg@uni-bielefeld.de,
tenbrink@sfbtr8.uni-bremen.de

Abstract. The ways in which objects are referred to by using spatial language depend on many factors, including the spatial configuration and the discourse context. We present the results of a web experiment in which speakers were asked to either describe *where* a specified item was located in a picture containing several items, or *which* item was specified. Furthermore, conditions differed as to whether the first six configurations were specifically simple or specifically complex. Results show that speakers' spatial descriptions are more detailed if the question is *where* rather than *which*, mirroring the fact that contrasting the target item from the others in *which* tasks may not always require an equally detailed spatial description as in *where* tasks. Furthermore, speakers are influenced by the complexity of initial configurations in intricate ways: on the one hand, individual speakers tend to self-align with respect to their earlier linguistic strategies; however, also a contrast effect could be identified with respect to the usage of combined projective terms.

1 Introduction

The investigation of spatial language usage encompasses a broad range of interesting research areas, including object reference and resolution on the basis of spatial position, spatial cognitive processes, mappings of language and other representations with the real world, formalizations, psycholinguistic approaches, and discourse analytic perspectives. Our approach focuses on a simple 2D situation shown on a computer screen, containing only a small number of similar items of two types, circles and squares. Our interest lies in how people refer to one (previously specified) object in such a scene depending on the discourse context, focusing on two crucial aspects which may affect the level of specificity in speakers' descriptions: first, the linguistic representations may be influenced by the given discourse task, such as describing *where* an object is versus specifying *which* object is intended. Second, there may be an effect of the discourse history: if speakers start out with complex configurations and – accordingly – complex descriptions, do they stick to the same level of detail, or do they rather switch to simpler descriptions in subsequent utterances? We investigated spatial descriptions in a web experiment in which these two aspects were systematically varied.

Previous research of the usage of spatial language revealed many aspects that influence the choice of spatial terms with respect to a given situation. For instance, the perspective chosen to describe a scene depends on various factors, such as the speaker's estimation of the cognitive abilities of the listener (Herrmann & Grabowski 1994), locatedness at a focal axis (Vorwerg & Sudheimer 2006), or a functional relation between two objects (Carlson 2000), and it may change in the course of a speaker's description (Tversky 1999). Aspects of the task that may influence perspective choice are the presence of interaction partners (Schober 1998) and the relative distance of the target object to the speaker vs. the addressee (Mainwaring et al. 2003), at least if the speaker chooses spatial terms denoting distance.[1] Furthermore, notions of proximity and accessibility play a role in the application of spatial terms; according to Miller & Johnson-Laird (1976), objects are conceptualized to be in a spatial relation if their *regions of interaction* overlap with each other. Pribbenow (1991) proposes a number of functional criteria for this notion, such as visual access between the two objects, and primacy of the relation between these two objects compared to any other objects present which could serve as relatum. Weiß (2005) observes that although object size influences the extension of overlapping regions, the determination of object regions depends on a complex interaction of different situation-specific factors.

In the present work, we focus on a scenario in which projective terms, such as *left of, right of, above, below*[2] are most likely to be used, because other terms are not discriminative (for instance, all objects have the same distance to each other). These terms have been investigated with respect to their semantics (e.g., Herskovits 1986, Grabowski 1998), their geometric and functional features (Eschenbach 2005, Coventry et al. 2001 and others), some extensions to non-spatial usage (e.g., Tyler & Evans 2003), their application in various reference frames (Herrmann 1990, Levinson 2003, and others), and many other aspects that will not be considered in detail here. In the following, we briefly examine the two criteria that are decisive for our research: first, the distinction between *where* and *which* tasks, and second, the influence of the discourse history.

1.1 Discourse Task

With respect to the first criterion, most psycholinguistic approaches to the applicability of projective terms have focused on a task in which speakers were asked to locate one object with respect to another; in other words, they were asked a *where* question. This kind of research has revealed that there is a graded structure underlying the applicability of projective terms. Each term denotes the relationship of an entity to another on or close to a 'focal' spatial axis; the closer the object is to the axis, the better the fit to the projective terms. For instance, the expressions *left* and *behind* can be used straightforwardly for 90° and 180° angles, respectively. However, the more the angles between the target object and the relatum depart from these focal axes, the more linguistic modifications are used for specifying the spatial relation. Simple,

---

[1] In this case, it may be advantageous to refer to the allocation of a *relatum* rather than perspective, since distance terms do not depend on a specific view direction.

[2] More specifically, in our work projective terms are those that are used to express relative location on a specific spatial axis as determined by a relatum and view direction (rather than a cardinal direction, for example). This excludes terms like *beside* (and German *neben*) because they are unspecific with respect to direction; see Tenbrink (2005c) for a thorough discussion.

unmodified spatial expressions are acceptable and applicable in a certain range; outside this range compounds or modifiers such as *left front* or *a little bit to the left* are more typical (Vorwerg & Rickheit 1999, Zimmer et al. 1998). The largest area that can be assumed (and has been proposed) for applicability of a projective term is a half-plane (e.g., Herskovits 1986:181f., Eshuis 2003). Thus, the terms are not discrete and mutually exclusive, but applicable to different degrees depending on the distance to the focal axis. These effects are captured in the notion of *spatial template* as used, for instance, by Carlson-Radvansky & Logan (1997). A comprehensive and thorough overview of previous insights as well as new experimental results in this regard is presented in Vorwerg (2001). Gradedness effects are not restricted to language, they can also be observed in non-linguistic tasks (e.g., Franklin et al. 1995, von Wolff 2001).

With respect to the other kind of discourse task addressed here, only few studies have addressed the identification of an object in a configuration of similar objects by using spatial language. Such a situation occurs whenever one object needs to be singled out via language (if other means such as gestures are not available), and the objects cannot be differentiated on the basis of their features such as colour, shape, and so on. Then, spatial terms – especially projective terms such as *left, right, behind* – are the most feasible means to achieve discrimination, since they can be used in nearly any kind of configuration (cf. Moratz & Tenbrink, subm.) According to Gorniak & Roy (2004) a frequent strategy in such a situation is "to refer to spatial extremes within groups of objects and to spatial regions in the scene" (p. 439). This is done, for example, by terms of distance such as *closest*, or by projective terms such as *in the front* or *on the left side*. As indicated by the authors' interpretation of the latter terms as "spatial extremes", such expressions, though linguistically unmodified, refer to the object that is situated at the most extreme position as compared to other objects that may also be situated in the same spatial region, e.g., within the left half of the picture. Another possibility is to refer to the extreme position using a projective superlative such as *leftmost*.

Further ideas on how reference is achieved by using spatial language can be gained from the study of object reference on a broader scale. Based on controlled psycholinguistic experimentation, Herrmann & Deutsch (1976) formulate a number of principles speakers use in order to distinguish the intended object from others present, in scenarios where the objects differ in more than one respect. Following the *principle of greatest distance*, speakers analyze the target object with respect to properties that can establish a (maximum) contrast to competing objects. Furthermore, speakers encode as many properties as needed for unambiguous object reference, but usually not more, being economic (cf. Grice's maxim of quantity, Grice 1975). But, according to the *principle of redundant verbalization*, they may encode more properties than needed if the situation is specifically complex or if several object properties are equally suitable for achieving reference. Herrmann & Deutsch's results are based on studies which explicitly exclude spatial reference. However, our own results (Tenbrink 2005b,c) reflect similar principles for application contexts in which only spatial localization is available for establishing contrast. In that work, data collected in a web-based study were analyzed qualitatively using discourse analytic methods. The results support a hypothesis posed already by Herskovits (1986:182) and also supported by Gorniak & Roy (2004) concerning the usage of unmodified spatial terms: They can suffice for a description even if the located object is not positioned on or even near to the focal axis with respect to the relatum *if the spatial context does not allow for a different interpretation*. In other words, if there is only one object situated within the left half plane of another object, the unmodified utterance *to the left* serves the purpose of identifying the target object even if it is far away from the left (90°) axis. This result leads to the conclusion that the graded structure of projective terms – as identified for *where* discourse tasks – does not apply to the same degree for their usage in *which* tasks. This idea does not contradict findings on applicability structure, but rather indicates that applicability structure does not need to be expressed linguistically. The actual application of a spatial term is a matter of discourse context, rather than of the semantic structure of the spatial terms themselves.

However, there is a caveat to this idea. It could be the case that *where* questions are context dependent in a similar way as *which* questions, depending on the current purpose. In studies designed for identifying the graded applicability structure, the task was naturally formulated in a way that triggered descriptions that were as precise as possible. For example, in Zimmer et al. (1998) participants were asked to produce utterances that enabled the listener to find a hidden object. In other cases (e.g., Vorwerg 2001) the best fitting expression was chosen, applicability was rated, or the object was positioned according to a spatial description. All of these discourse situations enhance precision to a high degree and are not directly comparable to a task in which an object simply should be identified by using spatial reference.

Furthermore, in most studies there are no further objects within the scenario which might influence the level of precision. It is likely that some differences in the results are influenced not directly by features of a scenario within each trial, but indirectly by the contrast set, i.e., by the contrast that is effected by comparing the present trial with the previous ones. This idea is pointed out, for instance, by Carlson & Logan (2001), who used acceptability ratings and a sentence-picture verification task to assess whether a distractor object influences the apprehension of spatial terms. Their results show that distractors consistently influence apprehension, notably in a way that is independent of the actual placement of the distractor. However, their study did not involve objects of the same class, nor did it involve free production of spatial language. Therefore, the interplay of the discourse task with the graded applicability structure for spatial terms in scenarios containing both same-class and different-class objects is still largely unexplored.

Our study sets out to directly address this issue by systematically comparing speakers' spontaneous linguistic choices in tasks differing only in the question posed to the participants, namely, *which* versus *where*. The scenarios contain varying numbers of objects of either similar or different object classes. The resulting spatial descriptions are analyzed with respect to their level of specificity. We started from the approach taken already in Tenbrink (2005b), but designed the study in a way that allowed both for the investigation from a discourse analytic perspective as before, and for the analysis by psycholinguistic methods involving statistical analyses. This combination was possible because the previous study already allowed for an estimation of the variability and features of speakers' spontaneous utterances, so that concrete hypotheses could be formulated.

1.2 Discourse History

If speakers produce a sequence of localizing utterances, such as in room descriptions or when answering a number of questions, often an intraindividually consistent use of reference frames can be observed (Ehrich, 1985; Ehrich & Koster, 1983; Levelt, 1982; Vorwerg, 2001). In some instances, the consistent use of a spatial perspective may be the result of resorting to a default frame of reference. However, if *intra*individual consistency is coupled with *inter*individual variability, a default frame cannot account for the effect. A possible explanation for such a pattern of results is the assumption of different "cognitive styles" (cf. Levelt, 1982). According to this account, speakers might keep their own strategies, their habitual, possibly in part genetically determined, ways to use projective terms.

However, speakers might as well just be influenced by what they encountered in the beginning. One way to test this idea may be to include different starting conditions that lead to interindividually varying patterns. Results on self-alignment in successive localizing utterances point to the conclusion that speakers may stick to one referential strategy, triggered by the beginnings of the discourse. In one study, the probability of a particular frame of reference to be chosen was a function of its initial choice within the sequence which in turn was influenced by a parameter of the configuration (Vorwerg & Sudheimer 2006). In another study, participants starting with a prepositional expression (preposition or prepositional adverb) tended to use it throughout the sequence, whereas participants starting with an adverbial description most often stuck to that (Vorwerg 2005). These effects can be interpreted as resulting from self-alignment with earlier utterances due to a pre-activation of cognitive representations by these self-produced utterances, similar to Pickering and Garrod's (2004) proposal for interactive alignment processes in dialogue. According to Pickering and Garrod the processes employed in language production and in language understanding draw basically upon the same representations. Therefore similar pre-activation processes might be expected for interactive alignment and self-alignment.

The question addressed here is whether the complexity of projective expressions in terms of compound or modified specifications of direction may be subject to processes of self-alignment. Specifically, we set out to study the effect of different starting conditions in terms of configuration complexity on the complexity of utterances at later points in time. We hypothesized that the level of detail chosen for spatial descriptions – although it depends very much on the type of spatial configuration – may be one of those linguistic choices that are kept relatively stable throughout a discourse, because it is a kind of conceptual choice.

Furthermore, provided that the starting condition is related to the complexity of later descriptions, another question is whether this effect interacts with the discourse task (as described in the previous section). For example, it is conceivable that the complexity of a projective term is more dependent on context for a "which" as opposed to a "where" question. In may also be the case that the effect of self-alignment processes (strength of pre-activation) interferes with the impact of the discourse question.

To investigate the influence of earlier utterances on later ones, the complexity of the initially presented spatial configurations was manipulated between participants answering *which* or *where* questions.

2 Method

2.1 Participants

As in a previous study (Tenbrink 2005b), we recruited participants for our web study by advertising in mailing lists and spreading the information in other ways. Participation was entirely voluntary and not rewarded. We automatically eliminated all results of participants who did not complete the experiment. The remaining results were consistent with the task formulation to such a high degree that we can conclude that the participants understood what they were required to do, in spite of the obvious disadvantage of a web study that clarification questions cannot be asked. Furthermore, the range of variability in the results overwhelmingly concerns aspects that can be directly related to the discourse task itself, so that the lack of control exhibited by web studies (cf. Reips 2002) also does not seem to diminish the results substantially. Among the aspects that are difficult to control in this kind of study are truthfulness of information about age, gender, native tongue, and continuity of participation; however, these factors do not seem to be crucial to our investigation at present.

Altogether, the results of 312 participants were stored and analyzed. 136 of these were English native speakers, and 176 were German native speakers. Their age ranged from 14 to 73 (*median* = 34, *interquartile range* = 14). Fifty-five percent of the participants were female. The overall number of utterances collected in this experiment is 7488; namely, 3264 English and 4224 German.

Table 1. Number of participants assigned to each experimental condition

Initial items	English		German	
	Where	Which	Where	Which
Simple	35	33	43	43
Complex	34	34	42	48

2.2 Design

The experimental design was a 2 (*discourse question:* "where" vs. "which") x 2 (*starting condition:* simple vs. complex start configurations) x 2 (*language:* English vs. German) between-subjects design. The subjects were assigned randomly to the conditions in their native language (see Table 1). Additionally, *section* (first vs. second vs. third part of the utterance sequence) was introduced as a within-subject factor.

2.3 Procedure

The first question that each participant received when opening the experimental web page was, "Are you a native speaker of English?" (for English, mutatis mutandis for German). Participants answering "No" immediately received a message that they could not participate. Although this method does not entirely guarantee the participation of native speakers only, it does preclude visitors from participating inadvertently because they were not sufficiently aware that they should be native speakers. Participants answering "Yes" received the following instruction:

"We are interested in how people talk about spatial arrangements. In the following, we present pictures of possible arrangements along with a question that we would like you to answer. There are no wrong or right answers, but please try to avoid using numbers like "first", "second" and so on.

We would like to ask you to work through all the tasks one after the other - without using the 'back' button of your browser - and press the 'submit' button after answering the last question. Otherwise, your data might be lost. Altogether, there will be 24 easy tasks for you."

They were then asked whether they had participated in a web study on spatial language before. After that, they immediately started with the first of 24 pictures. In the "Which" group, the question they read along with each picture was, "Which element

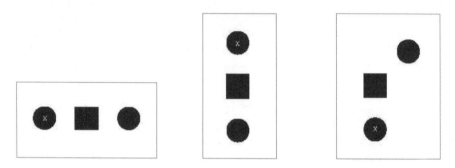

Fig. 1. Example start pictures in the starting-simple condition

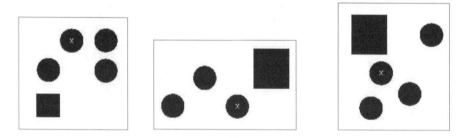

Fig. 2. Example start pictures in the starting-complex condition

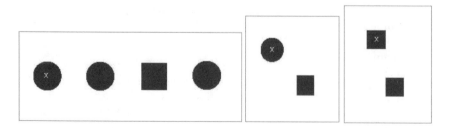

Fig. 3. Example pictures in positions 7–24

of the picture is marked by an X?", whereas in the "Where" group, the question was always, "Where is the element that is marked by an X?" The participants could type their answers into an input field. The first six pictures shown always appeared in the same order, whereas pictures no. 7–24 were distributed randomly. In the starting-simple condition, the first six pictures were specifically simple, such as the ones shown in Fig. 1. In the starting-complex condition, the first six pictures were complex (see Fig. 2). The difference concerns number of items as well as position of the marked object with respect to the focal axis. A pilot study had shown that speakers reliably produce more complex descriptions with the complex pictures, as intended. Our interest now lies in whether these initial complex descriptions would affect later utterances. After the initial pictures, the participants received 18 further pictures that differed with respect to the number of items, the presence of an object of a different class (a square in two different sizes), and positions in relation to the focal axes of the other objects; examples are shown in Fig. 3. Four of these 18 pictures contained only two objects of different classes (a circle and a square), which enables speakers to achieve discrimination by simple mention of the class name. This kind of configuration was included because it resembles the configurations used in earlier studies to elicit answers to "Where" questions to a higher degree than the more complex configurations do. Generally, the configurations of the pictures were specifically designed to test a range of further aspects which will not be in focus in the present analysis. Here, we address the main effects of conditions rather than configurations, and therefore abstract from most of the differences (apart from the same/different class distinction) between the pictures that appeared – in a random order – after the first six pictures.

After finishing the 24 descriptions, participants were then asked to fill in a questionnaire asking about age, gender, and other aspects that could potentially be of interest. Finally, they could ask to be informed about the results of the study.

2.4 Analysis

The simplest and at the same time informative way to talk about a spatial configuration in visual space may be the use of a projective term (such as *on the right*). This may be particularly true when only one possible spatial perspective is involved. More complex spatial utterances arise by adding either a second projective term (such as *above and left of the square*), a modifier used for precisification or qualification (such as *the circle directly above the square* or *slightly to the left of the square*), or both (such as *above and slightly to the left of the square*). Such linguistic hedges may be used by speakers to communicate more precisely the degree of applicability of a projective term or the degree of deviation from the focal axis (Franklin, Henkel, & Zangas, 1995; Vorwerg & Rickheit, 1999). On the other hand, a speaker may as well produce an utterance without any projective term (such as *between the two circles* or *in the middle of the arrangement*).

Each utterance was coded with respect to the following criteria:

- **Occurrence of a projective term:** We distinguish between utterances not containing a projective term at all, because participants used other solutions to the task, and utterances containing projective terms.
- **Second projective term:** For those utterances containing a projective term, it was coded whether a second project term was present or not. The frequently occurring

adverb "schräg"/"diagonal(ly)" in both German and English is semantically in between a linguistic modifier and the combination of two projective terms, in that it syntactically modifies a spatial description but semantically contributes the notion also conveyed by a combination, namely, the spatial position of the object between two focal axes. Here, we analyzed this item as additional projective term even if no second spatial direction was explicitly mentioned.
- **Precisification:** For those utterances containing a projective term, it was coded whether a modifier expressing a qualification or intensification was present or not. This variable is defined in terms of the linguistic ways of "precisifying" the spatial position, i.e., making the position more precise compared to the vague meaning of a projective term encompassing a whole category of possible spatial directions.

These criteria were motivated directly from the previous discourse-analytic results gained from the study reported in Tenbrink (2005b,c), in which it became obvious that the presence of projective terms is an indicator of the overall strategy taken, the combination of two projective terms often indicates that two spatial dimensions are equally suitable for the description, whereas the presence of linguistic modifiers is a good indicator of the level of detail chosen in the utterance. In the analysis below, we take the occurrence of exactly one projective term as the expected default case (since, as discussed above, it is a simple and flexible way of indicating an object's position) and investigate deviations from this option.

Altogether, 88.8 % of utterances contained a projective expression, and about one third of utterances (34.3 %) contained only one projective expression with neither modifier nor compound direction.

We subdivided the whole series of utterances produced by each participant into four sections comprising six answers each. To investigate which factors contribute to the tendency to produce non-projective expressions instead of projective terms, or to produce more complex projective utterances in terms of either using compound directions (two projective terms) or modifiers (precisifications), we then computed the number of utterances containing (1) no projective expression, or (2) an additional projective term, or (3) a precisification for each section, and analyzed the results with respect to the three between-subjects factors *discourse question* ("where" vs. "which"), *starting condition* (simple vs. complex start configurations), and *language* (English vs. German).

The first quarter of the series corresponded to the six start configurations. This section is analyzed separately because participants received different configurations according to starting condition (simple vs. complex). The remaining 18 configurations were identical across conditions and presented to each participant in random order. These were also subdivided into sections of six utterances each, and provided the data for the main analysis. The subdivision of the main part of the discourse was made in order to provide for direct comparability with the start items, and –more importantly– to study whether the three dependent variables and their dependence on initial items changed over time.

2.5 Immediate Effects of Start Configurations

As a basis for determining effects of discourse history (in the main analysis, see Results section), the complexity of the six initial utterances depending on the *starting condition* (simple vs. complex configurations) was analyzed. Only if the manipulated complexity of these first six spatial arrangements affects the complexity of the associated utterances, can there be an effect of early discourse history on subsequent utterances. Additionally, *discourse question* and *language* were included as factors in order to ascertain whether there are interaction effects and to provide a basis for close comparison with the main results.

We subjected the data on (1) number of utterances without any projective term, (2) number of utterances containing a second projective, and (3) number of utterances containing a precisification to an ANOVA each, with *starting condition*, *discourse question*, and *language* as factors.

Univariate ANOVAs showed significant effects of *starting condition* (complexity of configurations) on all dependent measures: number of non-projective utterances [$F(1,304)=116.3$, $p<.001$, $\eta_p^2=.277$), number of utterances containing a second projective term [$F(1,304)=351.0$, $p<.001$, $\eta_p^2=.536$], and the number of utterances containing a precisification [$F(1,304)=139.5$, $p<.001$, $\eta_p^2=.315$]. Means for all three measures can be seen in Table 2. The ANOVAs indicated no other significant main effects or interactions (all $Fs \leq 1.4$). A significance level of .05 has been adopted throughout the paper.

Table 2. Mean number of utterances per speaker containing (1) no projective term, (2) a second projective term, (3) a precisifying modifier. (SD are given in brackets. A * indicates significant differences.) These start configurations form the starting condition for subsequent items.

	Start configurations	
	Simple	**Complex**
'No projective term'*	0.06 (0.07)	1.12 (0.07)
'Second projective term'*	0.63 (0.09)	2.92 (0.09)
'Precisification'*	0.69 (0.11)	2.57 (0.11)

The data show that all three measures differ between both conditions, with participants subjected to the starting-complex condition producing more alternative utterance without a projective term, more utterances with projective-term combinations, as well as more utterances containing precisifying modifiers. As indicated by the partial-eta-squared values, the effect size is largest for the second-projective-term variable.

Whether this initial difference in the complexity of utterances for complex versus simple configurations affected later utterances – although later configurations did not differ between both conditions – was determined in the following analyses. All following analyses concern positions 7–24.

2.6 Results

All utterances produced after the first six initial utterances (items 7-24) were in the main focus of our analysis. These 18 configurations were identical for all participants, with the order of presentation being randomized.

Mixed-model analyses of variance were performed on the three between-subject factors (*starting condition*, *discourse question*, and *language*) and the within-subject factor (*section of discourse*) for each of the dependent variables (unstructured covariance model for each of them).

For the *number of alternative utterances without projective terms*, the ANOVA revealed significant main effects of *discourse question* [$F(1,304.0)=66.5$, $p<.001$] and *starting condition* [$F(1,304.0)=6.3$, $p=.013$]. Participants produced a higher number of "non-projective" utterances in the starting-complex condition and answering a "which" question. There was no significant effect of *language* [$F=2.3$]. No significant interactions were observed [all $Fs \leq 1.5$]. There was also no main effect of *section of discourse* [$F<1$]. Estimated means for the main factors are given in Table 3.

For the *number of utterances containing a second projective term*, again significant main effects of *discourse question* [$F(1,304.0)=139.2$, $p<.001$] and *starting condition* [$F(1,304.0)=9.5$, $p=.002$] were obtained. These main effects reveal that participants produced significantly more compound projective terms in the starting-simple condition and answering a "where" question (see Table 3). These main effects are qualified

Table 3. Mean number of utterances per speaker containing (1) no projective term, (2) a second projective term, (3) a precisifying modifier. (SD are given in brackets. A * indicates significant differences).

	Starting Condition	
	Simple	**Complex**
'No projective term'*	0.60 (0.06)	0.81 (0.06)
'Second projective term'*	2.50 (0.08)	2.14 (0.08)
'Precisification'	1.56 (0.09)	1.42 (0.09)

	Discourse question	
	Where	**Which**
'No projective term'*	0.36 (0.06)	1.05 (0.06)
'Second projective term'*	3.02 (0.08)	1.62 (0.08)
'Precisification'*	1.69 (0.09)	1.29 (0.09)

	Language	
	English	**German**
'No projective term'	0.77 (0.06)	0.64 (0.05)
'Second projective term'	2.37 (0.09)	2.27 (0.08)
'Precisification'*	1.77 (0.09)	1.21 (0.08)

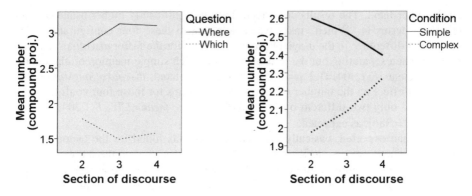

Fig. 4. Significant interactions for Number of compound projective terms: *Starting condition* x *section of discourse* and *discourse question* x *section of discourse*

by significant interactions with *section of discourse* for both *discourse question* [$F(2,304.0)=6.7$, $p=.001$] and *starting condition* [$F(2,304.0)=3.1$, $p<.05$]. As Fig. 4 shows, the difference between those participants having started simple and those having started complex gets smaller over time, whereas the difference between those speakers answering "which" questions and those answering "where" questions gets larger from the second to the third section of discourse. (The first section comprised the six start items.) There were no main effects for *language* or *section of discourse* [$Fs<1$]. No other significant interactions occurred; there was only a non-significant tendency for a larger difference between "where" and "which" in English than in German [$F(1,304.0)=3.0$, $p=.08$; all other $Fs\leq1.7$].

For the *number of utterances containing a precisification*, the ANOVA revealed significant main effects of *language* [$F(1,304.0)=19.2$, $p<.001$] and *discourse question* [$F(1,304.0)=10.2$, $p=.002$]. Participants produced more utterances with precisifications of projective terms in English than in German and when answering a "where" question than when answering a "which" question (see Table 3). There were no main effects of *starting condition* (simple vs. complex) or *section of discourse* ($Fs\leq1.5$). Neither were there any significant interactions; there was only a non-significant tendency for a slightly stronger decline in the number of precisifications from the second to the third section of discourse in English than in German [$F(2,304.0)=2.9$, $p=.055$; all other $Fs\leq2.3$].

In order to explore further whether the significant difference in the employment of projective terms with respect to *starting condition* and *discourse question* depended on the type of configuration, we carried out a further analysis with respect to a subgroup of the configurations, namely, those that only consisted of two objects of a different class (4 out of 18 configurations). We hypothesized that, in a "Which" task, discriminative object reference could be achieved simply by mentioning the class name in these cases. Indeed, a closer look at the data revealed that, in these four situations, almost all utterances (one exception) in the "Which" condition that did not contain a projective term did not contain a spatial description at all; only the class name was mentioned. Therefore, we subjected the data to separate univariate ANOVAs for this subgroup of configurations (4) on the one hand and all other configurations (14) on the other hand. The results confirmed that the significantly higher usage of non-projective terms in "Which" tasks was solely due to these four configurations. No significant difference in the usage of projective terms for the factor *discourse question* remains when separating out the four pictures in which simple mention of class name was sufficient [$F(1,304)=1.2$, $p=.265$]. On the other hand, *discourse question* had a significant effect on the number of non-projective terms for those four configurations made up of only two different objects [*mean*=0.56 vs. *mean*=2.78; $F(1,304)=191.0$, $p<.001$, $\eta_p^2=.386$], as expected.

Other than expected, basically the same pattern was found for the factor *starting condition*: There was a significant effect of *starting condition* on the number of non-projective terms for those four configurations made up of only two differently-shaped objects [*mean*=1.42 vs. *mean*=1.91<; $F(1,304)=9.4$, $p=.002$, $\eta_p^2=.03$], but not for the other 14 configurations [$F(1,304)=0.7$, $p=.391$]. However, the effect of *discourse question* was much larger than the effect of *starting condition*. No other main effects or interactions proved significant in either of these analyses [$Fs \leq 2.0$]. This pattern of results also was confirmed by nonparametric Mann-Whitney U tests.

As there was no significant main effect of *starting condition* on the number of precisifications, we wondered whether the total numbers of precisifications produced by a participant in the main part of the experiment (items 7-24) may be related to their corresponding numbers associated with the start configurations (items 1-6). This was indeed confirmed by rank correlation tests for each subgroup of "*discourse question* x *starting condition*" combination: "where/simple" ($\rho=.48$, $p<.001$), "where/complex" ($\rho=.57$, $p<.001$), "which/simple" ($\rho=.26$, $p=.01$), "which/complex" ($\rho=.43$, $p<.001$). Thus, although we observed no significant effect of *starting condition*, individual participants tended to re-use their own individual strategies with respect to the usage of modifiers.

By contrast, this is not the case for the usage of compound projectives. As stated above we found an overall trend towards using more compound terms in reaction to the starting-simple condition. But none of the four "*discourse question* x *starting condition*" subgroups showed a significant rank correlation between the number of compound projectives in the initial and in the main part of the experiment; although the "where/complex" subgroup ($\rho=.19$) only narrowly failed to reach significance ($p<.055$; all other $\rho s \leq .11$). These results lead to the conclusion that the negative correlation found between initial number of compound projectives and projectives in the main part is based completely on relating subjects from different subgroups.

This finding as well as the lack of a starting-condition effect for number of precisifications despite existing individual correlations lead to the hypothesis that interrelations between the three different measures obtained might be responsible for the partially missing consistency effects. If the probability of a participant to modify a projective term or to combine it with another one is not independent of each other or of the frequency of nonprojective utterances, some dependencies on initial language usage may not show up in the data. This is indeed what we find: The tendencies to use a projective term with a precisification or with a second projective term are positively correlated with each other ($r=.31$, $p<.001$). However, the probability to combine a projective term with another one is negatively associated with the tendency to produce utterances without any projective terms ($r=-.55$, $p<.001$). Importantly, the same pattern of results has been observed within all four *condition* x *question* subgroups.

Furthermore, speakers who started out using the term *schräg* (in German) or *diagonal* (in English) in the starting-complex condition showed a significant tendency to do so in later utterances as well (for *where*: $r=.70$, $p<.001$; for *which*: $r=.43$, $p<.001$). In the starting-simple condition this term does not occur often enough to allow for correlation tests. These results confirm that there are different patterns of language use, whose parameters interact with each other.

3 Discussion

The results of our experiment clearly show effects with respect to both criteria addressed, namely, discourse task and discourse history. These effects are consistent in both languages under analysis, English and German, with only a few language-specific differences which are not analyzed here because no specific predictions were made (but see Tenbrink 2005a for a detailed qualitative analysis of linguistic options and their usage in the two languages). Most importantly in this context, the results concerning discourse history and discourse task generalize to both languages.

With respect to the discourse task, speakers attend very much to whether they are asked to describe a spatial relation (by a *where* question) or whether the recipient should be enabled to identify an object out of several competing ones (by a *which* question). For the latter task, it is sufficient to point to an obvious contrast between the target object and the other objects present. In our scenarios, this may happen in one of two ways. First, speakers may use only the class name if it is discriminative. This is the case in the four (of 18) situations in which the competing object belonged to a different class, and many speakers did indeed make use of this option, leading to the significant difference that we found between the *which* and *where* questions concerning the occurrence of projective terms. We did not expect simple mention of class name to be an option in *where* questions; however, a few speakers actually did use this kind of simple answer, especially in relation to the four configurations in which only two different objects were shown. It seems that these configurations lend themselves more easily to descriptions referring to just the shape (or another identifying feature). It may sometimes be appropriate for a speaker to answer just "It's the circle", when answering the question "Where is the element that is marked by an X?", although this is much less frequent than with "Which" questions. Whether one does so or not, however, is influenced by earlier language choices, as our results show.

Second, speakers may use a spatial description that is suitable for discriminating the target object from the competing candidate(s). This is a convenient option whenever several identical objects are present, which was the case in most of the configurations we used (14 of 18). Thus, although a *which* question is not inherently spatial, a spatial utterance could be expected in these cases as a suitable reaction. However, for the purpose of disambiguation the spatial description does not require much detail. This fact is reflected clearly in our data by the fact that precisifications and compound projective terms were used significantly less frequently when a *which* question was asked than when a *where* question was asked. Furthermore, data suggest that the difference between both discourse tasks grows with a decrease of starting-condition influence (cf. Fig. 4).

Taken together, therefore, with respect to referential identification (*which*) tasks, the results of the present experiment statistically confirm the qualitative findings in Tenbrink (2005b,c), and they also support the general principles proposed by Herskovits (1986) and Herrmann & Deutsch (1976). Their findings suggest that suitable dimensions are chosen for discrimination of objects in a scenario, typically without providing more detailed information than is needed for establishing a contrast to the competing objects.

In principle, spatial descriptions in answer to a *where* task do not have to be particularly precise either. There is no indication in the literature on the gradedness structure of projective terms that a deviation from the focal axis must necessarily be marked linguistically. However, as our results show, speakers systematically provide more details in their description if the spatial relationship itself is in focus. We can conclude from this direct comparison that the discourse task matters for speakers' linguistic choices even if other context factors remain constant, and that different principles for communication and spatial description are at work if the spatial description is itself in focus, as opposed to being a vehicle for the discourse task of discriminative reference. The results on the application of spatial expressions gathered in psycholinguistic experimentation can therefore not straightforwardly be generalized to other kinds of discourse task.

These results may provide an explanation for a discrepancy between Herskovits' "shifting contrast principle" (Herskovits 1986:81) and the results of Carlson & Logan (2001). Herskovits' principle is formulated as follows.

> "If two objects, A and B, are placed in a relation to a reference object in such a way that the ideal meaning of a preposition (...) is truer of A than of B, then one can use that preposition to discriminate A from B so that the locative phrase will be assumed true of A but not of B."

This claim seems to imply that the location of an object that is placed somewhat away from the focal axis cannot be described by using the corresponding projective term if there is another object in the scenario that is closer to the focal axis. Carlson & Logan (2001) set out to test whether a distractor object indeed influences apprehension of spatial terms in this way, and found that – contrary to Herskovits' suggestions – "performance did not vary depending on whether the distractor instantiated a better example of the spatial term than an equally good example" (Carlson & Logan 2001:889). However, their discourse task focused on the acceptability or verification of a description for a spatial relation, i.e., it resembled a *where* task without involving free production of language. Herskovits' claim becomes more sensible if applied for *which* tasks: If a spatial term is applied to identify an object out of several similar ones, it does not seem to be a good solution to use a spatial term to refer to an object that is farther away from the relevant focal axis than a competing object (especially, an object of the same kind) is. Since Herskovits uses the formulation "to discriminate A from B", this is probably the kind of scenario she had in mind. Her example in Herskovits (1986:183; Fig. 10.16) supports this interpretation. While our analysis presented here only points to a general difference between *where* and *which* task as regards precision, our earlier results reported in Tenbrink (2005b) indicate that competing objects do indeed influence speakers' usage of modifications together with projective terms in *which* tasks. We will investigate these effects in our present data more closely in future analyses. In this way, the influence of further objects in the scenario on speakers' free production of spatial language can be investigated directly for both kinds of tasks. Possibly, not only the formulation of the discourse task itself, but also participants' flexibility in choosing their own linguistic strategies influences the ways in which context objects come into play.

With respect to our second major research issue, discourse history, our results suggest that speakers are influenced by their own earlier contributions in intricate ways. On the one hand, the group data show a contrast effect: Generally, speakers react to the contrast between simpler to more complex, or complex to simpler configurations by over-adjusting their spatial descriptions, i.e., using more compound projective terms if later scenarios are more complex than the initial ones, and using fewer compound terms if (the same) later scenarios are less complex than the initial ones. Possibly, the same configurations are *perceived* as more or less complex, given the impression of the first six configurations, and are therefore described accordingly. This effect gradually diminishes throughout the subsequent discourse. On the other hand, there was no such effect for the other two variables. There was a higher frequency of utterances without projective terms in the starting-complex condition that was consistent between initial items and the subsequent sections, whereas the frequency of utterances containing a precisifying modifier was not significantly related to the starting condition at all.

To account for these results, individual strategies need to be examined. Participants in the starting-complex condition often used creative solutions to the task because of the complexity of the initial configurations, and then stuck to this strategy. Example answers for this creativity are the following (in reference to the first picture shown in Fig. 2, which was the very first picture that these participants saw):

"It is in the center row standing alone."

"NNE of the square element and about 0.5 of the extent of the region from the square"

Actually, the participant producing the latter utterance consistently stuck to this strategy of using cardinal directions (typically together with quantitative measures) throughout the whole study. Other participants chose to simply mention the class name, and then re-used this strategy (as discussed above). These examples illustrate our statistical results, which confirm that, with respect to the usage of projective terms, speakers do indeed re-use their own initial strategies. This effect could be shown to be influenced by the starting condition.

With respect to precisifying modifiers and combined terms, the results do not confirm an influence of the starting condition in the way we expected. However, the analysis of individual correlations nevertheless points to a consistency in speakers' strategies at least for precisifications: Individuals who started out using modifiers tended to do so throughout the whole discourse. This result may or may not be due to a habitual way of using (spatial) language. The correlation data on compound terms, on the other hand, neither show a contrast effect as with the global group data, nor do they show unequivocally that speakers are intra-individually consistent in their usage of combined projective terms. The observed contrast effect for the group can therefore not be accounted for by individual language usage. However, an interaction between different parameters of language use has been observed, which shows, for example, that the tendency to use compound projectives is negatively related to the tendency to produce utterances without any projectives, but positively related to the probability to use a projective term with a modifier. That is, some people may use compounds and modifiers interchangeably as ways of expressing complexity in their spatial descriptions, while others choose not to use projective terms at all, and are then rather consistent in this strategy.

In our data, therefore, the choice of using projective terms at all as opposed to a different strategy seems to be the only way in which speakers can be said to "self-align" in the sense that we predicted. Obviously, other factors also come into play in addition to self-alignment, resulting in the observed general contrast effect. Furthermore, the detected general individual consistency with respect to precisifying modifiers may or may not be explained in terms of self-alignment.

It can be concluded from these results that conceptual choices with regard to using spatial terms may influence subsequent utterances within a discourse, presumably based on a pre-activation of conceptual representations. However, further factors contribute to speakers' actual choices in the employment of projective terms, resulting in a complex interplay of factors related to the task, the discourse history, and the scenarios presented in an experimental trial sequence. This result certainly calls for further detailed and focused experimentation.

4 Conclusion and Outlook

We have presented the results of a web experiment testing whether speakers attend to the discourse task by providing more or less detailed spatial descriptions as required in each case, and whether speakers are influenced in later linguistic choices depending on the first configurations they were presented with. Both hypotheses were confirmed at least in part, showing a) that the iterative procedure of first analyzing free-production data qualitatively and then turning to more controlled quantitative analysis leads to reliable results, b) that psycholinguistic results may be confined to the exact discourse task suggested by the experimental design, and c) that the level of detail in a spatial description is influenced systematically by a variety of discourse factors such as the current task and the discourse history.

As a next step, the data will be analyzed with respect to the impact of the precise configurations with respect to which the spatial descriptions were produced. The analysis of our initial "simple" versus "complex" configurations already shows a consistent effect of the complexity of the spatial situation itself; further details will be derived from a close analysis by configuration. Here, as well, the results of the earlier qualitative discourse-analytic study will serve as a starting point for the examination of hypotheses as to speakers' spontaneous linguistic choices.

Furthermore, as our results point to the importance of various discourse factors, future work should be geared towards the integration of results gained in the area of discourse analysis with the specific findings collected in psycholinguistic laboratories. In this way, combined approaches should increasingly highlight generalizable discourse factors that influence speakers' spontaneous choices in settings that are decidedly more natural than that which can be provided by a simplistic web study. Also, a more natural discourse context would provide an opportunity to examine the relationship between situational, discourse-related factors and the effects of functional relationships between objects.

References

Carlson, L.A.: Object use and object location: The effect of function on spatial relations. In: van der Zee, E., Nikanne, U. (eds.) Cognitive interfaces: Constraints on linking cognitive information, pp. 94–115. Oxford University Press, Oxford (2000)

Carlson, L., Logan, G.D.: Using spatial terms to select an object. Memory & Cognition 29, 883–892 (2001)

Carlson-Radvansky, L.A., Logan, G.D.: The influence of reference frame selection on spatial template construction. Journal of Memory and Language 37, 411–437 (1997)

Coventry, K.R., Prat-Sala, M., Richards, L.V.: The interplay between geometry and function in the comprehension of 'over', 'under', 'above' and 'below'. Journal of Memory and Language 44, 376–398 (2001)

Eschenbach, C.: Contextual, functional, and geometric components in the semantics of projective terms. In: Carlson, L., van der Zee, E. (eds.) Functional features in language and space: Insights from perception, categorization, and development, pp. 71–91. Oxford University Press, Oxford (2005)

Ehrich, V.: Zur Linguistik und Psycholinguistik der sekundären Raumdeixis [Linguistics and psycholinguistics of secondary spatial deixis]. In: Schweizer, H. (ed.) Sprache und Raum. Psychologische und linguistische Aspekte der Aneignung und Verarbeitung von Räumlichkeit. Ein Arbeitsbuch für das Lehren von Forschung. Stuttgart: Metzler (1985)

Ehrich, V., Koster, C.: Discourse organization and sentence form: The structure of room descriptions in Dutch. Discourse Processes 6, 169–195 (1983)

Eshuis, R.: Memory for locations relative to objects: Axes and the categorization of regions. In: van der Zee, E., Slack, J. (eds.) Representing direction in language and space, pp. 226–254. Oxford University Press, Oxford (2003)

Franklin, N., Henkel, L.A., Zangas, T.: Parsing surrounding space into regions. Memory and Cognition 23, 397–407 (1995)

Gorniak, P., Roy, D.: Grounded semantic composition for visual scenes. Journal of Artificial Intelligence Research 21, 429–470 (2004)

Grabowski, J.: Ein psychologisch-anthropologisches Modell der einheitlichen semantischen Beschreibung dimensionaler Präpositionen [A psychological-anthropological model for the unitary semantic description of dimensional prepositions]. In: Ludewig, P., Geurts, B. (eds.) Lexikalische Semantik aus kognitiver Sicht – Perspektiven im Spannungsfeld linguistischer und psychologischer Modellierungen, Tübingen, Narr, pp. 11–40 (1998)

Grice, H.P.: Logic and conversation. In: Cole, Peter, Morgan, Jerry (eds.) Syntax and semantics, New York, San Francisco, London (1975)

Herrmann, T.: Vor, hinter, rechts und links: das 6H-Modell. Psychologische Studien zum sprachlichen Lokalisieren [In-front, behind, right and left: the 6M model]. Zeitschrift für Literaturwissenschaft und Linguistik 78, 117–140 (1990)

Herrmann, T., Deutsch, W.: Psychologie der Objektbenennung [Psychology of object naming]. Bern, Hans Huber (1976)

Herrmann, T., Grabowski, J.: Sprechen: Psychologie der Sprachproduktion [Speaking: Psychology of speech production]. Spektrum, Heidelberg (1994)

Herskovits, A.: Language and spatial cognition. Cambridge University Press, Cambridge (1986)

Levelt, W.J.M.: Cognitive styles in the use of spatial direction terms. In: Jarvella, R.J., Klein, W. (eds.) Speech, place, and action, pp. 251–268. Wiley, Chichester (1982)

Levinson, S.C.: Space in language and cognition. Cambridge University Press, Cambridge (2003)

Mainwaring, S.D., Tversky, B., Ohgishi, M., Schiano, D.J.: Descriptions of simple spatial scenes in English and Japanese. Spatial Cognition and Computation 3(1), 3–42 (2003)

Miller, G.A., Johnson-Laird, P.N.: Language and perception. Cambridge University Press, Cambridge (1976)

Moratz, R., Tenbrink, T. (subm). Affordance-based human-robot interaction. In: Moratz, R. (ed.) Proc. Dagstuhl Seminar 06231: Towards Affordance-Based Robot Control. LNCS (LNAI), Springer, Heidelberg (2006)

Pribbenow, S.: Zur Verarbeitung von Lokalisierungsausdrücken in einem hybriden System [On the processing of localizing expressions in a hybrid system]. Dissertation, Fachbereich Informatik der Universität Hamburg (1991)

Reips, U.-D.: Theory and techniques of Web experimenting. In: Batinic, B., Reips, U.-D., Bosnjak, M. (eds.) Online Social Sciences, Hogrefe & Huber, Seattle (2002)

Schober, M.F.: How addressees affect spatial perspective choice in dialogue. In: Olivier, P.L., Gapp, K.-P. (eds.) Representation and processing of spatial expressions, pp. 231–245. Lawrence Erlbaum, Mahwah, NJ (1998)

Tenbrink, T.: Identifying objects in English and German: A contrastive linguistic analysis of spatial reference. In: Proc. Workshop on Spatial Language and Dialogue (5th Workshop on Language and Space), October 23-25, 2005, Delmenhorst, Germany (2005a)

Tenbrink, T.: Identifying objects on the basis of spatial contrast: An empirical study. In: Freksa, C., Knauff, M., Krieg-Brückner, B., Nebel, B., Barkowsky, T. (eds.) Spatial cognition IV: Reasoning, action, interaction, pp. 124–146. Springer, Berlin (2005b)

Tenbrink, T.: Localising objects and events: Discoursal applicability conditions for spatiotemporal expressions in English and German. Dissertation, University of Bremen (2005c)

Tversky, B.: Spatial perspective in descriptions. In: Bloom, P., Peterson, M.A., Nadel, L., Garrett, M.F. (eds.) Language and space, pp. 109–169. MIT Press, Cambridge, MA (1999)

Tyler, A., Evans, V.: The semantics of English prepositions: Spatial sciences, embodied meaning, and cognition. Cambridge University Press, Cambridge (2003)

Vorwerg, C.: Raumrelationen in Wahrnehmung und Sprache: Kategorisierungsprozesse bei der Benennung visueller Richtungsrelationen [Spatial relations in perception and language: Categorization processes in naming visual directional relations]. Wiesbaden, DUV (2001)

Vorwerg, C.: Consistency in successive spatial utterances. In: Proc. Workshop on Spatial Language and Dialogue (5th Workshop on Language and Speech), October 23-25, 2005, Delmenhorst, Germany (2005)

Vorwerg, C., Rickheit, G.: Richtungsausdrücke und Heckenbildung beim sprachlichen Lokalisieren von Objekten im visuellen Raum [Direction terms and hedges in the verbal localization of objects in visual space]. Linguistische Berichte 178, 152–204 (1999)

Vorwerg, C., Sudheimer, J.: Saliency impact on initial frame-of-reference selection, and consistency in verbal localization. Manuscript submitted for publication (2006)

von Wolff, A.: Transformation und Inspektion mentaler Umraumrepräsentationen: Modell und Empirie. GeoInfo Series, Vienna (2001)

Weiß, P.: Raumrelation und Objekt-Regionen. Psycholinguistische Überlegungen zur Bildung lokalisationsspezifischer Teilräume [Spatial relations and object regions: Psycholinguistic considerations on the formation of localization-specific partial spaces]. Wiesbaden, DUV (2005)

Zimmer, H.D., Speiser, H.R., Baus, J., Blocher, A., Stopp, E.: The use of locative expressions in dependence of the spatial relation between target and reference object in two-dimensional layouts. In: Freksa, C., Habel, C., Wender, K.F. (eds.) Spatial cognition. An interdisciplinary approach to representing and processing spatial knowledge, pp. 223–240. Springer, Berlin (1998)

Autobahn People: Distance Estimations Between German Cities Biased by Social Factors and the Autobahn

Claus-Christian Carbon

University of Vienna, Faculty of Psychology,
Department of Psychological Basic Research,
Liebiggasse 5, A-1010 Vienna, Austria
Tel.: +43-1-4277 47921
ccc@experimental-psychology.com

Abstract. Re-analyses of a study on distance estimations between German cities [1] revealed that conclusions drawn from aggregated data sets can also be found on the individuals' data level. It could be shown that mental barriers, here the former iron curtain which has been physically absent for 15 years, while still assuming a significant role in social interactions, are powerful predictors for psychological distance estimations. Moreover, by integrating new demographical information about the participants, evidence can be found that social attitudes are often much stronger predictions for distance estimations than implicit or explicit geographical knowledge. For instance, it was revealed that the attitude towards German reunification plays a crucial role in estimating distances between cities crossing the former iron curtain: the more negative the attitude the more pronounced was the overestimation of distances. This trend was impenetrable by higher route knowledge measured implicitly by the extent of traveling experience and explicitly by ratings of geographical knowledge about Germany. Furthermore, participants appeared to base their estimations of direct (air) distances between German cities on distances resulting from their experience with the German Autobahn system.

Keywords: Mental map, cognitive map, distance estimation, distortion, social factors, representation, road, route, highway.

1 Introduction

Whenever we have to estimate direct distances between cities "as the crow flies", we are confronted with a task that we are not used to. Everyday experience on distances is mainly obtained by looking at signs, getting information from (e.g., GPS based) navigation systems or by implicitly calculating distances via the temporal length of journeys. All these procedures are based on the given road structures. An alternative way for estimating distances is given by imaging of cartographic maps, which is supposed to provide an overview of the whole geographic structure.

The present paper reinvestigates a data set provided by Carbon and Leder [1] in order to analyze whether German participants base distance estimations between German cities on map-related knowledge, precisely whether they can estimate direct air distances, or whether they use the prominent German Autobahn structure as a heuristic to solve this task. Furthermore, it was tested whether the former iron curtain, which was found to be a relevant factor in systematically overestimating distances across this historical border [1], can also be identified as a biasing factor when measured with alternative measures other than those provided by Carbon and Leder [1]. Ultimately, it was analyzed whether social factors, such as the participant's attitude towards German reunification, which strongly biases the estimation of distances crossing the former iron curtain, can be modified by traveling experience and geographical knowledge.

1.1 Distance Estimations and Distance Distortions

In the way finding and cognitive map literature, two major coding formats have been proposed for the mental representation of environmental information [2]: Egocentric information in terms of uni-directed routes with a specific starting point and destination, and allocentric information in terms of precise mental maps with distances coded in a bird's eye view [3]. Neither of the two modes of representation are commonly seen as competitive accounts for coding geographical information, but the allocentric mode is assumed to evolve from long-term practice of egocentric information. According to this developmental approach [4], mental navigation starts in the first stage with simple landmarks (highly distinct, and stable large-scale objects, such as buildings, signs, trees, etc.) that are loosely interconnected. Over time these landmarks get more and more interlinked to form a coherent travel route, which is only represented in a sequential way, thus, requiring a step by step activation procedure to enable way finding [5]. Only after extensive practice, a final step can be reached where topographical knowledge can be formed.

However, whether this final stage can ever be reached adequately remains an open question. In contrast to the predictions of representations based on accurate mental maps which have a structure similar to cartographic maps [3], a number of studies have demonstrated mild to severe deviations from such ideal mental maps. In general, Stevens and Galanter [6] showed that psychological distances and physical distances are related by a power function versus a linear relationship. However, it has only been demonstrated recently that the appropriateness of a power function depended on the specific regional structure to which the cities pertained [7]. Ekman and Bratfisch [8] showed systematic overestimations between cities that participants did not attach much emotional involvement[1]. Moreover, mental maps seem to be organized as complex hierarchical structures [9-12]. Last but not least, Carbon and Leder [1] revealed social attitudes as powerful enough to distort mental maps in a systematic way. Taking all these types of distortion factors into consideration, Tversky [13] concluded that there is no simple way to account for all of these factors, but that our mental maps are systematically distorted by a variety of influences.

1.2 The Specific Situation of Germany

For four decades, Germany was divided into two parts belonging to diametrically opposed political systems. Whereas the Federal Republic of Germany ("West

---

[1] Emotional involvement was measured by imaging something important happening in the cities and estimating their degree of emotional involvement in what might happen.

Germany") was integrated in the NATO and European (Economic) Community, the German Democratic Republic ("East Germany") was a key member of the Warsaw Pact and the COMECON. Moreover, West Germany had a free market economy, whereas East Germany adhered to a socialist economy. Although both countries entertained permanent diplomatic relations with each other, in fact, both countries were strong opponents in these days. This antagonistic situation was even intensified by the construction of the Berlin Wall in 1961, which back then represented a visible sign of the political separation of both countries.

Germany has been reunified since 1990, merging these formerly sovereign countries. The reunified conglomerate consists of ten federal states in the former western part and six states in the former eastern part including one state, Berlin, which was formerly divided and pertained to both countries (West-Berlin and East-Berlin). In an experimental study, Carbon and Leder [1] investigated whether the former border between both countries still exists as some sort of "mental wall". They used a distance estimation paradigm based on the idea of Ekman and Bratfisch [8] that distances between two places with low emotional involvement will be systematically overestimated. When additionally taking the attitude towards German reunification into account, it could be demonstrated that those participants who had a negative attitude towards the reunification strongly overestimated distances across the former iron curtain. Interestingly, this was not found for people with a positive attitude towards the German reunification. After the Berlin Wall had been demolished more than 15 years ago, streets and bridges between both parts of Germany have been re-activated and today Germany does not have any physical borders within the country. Thus, this systematic distortion indicates a "mental wall" [1], which is rather resistant to political change. The authors concluded that this mental wall is a sign for the failure, at least in some respect, of the German (re-)unification.

1.3 The Aim of the Present Study

This study is based on the experimental data obtained by the study of Carbon and Leder [1]. In the original study, only *group* data, sampled over participants belonging to different levels of the social attitude towards German reunification, was analyzed. Further demographical variables, such as traveling experience or geographical knowledge about Germany were not integrated in the presented analyses. Furthermore, the authors did only test hypotheses with the psychological distance data, ignoring response[2] time data [14]. In addition, the authors only included the air distances as predictors, but did not test whether the participants based their estimations on relevant distances found in reality, such as kilometers on national highways (Autobahns).

The present paper aims to address all these points in order to facilitate a more comprehensive view towards distance estimation strategies in general and towards the mental wall between West and East Germany in particular.

---

[2] Throughout the paper, the term *response time* is used instead of *reaction time* according to the terminology proposed by Luce [14] in which the term reaction time is only used for experiments for which measuring time aspects are the first priority. Here, in contrast, the major variable was the estimation of distances; therefore the term *response time* seems to be more adequate within this context.

2 Study

This study is a re-evaluation and extension of the analyses of Carbon and Leder [1] based on a distance estimation paradigm between German cities. To minimize redundancy, only key facts of the original study and newly calculated data and tested hypotheses are presented in the following sections.

2.1 Method

2.1.1 Participants
Eighty-three participants (54 female; 32 raised in the former western part of Germany; mean age: 25.9 years) participated in the study as volunteers. They were all naive to the purpose of the study.

2.1.2 Materials
11 well-known German cities, including Berlin, were selected as anchor points for following distance estimations. No federal state was represented by more than one of these cities. The 11 cities included five cities that were located in the western part of Germany; five in the eastern part of Germany and the 11th city was Berlin, the capital of Germany. Given this set of cities, 25 distances across the former iron curtain (across distances) and 20 distances within one of the two former parts of Germany (within distances) can be rated.

2.1.3 Procedure
First the participants were asked to provide demographical information (attitude towards German reunification, traveling experience within Germany, geographical knowledge about Germany, etc.). The attitude towards German reunification was measured on a four-point rating scale from 1 to 4 (1 --> very negative; 2 --> negative; 3 --> neutral; 4 --> positive). Then the participants had to estimate all possible air distances ("as the crow flies") between the 11 presented German cities in a sequential way by typing the exact distance in kilometers, thus realizing a spatial resolution of 1 km. The order of all 110 distance estimations (including Berlin and both directions, e.g. Berlin-Nuremberg and Nuremberg-Berlin) was randomized via an experimental control software [15] which also registered response times from the onset of the stimulus with a resolution of 16 ms.

2.2 Results and Discussion

In the original study, only sampled group data based on distance estimations were presented and contrasted with direct air distances. Here, analyses based on individual participant's data will be provided, which are also contrasted with relevant distances measured on German Autobahns. Furthermore, additional demographical data will be considered. Additionally, response time data will complement the data analysis of psychological distances. For all following analyses, people with a "*very* negative" and a "negative" attitude towards the German reunification were collapsed into one single group later on labeled as "negative" due to the low number of only 5 members pertaining to the "very negative" group.

2.2.1 Autobahn Distances Versus Air Distances

In everyday life, we use the main routes to travel along long distances between cities. Although these routes are planned and realized as straight-lined as possible to minimize distances, political, geographical and topographical facts prevent perfect straightness. Thus, people might not base their distance estimations on direct air distances but rather on the given infrastructure. In Germany, this infrastructure mainly manifests itself in the so-called Autobahn system (the "Bundesautobahn" based on the former "Reichsautobahn"). As this system is strongly interconnected with a total length of approximately 12.000 km, its usage is toll-free and there is no speed limit, the Autobahn is a very popular infrastructure which might strongly bias people's cognitive maps. We analyzed this hypothesis by comparing air distances and distances on Autobahns with psychological distances. In order to increase the clarity of data (a) psychological data were related to both types of physical distances. The closer the index to the ideal ratio of 1.0 (i.e., psychological distance = physical distance), the more probable the estimation is based on this type of physical distance. (b) Only the sampled 45 undirected distances (all distances without Berlin as one of the anchors) were included in all of the following analyses.

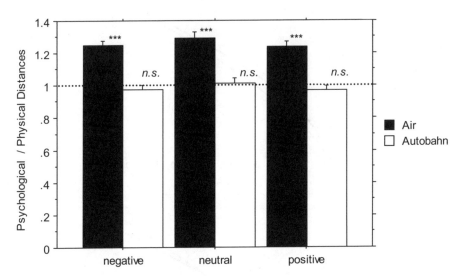

Fig. 1. Psychological distances compared with air distances vs. Autobahn distances. For each data bar, error bars as SEs of the mean and p-values of one-sample t-tests (two-tailed) against the hypothesized mean of 1.0 are shown (p < .001 equals ***).

Figure 1 clearly demonstrates that participants estimated the distances very close to Autobahn distances versus air distances, although both physical measures correlated very highly (r^2 = .9734, p < .0001). It is important to stress that the participants were forced to base their estimations on *air distances*. Nevertheless, the dissociate outcome of six independent one-sample *t*-tests against a hypothesized mean of 1.0 strongly favors

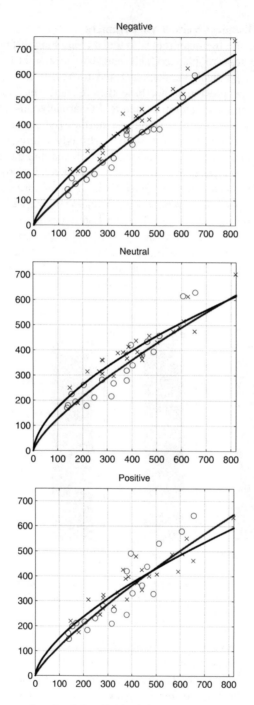

Fig. 2. Curve plots (power functions) for all attitude groups (negative, neutral, positive) split for within (blue circles) and across (black crosses) distances. On the x-axis the air distances, on the y-axis the psychological distances are given.

the idea that the participants based their distance estimations on distances measured on the Autobahn which mirror their everyday life experiences (see *p*-values in Figure 1).

The high fit between psychological and Autobahn distances (see Figure 2) was also revealed by regression analyses of the 25 across and 20 within distances for each attitude group (negative, neutral or positive attitude towards German reunification).

Table 1 outlines the curve parameters and the fits as the correlations between estimated and modeled distances. None of the correlations were below .90.

Table 1. Fit functions for the group data level split by the *attitude* towards the German reunification and the *distance type*

Attitude	Distance Type	Fit Function	Correlation
Negative	Across	$y = 5.64 \cdot x^{0.71}$	.9694
Negative	Within	$y = 1.98 \cdot x^{0.86}$	.9685
Neutral	Across	$y = 10.73 \cdot x^{0.60}$	.9237
Neutral	Within	$y = 3.98 \cdot x^{0.75}$	.9286
Positive	Across	$y = 7.76 \cdot x^{0.65}$	.9093
Positive	Within	$y = 2.92 \cdot x^{0.80}$	.9009

Thus, it is highly plausible that the most important basis for estimations of *air distances* in the present study was the German Autobahn system.

2.2.2 The Mental Wall Between East and West

Carbon and Leder [1] based their main argument for the mental wall between East and West Germany on (a) regression analyses and (b) 6 pairs of distances which were comparable in length. Here, also evidence for the mental wall on the basis of the relation between the psychological distance and the physical (Autobahn) distance will be presented.

A mixed-design ANOVA with the between-subjects factor *attitude* (negative, neutral, positive) and the within-subjects factor *distance type* (across, within) revealed a main effect of *distance type*, $F(1, 80) = 13.00$, $p < .0005$. Most importantly, there was a significant interaction between both factors, $F(2, 80) = 6.15$, $p = .0033$, illustrated in Figure 3.

Further analyses of the simple main effects (see Figure 3) showed that *across* distances were particularly overestimated in comparison to *within* distances for the negative attitude group. For the neutral group there was, in accordance with the original findings, also a weak, but reliable, effect of *distance type*. Nevertheless, there was no effect for the positive attitude group.

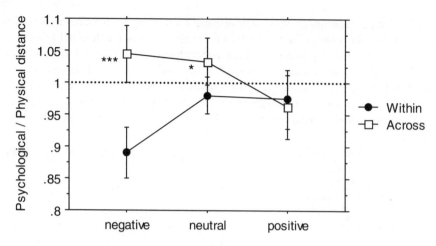

Fig. 3. The effect of *distance type* (across vs. within) on the relation between psychological distances and physical (Autobahn) distances. For each data bar, error bars as *SE*s of the mean and *p*-values of the simple main effects for each *attitude* group are shown ($p < .001$ equals ***, $p < .05$ equals *).

2.2.3 Analyses Based on the Individual's Level

The main argument of Carbon and Leder [1] for claiming the existence of a mental wall between the western and the eastern part of Germany is based on sampled group data. However, on the basis of group data it is not possible to test the slopes of the power functions of across and within distances for statistical differences. Here, curve fittings for both *distance types* (across vs. within) were conducted for all participants on an individual level.

Table 2. Fit functions for the individual data level split by the *attitude* towards the German reunification and the *distance type*

Attitude	Distance Type	Fit Function	Correlation
Negative	Across	$y = 42.05 \cdot x^{0.70}$	.8386
Negative	Within	$y = 3.60 \cdot x^{0.87}$	.8734
Neutral	Across	$y = 24.12 \cdot x^{0.64}$	.6745
Neutral	Within	$y = 11.84 \cdot x^{0.79}$	.7483
Positive	Across	$y = 12.46 \cdot x^{0.69}$	.6983
Positive	Within	$y = 8.52 \cdot x^{0.82}$	.7633

For each participant both curve parameters were calculated to define the power function and the correlation (see Table 2). Both curve parameters were averaged across the 3 *attitude* groups and the 2 *distance types* by the arithmetic means. Due to the steep distribution of correlations, all correlations were Fisher-Z transformed in order to further analyze them. The arithmetic means of these Z-values were then reconverted by inverted Fisher-Z transformation following the advice of R. A. Fisher. Again, the mean correlations, although based on fuzzy individual data, were rather high, none of them below .67. Furthermore, referring to the curve parameters, two-tailed t-tests revealed that multiplicands of the functions for across vs. within distances differed only significantly for the neutral attitude group ($p = .0320$) and as a trend for the negative group ($p = .0758$, n.s.). The exponents differed for the negative ($p = .0094$) and the neutral group ($p = .0044$), but only as a trend for the positive group ($p = .0509$, n.s.). Thus, the data on the individual level is quite compatible with the idea of a mental wall between the former western and the eastern parts of Germany proposed by Carbon and Leder [1], especially when participants had a negative attitude towards German reunification.

2.2.4 Integration of Further Demographical Information

Due to the specific social psychological approach adhered to in the original Carbon and Leder [1] study, only alterations of the mental map on the basis of the attitudes towards the German reunification had been analyzed. In order to generate a more complete picture of the effects, here, further variables were included in our calculations which might influence the "width" of the mental wall. These variables reflect the extent of geographical knowledge on Germany and the level of traveling experience throughout Germany. Both variables were assessed by 4-level ratings, whose levels were collapsed in such a way that 2-level variables were obtained which were comparable in size. Geographical knowledge was split in a low (bad and rather bad) and high (rather good and good) level of knowledge; traveling experience was split in a low (less than 10.000 km per year) and high (more than 10.000 km per year) level of experience.

Both variables were tested in two separate mixed-design ANOVAs due to the unequal number of participants per level. Both ANOVAs comprised one demographical between-subjects factor (*geographical knowledge* or *traveling experience*, respectively), the between-subjects factor *attitude* (negative, neutral, positive) and the within-subjects factor *distance type* (across vs. within). In accordance with the analyzes presented above, the relation between psychological distances and physical distances based on Autobahn kilometers, was used as dependent variable. Both analyses did not show any effect of any of the two other demographical factors nor an interaction with them, $Fs < 1.43, p > .2355$, n.s. Thus, the width of the mental wall was *not* influenced by implicit (traveling experience) or explicit knowledge (geographical knowledge) about Germany's geography!

2.2.5 Analyses of Response Time Data

For the original study only distance estimations but no response times were reported. Here we also present response times which are valuable as indirect distance measures in contrast to direct measures, such as distance or bearing estimations [16]. First, the quality of response times as an indirect distance measure will be investigated by correlating response time data with physical (Autobahn) distances. Second, an index will be presented which relates response times with psychological distances.

For all 3 *attitude* groups (negative, neutral, positive) and both *distance types* the correlations between response times and Autobahn distances were calculated. The only significant correlations were obtained for the *within* distances in the neutral group ($r^2 = .3469$, $p = .0090$), and for the *within* ($r^2 = .4493$, $p = .0015$) and the *across* distances ($r^2 = .1939$, $p = .0283$) in the positive group. This indicates that the distance estimations of the negative attitude group followed more complex cognitive mechanisms, probably social psychologically relevant ones that are related to the negative attitude towards German reunification, but cannot be described easily by linear functions of response times.

The response time data were further analyzed by calculating an index which relates response times to psychological distances. This index is a quasi measure for the estimation effort in milliseconds for each estimated km. A mixed-design ANOVA with the between-subjects factor *attitude* (negative, neutral, positive) and the within-subjects factor *distance type* (across, within) revealed main effects of *attitude*, $F(2, 80) = 4.48$, $p = .0144$, and *distance type*, $F(1, 80) = 67.40$, $p < .0001$. Most importantly, there was a significant interaction between both factors, $F(2, 80) = 6.42$, $p = .0026$, as illustrated in Figure 4.

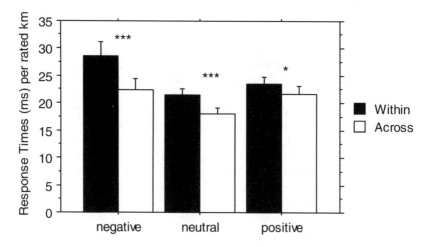

Fig. 4. The effect of *distance type* (across vs. within) on the relation response times and psychological distances. For each data bar, error bars as *SE*s of the mean and *p*-values of the simple main effects for each *attitude* group are shown ($p < .001$ equals ***, $p < .05$ equals *).

Further analyses of simple main effects of *distance types* on the different levels of *attitude* revealed large effects for the negative and neutral attitude group and a small effect for the positive group. Interestingly, the crossing of the former border did not cause an additional time factor, but systematically *decreased* response times. This might be an indication for generally different underlying cognitive processes for estimating distances *across* or *within*, probably based on simpler heuristics for distance estimations *across* the former border. Once again, these findings are in accordance with the idea of a mental wall between the former western and eastern parts of Germany.

4   General Discussion

The present study re-evaluated and extended the findings of distance distortions by social factors provided by the data of Carbon and Leder [1].

The participants in the present experiment were asked to estimate direct *air* distances. Nevertheless, comparisons between the physical distances of air distances and Autobahn distances indicated that they based their estimations not on *air* but on *Autobahn* distances. This indicates strong influences of experiential information on cognitive processes.

In accordance with the original analyses, it was shown that distances across the former iron curtain were systematically overestimated, particularly when the attitude towards the German reunification was negative. This "mental wall" [1] was *not* existent in terms of overestimations when participants had a positive attitude towards the reunification. These main conclusions could be drawn not only on the basis of sampled group data with reference to air distances but also with reference to Autobahn distances. Moreover, curve fittings on the basis of individuals' data also favored the idea of a mental wall which is mainly existent for the negative attitude group.

When response times were additionally taken into account, there were large effects in accordance with the idea of a mental wall only for the negative and the neutral attitude group. This demonstrates that taking response time data into account can be a valuable additional variable to test specific hypotheses in the field of spatial cognition.

Importantly, the reliable effects of distance overestimations which cross the former iron curtain were impenetrable both by implicit knowledge about German geography operationalized by participants' traveling experience, as well as by explicit knowledge about German geography operationalized by participants' geographical knowledge. Again, this underlines the strong impact of social attitudes on cognitive processing.

In short: we cannot understand human cognition without taking into consideration the complex interplay of social attitudes and experiential information [17].

Acknowledgement

We thank Andrea Lyman for proofreading the text and two anonymous reviewers for giving valuable comments on an earlier version of this paper. This paper is based on a re-evaluation and further analyses of a data set initially presented in "Carbon, C.C., Leder, H.: The Wall inside the Brain: Overestimation of distances crossing the former iron curtain. Psychon B Rev 12 (2005) 746-750".

References

1. Carbon, C.C., Leder, H.: The Wall inside the Brain: Overestimation of distances crossing the former iron curtain. Psychon B Rev 12, 746–750 (2005)
2. Tversky, B., Schiano, D.J.: Perceptual and conceptual factors in distortions in memory for graphs and maps. J. Exp. Psychol. Gen. 118, 387–398 (1989)
3. Tolman, E.C.: Cognitive maps in rats and men. Psychol. Rev. 55, 189–208 (1948)
4. Siegel, A.W., White, S.H.: The development of spatial representations of large-scale environments. Adv. Child Dev. Behav. 10, 9–55 (1975)

5. Cousins, J.H., Siegel, A.W., Maxwell, S.E.: Way Finding and Cognitive Mapping in Large-Scale Environments - a Test of a Developmental Model. J. Exp. Child Psychol. 35, 1–20 (1983)
6. Stevens, S.S., Galanter, E.H.: Ratio scales and category scales for a dozen perceptual continua. J. Exp. Psychol. 54, 377–411 (1957)
7. Friedman, A., Montello, D.R.: Global-Scale Location and Distance Estimates: Common Representations and Strategies in Absolute and Relative Judgments. Journal of Experimental Psychology: Learning, Memory, & Cognition (in press)
8. Ekman, G., Bratfisch, O.: Subjective distance and emotional involvement. A psychological mechanism. Acta Psychol (Amst) 24, 430–437 (1965)
9. Stevens, A., Coupe, P.: Distortions in judged spatial relations. Cognit. Psychol. 10, 422–437 (1978)
10. Hirtle, S.C., Jonides, J.: Evidence of hierarchies in cognitive maps. Mem. Cognit. 13, 208–217 (1985)
11. McNamara, T.P.: Mental representations of spatial relations. Cognit. Psychol. 18, 87–121 (1986)
12. Voicu, H.: Hierarchical cognitive maps. Neural Networks 16, 569–576 (2003)
13. Tversky, B.: Distortions in Cognitive Maps. Geoforum 23, 131–138 (1992)
14. Luce, D.: Response time: Their role in inferring elementary mental organizations. Oxford University Press, New York (1986)
15. Cohen, J.D., MacWhinney, B., Flatt, M., Provost, J.: PsyScope: a new graphic interactive environment for designing psychology experiments. Behavioral Research Methods, Instruments, and Computers 25, 257–271 (1993)
16. Wender, K.F., WagenerWender, M., Rothkegel, R.: Measures of spatial memory and routes of learning. Psychological Research-Psychologische Forschung 59, 269–278 (1997)
17. Schwarz, N.: Social judgment and attitudes: warmer, more social, and less conscious. European Journal of Social Psychology 30, 149–176 (2000)

Author Index

Allen, Gary L.   59
Avraamides, Marios N.   270

Barkowsky, Thomas   191
Bateman, John A.   107
Bertel, Sven   191
Bishop, Ian D.   156
Bogaert, Peter   20
Brösamle, Martin   381
Büchner, Simon J.   365, 381

Carbon, Claus-Christian   489
Cohn, Anthony G.   20

Davies, Clare   439
Davies, Jim   127
De Maeyer, Philippe   20
Donikian, Stéphane   421
Dylla, Frank   39

Fangmeier, Thomas   175
Fischer, Kerstin   76
Fitting, Sylvia   59
Freksa, Christian   39
Frommberger, Lutz   39

Garsoffky, Bärbel   140
Glasgow, Janice   127
Gunzelmann, Glenn   288

Haque, Shazia   308
Hirtle, Stephen C.   346
Hois, Joana   107
Hölscher, Christoph   365, 381
Hornsby, Kathleen Stewart   156
Huff, Markus   140

Klippel, Alexander   308, 453
Knauff, Markus   175
Kulik, Lars   308
Kuo, Tony   127
Kyranidou, Melina-Nicole   270

Lautenschütz, Anna-Katharina   439
Lyon, Don R.   288

Mandel, Christian   327
Mau, Inga   156
McNamara, Timothy P.   249
Meilinger, Tobias   365, 381
Mukerjee, Amitabha   210

Pederson, Eric   439

Ragni, Marco   175
Raubal, Martin   439
Richter, Kai-Florian   453
Röfer, Thomas   107
Ross, Robert J.   327
Rump, Björn   249

Sarkar, Mausoom   210
Schlick, Christopher M.   229
Schultheis, Holger   191
Schwan, Stephan   140
Schwering, Angela   439
Scivos, Alexander   1
Seifert, Inessa   90, 191
Shi, Hui   327
Srinivas, Samvith   346
Strube, Gerhard   365

Tenbrink, Thora   401, 470
Thomas, Romain   421

Van de Weghe, Nico   20
Vorwerg, Constanze   470

Wallgrün, Jan Oliver   39
Webber, Lara   175
Wedell, Douglas H.   59
Wiener, Jan M.   401
Winkelholz, Carsten   229
Witlox, Frank   20
Wolter, Diedrich   39
Wünstel, Michael   107

Printing: Mercedes-Druck, Berlin
Binding: Stein+Lehmann, Berlin